Mechanical Vibrations

J. P. DEN HARTOG

Professor Emeritus of Mechanical Engineering
Massachusetts Institute of Technology

DOVER PUBLICATIONS, INC.
NEW YORK

This Dover edition, first published in 1985, is an unabridged, slightly corrected republication of the fourth edition (1956) of the work first published by the McGraw-Hill Book Company, Inc., New York, in 1934. A brief Preface has been added to this edition.

Library of Congress Cataloging-in-Publication Data

Den Hartog, J. P. (Jacob Pieter), 1901–
    Mechanical vibrations.

    Reprint. Originally published : 4th ed. New York : McGraw-Hill, 1956. With a new preface.
    Includes index.
    1. Vibration. 2. Mechanics, Applied. I. Title.
TA355.D4 1985   620.3   84-18806
ISBN-13: 978-0-486-64785-2 (pbk.)
ISBN-10: 0-486-64785-4 (pbk.)

Manufactured in the United States by LSC Communications
64785420   2018
www.doverpublications.com

PREFACE TO THE FOURTH EDITION

This book grew from a course of lectures given to students in the Design School of the Westinghouse Company in Pittsburgh, Pa., in the period from 1926 to 1932, when the subject had not yet been introduced into the curriculum of our technical schools. From 1932 until the beginning of the war, it became a regular course at the Harvard Engineering School, and the book was written for the purpose of facilitating that course, being first published in 1934. In its first edition, it was influenced entirely by the author's industrial experience at Westinghouse; the later editions have brought modifications and additions suggested by actual problems published in the literature, by private consulting practice, and by service during the war in the Bureau of Ships of the U.S. Navy.

The book aims to be as simple as is compatible with a reasonably complete treatment of the subject. Mathematics has not been avoided, but in all cases the mathematical approach used is the simplest one available.

In the fourth edition the number of problems again has been increased substantially, rising from 81 in the first edition to 116 and 131 in the second and third, and to 230 in this present book. Changes in the text have been made in every chapter to bring the subject up to date; in order to keep the size of the volume within bounds these changes consisted of deletions as well as additions.

During the life of this book, from 1934 on, the art and science of engineering has grown at an astonishing rate and the subject of vibration has expanded with it. While in 1934 it could be said that the book covered more or less what was known and technically important, no such claim can be made for this fourth edition. In these twenty years our subject has become the parent of three vigorously growing children, each of which now stands on its own feet and is represented by a large body of literature. They are (1) electronic measuring instruments and the theory and practice of instrumentation, (2) servomechanisms and control or systems engineering, (3) aircraft flutter theory or "aeroelasticity."

No attempt has been made to cover these three subjects, since even a superficial treatment would have made the book several times thicker. However, all three subjects are offshoots of the theory of vibration and

v

cannot be studied without a knowledge of that theory. While in 1934 a mechanical engineer was considered well-educated without knowing anything about vibration, now such knowledge is an important requirement. Thus, although in its first edition this book more or less presented the newest developments in mechanical engineering, it now has come to cover subject matter which is considered a necessary tool for almost every mechanical engineer.

As in previous editions, the author gratefully remembers readers who have sent in comments and reported errors, and expresses the hope that those who work with the present edition will do likewise. He is indebted to Prof. Alve J. Erickson for checking the problems and reading the proof.

J. P. DEN HARTOG

PREFACE TO THE DOVER EDITION

The first edition of *Mechanical Vibrations*, published in 1934 by McGraw-Hill, was followed by three other editions in English and by translations into eleven different languages. It has been out of print in English but has remained available in four of these translations. This Dover reprint, an unaltered republication of the fourth edition, can be termed a half-century edition. Mr. Hayward Cirker, President of Dover Publications, Inc., paid me a very great and highly appreciated compliment by offering to reprint the book at this time. And now I say, "Thank you kindly, Mr. Dover."

JACOB P. DEN HARTOG

Concord, Massachusetts
July, 1984

CONTENTS

LIST OF SYMBOLS

a, A = cross-sectional area.
a_0 = amplitude of support.
a_n = Fourier coefficient of sin $n\omega t$.
b_n = Fourier coefficient of cos $n\omega t$.
c = damping constant, either linear (lb. in.$^{-1}$ sec.) or torsional (lb. in. rad.$^{-1}$).
C = condenser capacity.
c_c = critical damping constant, Eq. (2.16).
C_1, C_2 = constants.
d, D = diameters.
D = aerodynamic drag.
e = eccentricity.
e = amplitude of pendulum support (Sec. 8.4 only).
E = modulus of elasticity.
E_0 = maximum voltage, E_0 sin ωt.
f = frequency = $\omega/2\pi$.
f_n = natural frequency.
f and g = numerical factors used in the same sense in one section only as follows:
Sec. 3.3 as defined by Eq. (3.23), page 96. Sec. 4.3 as defined by Eq. (4.19), page 133.
F = force in general or dry friction force in particular.
F = frequency function [Eq. (4.7), page 125].
g = acceleration of gravity.
g = See f.
G = modulus of shear.
h = height in general; metacentric height in particular (page 106).
i = electric current.
I = moment of inertia.
$j = \sqrt{-1}$ = imaginary unit.
k, K = spring constants.
Kin = kinetic energy.
Δk = variation in spring constant (page 337).
l = length in general; length of connecting rod in Chap. 5.
l_n = distance from nth crank to first crank (Sec. 5.3).
L = inductance.
L = aerodynamic lift.
m, M = mass.
M = moment or torque.
$\overline{\mathfrak{M}}$ = angular momentum vector.
$\mathfrak{M}$ = magnitude of angular momentum.
n = a number in general; a gear ratio in particular (page 29).
p = real part of complex frequency s (page 133).
p = pressure.
p_1, p_2 = (in Sec. 8.3 only) defined by Eqs. (8.17) and (8.18), page 344.

x

P_0 = maximum force, $P_0 \sin \omega t$.

Pot = potential energy.

q = natural frequency of damped vibration (pages 39 and 133).

q = load per unit length on beam (page 148).

Q = condenser charge.

r, R = radius of circle.

R = electrical resistance.

s = complex frequency = $\pm p \pm jq$ (page 133).

$\mathbf{s}$ = (in Sec. 8.3 only) multiplication factor.

t = time.

T = period of vibration = $1/f$.

T_0 = maximum torque $T_0 \sin \omega t$.

$\mathbf{T}$ = tension in string.

v, V = velocity.

$\mathbf{v}, \mathbf{V}$ = volume.

W = work or work per cycle.

$\mathbf{W}$ = weight.

x = displacement.

x_0 = maximum amplitude.

x_{st} = static deflection, usually = P_0/k.

$y = y_0 \sin \omega t$ = amplitude of relative motion.

y = lateral deflection of string or bar.

Z = impedance.

α = angle in general; angle of attack of airfoil.

α_n = nth crank angle in reciprocating engine.

α_{mn} = influence number, deflection at m caused by unit force at n.

β_n = angular amplitude of vibration of nth crank (Chap. 5).

$\bar{\beta}_n$ = vector representing β_n.

δ = small length or small quantity in general.

δ_{st} = static deflection.

ϵ = parameter defined in Eq. (8.35), page 364.

λ = a length.

μ = mass ratio m/M (Secs. 3.2 and 3.3).

μ_1 = mass per unit length of strings, bars, etc.

ξ = longitudinal displacement of particle along beam (page 136).

ρ = radius of gyration.

φ, ϕ = phase angle or some other angle.

φ_n = phase angle between vibration of nth crank and first crank (Chap. 5).

ψ = an angle.

ω = circular frequency = $2\pi f$.

ω = angular velocity.

Ω = large angular velocity.

ω_n, Ω_n = natural circular frequencies.

Vector quantities are letters with superposed bar, $\bar{a}$, $\bar{V}$, $\bar{M}$, etc.

Scalar quantities are letters without bar, a, T, $\mathbf{T}$, $\mathbf{M}$, etc. Note especially that **bold-face** type does not denote a vector, but is used merely for avoiding confusion. For example, $\mathbf{V}$ denotes volume and V velocity.

Subscripts used are the following: a = absorber; c = critical, e = engine, f = friction, g = governor or gyroscope, k = variation in spring constant k, p = propeller, s = ship, st = statical, w = water.

CHAPTER 1

KINEMATICS OF VIBRATION

1.1 Definitions. A vibration in its general sense is a periodic motion, *i.e.*, a motion which repeats itself in all its particulars after a certain interval of time, called the *period* of the vibration and usually designated by the symbol T. A plot of the displacement x against the time t may be a curve of considerable complication. As an example, Fig. 1.1a shows the motion curve observed on the bearing pedestal of a steam turbine.

The simplest kind of periodic motion is a *harmonic motion;* in it the relation between x and t may be expressed by

$$x = x_0 \sin \omega t \qquad (1.1)$$

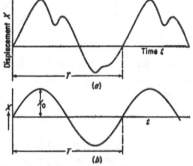

as shown in Fig. 1.1b, representing the small oscillations of a simple pendulum. The maximum value of the displacement is x_0, called the *amplitude* of the vibration.

The period T usually is measured in seconds; its reciprocal $f = 1/T$

Fig. 1.1. A periodic and a harmonic function, showing the period T and the amplitude x_0.

is the *frequency* of the vibration, measured in *cycles per second*. In some publications this is abbreviated as *cyps* and pronounced as it is written. In the German literature cycles per second are generally called *Hertz* in honor of the first experimenter with radio waves (which are electric vibrations).

In Eq. (1.1) there appears the symbol ω, which is known as the *circular frequency* and is measured in radians per second. This rather unfortunate name has become familiar on account of the properties of the vector representation, which will be discussed in the next section. The relations between ω, f, and T are as follows. From Eq. (1.1) and Fig. 1.1b it is clear that a full cycle of the vibration takes place when ωt has passed through 360 deg. or 2π radians. Then the sine function resumes its previous values. Thus, when $\omega t = 2\pi$, the time interval t is equal to

1

the period T or

$$T = \frac{2\pi}{\omega} \text{ sec.} \tag{1.2}$$

Since f is the reciprocal of T,

$$f = \frac{\omega}{2\pi} \text{ cycles per second} \tag{1.3}$$

For rotating machinery the frequency is often expressed in vibrations per minute, denoted as v.p.m. $= 30\omega/\pi$.

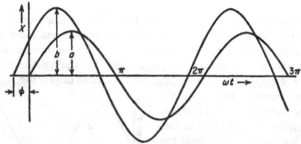

FIG. 1.2. Two harmonic motions including the phase angle φ.

In a harmonic motion for which the displacement is given by $x = x_0 \sin \omega t$, the velocity is found by differentiating the displacement with respect to time,

$$\frac{dx}{dt} = \dot{x} = x_0\omega \cdot \cos \omega t \tag{1.4}$$

so that the velocity is also harmonic and has a maximum value ωx_0.

The acceleration is

$$\frac{d^2x}{dt^2} = \ddot{x} = -x_0\omega^2 \sin \omega t \tag{1.5}$$

also harmonic and with the maximum value $\omega^2 x_0$.

Consider two vibrations given by the expressions $x_1 = a \sin \omega t$ and $x_2 = b \sin (\omega t + \varphi)$ which are shown in Fig. 1.2, plotted against ωt as abscissa. Owing to the presence of the quantity φ, the two vibrations do not attain their maximum displacements at the same time, but the one is φ/ω sec. behind the other. The quantity φ is known as the *phase angle* or *phase difference* between the two vibrations. It is seen that the two motions have the same ω and consequently the same frequency f. A phase angle has meaning only for two motions of the same frequency: if the frequencies are different, phase angle is meaningless.

Example: A body, suspended from a spring, vibrates vertically up and down between two positions 1 and 1½ in. above the ground. During each second it reaches the top position (1½ in. above ground) twenty times. What are T, f, ω, and x_0?

Solution: x_0 = ¼ in., T = ¹⁄₂₀ sec, f = 20 cycles per second, and ω = $2\pi f$ = 126 radians per second.

1.2. The Vector Method of Representing Vibrations. The motion of a vibrating particle can be conveniently represented by means of a rotating vector. Let the vector d (Fig. 1.3) rotate with uniform angular velocity ω in a counterclockwise direction. When time is reckoned from the horizontal position of the vector as a starting point, the horizontal projection of the vector can be written as

$$a \cos \omega t$$

and the vertical projection as

$$a \sin \omega t$$

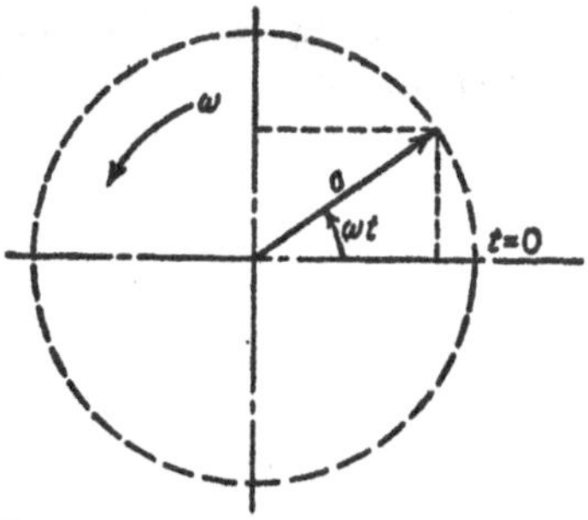

Either projection can be taken to represent a reciprocating motion; in

FIG. 1.3. A harmonic vibration represented by the horizontal projection of a rotating vector.

the following discussion, however, we shall consider only the *horizontal* projection.

This representation has given rise to the name *circular frequency* for ω. The quantity ω, being the angular speed of the vector, is measured in *radians per second;* the frequency f in this case is measured in *revolutions per second.* Thus it can be seen immediately that

$$\omega = 2\pi f.$$

The *velocity* of the motion $x = a \cos \omega t$ is

$$\dot{x} = -a\omega \sin \omega t$$

and can be represented by (the horizontal projection of) a vector of length $a\omega$, rotating with the same angular velocity ω as the displacement vector but situated always 90 deg. ahead of that vector.

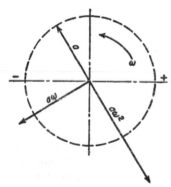

FIG. 1.4. Displacement, velocity, and acceleration are perpendicular vectors.

The *acceleration* is $-a\omega^2 \cos \omega t$ and is represented by (the horizontal projection of) a vector of length $a\omega^2$ rotating with the same angular speed ω and 180 deg. ahead of the position or displacement vector or 90 deg. ahead of the velocity vector (Fig. 1.4). The truth of these statements can be easily verified by following the various vectors through one complete revolution.

This vector method of visualizing reciprocating motions is very convenient. For example, if a point is simultaneously subjected to two motions of the same frequency which differ by the phase angle φ, namely, $a \cos \omega t$ and $b \cos (\omega t - \varphi)$, the addition of these two expressions by the methods of trigonometry is wearisome. However, the two vectors are easily drawn up, and the total motion is represented by the geometric sum of the two vectors as shown in the upper part of Fig. 1.5. Again

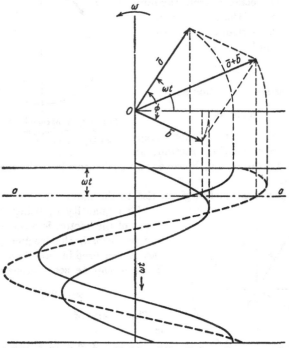

FIG. 1.5. Two vibrations are added by adding their vectors geometrically.

the entire parallelogram $\bar{a}, \bar{b}$ is considered to rotate in a counterclockwise direction with the uniform angular velocity ω, and the horizontal projections of the various vectors represent the displacements as a function of time. This is shown in the lower part of Fig. 1.5. The line a-a represents the particular instant of time for which the vector diagram is drawn. It is readily seen that the displacement of the sum (dotted line) is actually the sum of the two ordinates for $\bar{a}$ and $\bar{b}$.

That this vector addition gives correct results is evident, because $a \cos \omega t$ is the horizontal projection of the $\bar{a}$-vector and $b \cos (\omega t - \varphi)$ is the horizontal projection of the $\bar{b}$-vector. The horizontal projection of the geometric sum of these two vectors is evidently equal to the sum of the horizontal projections of the two component vectors, which is exactly what is wanted.

Addition of two vectors is permissible only if the vibrations are of the same frequency. The motions $a \sin \omega t$ and $a \sin 2\omega t$ can be represented by two vectors, the first of which rotates with an angular speed ω and the second with twice this speed, i.e., with 2ω. The relative position of these two vectors in the diagram is changing continuously, and consequently a geometric addition of them has no meaning.

A special case of the vector addition of Fig. 1.5, which occurs rather often in the subsequent chapters, is the addition of a sine and a cosine wave of different amplitudes: $a \sin \omega t$ and $b \cos \omega t$. For this case the two vectors are perpendicular, so that from the diagram of Fig. 1.6 it is seen at once that

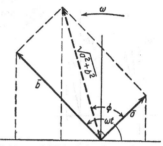

Fig. 1.6. Addition of a sine and cosine wave of different amplitudes.

$$a \sin \omega t + b \cos \omega t = \sqrt{a^2 + b^2} \sin (\omega t + \varphi) \tag{1.6}$$

where $\qquad\qquad\qquad\qquad \tan \varphi = b/a.$

Example: What is the maximum amplitude of the sum of the two motions

$$x_1 = 5 \sin 25t \text{ in.} \qquad \text{and} \qquad x_2 = 10 \sin (25t + 1) \text{ in.?}$$

Solution: The first motion is represented by a vector 5 in. long which may be drawn vertically and pointing downward. Since in this position the vector has no horizontal projection, it represents the first motion at the instant $t = 0$. At that instant the second motion is $x_2 = 10 \sin 1$, which is represented by a vector of 10 in. length turned 1 radian (57 deg.) in a counter-clockwise direction with respect to the first vector. The graphical vector addition shows the sum vector to be 13.4 in. long.

1.3. Beats. If the displacement of a point moving back and forth along a straight line can be expressed as the sum of two terms, $a \sin \omega_1 t + b \sin \omega_2 t$, where $\omega_1 \neq \omega_2$, the motion is said to be the "superposition" of two vibrations of different frequencies. It is clear that such a motion is not itself sinusoidal. An interesting special case occurs when the two frequencies ω_1 and ω_2 are nearly equal to each other. The first vibration can be represented by a vector $\bar{a}$ rotating at a speed ω_1, while the $\bar{b}$-vector rotates with ω_2. If ω_1 is nearly equal to ω_2, the two vectors will retain sensibly the same relative position during one revolution, i.e., the angle included between them will change only slightly. Thus the vectors can be added geometrically, and during one revolution of the two vectors the motion will be practically a sine wave of frequency $\omega_1 \approx \omega_2$ and amplitude $\bar{c}$ (Fig. 1.7). During a large number of cycles, however, the relative position of $\bar{a}$ and $\bar{b}$ varies, because ω_1 is not exactly equal to ω_2, so that the magnitude of the sum vector $\bar{c}$ changes. Therefore the resulting motion can be described approximately as a sine wave with a frequency ω_1 and an amplitude varying slowly between $(b + a)$ and $(b - a)$, or, if $b = a$, between $2a$ and 0 (Figs. 1.7 and 1.8).

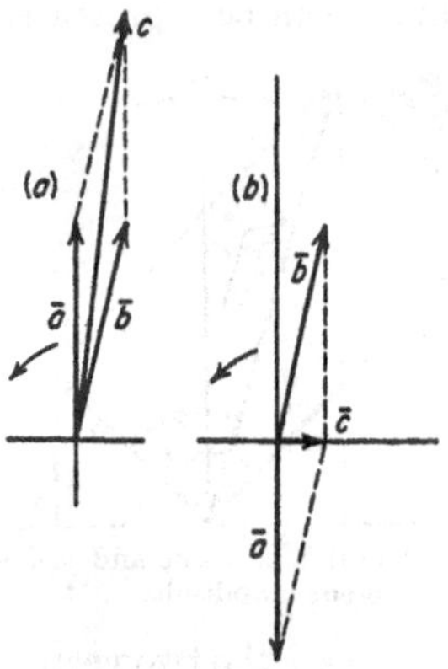

This phenomenon is known as *beats*. The beat frequency is the number of times per second the amplitude passes from a minimum through a maximum to the next minimum (A to B in Fig. 1.8). The period of one beat evidently corresponds to the time required for a full revolution of the $\bar{b}$-vector with respect to the $\bar{a}$-vector. Thus the beat frequency is seen to be $\omega_1 - \omega_2$.

Example: A body describes simultaneously two vibrations, $x_1 = 3 \sin 40t$ and $x_2 = 4 \sin 41t$, the units being inches and seconds. What is the maximum and minimum amplitude of the combined motion and what is the beat frequency?

FIG. 1.7. Vector diagrams illustrating the mechanism of beats.

Solution: The maximum amplitude is $3 + 4 = 7$ in.; the minimum is $4 - 3 = 1$ in. The circular frequency of the beats $\omega_b = 41 - 40 = 1$ radian per second. Thus $f_b = \omega_b/2\pi = 1/2\pi$ cycles per second. The period T_b or duration of one full beat is $T_b = 1/f_b = 6.28$ sec.

The phenomenon can be observed in a great many cases (pages 84, 332). For audio or sound vibrations it is especially notable. Two tones of

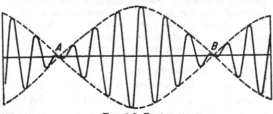

FIG. 1.8. Beats.

slightly different pitch and of approximately the same intensity cause fluctuations in the total intensity with a frequency equal to the difference of the frequencies of the two tones. For example, beats can be heard in electric power houses when a generator is started. An electric machine has a "magnetic hum," of which the main pitch is equal to twice the frequency of the current or voltage, usually 120 cycles per second. Just before a generator is connected to the line the electric frequency of the generator is slightly different from the line frequency. Thus the hum of the generator and the hum of the line (other generators or transformers) are of different pitch, and beats can be heard.

The existence of beats can be shown also by trigonometry. Let the two vibrations be $a \sin \omega_1 t$ and $b \sin \omega_2 t$, where ω_1 and ω_2 are nearly equal and $\omega_2 - \omega_1 = \Delta\omega$.

Then
$$a \sin \omega_1 t + b \sin \omega_2 t$$
$$= a \sin \omega_1 t + b(\sin \omega_1 t \cos \Delta\omega t + \cos \omega_1 t \sin \Delta\omega t)$$
$$= (a + b \cos \Delta\omega t) \sin \omega_1 t + b \sin \Delta\omega t \cos \omega_1 t$$

Applying formula 1.6 the resultant vibration is

$$\sqrt{(a + b \cos \Delta\omega t)^2 + b^2 \sin^2 \Delta\omega t} \cdot \sin (\omega_1 t + \varphi)$$

where the phase angle φ can be calculated but is of no interest in this case. The amplitude, given by the radical, can be written

$$\sqrt{a^2 + b^2(\cos^2 \Delta\omega t + \sin^2 \Delta\omega t) + 2ab \cos \Delta\omega t}$$
$$= \sqrt{a^2 + b^2 + 2ab \cos \Delta\omega t}$$

which expression is seen to vary between $(a + b)$ and $(a - b)$ with a frequency $\Delta\omega$.

1.4. A Case of Hydraulic-turbine Penstock Vibration. A direct application of the vector concept of vibration to the solution of an actual problem is the following.

In a water-power generating station the penstocks, *i.e.*, the pipe lines conducting the water to the hydraulic turbines, were found to be vibrating so violently that the safety of the brick building structure was questioned. The frequency of the vibration was found to be $113\frac{1}{3}$ cycles per second, coinciding with the product of the speed (400 r.pm.) and the number of buckets (17) in the rotating part of the (Francis) turbine. The penstocks emitted a loud hum which could be heard several miles away. Incidentally, when standing close to the electric transformers of the station, the $6\frac{2}{3}$ cyps. beat between the penstock and transformer hums could be plainly heard. The essential parts of the turbine are shown schematically in Fig. 1.9, which is drawn in a horizontal plane, the turbine shaft being vertical. The water enters from the penstock I into the "spiral case" II; there the main stream splits into 18 partial streams on account of the 18 stationary, non-rotating guide vanes. The water then enters the 17 buckets of the runner and finally turns through an angle of 90 deg. to disappear into the vertical draft tube III.

Two of the 18 partial streams into which the main stream divides are shown in the figure. Fixing our attention on one of these, we see that for each revolution of the runner, 17 buckets pass by the stream, which thus is subjected to 17 impulses. In total, $113\frac{1}{3}$ buckets are passing per second, giving as many impulses per second, which are transmitted back through the water into the penstock. This happens not only in stream a but in each of the other partial streams as well, so that there arrive into the penstock 18 impulses of different origins, all having the same frequency of 113⅓ cycles per second. If all these impulses had the same *phase*, they would all add up arithmetically and give a very strong disturbance in the penstock.

Assume that stream *a* experiences the maximum value of its impulse when the two vanes 1 and 1 line up. Then the maximum value of the impulse in stream *b* takes place somewhat *earlier* (to be exact, 1/(17 × 18)th revolution earlier, at the instant that the two vanes 2 and 2 are lined up).

The impulse of stream *a* travels back into the penstock with the velocity of sound in water (about 4,000 ft./sec.)† and the same is true for the

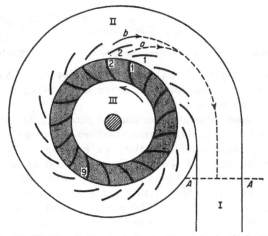

FIG. 1.9. Vibration in the penstock of a Francis hydraulic turbine.

impulse of stream *b*. However, the path traveled by the impulse of *b* is somewhat longer than the path for *a*, the difference being approximately one-eighteenth part of the circumference of the spiral case. Because of this fact, the impulse *b* will arrive in the penstock *later* than the impulse *a*.

In the machine in question it happened that these two effects just canceled each other so that the two impulses *a* and *b* arrived at the cross section *AA* of the penstock simultaneously, *i.e.*, in the same phase. This of course is true not only for *a* and *b* but for all the 18 partial streams. In the vector representation the impulses behave as shown in Fig. 1.10a, the total impulse at *AA* being very large.

In order to eliminate the trouble, the existing 17-bucket runner was

† The general streaming velocity of the water is small in comparison to the velocity of sound, so that its effect can be neglected.

removed from the turbine and replaced by a 16-bucket runner. This does not affect the time difference caused by the different lengths of the paths a, b, etc., but it does change the interval of time between the impulses of two adjacent guide vanes. In particular, the circumferential distance between the bucket 2 and guide vane 2 becomes twice as large after the change. In fact, at the instant that rotating bucket 1 gives its impulse, bucket 9 also gives its impulse, whereas in the old construction bucket 9 was midway between two stationary vanes (Fig. 1.9).

It was a fortunate coincidence that half the circumference of the spiral case was traversed by a sound wave in about $\frac{1}{2} \times \frac{1}{113}$ sec., so that the two impulses due to buckets 1 and 9 arrived in the cross section AA in phase opposition (Fig. 1.10b). The phase angle between the impulses at section AA of two adjacent partial streams is thus one-ninth of 180 degrees, and the 18 partial impulses arrange themselves in a circular diagram with a zero resultant.

The analysis as given would indicate that after the change in the runner had been made the vibration would be totally absent. However, this is not to be expected, since the reasoning given is only approximate, and many effects have not been considered (the spiral case has been replaced by a narrow channel, thus neglecting curvature of the wave front, reflection of the waves against the various obstacles, and effect of damping). In the actual case the amplitude of the vibration on the penstock was reduced to one-third of its previous value, which constituted a satisfactory solution of the problem.

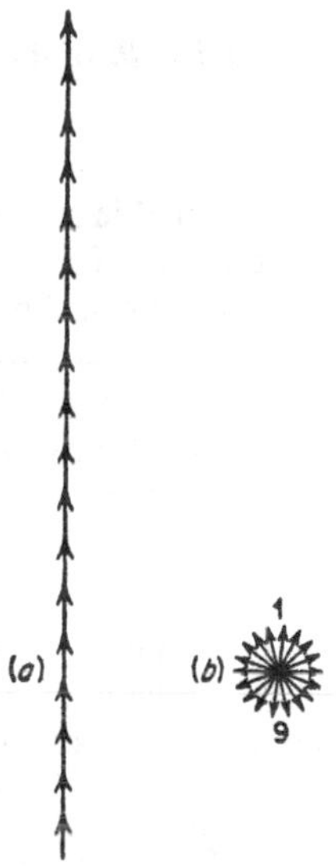

Fig. 1.10. The 18 partial impulses at the section AA of Fig. 1.9 for a 17-bucket runner (a) and for a 16-bucket runner (b).

1.5. Representation by Complex Numbers. It was shown in the previous sections that rotating vectors can represent harmonic motions, that the geometric addition of two vectors corresponds to the addition of two harmonic motions of the same frequency, and that a differentiation of such a motion with respect to time can be understood as a multiplication by ω and a forward turning through 90 deg. of the representative vector. These vectors, after a little practice, afford a method of visualizing harmonic motions which is much simpler than the consideration of the sine waves themselves.

For *numerical* calculations, however, the vector method is not well adapted. since it becomes necessary to resolve the vectors into their horizontal and vertical components. For instance, if two motions have to be added as in Fig. 1.5, we write

$$\bar{c} = \bar{a} + \bar{b}$$

meaning *geometric* addition.   To calculate the length of $\bar{c}$, *i.e.*, the amplitude of the sum motion, we write

$$\bar{a} = a_x + a_y$$

which means that $\bar{a}$ is the geometric sum of a_x in the x-direction and a_y in the y-direction.   Then

$$\bar{c} = a_x + a_y + b_x + b_y = (a_x + b_x) + (a_y + b_y)$$

and the *length* of $\bar{c}$ is consequently

$$c = \sqrt{(a_x + b_x)^2 + (a_y + b_y)^2}$$

This method is rather lengthy and loses most of the advantage due to the introduction of vectors.

There exists, however, a simpler method of handling the vectors numerically, employing imaginary numbers.   A complex number can be represented graphically by a point in a plane where the *real* numbers 1, 2, 3, etc., are plotted horizontally and the *imaginary* numbers are plotted vertically.   With the notation

$$j = \sqrt{-1}$$

these imaginary numbers are j, $2j$, $3j$, etc.   In Fig. 1.11, for example, the point $3 + 2j$ is shown.   In joining that point with the origin, the complex number can be made to represent a vector.   If the angle of the vector with the horizontal axis is α and the length of the vector is a, it can be written as

$$a(\cos \alpha + j \sin \alpha)$$

Harmonic motions are represented by *rotating* vectors.   A substitution of the variable angle ωt for the fixed angle α in the last equation leads to

$$a(\cos \omega t + j \sin \omega t) \qquad (1.7)$$

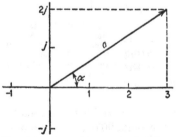

FIG. 1.11. A vector represented by a point in the complex plane.

representing a rotating vector, the horizontal projection of which is a harmonic motion.   But this horizontal projection is also the real part of (1.7).   Thus if we say that a "vector represents a harmonic motion," we mean that the *horizontal projection of the rotating* vector represents that motion.   Similarly if we state that a "complex number represents a harmonic motion," we imply that the *real part of such a number, written in the form* (1.7) represents that motion.

Example: Solve the example of page 5 by means of the complex method.
Solution: The first vector is represented by $-5j$ and the second one by $-10j \cos 57°$ $+ 10 \sin 57° = -5.4j + 8.4$. The sum of the two is $8.4 - 10.4j$, which represents a vector of the length $\sqrt{(8.4)^2 + (10.4)^2} = 13.4$ in.

Differentiate (1.7), which gives the result

$$a(-\omega \sin \omega t + j\omega \cos \omega t) = j\omega \cdot a(\cos \omega t + j \sin \omega t)$$

since by definition of j we have $j^2 = -1$. It is thus seen that *differentiation of the complex number* (1.7) *is equivalent to multiplication by* $j\omega$.

In vector representation, differentiation multiplies the length of the vector by ω and turns it ahead by 90 deg. Thus we are led to the conclusion that multiplying a complex number by j is equivalent to moving it a quarter turn ahead without changing its absolute value. That this is so can be easily verified:

$$j(a + jb) = -b + ja$$

which Fig. 1.12 actually shows in the required position.

In making extended calculations with these complex numbers the ordinary rules of algebra are followed. With every step we may remember that the motion is represented by only the *real* part of what we are writing down. Usually, however, this is not done: the algebraic manipulations are performed without much recourse to their physical meaning and only the final answer is interpreted by considering its real part.

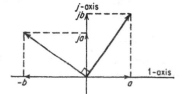

FIG. 1.12. Multiplying a complex number by j amounts to turning its vector ahead through 90 deg.

For simple problems it is hardly worth while to study the complex method, since the solution can be obtained just as easily without it. However, for more complicated problems, such as are treated in Sec. 3.3, for example, the labor-saving brought about by the use of the complex notation is substantial.

The expression (1.7) is sometimes written in a different form,

$$a(\cos \omega t + j \sin \omega t) = ae^{j\omega t}$$

or, if for simplicity $a = 1$ and $\omega t = \alpha$,

$$e^{j\alpha} = \cos \alpha + j \sin \alpha \tag{1.8}$$

The right-hand side of this equation is an ordinary complex number, but the left-hand side needs to be interpreted, as follows. The Maclaurin series development of e^x is

$$e^x = 1 + x + \frac{x^2}{2!} + \frac{x^3}{3!} + \cdots$$

Substituting $x = j\alpha$ this becomes

$$e^{j\alpha} = 1 + j\alpha - \frac{\alpha^2}{2!} - j\frac{\alpha^3}{3!} + \frac{\alpha^4}{4!} + j\frac{\alpha^5}{5!} - \cdots$$

$$= \left(1 - \frac{\alpha^2}{2!} + \frac{\alpha^4}{4!} - \cdots\right) + j\left(\alpha - \frac{\alpha^3}{3!} + \frac{\alpha^5}{5!} - \cdots\right)$$

The right-hand side is a complex number, which by definition is the meaning of $e^{j\alpha}$. But we recognize the brackets to be the Maclaurin developments of $\cos \alpha$ and $\sin \alpha$, so that formula (1.8) follows.

A simple graphical representation of the result can be made in the complex plane of Fig. 1.11 or 1.12. Consider the circle with unit radius in this plane. Each point on the circle has a horizontal projection $\cos \alpha$ and a vertical projection $\sin \alpha$ and thus represents the number, $\cos \alpha + j \sin \alpha = e^{j\alpha}$. Consequently the number $e^{j\alpha}$ is represented by a point on the unit circle, α radians away from the point $+1$. If α is now made equal to ωt, it is seen that $e^{j\omega t}$ represents the rotating unit vector of which the horizontal projection is a harmonic vibration of unit amplitude and frequency ω.

On page 39 we shall have occasion to make use of Eq. (1.8).

1.6. Work Done on Harmonic Motions. A concept of importance for many applications is that of the work done by a harmonically varying force upon a harmonic motion of the same frequency.

Let the force $P = P_0 \sin (\omega t + \varphi)$ be acting upon a body for which the motion is given by $x = x_0 \sin \omega t$. The work done by the force during a small displacement dx is $P\, dx$, which can be written as $P\frac{dx}{dt} \cdot dt$.

During one cycle of the vibration, ωt varies from 0 to 2π and consequently t varies from 0 to $2\pi/\omega$. The work done during one cycle is:

$$\int_0^{\frac{2\pi}{\omega}} P\frac{dx}{dt}\, dt = \frac{1}{\omega}\int_0^{2\pi} P\frac{dx}{dt}\, d(\omega t) = P_0 x_0 \int_0^{2\pi} \sin(\omega t + \varphi)\cos \omega t\, d(\omega t)$$

$$= P_0 x_0 \int_0^{2\pi} \cos \omega t[\sin \omega t \cos \varphi + \cos \omega t \sin \varphi]d(\omega t)$$

$$= P_0 x_0 \cos \varphi \int_0^{2\pi} \sin \omega t \cos \omega t\, d(\omega t) + P_0 x_0 \sin \varphi \int_0^{2\pi} \cos^2 \omega t\, d(\omega t)$$

A table of integrals will show that the first integral is zero and that the second one is π, so that the work per cycle is

$$W = \pi P_0 x_0 \sin \varphi \tag{1.9}$$

This result can also be obtained by a graphical method, which interprets the above calculations step by step, as follows.

The force and motion can be represented by the vectors $\bar{P}_0$ and $\bar{x}_0$

(Fig. 1.13). Now resolve the force into its components $P_0 \cos \varphi$ in phase with the motion, and $P_0 \sin \varphi$, 90 deg. ahead of the motion x_0. This is permissible for the same reason that geometric addition of vectors is allowed, as explained in Sec. 1.2. Thus the work done splits up into two parts, one part due to a force in phase with the motion and another part due to a force 90 deg. ahead of the motion.

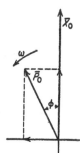

Consider the first part as shown in Fig. 1.14a, in which the ordinates are the displacement x and the "in phase" component of the force. Between A and B the force is positive, say upward, and the body is moving in an upward direction also; positive work is done. Between B and C, on the other hand, the body moves downward toward the equilibrium point while the force is still positive (upward, though of gradually diminishing intensity), so that negative work is done. The work between A and B cancels that between B and C, and over a whole cycle the work done is zero. *If a harmonic force acts on a body subjected to a harmonic motion of the same frequency, the component of the force in phase with the displacement does no work.*

Fig. 1.13. A force and a motion of the same frequency.

It was shown in Sec. 1.2 that the velocity is represented by a vector 90 deg. ahead of the displacement, so that the statement can also be worded as follows:

A force does work only with that component which is in phase with the velocity.

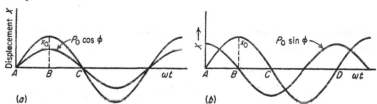

Fig. 1.14. A force in phase with a displacement does no work over a full cycle; a force 90 deg. out of phase with a displacement does a maximum amount of work.

Next we consider the other component of the force, which is shown in Fig. 1.14b. During the interval AB the displacement increases so that the motion is directed upward, the force is positive, and consequently upward also, so that positive work is done. In the interval BC the motion is directed downward, but the force points downward also, so that the work done is again positive. Since the whole diagram is symmetrical about a vertical line through B, it is clear that the work done during AB equals that done during BC. The work done during the whole cycle AD is four times that done during AB.

To calculate that amount it is necessary to turn to the definition of work:

$$W = \int P \, dx = \int P \frac{dx}{dt} \, dt = \int Pv \cdot dt$$

This shows that the work done during a cycle is the time integral of the product of *force* and *velocity*. The force is (Fig. 1.14b) $P = (P_0 \sin \varphi) \cdot \cos \omega t$ and the velocity is $v = x_0 \omega \cos \omega t$, so that the work per cycle is

$$\int_0^T P_0 \sin \varphi \cos \omega t \, x_0 \omega \cos \omega t \, dt = P_0 x_0 \sin \varphi \int_0^{2\pi} \cos^2 \omega t \, d(\omega t)$$

The value of the definite integral on the right-hand side can be deduced from Fig. 1.15, in which curve I represents $\cos \omega t$ and curve II represents $\cos^2 \omega t$. The curve $\cos^2 \omega t$ is sinusoidal about the dotted line AA as

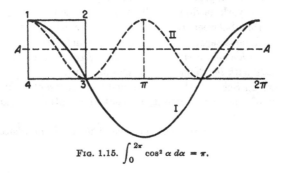

FIG. 1.15. $\int_0^{2\pi} \cos^2 \alpha \, d\alpha = \pi$.

center line and has twice the frequency of $\cos \omega t$, which can be easily verified by trigonometry:

$$\cos^2 \alpha = \tfrac{1}{2}(1 + \cos 2\alpha)$$

Consider the quadrangle 1-2-3-4 as cut in two pieces by the curve II and note that these two pieces have the same shape and the same area. The distance 1-4 is unity, while the distance 3-4 is $\pi/2$ radians or 90 deg. Thus the area of the entire quadrangle is $\pi/2$ and the area of the part under curve II is half of that. Consequently the value of our definite integral taken between the limits 0 and $T/4$ is $\pi/4$, and taken between the limits 0 and T it is π. Thus the work during a cycle is

$$W = \pi P_0 x_0 \sin \varphi \tag{1.9}$$

It will be seen in the next section that a periodic force as well as a periodic motion may be "impure," *i.e.*, it may contain "higher harmonics" in addition to the "fundamental harmonic." In this connection it is of importance to determine the work done by a harmonic force on a harmonic motion of a frequency *different* from that of the force. Let the force vary with a frequency which is an integer multiple of ω, say $n\omega$, and let the frequency of the motion be another integer multiple of ω, say $m\omega$. It will now be proved that the work done by such a force on such a motion during a full cycle of ω is zero.

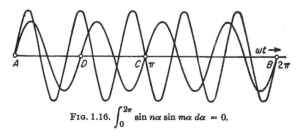

Fig. 1.16. $\int_0^{2\pi} \sin n\alpha \sin m\alpha \, d\alpha = 0.$

Let the force be $P = P_0 \sin n\omega t$ and let the corresponding motion be $x = x_0 \sin (m\omega t + \varphi)$. Then the work per cycle is

$$\int P \, dx = \int_0^T P \frac{dx}{dt} dt = \int_0^T P_0 \sin n\omega t \cdot x_0 m\omega \cdot \cos (m\omega t + \varphi) \, dt$$

Since $\qquad \cos (m\omega t + \varphi) = \cos m\omega t \cos \varphi - \sin m\omega t \sin \varphi$

and since φ is independent of the time and can be brought in front of the integral sign, the integral splits up into two parts of the form

$$\int_0^T \sin n\omega t \sin m\omega t \, dt \qquad \text{and} \qquad \int_0^T \sin n\omega t \cos m\omega t \, dt$$

Both these integrals are zero if n is different from m, which can be easily verified by transforming the integrands as follows:

$$\sin n\omega t \sin m\omega t = \tfrac{1}{2} \cos (n - m)\omega t - \tfrac{1}{2} \cos (n + m)\omega t$$
$$\sin n\omega t \cos m\omega t = \tfrac{1}{2} \sin (n + m)\omega t + \tfrac{1}{2} \sin (n - m)\omega t$$

Since the interval of integration is $T = 2\pi/\omega$, the sine and cosine functions are integrated over multiples of 2π, giving a zero result. In order to gain a physical understanding of this fact let us consider the first of the above two integrals with $n = 4$ and $m = 5$. This case is represented in Fig. 1.16, where the amplitudes of the two waves are drawn to different vertical scales in order to distinguish them more easily. The time interval over which the integration extends is the interval AB. The ordinates of the two curves have to be multiplied together and then integrated. Consider two points, one somewhat to the right of A and another at the same distance to the left of C. Near A both waves are positive; near C

one is positive and the other is negative, but the absolute *values* of the ordinates are the same as near A. Therefore the contribution to the integral of an element near A cancels the contribution of the corresponding element near C. This canceling holds true not only for elements very *near* to A and C but generally for two elements at equal distances to the left from C and to the right from A. Thus the integral over the region AD cancels that over CD. In the same way it can be shown that the integral over CB is zero.

It should be understood that the work is zero only over a *whole cycle*. Starting at A, both waves (the force and the velocity) are positive, so that positive work is done. This work, however, is returned later on (so that in the meantime it must have been stored in the form of potential or kinetic energy).

This graphical process can be repeated for any combination of integral values of m and n and also for integrals containing a cosine in the integrand. When m becomes equal to n, we have the case of equal frequencies as already considered. Even then there is no work done when the force and displacement are in phase. In case $m = n$ and the force and displacement are 90 deg. out of phase, the work per cycle of the nth harmonic is $\pi P_0 x_0$ as before, and since there are n of these cycles in *one* cycle of the fundamental frequency ω, the work per fundamental cycle is $n\pi P_0 x_0$.

The results thus obtained can be briefly summarized as follows:

1. *The work done by a harmonic force acting upon a harmonic displacement or velocity of a different frequency from that of the force is zero during a time interval comprising both an integer number of force cycles and a (different) integer number of velocity cycles.*

2. *The work done by a harmonic force 90 deg. out of phase with a harmonic velocity of the same frequency is zero during a whole cycle.*

3. *The work done by a harmonic force of amplitude P_0 and frequency ω, in phase with a harmonic velocity $v_0 = x_0\omega$ of the same frequency, is $\pi P_0 v_0/\omega = \pi P_0 x_0$ over a whole cycle.*

Example: A force $10 \sin 2\pi 60t$ (units are pounds and seconds) is acting on a displacement of $\frac{1}{10} \sin [2\pi 60t - 45°]$ (units are inches and seconds). What is the work done during the first second, and also during the first one-thousandth of a second?

Solution: The force is 45 deg. out of phase with the displacement and can be resolved into two components, each of amplitude $10/\sqrt{2}$ lb., being in phase and 90 deg. out of phase with the displacement. The component in phase with the displacement does no work. That 90 deg. out of phase with the displacement does per cycle $\pi P_0 x_0 = \pi \cdot \dfrac{10}{\sqrt{2}} \cdot \dfrac{1}{10} = 2.22$ in. lb. of work. During the first second there are 60 cycles so that the work performed is $60 \times 2.22 = 133$ in. lb.

During the first one-thousandth of a second there are $60/1,000 = 0.06$ cycle, so that the vectors in the diagram turn through only 0.06×360 deg. $= 21.6$ deg. Formula

(1.9) holds only for a full cycle. For part of a cycle the integration has to be performed in full:

$$W = \int P \, dx = \int P_0 \sin \omega t \cdot x_0 \omega \cos (\omega t - \varphi) \, dt$$

$$= P_0 x_0 \int_0^{21.6°} \sin (\omega t) \cos (\omega t - \varphi) d(\omega t)$$

$$= 10 \cdot \tfrac{1}{10} \int_0^{21.6°} \sin (\omega t)[\cos (\omega t) \cos \varphi + \sin (\omega t) \sin \varphi] d(\omega t)$$

$$= \cos \varphi \int_0^{21.6°} \sin (\omega t) \cos (\omega t) d(\omega t) + \sin \varphi \int_0^{21.6°} \sin^2 (\omega t) d(\omega t)$$

$$= \tfrac{1}{2} \cos \varphi \sin^2 (\omega t) + \sin \varphi[\tfrac{1}{2}\omega t - \tfrac{1}{4} \sin 2\omega t] \Big|_0^{21.6°}$$

$$= \frac{1}{2} \cos 45° \sin^2 21.6° + \frac{1}{2} \frac{21.6}{57.3} \sin 45° - \frac{1}{4} \sin 45° \sin 43.2°$$

$$= \frac{1}{2} \times 0.707 \times 0.368^2 + \frac{1}{2} \frac{21.6}{57.3} \times 0.707 - \frac{1}{4} \times 0.707 \times 0.685$$

$$= 0.048 + 0.133 - 0.121 = +0.060 \text{ in. lb.}$$

This is considerably less than one-thousandth part of the work performed in a whole second, because during this particular $1/1{,}000$ sec. the force is very small, varying between 0 and $0.368 P_0$.

1.7. Non-harmonic Periodic Motions. A "periodic" motion has the property of repeating itself in all details after a certain time interval, the "period" of the motion. All harmonic motions are periodic, but not every periodic motion is harmonic. For example, Fig. 1.17 represents the motion

$$x = a \sin \omega t + \frac{a}{2} \sin 2\omega t,$$

the superposition of two sine waves of different frequency. It is periodic but not harmonic. The mathematical theory shows that any periodic curve $f(t)$ of frequency ω can be split up into a series of sine curves of frequencies ω, 2ω, 3ω, 4ω, etc. Or

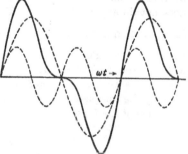

FIG. 1.17. The sum of two harmonic motions of different frequencies is not a harmonic motion.

$$f(t) = A_0 + A_1 \sin (\omega t + \varphi_1) + A_2 \sin (2\omega t + \varphi_2) + A_3 \sin (3\omega t + \varphi_3) + \cdots \quad (1.10)$$

provided that $f(t)$ repeats itself after each interval $T = 2\pi/\omega$. The amplitudes of the various waves A_1, A_2, . . . , and their phase angles φ_1, φ_2, . . . , can be determined analytically when $f(t)$ is given. The series (1.10) is known as a $Fourier\ series$.

The second term is called the fundamental or first harmonic of $f(t)$

and in general the $(n + 1)$st term of frequency $n\omega$ is known as the nth harmonic of $f(t)$. Since

$$\sin (n\omega t + \varphi_n) = \sin n\omega t \cos \varphi_n + \cos n\omega t \sin \varphi_n$$

the series can also be written as

$$
\begin{aligned}
f(t) = a_1 \sin \omega t + a_2 \sin 2\omega t + \cdots + a_n \sin n\omega t + \cdots + b_0 \\
+ b_1 \cos \omega t + b_2 \cos 2\omega t + \cdots + b_n \cos n\omega t + \cdots
\end{aligned} \quad (1.11)
$$

The constant term b_0 represents the "average" height of the curve $f(t)$ during a cycle. For a curve which is as much above the zero line during a cycle as it is below, the term b_0 is zero. The amplitudes $a_1 \ldots a_n$ $\ldots, b_1 \ldots b_n \ldots$ can be determined by applying the three energy theorems of page 16.

Consider for that purpose $f(t)$ to be a *force*, and let this nonharmonic force act on a point having the harmonic velocity $\sin n\omega t$. Now consider the force $f(t)$ as the sum of all the terms of its Fourier series and determine the work done by each harmonic term separately. All terms of the force except $a_n \sin n\omega t$ and $b_n \cos n\omega t$ are of a frequency different from that of the velocity $\sin n\omega t$, so that no work per cycle is done by them. Moreover, $b_n \cos n\omega t$ is 90 deg. out of phase with the velocity so that this term does not do any work either. Thus the total work done is that of the force $a_n \sin n\omega t$ on the velocity $\sin n\omega t$; it is $\dfrac{\pi \cdot a_n \cdot 1}{n\omega}$ per cycle of the $n\omega$-frequency. Per cycle of the fundamental frequency (which is n times as long), the work is $\pi a_n / \omega$.

Thus the amplitude a_n is found to be ω/π times as large as the work done by the complete non-harmonic force $f(t)$ on a velocity $\sin n\omega t$ during one cycle of the force. Or, mathematically

$$a_n = \frac{\omega}{\pi} \int_0^{\frac{2\pi}{\omega}} f(t) \sin n\omega t \, dt \quad (1.12a)$$

By assuming a velocity $\cos n\omega t$ instead of $\sin n\omega t$ and repeating the argument, the meaning of b_n is disclosed as

$$b_n = \frac{\omega}{\pi} \int_0^{\frac{2\pi}{\omega}} f(t) \cos n\omega t \, dt \quad (1.12b)$$

The relations between a_n, b_n and the quantities A_n, φ_n of Eq. (1.10) are as shown in Eq. (1.6), page 5, so that

$$A_n^2 = a_n^2 + b_n^2 \quad \text{and} \quad \tan \varphi_n = \frac{b_n}{a_n}.$$

Thus the work done by a non-harmonic force of frequency ω upon a harmonic velocity of frequency $n\omega$ is merely the work of the component of the nth harmonic of that force in phase with the velocity; the work of all other harmonics of the force is zero when integrated over a complete force cycle.

With the aid of the formulas (1.12) it is possible to find the a_n and b_n for any periodic curve which may be given. The branch of mathematics which is concerned with this problem is known as *harmonic analysis*.

Example 1: Consider the periodic "square-top wave," $f(t) = F_0 =$ constant for $0 < \omega t < \pi$ and $f(t) = -F_0$ for $\pi < \omega t < 2\pi$. The Fourier coefficients are found from Eqs. (1.12) as follows:

$$a_n = \frac{\omega}{\pi}\left(F_0 \int_0^{\frac{\pi}{\omega}} \sin n\omega t\, dt - F_0 \int_{\frac{\pi}{\omega}}^{\frac{2\pi}{\omega}} \sin n\omega t\, dt\right)$$

$$= \frac{F_0}{n\pi}\left(- \cos n\omega t \,\Big|_{t=0}^{t=\frac{\pi}{\omega}} + \cos n\omega t \,\Big|_{t=\frac{\pi}{\omega}}^{t=\frac{2\pi}{\omega}}\right)$$

$$= \frac{F_0}{n\pi}\left(- \cos n\pi + \cos 0 + \cos 2n\pi - \cos n\pi\right)$$

For even orders $n = 0, 2, 4, \ldots$, all angles are multiples of 360 deg. $= 2\pi$, and the four terms in the parentheses cancel each other. For odd orders $n = 1, 3, 5, \ldots$, we have $\cos n\pi = -1$, while $\cos 0 = \cos 2n\pi = +1$, so that the value in the parentheses is 4, and $a_n = 4F_0/n\pi$ ($n =$ odd).

$$b_n = \frac{\omega}{\pi}\left(F_0 \int_0^{\frac{\pi}{\omega}} \cos n\omega t\, dt - F_0 \int_{\frac{\pi}{\omega}}^{\frac{2\pi}{\omega}} \cos n\omega t\, dt\right)$$

$$= \frac{F_0}{n\pi}\left(\sin n\omega t\,\Big|_0^{\frac{\pi}{\omega}} - \sin n\omega t\,\Big|_{\frac{\pi}{\omega}}^{\frac{2\pi}{\omega}}\right)$$

$$= \frac{F_0}{n\pi}(0 - 0 - 0 + 0) = 0$$

Hence the "square wave" of height F_0 is

$$f(t) = \frac{4F_0}{\pi}\left(\sin \omega t + \frac{1}{3}\sin 3\omega t + \frac{1}{5}\sin 5\omega t + \cdots\right)$$

Example 2: The curve c of Fig. 8.17 (page 352) shows approximately the damping force caused by turbulent air on a body in harmonic motion. If the origin of coordinates of Fig. 8.17 is displaced one-quarter cycle to the left, the mathematical expression for the curve is

$$f(\omega t) = \sin^2 \omega t \quad \text{for } 0 < \omega t < \pi$$
$$f(\omega t) = -\sin^2 \omega t \quad \text{for } \pi < \omega t < 2\pi$$

Find the amplitudes of the various harmonics of this curve.

Solution: The curve to be analyzed is an "antisymmetric" one, *i.e.*, the values of $f(\omega t)$ are equal and opposite at two points $\pm \omega t$ at equal distances on both sides of the origin. Sine waves are antisymmetric and cosine waves are symmetric. An antisymmetric curve cannot have cosine components. Hence, all b_n are zero. This can be further verified by sketching the integrand of Eq. (1.12) in the manner of Fig. 1.16 and showing that the various contributions to the integral cancel each other. The constant term $b_0 = 0$, because the curve has no average height. For the sine components we find

$$a_n = \frac{\omega}{\pi} \int_0^{\frac{2\pi}{\omega}} f(\omega t) \sin n\omega t \, dt$$

$$= \frac{1}{\pi} \left[\int_0^{\pi} \sin^2 \omega t \sin n\omega t \, d(\omega t) - \int_{\pi}^{2\pi} \sin^2 \omega t \sin n\omega t \, d\omega t \right]$$

The integrands can be transformed by means of the last formula on page 15,

$$\sin^2 \omega t \sin n\omega t = (\tfrac{1}{2} - \tfrac{1}{2} \cos 2\omega t) \sin n\omega t$$
$$= \tfrac{1}{2} \sin n\omega t - \tfrac{1}{4} \sin (n + 2)\omega t - \tfrac{1}{4} \sin (n - 2)\omega t$$

The indefinite integral of this is

$$F(\omega t) = \frac{-1}{2n} \cos n\omega t + \frac{1}{4(n + 2)} \cos (n + 2)\omega t + \frac{1}{4(n - 2)} \cos (n - 2)\omega t$$

The harmonic a_n is $1/\pi$ times the definite integrals.
Since $F(2\pi) = F(0)$, we have

$$a_n = \frac{1}{\pi} \left[F(\pi) - F(0) - F(2\pi) + F(\pi) \right] = \frac{2}{\pi} \left[F(\pi) - F(0) \right]$$
$$= \frac{2}{\pi} (\cos n\pi - 1) \cdot \left[-\frac{1}{2n} + \frac{1}{4(n + 2)} + \frac{1}{4(n - 2)} \right] = \frac{4}{\pi} \frac{\cos n\pi - 1}{n(n^2 - 4)}$$

For even values of n the a_n thus is zero, while only for odd values of n the harmonic exists. In particular for $n = 1$, we have for the fundamental harmonic

$$a_1 = \frac{8}{3\pi} = 0.85$$

Thus the amplitude of the fundamental harmonic is 85 per cent of the maximum amplitude of the curve itself, and the Fourier series is

$$f(\omega t) = \frac{8}{3\pi} \left(\sin \omega t - \frac{1}{5} \sin 3\omega t - \frac{1}{35} \sin 5\omega t - \frac{1}{63} \sin 7\omega t - \cdots \right)$$

The evaluation of the integrals (1.12) by calculation can be done only for a few simple shapes of $f(t)$. If $f(t)$ is a curve taken from an actual vibration record or from an indicator diagram, we do not even possess a mathematical expression for it. However, with the aid of the curve so obtained the integrals can be determined either graphically or numerically or by means of a machine, known as a *harmonic analyzer*.

Such a harmonic analyzer operates on the same principle as Watt's steam-engine indicator. The indicator traces a closed curve of which the ordinate is the steam pressure (or piston force) and the abscissa is the piston displacement. The area of this closed curve is the work done by the piston force per cycle. The formulas (1.12) state that the coefficients a_n or b_n are ω/π times the work done per cycle by the force $f(t)$ on a certain displacement of which the velocity is expressed by sin $n\omega t$. To obtain complete correspondence between the two cases, we note that sin $n\omega t$ is the velocity of $\dfrac{-1}{n\omega}$ cos $n\omega t$, so that (1.12a) can be written in the modified form

$$a_n = \frac{-1}{n\pi} \int f(t)d(\cos n\omega t) = \frac{-1}{n\pi} \oint P\, ds$$

The symbol $\oint$ indicates that the integration extends over the closed curve described during one cycle of the force $f(t)$.

The machine is shown schematically in Fig. 1.18. Its output is a pen point D which scribes a curve on a piece of paper fastened to the table E. To complete the analogy with Watt's indicator the vertical motion of the pen D must follow the force $f(t)$, while the horizontal motion of

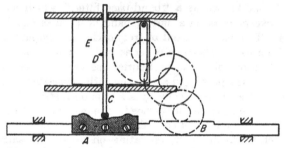

Fig. 1.18. The harmonic analyzer, an instrument operating on the same principle as Watt's steam-engine indicator.

the table E must follow the velocity cos $n\omega t$. The vertical motion of the pen D is ensured by making it ride on a template A, representing one cycle of the curve $f(t)$ which is to be analyzed. The template A is fastened to a rack and a pinion B, which is rotated by an electric motor. The arm C is guided so that it can move in its longitudinal direction only and is pressed lightly against the template by a spring. Thus the vertical motion of the pen D on the arm C is expressed by $f(t)$. The table E moves horizontally and is driven by a scotch crank and gear which is connected by suitable intermediate gears to B so that E oscillates n times harmonically, while A moves through the length of the diagram. The machine has with it a box of spare gears so that any gear ratio n from 1 to 30 can be obtained by replacing one gear in the train by another.

The horizontal motion of the table E is expressed by sin $n\omega t$ or by cos $n\omega t$, depending on the manner in which the gears are interlocked. The point D will thus trace a closed curve on the table, for which the area equals a_n or b_n (multiplied by a constant $1/n\pi$). Instead of actually tracing this curve, the instrument usually carries a planimeter of which one point is attached to E and the other end to D, so that the area is given directly by the planimeter reading.

Harmonic analyzers have been built on other principles as well. An interesting optical method using the sound tracks of motion picture films was invented by Wente and constructed by Montgomery, both of the Bell Telephone Laboratories.

Electrical harmonic analyzers giving an extremely rapid analysis of the total harmonics $A_n = \sqrt{a_n^2 + b_n^2}$ [Eqs. (1.10) and (1.11)], without giving information on the phase angles φ_n [or the ratios a_n/b_n, Eq. (1.10)], are available on the market. They have been developed by the Western Electric Company for sound or noise analysis and require the original curve to be available in the form of an electric voltage, varying with the time, such as results from an electric vibration pickup (page 62) or a microphone. This voltage, after proper amplification, is fed into an electric network known as a "band-pass filter," which suppresses all frequencies except those in a narrow band of a width equal to 5 cycles per second. This passing band of frequencies can be laid anywhere in the range from 10 to 10,000 cycles per second. When a periodic (steady-state) vibration or noise is to be Fourier-analyzed, a small motor automatically moves the pass band across the entire spectrum and the resulting analysis is drawn graphically by a stylus on a strip of waxed paper, giving the harmonic amplitudes vs. the frequency from 10 to 10,000 cycles per second, all in a few minutes. The record is immediately readable.

Another electrical analyzer, operating on about the same principle but without graphic recording, is marketed by the General Radio Company, Cambridge, Mass.

Problems 1 to 11.

CHAPTER 2

THE SINGLE-DEGREE-OF-FREEDOM SYSTEM

2.1. Degrees of Freedom. A mechanical system is said to have one degree of freedom if its geometrical position can be expressed at any instant by one number only. Take, for example, a piston moving in a cylinder; its position can be specified at any time by giving the distance from the cylinder end, and thus we have a system of one degree of freedom. A crank shaft in rigid bearings is another example. Here the position of the system is completely specified by the angle between any one crank and the vertical plane. A weight suspended from a spring in such a manner that it is constrained in guides to move in the up-and-down direction only is the classical single-degree-of-freedom vibrational system (Fig. 2.3).

Generally if it takes n numbers to specify the position of a mechanical system, that system is said to have n degrees of freedom. A disk moving in its plane without restraint has three degrees of freedom: the x- and y-displacements of the center of gravity and the angle of rotation about the center of gravity. A cylinder *rolling* down an inclined plane

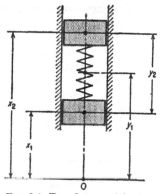

FIG. 2.1. Two degrees of freedom.

has one degree of freedom; if, on the other hand, it descends partly rolling and partly sliding, it has two degrees of freedom, the translation and the rotation.

A rigid body moving freely through space has six degrees of freedom, three translations and three rotations. Consequently it takes six numbers or "coordinates" to express its position. These coordinates are usually denoted as x, y, z, φ, ψ, χ. A system of two rigid bodies connected by springs or other ties in such a manner that each body can move only along a straight line and cannot rotate has two degrees of freedom (Fig. 2.1). The two quantities determining the position of such a system can be chosen rather arbitrarily. For instance, we may call the distance from a fixed point O to the first body x_1, and the distance from O to the second body x_2. Then x_1 and x_2 are the coordinates. However, we might also choose the distance from O to the center of gravity of the two bodies for one of the coordinates and call that y_1. For the other coordinate we might choose the distance between the two bodies, $y_2 = x_2 - x_1$. The pair of numbers x_1, x_2 describes the position completely, but the pair y_1, y_2 does it equally well. The latter choice has a certain practical advantage in this case, since usually we are not interested so much in the location of the system as a whole as in the stresses inside it. The stress in the spring of Fig. 2.1 is completely determined by y_2, so that for its calculation a knowledge of y_1 is not required. A suitable choice of the coordinates of a system of several degrees of freedom may simplify the calculations considerably.

It should not be supposed that a system of a single degree of freedom is always very simple. For example, a 12-cylinder gas engine, with a rigid crank shaft and a rigidly mounted cylinder block, has only one degree of freedom with all its moving pistons, rods, valves, cam shaft, etc. This is so because a single number (for instance, the angle through which the crank shaft has turned) determines completely the location of every moving part of the engine. However, if the cylinder block is mounted on flexible springs so that it can freely move in every direction (as is the case in many modern automobiles), the system has seven degrees of freedom, namely the six pertaining to the block as a rigid body in free space and the crank angle as the seventh coordinate.

A completely flexible system has an infinite number of degrees of freedom. Consider, for example, a flexible beam on two supports. By a suitable loading it is possible to bend this beam into a curve of any shape (Fig. 2.2). The description of this curve requires a function $y = f(x)$, which is equivalent to an infinite number of numbers. To each location

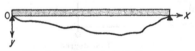

Fig. 2.2. A beam has an infinite number of degrees of freedom.

x along the beam, any deflection y can be given independent of the position of the other particles of the beam (within the limits of strength of the beam) and thus complete determination of the position requires as many values of y as there are points along the beam. As was the case in Fig. 2.1, the $y = f(x)$ is not the only set of numbers that can be taken to define the position. Another possible way of determining the deflection curve is by specifying the values of all its Fourier coefficients a_n and b_n [Eq. (1.11), page 18], which again are infinite in number.

2.2. Derivation of the Differential Equation. Consider a mass m suspended from a rigid ceiling by means of a spring, as shown in Fig. 2.3. The "stiffness" of the spring is denoted by its "spring constant" k, which by definition is *the number of pounds tension necessary to extend the spring 1 in.* Between the mass and the rigid wall there is also an oil or air dashpot mechanism. This is not supposed to transmit any force to the mass as long as it is at rest, but as soon as the mass moves, the "damping force" of the dashpot is $c\dot{x}$ or $c\,dx/dt$, i.e., proportional to the velocity and directed opposite to it. The quantity c is known as the *damping constant* or more at length as the *coefficient of viscous damping.*

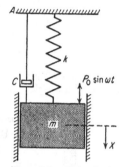

FIG. 2.3. The fundamental single-degree-of-freedom system.

The damping occurring in actual mechanical systems does not always follow a law so simple as this $c\dot{x}$-relation; more complicated cases often arise. Then, however, the mathematical theory becomes very involved (see Chap. 8, pages 361 and 373), whereas with "viscous" damping the analysis is comparatively simple.

Let an external alternating force $P_0 \sin \omega t$ be acting on the mass, produced by some mechanism which we need not specify in detail. For a mental picture assume that this force is brought about by somebody pushing and pulling on the mass by hand.

The problem consists in calculating the motions of the mass m, due to this external force. Or, in other words, if x be the distance between any instantaneous position of the mass during its motion and the equilibrium position, we have to find x as a function of time. The "equation of motion," which we are about to derive, is nothing but a mathematical expression of Newton's second law,

$$\text{Force} = \text{mass} \times \text{acceleration}$$

All forces acting on the mass will be considered positive when acting downward and negative when acting upward.

The spring force has the *magnitude* kx, since it is zero when there is no extension x. When $x = 1$ in., the spring force is k lb. by definition, and consequently the spring force for any other value of x (in inches) is kx (in pounds), because the spring follows Hooke's law of proportionality between force and extension.

The *sign* of the spring force is negative, because the spring pulls *upward* on the mass when the displacement is *downward*, or the spring force is negative when x is positive. Thus the spring force is expressed by $-kx$.

The damping force acting on the mass is also negative, being $-c\dot{x}$, because, since it is directed *against* the velocity $\dot{x}$, it acts *upward* (negative) while $\dot{x}$ is directed *downward* (positive). The three downward forces acting on the mass are

$$-kx - c\dot{x} + P_0 \sin \omega t$$

Newton's law gives

$$m\frac{d^2x}{dt^2} = m\ddot{x} = -kx - c\dot{x} + P_0 \sin \omega t,$$

or $$m\ddot{x} + c\dot{x} + kx = P_0 \sin \omega t \qquad (2.1)$$

This very important equation† is known as the *differential equation of motion of a single-degree-of-freedom system.* The four terms in Eq. (2.1)
are the inertia force, the damping force, the spring force, and the external force.

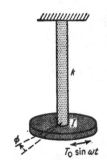

Before proceeding to a calculation of x from Eq. (2.1), *i.e.*, to a solution of the differential equation, it is well to consider some other problems that will lead to the same equation.

2.3. Other Cases. Figure 2.4 represents a disk of moment of inertia I attached to a shaft of torsional stiffness k, defined as *the torque in inch-pounds necessary to produce 1 radian twist at the disk.* Consider the twisting motion of the disk under the influence of an externally applied torque $T_0 \sin \omega t$. This again is a one-degree-of-freedom problem since the torsional displacement of the disk from its equilibrium position can be expressed by a single quantity, the angle φ. Newton's law for a rotating body states that

FIG. 2.4. The torsional one-degree-of-freedom system.

$$\text{Torque} = \text{moment of inertia} \times \text{angular acceleration} = I\frac{d^2\varphi}{dt^2} = I\ddot{\varphi}$$

As in the previous problem there are *three* torques acting on the disk: the spring torque, damping torque, and external torque. The spring torque is $-k\varphi$, where φ is measured in radians. The negative sign is evident for the same reason that the spring force in the previous case was $-kx$. The damping torque is $-c\dot{\varphi}$, caused by a dashpot mechanism not shown in the figure. The "damping constant" c in this problem is the *torque on the disk caused by an angular speed of rotation of 1 radian per second.*

† In the derivation, the effect of gravity has been omitted. The amplitude x was measured from the "equilibrium position," *i.e.*, from the position where the downward force mg is held in equilibrium by an upward spring force $k\delta$ (δ being the deflection of the spring due to gravity). It would have been possible to measure x_1 from the position of the *unstressed* spring, so that $x_1 = x + \delta$. In Eq. (2.1), then, x must be replaced by x_1, and on the right-hand side a force mg must be added. This leads to the same result (2.1).

The external torque is $T_0 \sin \omega t$, so that Newton's law leads to the differential equation

$$I\ddot{\varphi} + c\dot{\varphi} + k\varphi = T_0 \sin \omega t \qquad (2.2)$$

which has the same form as Eq. (2.1).

As a third example, consider an electric circuit with an alternating-current generator, a condenser C, resistance R, and inductance L all in series.   Instead of Newton's law, use the relation that the instantaneous voltage of the generator $e = E_0 \sin \omega t$ is equal to the sum of the three voltages across C, R, and L. Let i be the instantaneous value of the current in the circuit in the direction indicated in Fig. 2.5.   According

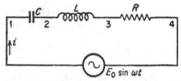

FIG. 2.5. The electrical single-degree-of-freedom circuit.

to Ohm's law, the voltage across the resistance is $V_3 - V_4 = Ri$. The voltage across the inductance is $V_2 - V_3 = L\dfrac{di}{dt}$. For the condenser, the relation $Q = CV$ holds, where Q is the charge, C the capacitance, and V the voltage. The charge Q can be expressed in terms of i, as follows: If the current i flows during a time element dt, the quantity of electricity transported through the circuit is $i\,dt$. This does not flow *through* the condenser but merely increases its charge so that

$$dQ = i\,dt\dagger$$

Hence $i = \dfrac{dQ}{dt} = \dot{Q}$ or $Q = \int i\,dt$

To show that this electric circuit behaves in the same manner as the vibrating mass of Fig. 2.3 it is better to work with the charge Q rather than with the more familiar current i.   The various voltage drops can be written

$$V_1 - V_2 = \frac{Q}{C}$$

$$V_2 - V_3 = L\frac{di}{dt} = L\frac{d^2Q}{dt^2} = L\ddot{Q}$$

$$V_3 - V_4 = Ri = R\frac{dQ}{dt} = R\dot{Q}$$

As the sum of these three voltage drops must equal the generator voltage,

† The letter i unfortunately is dotted. To avoid confusion it is agreed that i shall mean the current itself and that for its differential coefficient the Leibnitz notation di/dt will be used.

the differential equation is

$$L\ddot{Q} + R\dot{Q} + \frac{1}{C}Q = E_0 \sin \omega t \qquad (2.3)$$

which is of exactly the same form as Eq. (2.1).

Therefore, the linear, torsional, and electrical cases thus far discussed all lead to the same differential equation. The translation from one case to another follows directly from the table shown below.

All the mechanical statements made have their electrical analogues and *vice versa*. For example, it was stated that "the voltage across the inductance L is $L\frac{di}{dt}$." In mechanical language this would be expressed as "the force of the mass m is $m\frac{dv}{dt}$." A mechanical statement would be "The energy stored in the mass is $\frac{1}{2}mv^2$." The electrical analogue is "The energy stored in the inductance is $\frac{1}{2}Li^2$."

Linear		Torsional		Electrical	
Mass	m	Moment of inertia	I	Inductance	L
Stiffness	k	Torsional stiffness	k	1/Capacitance	$1/C$
Damping	c	Torsional damping	c	Resistance	R
Force	$P_0 \sin \omega t$	Torque	$T_0 \sin \omega t$	Voltage	$E_0 \sin \omega t$
Displacement	x	Angular displacement	φ	Condenser charge	Q
Velocity	$\dot{x} = v$	Angular velocity	$\dot{\varphi} = \omega$	Current	$\dot{Q} = i$

Nor are these three cases the only ones that are determined by Eq. (2.1). Any system with inertia, elasticity, and damping proportional to the velocity, for which the displacements can be described by a single quantity, belongs to this class. For example, consider two disks of moment of inertia I_1 and I_2, joined by a shaft of torsional stiffness k in-lb./radian (Fig. 2.6). On the first disk the torque $T_0 \sin \omega t$ is made to act, while there is a damping with constant c, proportional to the velocity of twist in the shaft. What will be the motion? There are two disks, each of which can assume an angular position independent of the other by twisting the shaft. Apparently, therefore, this is a "two-degree-of-freedom" system. However, the quantity in which the engineer is most interested is the angle of twist of the shaft, and it is possible to express the motion in terms of this quantity only. Let φ_1 and φ_2 be the angular displace-

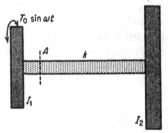

Fig. 2.6. Torsional vibrations of two disks on an elastic shaft.

ments of the two disks, then $\varphi_1 - \varphi_2$ is the shaft twist, $k(\varphi_1 - \varphi_2)$ is the shaft torque, and $c(\dot{\varphi}_1 - \dot{\varphi}_2)$ is the damping torque. Apply Newton's law to the first disk,

$$T_0 \sin \omega t = I_1 \ddot{\varphi}_1 + k(\varphi_1 - \varphi_2) + c(\dot{\varphi}_1 - \dot{\varphi}_2)$$

and to the second disk,

$$0 = I_2 \ddot{\varphi}_2 + k(\varphi_2 - \varphi_1) + c(\dot{\varphi}_2 - \dot{\varphi}_1)$$

Divide the first equation by I_1, the second by I_2, and subtract the results from each other:

$$\frac{T_0}{I_1} \sin \omega t = (\ddot{\varphi}_1 - \ddot{\varphi}_2) + \left(\frac{k}{I_1} + \frac{k}{I_2}\right)(\varphi_1 - \varphi_2) + \left(\frac{c}{I_1} + \frac{c}{I_2}\right)(\dot{\varphi}_1 - \dot{\varphi}_2)$$

Call the twist angle $\varphi_1 - \varphi_2 = \psi$, and multiply the whole equation by $I_1 I_2/(I_1 + I_2)$,

$$\frac{I_1 I_2}{I_1 + I_2} \ddot{\psi} + c\dot{\psi} + k\psi = \frac{I_2 T_0}{I_1 + I_2} \cdot \sin \omega t \qquad (2.4)$$

giving again an equation of the form (2.1). Of course, this equation, when solved, tells us only about the twist in the shaft or about the *relative* motion of the two disks with respect to each other. No information can be gained from it as to the motions of the disks individually.

A variant of Fig. 2.6 is shown in Fig. 2.7, in the shaft of which is inserted a gear-and-pinion system. Let the disks again have the moments of inertia I_1 and I_2, and assume the gears G and P to be without any inertia whatsoever. Also assume the gear teeth to be stiff, so that the torsional flexibility is limited to the shafts k_1 and k_2. The gear ratio is n.

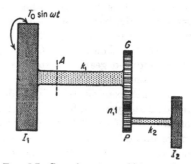

FIG. 2.7. Geared system which can be reduced to the system of Fig. 2.6.

The differential equation for Fig. 2.7 could be derived from Newton's law directly, but suppose we reduce Fig. 2.7 to Fig. 2.6 by omitting the gears and replacing k_2, I_2, and ψ by other "equivalent quantities" so that the differential equation (2.4) can be applied.

In Fig. 2.6 the elasticity k can be determined experimentally by clamping I_2 and applying a constant torque T_0 to I_1. This causes I_1 to deflect through an angle φ_0, so that $k = T_0/\varphi_0$. Repeat this experiment with Fig. 2.7., *i.e.*, clamp I_2 and apply T_0 to I_1. On account of the gears the torque in the shaft k_2 is T_0/n, and the angle of twist of k_2 is therefore

T_0/nk_2. Since I_2 is clamped, this is the angle of rotation of the pinion P. The angle of the gear G is n times smaller or T_0/n^2k_2. Add to this the angle T_0/k_1 for the shaft k_1 and we have the angular displacement of I_1. Thus the equivalent k is

$$\frac{1}{k} = \frac{\varphi}{T_0} = \frac{1}{k_1} + \frac{1}{n^2k_2}$$

Now consider the inertia. The inertia I_2 in Fig. 2.6 could be determined by the following hypothetical experiment. Give I_1 (or the whole shaft k) a *constant* angular acceleration α. Then the shaft at the section A would experience a torque $T_0 = \alpha I_2$ coming from the right. Thus, $I_2 = T_0/\alpha$. Repeat this experiment in Fig. 2.7. The acceleration α in k_1 and G becomes $n\alpha$ in k_2 Hence, the torque in k_2 is $n\alpha I_2$. This is also the torque at the pinion P. The gear G makes it n times larger, so that the torque at A is $n^2\alpha I_2$ and the equivalent of I_2 in the gearless system is n^2I_2. In general, therefore, a geared system (such as shown in Fig. 2.7) can be reduced to an equivalent non-geared system (Fig. 2.6) by the following rule:

Divide the system into separate parts each of which has the same speed within itself. (In Fig. 2.7 there are two such parts but in general there may be several.) Choose one of these parts as the base and assign numbers n to each of the other parts so that n is the speed ratio with respect to the base. ($n > 1$ for speeds higher than the base speed; the n of the base is unity.) Then, *remove all gears and multiply all spring constants k and all inertias I by the factors n^2*. The differential equation of the reduced gearless system is then the same as that of the original geared construction.

The last example to be considered resembles the first one in many respects and yet is different. Instead of having the force $P_0 \sin \omega t$ acting on the mass of Fig. 2.3, the upper end or ceiling A of the spring is made to move up and down with an amplitude a_0, the motion of A being determined by $a_0 \sin \omega t$. It will be shown that this *motion* of the top of the spring is completely equivalent to a *force* on the suspended mass.

Again let the downward displacement of the mass be x; then, since the top of the spring moves as $a_0 \sin \omega t$, the spring extension at any time will be $x - a_0 \sin \omega t$. The spring force is thus $-k(x - a_0 \sin \omega t)$ and the damping force is $-c(\dot{x} - a_0\omega \cos \omega t)$. Newton's law gives

$$m\ddot{x} + k(x - a_0 \sin \omega t) + c(\dot{x} - a_0\omega \cos \omega t) = 0$$

or $$m\ddot{x} + c\dot{x} + kx = ka_0 \sin \omega t + ca_0\omega \cos \omega t$$

By Eq. (1.6), page 5, the sum of a sine and a cosine wave of the same frequency is again a harmonic function, so that

$$m\ddot{x} + c\dot{x} + kx = \sqrt{(ka_0)^2 + (ca_0\omega)^2} \sin (\omega t + \varphi) \qquad (2.5)$$

Therefore, a motion of the top of the spring with amplitude a_0 is equivalent to a force on the mass with amplitude $\sqrt{(ka_0)^2 + (ca_0\omega)^2}$. The expressions ka_0 and $c\omega a_0$ in the radical are the maxima of the spring force and damping force, while the entire radical is the maximum value of the total force for the case where the mass is clamped, *i.e.*, where the x-motion is prevented.

Example: Find the differential equation of the *relative* motion y between the mass and the ceiling of Fig. 2.3, in which $P_0 = 0$ and in which the ceiling is moved harmonically up and down.

$$y = x - a_0 \sin \omega t$$

Solution: We have by differentiation:

$$x = y + a_0 \sin \omega t$$
$$\dot{x} = \dot{y} + a_0\omega \cos \omega t$$
$$\ddot{x} = \ddot{y} - a_0\omega^2 \sin \omega t$$

Substitute these into Eq. (2.5):

$$m\ddot{y} - ma_0\omega^2 \sin \omega t + c\dot{y} + ca_0\omega \cos \omega t + ky + ka_0 \sin \omega t$$
$$= ka_0 \sin \omega t + ca_0\omega \cos \omega t$$

or
$$m\ddot{y} + c\dot{y} + ky = ma_0\omega^2 \sin \omega t \tag{2.6}$$

Thus the relative motion between the mass and the moving ceiling acts in the same manner as the absolute motion of the mass with a ceiling at rest and with a force of amplitude $ma_0\omega^2$ acting on the mass. The right-hand side of (2.6) is the inertia force of the mass if it were moving at amplitude a_0; hence, it can be considered as the force that has to be exerted at the top of the spring if the spring is made stiff, *i.e.*, if the y-motion is prevented.

2.4. Free Vibrations without Damping. Before developing a solution of the general equation (2.1), it is useful to consider first some important simplified cases. If there is no external or impressed force $P_0 \sin \omega t$ and no damping ($c = 0$), the expression (2.1) reduces to

$$m\ddot{x} + kx = 0 \tag{2.7}$$

or
$$\ddot{x} = -\frac{k}{m}x$$

or, in words: *The deflection x is such a function of the time that when it is differentiated twice, the same function is again obtained, multiplied by a negative constant.* Even without a knowledge of differential equations, we may remember that such functions exist, *viz.*, sines and cosines, and a trial reveals that $\sin t \sqrt{k/m}$ and $\cos t \sqrt{k/m}$ are actually solutions of (2.7). The most general form in which the solution of (2.7) can be written is

$$x = C_1 \sin t \sqrt{\frac{k}{m}} + C_2 \cos t \sqrt{\frac{k}{m}} \tag{2.8}$$

where C_1 and C_2 are arbitrary constants. That (2.8) is a solution of (2.7) can be verified easily by differentiating (2.8) twice and then substituting in (2.7); that there are no solutions of (2.7) other than (2.8) need not be proved here; it is true and may be taken for granted.

Let us now interpret (2.8) physically. First, it is seen that the result as it stands is very indefinite; the constants C_1 and C_2 may have any value we care to assign to them. But the problem itself was never fully stated. The result (2.8) describes *all* the motions the system of mass and spring is capable of executing. One among others is the case for which $C_1 = C_2 = 0$, giving $x = 0$, which means that the mass remains permanently at rest.

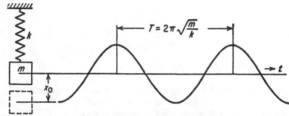

FIG. 2.8. Undamped free vibration starting from an initial displacement.

We now specify more definitely that the mass is pulled out of its equilibrium position to $x = x_0$ and then released without initial velocity. Measuring the time from the instant of release, the two conditions are

At $t = 0$, $x = x_0$ and $\dot{x} = 0$

The first condition substituted into (2.8) gives

$$x_0 = C_1 \cdot 0 + C_2 \cdot 1 \quad \text{or} \quad C_2 = x_0$$

For the second condition, Eq. (2.8) must be differentiated first and then we get

$$0 = C_1 \sqrt{\frac{k}{m}} \cdot 1 - C_2 \cdot \sqrt{\frac{k}{m}} \cdot 0 \quad \text{or} \quad C_1 = 0$$

Substitution of these results in (2.8) leads to the specific solution

$$x = x_0 \cos t \sqrt{\frac{k}{m}} \tag{2.8a}$$

This represents an *undamped vibration,* one cycle of which occurs when $t \sqrt{k/m}$ varies through 360 deg. or 2π radians (Fig. 2.8). Denoting the time of a cycle or the *period* by T, we thus have

$$\sqrt{\frac{k}{m}} \cdot T = 2\pi \quad \text{or} \quad T = 2\pi \sqrt{\frac{m}{k}} \tag{2.9}$$

It is customary to denote $\sqrt{k/m}$ by ω_n, called the "natural circular frequency." This value $\sqrt{k/m} = \omega_n$ is the angular velocity of the rotating vector which represents the vibrating motion (see page 3).

The reciprocal of T or the *natural frequency* f_n is

$$f_n = \frac{1}{T} = \frac{1}{2\pi}\sqrt{\frac{k}{m}} = \frac{\omega_n}{2\pi} \qquad (2.10)$$

measured in cycles per second. Hence it follows that if m is replaced by a mass twice as heavy, the vibration will be $\sqrt{2}$ times as slow as before. Also, if the spring is made twice as weak, other things being equal, the vibration will be $\sqrt{2}$ times as slow. On account of the absence of the impressed force $P_0 \sin \omega t$, this vibration is called a *free vibration*.

If we start with the assumption that the motion is harmonic, the frequency can be calculated in a very simple manner from an *energy consideration*. In the middle of a swing the mass has considerable kinetic energy, whereas in either extreme position it stands still for a moment and has no kinetic energy left. But then the spring is in a state of tension (or compression) and thus has elastic energy stored in it. At any position between the middle and the extreme, there is both elastic and kinetic energy, the sum of which is constant since external forces do no work on the system. Consequently, the kinetic energy in the middle of a stroke must be equal to the elastic energy in an extreme position.

We now proceed to calculate these energies. The spring force is kx, and the work done on increasing the displacement by dx is $kx \cdot dx$. The potential or elastic energy in the spring when stretched over a distance x is $\int_0^x kx \cdot dx = \frac{1}{2}kx^2$. The kinetic energy at any instant is $\frac{1}{2}mv^2$. Assume the motion to be $x = x_0 \sin \omega t$, then $v = x_0\omega \cos \omega t$. The potential energy in the extreme position is $\frac{1}{2}kx_0^2$, and the kinetic energy in the neutral position, where the velocity is maximum, is $\frac{1}{2}mv_{\max}^2 = \frac{1}{2}m\omega^2 x_0^2$.

Therefore, $\frac{1}{2}kx_0^2 = \frac{1}{2}m\omega^2 x_0^2$

from which $\omega^2 = k/m$, independent of the amplitude x_0. This "energy method" of calculating the frequency is of importance. In Chaps. 4 and 6, dealing with systems of greater complexity, it will be seen that a frequency determination from the differential equation often becomes so complicated as to be practically impossible. In such cases a generalized energy method, known as the method of *Rayleigh*, will lead to a result (see pages 141 to 155).

The formula $\omega_n = \sqrt{k/m}$ may be written in a somewhat different form. The weight of the mass m is mg, and the deflection of the spring caused by this weight is mg/k. It is called the static deflection δ_{st} or static sag of the spring under the weight.

$$\delta_{st} = \frac{mg}{k}$$

Hence, $$\frac{k}{m} = \frac{g}{\delta_{st}}$$

or $$\omega_n = \sqrt{\frac{g}{\delta_{st}}} \qquad (2.11)$$

If δ_{st} is expressed in inches, $g = 386$ in./sec.^{2}, and the frequency is

$$f_n = \frac{\sqrt{386}}{2\pi}\sqrt{\frac{1}{\delta_{st}}} = 3.14\sqrt{\frac{1}{\delta_{st}}} \text{ cycles per second}$$

$$f_n = 188\sqrt{\frac{1}{\delta_{st}}} \text{ cycles per minute} \qquad (2.11a)$$

This relationship, which is very useful for quickly estimating natural frequencies or critical speeds, is shown graphically in Fig. 2.9. It appears as a straight line on log-log paper.

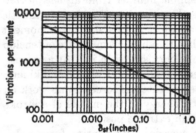

FIG. 2.9. Curve representing Eq. (2.11a) for the natural frequency of an undamped, single-degree system.

2.5. Examples. Consider some numerical examples of the application of the fundamental formula (2.10).

1. A steel bar of 1 by ½ in. cross section is clamped solidly in a vise at one end and carries a weight of 20 lb. at the other end (Fig. 2.10). (a) What is the frequency of the vibration if the distance between the weight and the vise is 30 in.? (b) What percentage change is made in the frequency by shortening the rod ¼ in.?

a. The weight of the bar itself is ½ by 1 by 30 cu. in. × 0.28 lb. per cubic inch or roughly 4 lb. The particles of the bar near the 20-lb. weight at its end vibrate with practically the same amplitude as that weight, whereas the particles near the clamped end vibrate hardly at all. This is taken account of by adding a fraction of the weight of the bar to the weight at its end. On page 155 it is shown that approximately one-quarter of the weight of the bar has to be thus added. Therefore the mass m in Eq. (2.10) is $21/g = {}^{21}\!/_{386}$ lb. in.$^{-1}$ sec.^{2}.

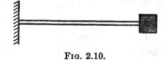

FIG. 2.10.

A force P at the end of a cantilever gives a deflection $\delta = Pl^3/3EI$. The spring constant by definition is

$$k = P/\delta = 3EI/l^3$$

The moment of inertia of the section is $I = \frac{1}{12}bh^3 = \frac{1}{24}$ (or $\frac{1}{96}$, depending upon whether the vibrations take place in the stiff or in the limber plane). The circular frequency is

$$\omega_n = \sqrt{\frac{k}{m}} = \sqrt{\frac{3 \cdot 30 \cdot 10^6 \cdot 386}{24 \cdot 30^3 \cdot 21}} = 50.4 \text{ radians per second}$$

The frequency $f_n = \omega_n/2\pi = 8.0$ cycles per second.

In case the bar vibrates in the direction of the weak side of the section, $I = \frac{1}{96}$, and f_n becomes one-half its former value, 4.0 cycles per second.

b. The question regarding the change in frequency due to a change in length can be answered as follows. The spring constant k is proportional to $1/l^3$, and the frequency consequently is proportional to $\sqrt{1/l^3} = l^{-\frac{3}{2}}$. Shortening the bar by 1 per cent will raise the frequency by $1\frac{1}{2}$ per cent. Thus the shortening of $\frac{1}{4}$ in. will increase f_n by $1\frac{1}{4}$ per cent.

2. As a second example consider a U-tube filled with water (Fig. 2.11). Let the total length of the water column be l, the tube cross section be A, and the mass of water per cubic inch be m_1. If the water oscillates back and forth, the mass in motion is $m_1 \cdot A \cdot l$. In this problem there is no specific "spring," but still the force of gravity tends to restore the water level to an equilibrium position. Thus we have a "gravity spring," of which the spring constant by definition is the force per unit deflection. Raise the level in one arm of the tube by 1 in., then it will fall in the other arm 1 in. This gives an unbalanced weight of 2 in. water column, causing a force of $(2m_1A) \cdot g$, which is the spring constant. Therefore the frequency is

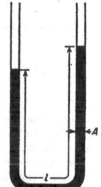

Fig. 2.11. Oscillations of a liquid column in a U-tube.

$$\omega_n = \sqrt{\frac{k}{m}} = \sqrt{\frac{2g}{l}}$$

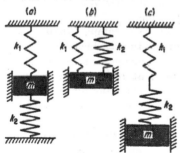

Fig. 2.12. Three systems with compound springs, which are equivalent to the system of Fig. 2.8. (a) and (b) have "parallel" springs; (c) has its springs "in series."

3. Consider the systems shown in Fig. 2.12, where a mass m is suspended from two springs k_1 and k_2 in three apparently different ways. However, the cases 2.12a and 2.12b are dynamically identical, because a downward deflection of 1 in. creates an upward force of $(k_1 + k_2)$ lb. in both cases.

Thus the natural frequency of such systems is

$$\omega_n = \sqrt{\frac{k_1 + k_2}{m}}$$

For Fig. 2.12c the situation is different. Let us pull downward on the mass with a force of 1 lb. This force will be transmitted through *both* springs in full strength. Their respective elongations are $1/k_1$ and $1/k_2$, the total elongation per pound being $\frac{1}{k_1} + \frac{1}{k_2}$. But, by definition, this is $1/k$, the reciprocal of the combined spring constant. Hence,

$$k = \frac{1}{\dfrac{1}{k_1} + \dfrac{1}{k_2}}$$

Rule: The combined spring constant of several "parallel" springs is $k = \Sigma k_n$; *for n springs "in series" the spring constant is found from* $1/k = \Sigma 1/k_n$.

For example, if a given coil spring of stiffness k is cut in two equal parts, each piece will have the stiffness $2k$. (It takes twice as much load to give to half the spring the same deflection as to the whole spring.) Putting the two half springs in series, we find, indeed, $\frac{1}{k} = \frac{1}{2k} + \frac{1}{2k}$.

It is of interest to note that this rule for compounding spring constants is exactly the same as that for finding the total conductance of series and parallel circuits in electrical engineering.

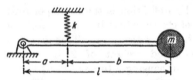

FIG. 2.13. The spring k as shown is equivalent to a fictitious spring of stiffness $k(a/l)^2$ placed at the mass m.

4. The last example to be discussed in this section is illustrated in Fig. 2.13. A massless, inflexible beam is hinged at one end and carries a mass m at the other end. At a distance a from the hinge there is a spring of stiffness k. What is the natural frequency of vibration of this system?

We shall consider the vibrations to be so small that the mass moves sensibly up and down only. In deriving the equation of motion on page 25, the spring force *on the mass* was equated to $m\ddot{x}$. In this case also we have to ask: What force has to be exerted *on the mass* in order to deflect it 1 in.? Let that force be F. Then from static equilibrium the force in the spring is $\frac{l}{a} \cdot F$. Since the deflection at the mass is 1 in., it is a/l in.

at the spring. This leads to a spring force $\frac{a}{l} \cdot k$. Hence

$$\frac{l}{a} \cdot F = \frac{a}{l} \cdot k \quad \text{or} \quad F = \left(\frac{a}{l}\right)^2 \cdot k$$

Therefore, the effective spring constant at the mass is $k \cdot (a/l)^2$. The effect of the stiffness of the spring is thus seen to diminish very fast when it is shifted to the left.

The frequency is

$$\omega_n = \frac{a}{l}\sqrt{\frac{k}{m}}$$

With the energy method of page 33 the calculation is as follows: Let the motion of the mass be $x = x_0 \sin \omega_n t$, where ω_n is as yet unknown. The amplitude of motion at the spring then is $x_0 a/l$ and the potential energy in the spring is $\frac{1}{2}k\delta^2 = \frac{1}{2}k(x_0 a/l)^2$. The kinetic energy of the mass is $\frac{1}{2}mv^2 = \frac{1}{2}m\omega_n^2 x_0^2$. Equating these two, the amplitude x_0 drops out and

$$\omega_n^2 = \frac{k}{m}\frac{a^2}{l^2}$$

Some of the problems at the end of this chapter can be solved more easily with the energy method than by a direct application of the formula involving $\sqrt{k/m}$.

2.6. Free Vibrations with Viscous Damping. It was seen that an undamped free vibration persists forever [Eq. (2.8) or (2.8a)]. Evidently this never occurs in nature; all free vibrations die down after a time. Therefore consider Eq. (2.1) with the damping term $c\dot{x}$ included, viz.:

$$m\ddot{x} + c\dot{x} + kx = 0 \tag{2.12}$$

The term "viscous damping" is usually associated with the expression $c\dot{x}$ since it represents fairly well the conditions of damping due to the viscosity of the oil in a dashpot. Other types of damping exist and will be discussed later (page 361). The solution of (2.12) cannot be found as simply as that of (2.7). However, if we consider the function $x = e^{st}$, where t is the time and s an unknown constant, it is seen that upon differentiation the same function results, but multiplied by a constant. This function, substituted in (2.12) permits us to divide by e^{st} and leads to an *algebraic* equation instead of a *differential* equation, which is a great simplification. Thus we assume that the solution is e^{st}. With this assumption, Eq. (2.12) becomes

$$(ms^2 + cs + k)e^{st} = 0 \tag{2.13}$$

If (2.13) can be satisfied, our assumption $x = e^{st}$ for the solution is correct. Since Eq. (2.13) is a quadratic in s, there are two values s_1 and s_2 that will make the left side of (2.13) equal to zero

$$s_{1,2} = -\frac{c}{2m} \pm \sqrt{\left(\frac{c}{2m}\right)^2 - \frac{k}{m}} \qquad (2.14)$$

so that $e^{s_1 t}$ and $e^{s_2 t}$ are both solutions of Eq. (2.12). The most general solution is

$$x = C_1 e^{s_1 t} + C_2 e^{s_2 t} \qquad (2.15)$$

where C_1 and C_2 are arbitrary constants.

In discussing the physical significance of this equation two cases have to be distinguished, depending upon whether the expressions for s in Eq. (2.14) are real or complex. Clearly for $(c/2m)^2 > k/m$, the expression under the radical is positive so that both values for s are real. Moreover, they are both negative because the square root is smaller than

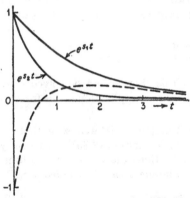

the first term $c/2m$. Thus (2.15) describes a solution consisting of the sum of two decreasing exponential curves, as shown in Fig. 2.14. As a representative example, the case $C_1 = 1$, $C_2 = -2$ is drawn as a dashed line.

Without analyzing any special cases by determining their values for C_1 and C_2, the figure shows that the motion is no "vibration" but rather a creeping back to the equilibrium position. This is due to the fact that for $(c/2m)^2 > k/m$ the damping c is extremely large. For smaller values of c, which pertain to more practical

FIG. 2.14. Motions of a single-degree system with damping greater than the critical damping c_c.

cases, (2.14) gives complex values for s, and the solution (2.15), as written, becomes meaningless. The damping c at which this transition occurs is called the *critical* damping c_c:

$$c_c = 2m \sqrt{\frac{k}{m}} = 2\sqrt{mk} = 2m\omega_n \qquad (2.16)$$

In case the damping is less than this, (2.14) can better be written as

$$s_{1,2} = -\frac{c}{2m} \pm j \sqrt{\frac{k}{m} - \left(\frac{c}{2m}\right)^2} = -\frac{c}{2m} \pm jq \qquad (2.17)$$

where $j = \sqrt{-1}$. Though the radical is now a real number both values of s contain j and consequently the solution (2.15) contains terms of the form e^{jat}, which have to be interpreted by means of Eq. (1.8), page 11. With (2.17) and (1.8), the solution (2.15) becomes

$$x = e^{-\frac{c}{2m}t} \left[C_1(\cos qt + j \sin qt) + C_2(\cos qt - j \sin qt) \right]$$

$$= e^{-\frac{c}{2m}t} \left[(C_1 + C_2) \cos qt + (jC_1 - jC_2) \sin qt \right] \qquad (2.18)$$

Since C_1 and C_2 were arbitrary constants, $(C_1 + C_2)$ and $(jC_1 - jC_2)$ are also arbitrary, so that for simplicity we may write them C_1' and C_2'. Thus

$$x = e^{-\frac{c}{2m}t} (C_1' \cos qt + C_2' \sin qt) \Bigg\}$$

where $\qquad q = \sqrt{\dfrac{k}{m} - \dfrac{c^2}{4m^2}}$ $\qquad (2.19a,\ b)$

This is the solution for a damping smaller than c_c. It consists of two factors, the first a decreasing exponential (Fig. 2.14) and the second a

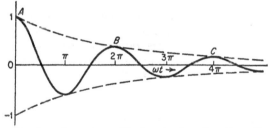

FIG. 2.15. Free vibration of a system with damping less than the critical damping of Eq. (2.16).

sine wave. The combined result is a "damped sine wave," lying in the space between the exponential curve and its mirrored image (Fig. 2.15). The smaller the damping constant c, the flatter will be the exponential curve and the more cycles it will take for the vibrations to die down.

The *rate* of this dying down is of interest and can be calculated in a simple manner by considering any two consecutive maxima of the curve: A-B, B-C, etc. During the time interval between two such maxima, *i.e.*, during $2\pi/q$ sec., the amplitude of the vibration (which at these maxima practically coincides with $e^{-\frac{c}{2m}t}$) diminishes from $e^{-\frac{c}{2m}t}$ to $e^{-\frac{c}{2m}\left(t+\frac{2\pi}{q}\right)}$. The latter of these two expressions is seen to be equal to the first one multiplied by the constant factor $e^{-\frac{\pi c}{mq}}$, which factor naturally is smaller than unity. It is seen that this factor is the same for *any* two consecutive maxima, independent of the amplitude of vibration or of the time.

The ratio between two consecutive maxima is constant; the amplitudes decrease in a geometric series.

If x_n is the nth maximum amplitude during a vibration and x_{n+1} is the next maximum, then we have seen that $x_{n+1} = x_n e^{-\pi c/mq}$ or also $\mathbf{\cdot}\log (x_n/x_{n+1}) = \pi c/mq = \delta$. This quantity δ is known as the logarithmic decrement. For small damping we have

$$\delta = \frac{\pi c}{mq} = 2\pi \frac{c}{c_c} \bigg/ \sqrt{1 - \left(\frac{c}{c_c}\right)^2} \approx \frac{2\pi c}{c_c}$$

and also $x_{n+1}/x_n = e^{-\delta} \approx 1 - \delta$, so that

$$\frac{x_n - x_{n+1}}{x_n} = \delta = \frac{2\pi c}{c_c} \tag{2.20}$$

The *frequency* of the vibration is seen to diminish with increasing damping according to (2.19b), which if written in a dimensionless form with the aid of (2.16) becomes

$$\frac{q}{\omega_n} = \sqrt{1 - \left(\frac{c}{c_c}\right)^2}$$

This relation is plotted in Fig. 2.16, where the ordinate q/ω_n is the ratio of the damped to the undamped natural frequency, while the abscissa is the ratio of the actual to the critical damp-

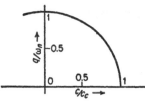

FIG. 2.16. The natural frequency of a damped single-degree-of-freedom system as a function of the damping; Eq. (2.19b).

ing constant. The figure is a circle; naturally for critical damping ($c = c_c$) the natural frequency q is zero. The diagram is drawn for negative values of c as well, the meaning of which will be explained later in Chap. 7 (page 283). On account of the horizontal tangent of the circle at $c = 0$, the natural frequency is practically constant and equal to $\sqrt{k/m}$ for all technical values of the damping ($c/c_c < 0.2$).

The undamped free vibration, being a harmonic motion, can be represented by a rotating vector, the end point of which describes a circle. With the present case of *damped* motion this graphical picture still holds, with the exception that the amplitude decreases with time. Thus, while revolving, the vector shrinks at a rate proportional to its length, giving a geometric series diminution. The end point of this vector describes a "logarithmic spiral" (Fig. 2.17). The amplitudes of a diagram like Fig. 2.15 can be derived from Fig. 2.17 by taking the horizontal projection of the vector, of which the end point lies on the spiral and which rotates with the uniform angular velocity q [Eq. (2.19)].

A special case of the foregoing occurs when the mass or inertia of the system is negligibly small, so that there remain only a spring and a dashpot.   We want to know the motion of the (massless) dashpot piston when it is released from an initial deflection x_0.   The differential equation is

$$c \frac{dx}{dt} + kx = 0$$

which can be solved directly by writing

$$\frac{c}{k} \frac{dx}{x} = - dt$$

$$t = - \frac{c}{k} \int \frac{dx}{x} = - \frac{c}{k} (\log x + \text{const.})$$

At $t = 0$ the deflection $x = x_0$, so that the constant is $- \log x_0$.   Hence

$$t = - \frac{c}{k} \log \frac{x}{x_0} \quad \text{and} \quad x = x_0 e^{-\frac{k}{c}t}, \qquad (2.21)$$

a relation represented by one of the solid curves of Fig. 2.14.   Evidently the exponent of the e-function is a dimensionless quantity, so that c/k

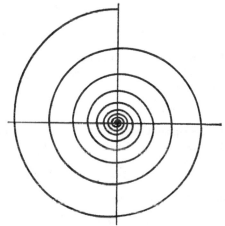

Fig. 2.17 Vector diagram of a damped free vibration.

must have the dimension of a time.   It is known as the *relaxation time*, which, by definition, is the time in which the deflection x_0 of the system "relaxes" to $1/e$th part of its original value.   On page 366 we shall have occasion to use this concept.

Example: In the system shown in Fig. 2.13, page 36, the mass weighs 1 oz.; the spring has a stiffness of 10 lb. per inch; $l = 4$ in.; $a = b = 2$ in.   Moreover, a dashpot mechanism is attached to the mid-point of the beam, *i.e.*, to the same point where the spring is fastened to it.   The dashpot produces a force of 0.001 lb. for a velocity of 1 in. per second.

a. What is the rate of decay of the free vibrations?

b. What would be the critical damping in the dashpot?

Solution: Let us first answer question (*b*) by means of Eq. (2.16). The undamped natural frequency is $\omega_n = \sqrt{k/m}$. On page 37 we found that the equivalent spring constant of Fig. 2.13 is ka^2/l^2 or $k/4 = 2.5$ lb. per inch. Thus

$$\omega_n = \sqrt{2.5 \times 16 \times 386} = 124 \text{ radians per second}$$

The critical damping constant of the system (*i.e.*, the critical damping of an imaginary dashpot *at the mass*) is, by Eq. (2.16)

$$2 \times \frac{1}{16 \times 386} \times 124 = 0.041 \text{ lb./in./sec.}$$

Since the dashpot is actually located at the mid-point of the beam, the dashpot must have a constant which is four times as great, for the same reason that the spring there must be taken four times as stiff as the "equivalent" spring (see page 37). Thus we find for the answer to question (*b*)

$$c_c = 0.164 \text{ lb./in./sec.}$$

a. The rate of decay is to be found from Eq. (2.20).

$$\frac{\Delta x}{x} = \delta = 2\pi \frac{c}{c_c} = 2\pi \frac{0.001}{0.164} = 0.038$$

or $$\frac{x_{n+1}}{x_n} = 1 - 0.038 = 0.962$$

2.7. Forced Vibrations without Damping. Another important particular case of Eq. (2.1) is the one where the damping term $c\dot{x}$ is made zero, while everything else is retained:

$$m\ddot{x} + kx = P_0 \sin \omega t \qquad (2.22)$$

It is reasonable to suspect that a function $x = x_0 \sin \omega t$ may satisfy this equation. Indeed, on substitution of this function Eq. (2.22) becomes

$$-m\omega^2 x_0 \sin \omega t + kx_0 \sin \omega t = P_0 \sin \omega t$$

which can be divided throughout by $\sin \omega t$, so that

$$x_0(k - m\omega^2) = P_0$$

or $$x_0 = \frac{P_0}{k - m\omega^2} = \frac{P_0/k}{1 - m\omega^2/k} = \frac{P_0/k}{1 - (\omega/\omega_n)^2}$$

and $$x = \frac{P_0/k}{1 - (\omega/\omega_n)^2} \cdot \sin \omega t \qquad (2.23)$$

is a solution of (2.22). The expression P_0/k in the numerator has a simple physical significance: it is the static deflection of the spring under the (constant) load P_0. We therefore write

$$\frac{P_0}{k} = x_{st}$$

and with this the solution becomes

$$\frac{x}{x_{st}} = \frac{1}{1 - (\omega/\omega_n)^2} \cdot \sin \omega t \qquad (2.24)$$

Although it is true that this is "a" solution of (2.22), it cannot be the most general solution, which must contain two integration constants. It can be easily verified, by substitution, that

$$x = C_1 \sin \omega_n t + C_2 \cos \omega_n t + \frac{x_{st}}{1 - (\omega/\omega_n)^2} \cdot \sin \omega t \qquad (2.25)$$

satisfies (2.22). The first two terms are the undamped *free* vibration; the third term is the undamped *forced* vibration. This is a manifestation of a general mathematical property of differential equations of this type, as stated in the following theorem:

Theorem: The general solution (2.25) of the complete differential equation (2.22) is the sum of the general solution (2.8) of the equation with zero right-hand member (2.7), *and* a particular solution (2.23) of the complete equation (2.22).

It is soon that the first two terms of (2.25) (the free vibration) form a sine wave having the free or natural frequency ω_n, whereas the forced vibration (the third term) is a wave having the forced frequency ω. Since we are at liberty to make ω what we please, it is clear that ω and ω_n are entirely independent of each other. The solution (2.25), being the sum of two sine waves of different frequencies, is itself *not* a harmonic motion (see Fig. 2.25c, page 54).

It is of interest now to examine more closely the implications of the result (2.24). Evidently x/x_{st} is a sine wave with an amplitude $1/[1 - (\omega/\omega_n)^2]$, depending on the frequency ratio ω/ω_n. Figure 2.18 represents this relation.

From formula (2.24) it follows immediately that for $\omega/\omega_n < 1$ the amplitudes or ordinates are positive, while for $\omega/\omega_n > 1$ they are negative. In order to understand the meaning of these negative amplitudes we return to Eq. (2.22) and the assumption $x_0 \sin \omega t$ for the solution made immediately thereafter. It appears that in the region $\omega/\omega_n > 1$ the results for x_0 are negative. But we can write

$$-x_0 \sin \omega t = +x_0 \sin (\omega t + 180 \text{ deg.})$$

which shows that a "negative amplitude" is equivalent to a positive amplitude of a wave which is merely 180 deg. out of phase with (in opposition to) the original wave. Physically this means that, while for $\omega/\omega_n < 1$ force and motion are in phase, they are in opposition for $\omega/\omega_n > 1$. Whereas for $\omega/\omega_n < 1$ the mass is below the equilibrium position when the force pushes downward, we find that for $\omega/\omega_n > 1$ the mass is above the equilibrium position while the force is pushing downward.

Usually this phase relation is considered as of slight interest, while the amplitude is vitally important; therefore, the negative sign may be disregarded and the dashed line in Fig. 2.18 appears.

There are three important points, A, B, and C in Fig. 2.18 at which it is possible to deduce the value of the ordinate from purely physical reason-

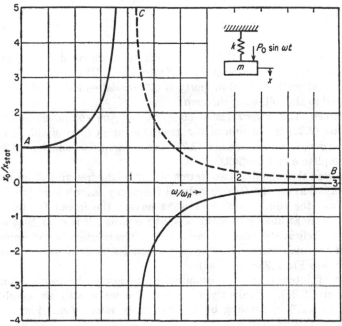

FIG. 2.18. Resonance diagram for the absolute motion of a system of which the mass is subjected to a force of constant amplitude and variable frequency; Eq. (2.23). This diagram is different from Fig. 2.20.

ing. First consider the point A, very close to $\omega = 0$; the force frequency is extremely slow, and the mass will be deflected by the force to the amount of its static deflection only. This is physically clear, and thus the amplitudes of the curve near the point A must be nearly equal to unity. On the other hand, for very high frequencies $\omega/\omega_n \gg 1$, the force moves up and down so fast that the mass simply has no time to follow, and the amplitude is very small (point B).

But the most interesting thing happens at point C, where the amplitude becomes infinitely large. This can also be understood physically. At $\omega/\omega_n = 1$, the forced frequency coincides exactly with the natural frequency. The force then can push the mass always at the right time in the right direction, and the amplitude can increase indefinitely. It is the case of a pendulum which is pushed slightly in the direction of its motion every time it swings: a comparatively small force can make the amplitude very large. This important phenomenon is known as "resonance," and the natural frequency is sometimes called also the "resonant frequency."

Thus far the theory has dealt with an impressed force of which the amplitude P_0 is *independent* of the frequency ω. Another technically important case is where P_0 is proportional to ω^2. For example, Fig. 2.19 represents a beam on two supports and carrying an unbalanced motor in the middle. While running, the motor axle experiences a rotating centrifugal force $m_1\omega^2 r$, where m_1 is the mass of the unbalance and r its distance from the center of the shaft. This rotating force can be resolved into a vertical component $m_1\omega^2 r \sin \omega t$ and a horizontal component $m_1\omega^2 r \cos \omega t$. Assume that the beam is very stiff against horizontal displacements but not so stiff against vertical ones. Then we have a single-

Fig. 2.19. Unbalanced motor giving a force $m\omega^2 a_0$ leading to the resonance diagram of Fig. 2.20.

degree-of-freedom system with a mass m (the motor), and a spring $k = 48EI/l^3$ (the beam), acted upon by a vertical disturbing force of amplitude $m_1\omega^2 r$, which is *dependent* on the frequency.

Another example of this type was discussed on page 31. There it was seen that the "relative motion" y between the mass and the support of Fig. 2.3 (where the support moves as $a_0 \sin \omega t$ and the force P_0 is absent) acts as if a force $ma_0\omega^2$ were acting on the mass. Incidentally, this case is of great importance since most vibration-recording instruments (vibrographs) are built on this principle (see page 57).

The resonance curve for the two cases just mentioned can be found directly from Eq. (2.23) by substituting $m\omega^2 a_0$ for P_0. Then

$$y_0 = \frac{m\omega^2 a_0/k}{1 - (\omega/\omega_n)^2} = a_0 \frac{(\omega/\omega_n)^2}{1 - (\omega/\omega_n)^2}$$

or
$$\frac{y_0}{a_0} = \frac{(\omega/\omega_n)^2}{1 - (\omega/\omega_n)^2} \tag{2.26}$$

It is to be remembered that a_0 is the amplitude of motion at the top of the spring, while y_0 is the relative motion between the mass and the top of the spring, or the extension of the spring, which is the same thing. The ordinates of the three points A, B, and C of Fig. 2.20, representing (2.26), can again be understood physically. At A the frequency ω is nearly zero; the top of the spring is moved up and down at a very slow rate; the mass follows this motion and the spring does not extend: $y_0 = 0$. At B the motion of the top of the spring is very rapid, so that the mass cannot follow and stands still in space. Then the relative motion is equal to the motion of the top and $y_0/a_0 = 1$. At the point C there is resonance, as before, so that the extensions of the spring become theoretically infinitely large.

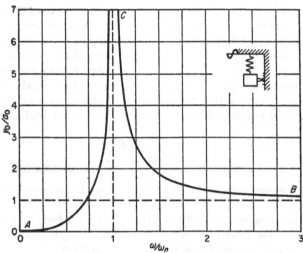

Fig. 2.20. Resonance diagram of Eq. (2.26) showing (a) the relative motion of a system in which the end of the spring is subjected to an alternating motion of constant amplitude a_0, and (b) the absolute motion of a system in which the mass experiences a force of variable amplitude $m\omega^2a_0$.

This last result is obviously not in agreement with actual observations, and it is necessary therefore to consider damping, which is done in Sec. 2.8.

Example: A motor generator set consists of a 25-cycle induction motor coupled to a direct-current generator. The set is rated at 200 hp. and 725 r.p.m. The connecting shaft has a diameter of $3\frac{5}{16}$ in. and a length of 14 in. The moment of inertia of the motor rotor is 150 lb. in. sec.[2] and that of the generator is 600 lb. in sec.[2]. The driving torque of the induction motor is not constant (see page 72) but varies between zero and twice the full-load torque T_0 at twice the frequency of the current, *i.e.*, 50 cycles per second, thus:

$$T_0 + T_0 \sin (2\pi \cdot 50t)$$

while the counter torque of the direct-current generator is constant in time. Find the maximum stress in the shaft at full load.

Solution: First find the torsional spring constant of the shaft.

$$k = \frac{\text{torque}}{\text{angle}} = \frac{GI_p}{l} = \frac{G \cdot \frac{\pi}{32} d^4}{l} = \frac{12 \cdot 10^6 \frac{\pi}{32} (3\frac{5}{16})^4}{14} = 10.20 \times 10^6 \text{ in. lb./rad.}$$

The system is idealized in Fig. 2.6 (page 28), and its differential equation is (2.4). The natural circular frequency is

$$\omega_n = \sqrt{\frac{k}{\dfrac{I_1 I_2}{I_1 + I_2}}} = \sqrt{\frac{10.2 \times 10^6 \times 750}{150 \times 600}} = 290 \text{ radians per second}$$

The forced frequency is 50 cycles per second, or

$$\omega = 2\pi f = 314 \text{ radians per second}$$

Apparently the system is excited at $^{314}\!/_{290} = 1.08$ times resonance, so that by Fig. 2.18 or Eq. (2.23) the effect of the torque is magnified by a factor

$$\frac{1}{1 - (1.08)^2} = 6.0$$

From Eq. (2.4) we see that the torque in question is $^{600}\!/_{750} T_0$, or four-fifths of the amplitude of the alternating component of the torque. As stated, the torque consists of a steady part T_0 and an alternating part of the same amplitude T_0. The maximum torque in the shaft thus is

$$T_0 + 6.0 \times \tfrac{4}{5} T_0 = 5.80 T_0$$

The steady torque T_0 can be found from the speed and horse power thus:

$$T_0 = \frac{\text{hp.}}{\omega} = \frac{200 \times 33,000}{725 \times 2\pi} = 1,450 \text{ ft. lb.} = 17,400 \text{ in. lb.}$$

The shear stress in the shaft due to this steady torque is

$$S_s = \frac{T_0 r}{I_p} = \frac{T_0 d/2}{\pi d^4/32} = 5 \frac{T_0}{d^3} = \frac{5 \times 17,400}{(3\tfrac{5}{16})^3} = 2,500 \text{ lb./in.}^2$$

On account of the proximity to resonance, this stress is multiplied by 5.80, so that the total maximum *shear* stress is 14,500 lb./in.². The "fatigue strength" of a steel, as listed, is derived from a tensile test, where the tensile stress is twice the shear stress. The fatigue limit of usual shaft steels is lower than 29,000 lb./in.², so that the shaft is expected to fail. The design can be improved by reducing the shaft diameter to $2\tfrac{1}{2}$ in. Then the natural frequency becomes 171 radians per second and the magnification factor 0.42. The new maximum tensile stress becomes 6,200 lb./in.², which is safe.

2.8. Forced Vibrations with Viscous Damping.

Finally, the complete Eq. (2.1),

$$m\ddot{x} + c\dot{x} + kx = P_0 \sin \omega t \tag{2.1}$$

will be considered. It can be verified that the theorem of page 43 holds here also. According to that theorem, the complete solution of (2.1) consists of the sum of the complete solution of the Eq. (2.12), which is (2.1) with the right-hand side zero, *and* a particular solution of the whole Eq. (2.1). But the solution of the equation with the zero right-hand side has already been obtained (Eq. 2.19), so that

$$x = e^{-\frac{c}{2m}t}(C_1 \sin qt + C_2 \cos qt) + \text{particular solution} \qquad (2.27)$$

It is therefore necessary merely to find the particular solution. Analogous to the case of Sec. 2.7, we might assume $x = x_0 \sin \omega t$, but then the term $c\dot{x}$ would give $\cos \omega t$, so that this assumption is evidently incorrect. It is possible to assume

$$x = A \sin \omega t + B \cos \omega t$$

and to substitute this in (2.1). In this case, only terms with $\sin \omega t$ and $\cos \omega t$ occur, but there are *two* constants A and B at our disposal. By solving for A and B algebraically, a particular solution can be obtained. Here we shall derive the result in a somewhat different manner, in order to give a clearer physical understanding of the phenomenon.

Let it be assumed that the solution is a sine wave with the forced frequency ω. Then all the four forces of Eq. (2.1) are sine waves of this frequency and can be represented by vectors. A differentiation is equivalent to a multiplication of the length of the vector with ω and a forward rotation through 90 deg., as explained on page 3.

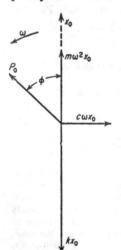

Let the displacement be represented by

$$x = x_0 \sin (\omega t - \varphi),$$

where x_0 and φ are as yet unknown, and draw this displacement as a vertical upward vector (dotted) in the diagram of Fig. 2.21. The spring force $-kx$ has an amplitude kx_0 and is directed downward in the diagram. The damping force $-c\dot{x}$ has an amplitude $c\omega x_0$ and is 90 deg. ahead of the spring force. The inertia force $-m\ddot{x}$ is 90 deg. ahead of the damping force and has an amplitude $m\omega^2 x_0$. The external force $P_0 \sin \omega t$ is φ deg. ahead of the displacement $x_0 \sin (\omega t - \varphi)$. Thus the complete diagram in Fig. 2.21 is obtained (x_0 and φ being unknown).

Fig. 2.21. Vector diagram from which Fig. 2.22 can be deduced.

Newton's law [or Eq. (2.1), which is the same thing] requires that the sum of the four forces be zero at all times. This means that the geometric sum of the four vectors in Fig. 2.21 must be zero, which again implies that the horizontal as well as the vertical component of this resultant must be zero. Expressed mathematically:

Vertical component: $kx_0 - m\omega^2 x_0 - P_0 \cos \varphi = 0$

Horizontal component: $c\omega x_0 - P_0 \sin \varphi = 0$

From these two equations the unknowns x_0 and φ are solved, with the result that

$$x_0 = \frac{P_0}{\sqrt{(c\omega)^2 + (k - m\omega^2)^2}} = \frac{\dfrac{P_0}{k}}{\sqrt{\left(1 - \dfrac{\omega^2}{\omega_n^2}\right)^2 + \left(2\dfrac{c}{c_c} \cdot \dfrac{\omega}{\omega_n}\right)^2}} \quad (2.28a)$$

$$\tan \varphi = \frac{c\omega}{k - m\omega^2} = \frac{2\dfrac{c}{c_c} \cdot \dfrac{\omega}{\omega_n}}{1 - (\omega^2/\omega_n^2)}. \quad (2.28b)$$

With the aid of the mechanical-electrical glossary of page 28, this can be translated into

$$Q_0 = \frac{E_0}{\sqrt{(R\omega)^2 + \left(\dfrac{1}{C} - L\omega^2\right)^2}}$$

or

$$Q_0\omega = \frac{E_0}{\sqrt{R^2 + \left(L\omega - \dfrac{1}{\omega C}\right)^2}} \quad (2.29)$$

Since $i = dQ/dt$, and $Q = Q_0 \sin \omega t$, the current is $i = Q_0\omega \cos \omega t$. The left-hand side of Eq. (2.29) is the maximum value of the current. The square root in the denominator to the right is known as the "impedance," a familiar element in electrical engineering.

The expressions (2.28a, b) for the amplitude x_0 and for the phase angle φ are in terms of "dimensionless quantities" or ratios only. There appear the frequency ratio ω/ω_n and the damping ratio c/c_c, where c_c is the "critical damping" of formula (2.16). P_0/k can be interpreted as the deflection of the spring under a load P_0; it is sometimes called the "static deflection" x_{st}.

These relations are plotted in Fig. 2.22a and b. The amplitude diagram contains a family of curves, one for each value of the damping c. All curves lie below the one for zero damping, which is of course the same curve as that of Fig. 2.18. Thus we see that the amplitude of forced vibration is diminished by damping. Another interesting property of the figure is that the maxima of the various curves do not occur any longer at $\omega/\omega_n = 1$ but at a somewhat smaller frequency. In fact, in the case of damped vibrations *three* different frequencies have to be distinguished, all of which coincide for $c = 0$, viz.,

(1) $\omega_n = \sqrt{\dfrac{k}{m}} = $ the "undamped natural frequency"

(2) $q = \sqrt{\dfrac{k}{m} - \left(\dfrac{c}{2m}\right)^2} = $ the "damped natural frequency"

(3) The "frequency of maximum forced amplitude," sometimes referred to as the "resonant frequency."

For small values of the damping these three frequencies are very close together.

The phase-angle diagram Fig. 2.22b also is of considerable interest. For no damping, it was seen that below resonance the force and the displacement are in phase ($\varphi = 0$), while above resonance they are 180 deg. out of phase. The phase-angle curve therefore shows a discontinuous jump at the resonance point. This can also be seen from Eq. (2.28b) by imagining the damping c very small. Below resonance, the denominator is positive so that tan φ is a very small positive number. Above resonance, tan φ is a very small negative number. Thus the angle φ itself is either close to 0 deg. or slightly smaller than 180 deg. Make the damping *equal* to zero, and φ becomes exactly 0 deg. or exactly 180 deg.

For dampings different from zero the other curves of Fig. 2.22b represent the phase angle. It is seen that in general the damping tends to smooth out the sharpness of the undamped diagrams for the amplitude as well as for the phase.

It is instructive to go back to the vector diagram of Fig. 2.21 and visualize how the amplitude and phase angle vary with the frequency. For very slow vibrations ($\omega \approx 0$) the damping and inertia forces are negligible and $P_0 = kx_0$, with $\varphi = 0$. With increasing frequency the damping vector grows, but the inertia force grows still faster. The phase angle cannot be zero any longer since P_0 must have a horizontal component to the left to balance $c\omega x_0$. The inertia-force vector will grow till it becomes as large as the spring force. Then φ must be 90 deg. and $P_0 = c\omega x_0$. This happens at resonance, because $m\omega^2 x_0 = kx_0$ or $\omega^2 = k/m$. Thus at resonance the phase angle is 90 deg., independent of damping. Above this frequency $m\omega^2 x_0$ will grow larger than kx_0, so that P_0 dips downward and φ is larger than 90 deg. For very high frequencies kx_0 is insignificant with respect to $m\omega^2 x_0$, so that P_0 is used up to balance the inertia force and $\varphi = 180$ deg.

At slow speeds the external force overcomes the spring force; at high speeds the external force overcomes inertia, while at resonance it balances the damping force.

The *energy relations* involved in this process also serve to give a better physical understanding. For *very slow* motions $\varphi = 0$, and it was shown on page 12 that no work is done over a whole cycle. In other words, no mechanical energy is transformed into heat during a cycle. Starting from the equilibrium position, the external force moves through a certain distance before reaching the extreme position. It certainly does work then. But that work is merely converted into potential or elastic energy stored in the spring. During the next quarter cycle the motion goes *against* the external force and the spring gives up its stored energy. At slow speeds, therefore, the work of the external force is thrown into elastic

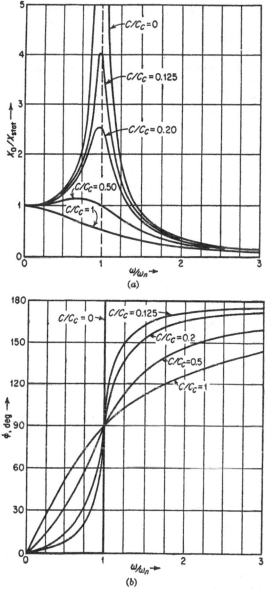

FIG. 2.22. (a) Amplitudes of forced vibration of any of Figs. 2.3 to 2.7 for various degrees of damping. (b) The phase angle between force and displacement as a function of the frequency for various values of the damping.

energy and nothing is converted into heat. At the resonant frequency, $\varphi = 90$ deg., and the work dissipated per cycle is $\pi P_0 x_0$ (page 12). The external force is equal and opposite to the damping force in this case, so that the work is dissipated in damping. The spring force and the inertia force are equal and opposite, and also in phase with the displacement. Each of these forces individually *does* perform work during a quarter cycle, but stores the energy, which is returned during the next quarter cycle. The work is stored periodically as elastic energy in the spring and as kinetic energy of motion of the mass.

Incidentally these energy relations can be used for calculating the "resonant amplitude." The damping force has the amplitude $c(\dot{x})_{max} = c\omega x_0$ and is 90 deg. out of phase with the displacement x_0. Consequently the work dissipated in damping per cycle is $\pi c\omega x_0^2$. The work done per cycle by the external force is $\pi P_0 x_0$ which must equal the dissipation of damping:

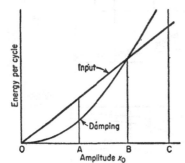

FIG. 2.23. Work per cycle performed by a harmonic force and by a viscous damping force for various amplitudes.

$$\pi P_0 x_0 = \pi c\omega x_0^2 \qquad (2.30)$$

This relation is illustrated by Fig. 2.23 in which the work per cycle done by the force P_0 at resonance and also that by the damping force are plotted against the amplitude of motion. Where the two curves intersect, we have energy equilibrium and *this* amplitude x_0 is the one that will establish itself. If at some instant the amplitude were greater, the energy dissipation would be greater than the input, which would gradually diminish the kinetic energy of the system until the equilibrium amplitude is reached.

Solving (2.30) for x_0, we obtain

$$(x_0)_{resonance} = \frac{P_0}{c\omega} \qquad (2.31)$$

Strictly speaking, this is the amplitude at the frequency where the phase angle is 90 deg., which is not exactly the frequency of maximum amplitude. However, these two frequencies are so close together that a very good approximation of the maximum amplitude can be obtained by equating the work done by the external force to the work dissipated by damping. For the single-degree-of-freedom system this method of calculating the resonant amplitude is of no great interest, but later we shall consider more complicated cases where an exact calculation is too laborious and where the approximate method of Eq. (2.30) and Fig. 2.23 gives acceptable results (page 202).

Equations (2.28a) and (2.28b) are the most important ones in this book. It is of interest to see what becomes of Eq. (2.28a), the magnification factor, for the case of resonance $\omega = \omega_n$. The magnification factor then becomes very simply $1/2c/c_c$. On page 40 [Eq. (2.20)] we saw that the percentage decay in amplitude of the *free* vibration per cycle is $\Delta x/x = 2\pi c/c_c$. Putting the two together, we find

$$\text{Magnifier at resonance} = \frac{\pi}{\text{percentage amplitude decay}}$$

This relation is plotted in Fig. 2.24.

Finally we return to the expression (2.27) on page 48 and remember that everything stated in the previous pages pertains to the "particular solution" or "forced vibration" only. The general solution consists of the damped free vibration superposed on the forced vibration. After a short time the damped free vibration disappears and the forced vibration alone persists. Therefore, the forced vibration is also called the "sustained vibration," while the free vibration is known as the "transient." The values of the constants C_1 and C_2 depend on the conditions at the start and can be calculated from these conditions by an analytical process similar to that

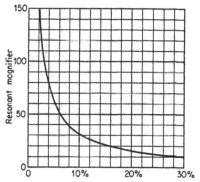

FIG. 2.24. Magnification factor at resonance as a function of percentage amplitude decay per cycle of free vibration.

performed on page 32. However, it is possible to construct the whole motion by physical reasoning only. As an example, consider the following problem:

A spring-suspended mass is acted on by an external harmonic force having a frequency eight times as slow as the natural frequency of the system. The mass is held tight with a clamp, while the external force is acting. Suddenly the clamp is removed. What is the ensuing motion if the damping in the system is such that the free vibration decreases by 10 per cent for each cycle?

In solving this problem, it is first to be noted that its statement is ambiguous, since it was not mentioned at what instant during the force cycle the mass was released. To make the problem definite, assume the release to occur at the moment that the forced vibration would just have its maximum amplitude. From the initial conditions of the problem it follows that at the instant of release the mass has no deflection and no velocity. We have prescribed the *forced* vibration to start with $x = x_0$

and $\dot{x} = 0$.   These two conditions can be satisfied only by starting a *free* vibration with $x = -x_0$ and $\dot{x} = 0$.   Then the *combined* or *total* motion will start at zero with zero velocity.   Figure 2.25a shows the free vibration, 2.25b the forced vibration, and 2.25c the combined motion.

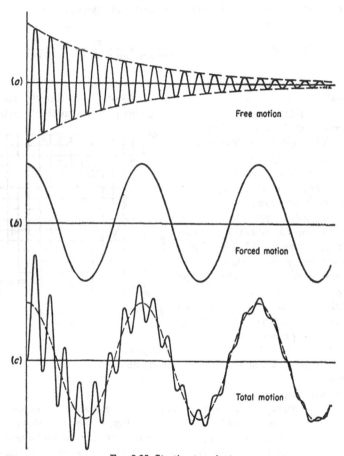

(a)

Free motion

(b)

Forced motion

(c)

Total motion

FIG. 2.25. Starting transient.

It is seen that the transient disappears quickly and that the maximum amplitude at the start is nearly twice as great as the sustained final amplitude.   If the difference between the free and forced frequencies is small and if the damping is also small, the diagram shows "beats" between the two frequencies (see page 5).   Because of damping such beats will disappear after some time.   In order to have sustained beats it is necessary to have two sustained or forced vibrations.

Example: An automobile has a body weighing 3,000 lb. mounted on four equal springs which sag 9 in. under the weight of the body. Each one of the four shock absorbers has a damping coefficient of 7 lb. for a velocity of 1 in. per second. The car is placed with all four wheels on a test platform which is moved up and down at resonant speed with an amplitude of 1 in. Find the amplitude of the car body on its springs, assuming the center of gravity to be in the center of the wheel base.

Solution: From Eq. (2.11) the natural frequency is

$$\omega_n = 2\pi f_n = \sqrt{386/\delta_{st}} = \sqrt{386/9} = 6.6 \text{ radians per second}$$

The damping of the system (four shock absorbers) is

$$c = 4 \times 7 = 28 \text{ lb./in./sec.}$$

The differential equation governing the motion is (2.5) of page 30. At resonance the disturbing force is

$$\sqrt{(ka_0)^2 + (ca_0\omega)^2}$$

Here $k = \dfrac{3,000 \text{ lb.}}{9 \text{ in.}} = 333$ lb./in.; $a_0 = 1$ in.; $c = 28$ lb./in./sec., and $\omega = \omega_n = 6.6$ radians per second.

$$\sqrt{(ka_0)^2 + (ca_0\omega)^2} = \sqrt{(333)^2 + (185)^2} = 380 \text{ lb.}$$

From Eq. (2.31) the amplitude of the car body is found,

$$x_0 = \frac{P_0}{c\omega} = \frac{380}{28 \times 6.6} = 2.06 \text{ in.}$$

2.9. Frequency-measuring Instruments.

Figure 2.20 is the key to the understanding of most vibration-measuring instruments. A vibration is sometimes a wave of rather complicated shape. When this wave has been traced on paper, everything regarding the vibration is known, but in many cases such complete knowledge is not necessary. We may want to know only the frequency or the amplitude of the motion or its acceleration. For such partial requirements, instruments can be made very much simpler and cheaper than if a record of the complete wave shape were demanded.

First, consider the methods of measuring *frequency* only. In many cases the vibration is fairly pure, *i.e.*, the fundamental harmonic has a much greater amplitude than any of the higher harmonics. In such cases a measurement of the frequency is usually easily made, and the result may give a hint of the cause of the vibration. Frequency meters are based nearly always on the resonance principle. For frequencies below about 100 cycles per second, *reed tachometers* are useful. There are two types of these: with a single reed and with a great many reeds.

The *single-reed frequency meter* consists of a cantilever strip of spring steel held in a clamp at one end, the other end being free. The length of the free portion of the strip can be adjusted by turning a knob, operating a screw mechanism in the clamp. Thus the natural frequency of the strip can be adjusted at will, and for each length the natural frequency in cycles per second is marked on the reed (see Fig. 4.28 on page 153). In use, the clamped end is pressed firmly against the vibrating object, so that the base of the reed partakes of the vibration to be measured. The screw is then turned slowly, varying the free length of the reed, until at one particular length it is in resonance with the impressed vibration and shows a large amplitude at the free end. The frequency is then read. Such an instrument is made and marketed by the Westinghouse Corporation.

Example: A variable-length, single-reed frequency meter consists of a strip of spring steel of cross section 0.200 by 0.020 in. and carries a weight of $\frac{1}{4}$ oz. at its end. What should be the maximum free length of the cantilever if the instrument is to be designed for measuring frequencies from 6 cycles per second to 60 cycles per second?

Solution: The spring constant of a cantilever beam is $3EI/l^3$. The moment of inertia of the cross section is $I = \frac{1}{12}bh^3 = \frac{1}{12} \times 0.2 \times (0.02)^3 = \frac{4}{3} \cdot 10^{-7}$ in.^{4}. The bending stiffness EI thus is $30 \cdot 10^6 \times \frac{4}{3} \cdot 10^{-7} = 4$ lb. in.^{2}, and the spring constant $k = 12/l^3$. The mass at the end is $m = 1/(4 \times 16 \times 386) = 4.05 \cdot 10^{-5}$ lb. in.$^{-1}$ sec.^{2}. The mass per inch of strip is $\mu_1 = 0.004 \times 0.28/386 = 0.29 \cdot 10^{-5}$ lb. in.$^{-2}$ sec.^{2}. Since about one-quarter of the strip length is effective as mass (see page 155), we have in total

$$m + \frac{\mu_1 l}{4} = (4.05 + 0.07l)10^{-5}$$

The frequency of maximum length is 6 cycles per second, or $\omega^2 = (2\pi \cdot 6)^2 = 1,420$ rad.^{2}/sec.^{2}.

Applying Eq. (2.10),

$$1,420 = \frac{12 \cdot 10^5}{l^3(4.05 + 0.07l)}$$

or
$$l^3(1 + 0.017l) = 206$$

This equation can be solved by trial and error. Since the second term in the parentheses (due to the mass of the strip) is small with respect to the first term (due to the $\frac{1}{4}$-oz. mass), we neglect the second term as a first guess.

$$l^3 = 206 \qquad \text{or} \qquad l = 5.9$$

With this, the parenthesis becomes $1 + 5.9 \times 0.017 = 1.10$, so that

$$l^3 = \frac{206}{1.10} = 187 \text{ in.}^3$$

and
$$l = 5.72 \text{ in.}$$
which is sufficiently accurate.

The other type of frequency meter employs a great number of reeds and is known as *Frahm's* tachometer. It consists of a light box b containing many small cantilever spring-steel strips a placed in one or more rows. Each reed has a slightly higher natural frequency than its left-hand neighbor, so that a whole range of natural frequencies is covered. In use, the box is placed on the vibrating machine with the result that most of the reeds hardly move at all. However, one or two of them for which the natural frequency is very close to that of the impressed vibration will swing with considerable amplitude.   This is made clearly visible by painting the inside of the box dull black and giving white tips c to the free ends of the reeds (Fig. 2.26).   Tachometers of this type are widely used.

The same instrument is also used for indicating the frequency of an alternating electric current.   The mechanical excitation of an impressed force is replaced by an electric excitation.   To this end one or more coils are placed in the box under the reeds.   The current flowing through these coils produces an alternating magnetic force on the reeds.

2.10. Seismic Instruments.   For measurement of the *amplitude* of the vibration a "seismic" instrument is ordinarily used, consisting of a mass mounted on springs inside a box.   The box is then placed on the vibrating machine, and the amplitude of the relative motion between the box and the mass follows the diagram of Fig. 2.20 for the various frequencies of

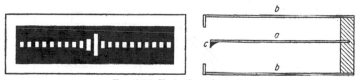

FIG. 2.26. Frahm's tachometer.

the motion to be recorded.   It is seen that, when the disturbing frequency is large with respect to the natural frequency of the instrument, the recorded amplitude y_0 is practically the same as that of the motion a_0. Thus *to get a displacement-measuring device or "vibrometer" it is necessary to give the instrument a natural frequency at least twice as slow as the slowest vibration to be recorded.*   In case the motion is impure, e.g., contains higher harmonics, this does not present any difficulty, since any higher harmonic has a higher frequency than the fundamental and will be recorded still more precisely.

A seismic mass on springs is capable of recording *accelerations* also. If the motion be $a_0 \sin \omega t$, the corresponding acceleration is $-a_0\omega^2 \sin \omega t$, with the amplitude $a_0\omega^2$.   Now, the left-hand branch of Fig. 2.20 (from $\omega/\omega_n = 0$ to $\omega/\omega_n = \frac{1}{2}$) has practically this $a_0\omega^2$ characteristic.   The equation of Fig. 2.20 is (page 45)

$$\frac{y_0}{a_0} = \frac{(\omega/\omega_n)^2}{1 - (\omega/\omega_n)^2} \tag{2.26}$$

For small values of ω/ω_n, the denominator differs only slightly from unity, so that the equation becomes approximately

$$\frac{y_0}{a_0} = \left(\frac{\omega}{\omega_n}\right)^2 \qquad \text{or} \qquad y_0 = \frac{1}{\omega_n^2} \cdot a_0 \omega^2$$

Here $1/\omega_n^2$ is a constant of the instrument, independent of the frequency of the external vibration. Hence the extreme left-hand part of Fig. 2.20 actually represents the accelerations at various frequencies.

An accelerometer is a seismic instrument with a natural frequency at least twice as high as the highest frequency of the accelerations to be recorded. This statement carries the possibility of a real difficulty, because an impure motion contains harmonics of frequencies higher than the fun-

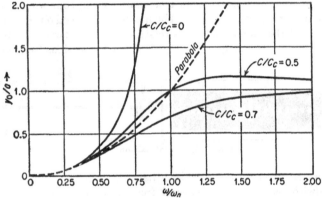

FIG. 2.27. Resonance curves with various amounts of damping compared with the parabolic curve of an ideal accelerometer.

damental and it may well be that one of these frequencies is very close to the natural frequency of the instrument. This trouble is peculiar to the accelerometer. A vibrometer is free from it since the harmonics in a wave are always *higher* in frequency than the main or fundamental wave, so that there is danger of resonance only when the recorded main frequency is *lower* than the natural frequency of the instrument. In order to avoid this particular difficulty, it is necessary to introduce damping in the accelerometer.

Figure 2.27 shows four curves: the parabola required of an ideal accelerometer and the three response curves for three different amounts of damping. The curves for 0.5 or 0.7 critical damping lie even closer to the desired parabola than does the undamped characteristic. Moreover, no resonance is to be feared. An accelerometer, therefore, with damping between half and 0.7 critical value will record accelerations up to three-quarters of the instrument frequency without appreciable error, while higher harmonics in the acceleration are diminished or, if their frequency is sufficiently high, they are practically suppressed.

The calculation of the curves of Fig. 2.27 is as follows: The differential equation (2.4), page 29, applies. Its solution [Eq. (2.28a), page 49] can be used immediately, after replacing P_0 by $m\omega^2 a_0$. Thus

$$\frac{y_0}{a_0} = \frac{\omega^2/\omega_n^2}{\sqrt{\left(1 - \frac{\omega^2}{\omega_n^2}\right)^2 + \left(2\frac{c}{c_c}\frac{\omega}{\omega_n}\right)^2}}$$

is the equation of Fig. 2.27. The reader would do well to check the formula with the figure for a few points.

The phase-angle formula (2.28b) and the corresponding figure 2.22b can be applied to this case without any change at all. It is interesting to note that for a damping between 0.5 and 0.7 critical the phase characteristic Fig. 2.22b differs but slightly from a straight diagonal line in the region below resonance. This has the advantage of avoiding an error known as "phase distortion." For each harmonic of an impure wave the damped instrument shows a different phase angle between the actual wave and its record. If this angle is proportional to the frequency, all the recorded waves form the same combined pattern as the actual waves.

Historically, the oldest seismic instruments are the *seismographs* for the recording of earthquake vibrations. The elastically suspended mass in these devices is sometimes very large, weighing a ton or more. The natural frequency is very low, of the order of a single vibration per 10 sec.

For *technical* applications a great variety of portable instruments are on the market, weighing from about 20 lb. for general use to an ounce or less for airplane work. The main difference among the various instruments lies in the manner of recording. In the most simple ones a dial gage is attached to the frame of the instrument and rests with its foot on the seismic mass. Figure 2.28 shows such an arrangement with one gage for horizontal and one for vertical vibrations. The vibratory motion is usually so rapid that the pointer of the

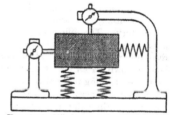

FIG. 2.28. Vibrometer for horizontal and vertical motions.

gage is seen as two pointers with a blurred region between them; twice the amplitude of the vibration is then the distance between the two positions of the pointer.

In a variation of this scheme the dial gage is replaced by a tiny mirror which is given a rocking motion by the vibration. The light of a small automobile headlight passes through a slit and is then reflected from the rocking mirror on a strip of ground glass. With the mirror standing still the image is a line, which broadens into a band due to the vibration. All instruments of these types, where no permanent record is made, are called *vibrometers.* The more elaborate *vibrographs* contain a recording mechanism, which usually is larger than the seismic part of the instrument. Some have a pen recording on a band of paper, which is moved by clockwork; some scratch the record on celluloid or glass, which is examined subsequently under the microscope, and some throw a light beam on a moving photographic film. Vibrographs sometimes are built without special damping devices. These devices *do* appear in accelerometers, sometimes as dashpots with either air or oil, or in the form of magnetic damping, where the seismic mass carries a tongue or thin copper plate moving parallel to its own plane in the narrow slit between the two poles of a powerful electromagnet. The motion of the tongue induces eddy currents in itself, and these currents develop a damping force proportional to the velocity.

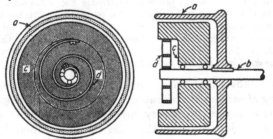

FIG. 2.29. Seismic part of a torsiograph.

Example 1: The vibrograph is sometimes used without the seismic part at all, *i.e.,* as a mere recording device. In that case the instrument is mounted in a place free from vibration, for example it is placed on a mass which is suspended from a crane in the factory. The only connection with the vibrating object is a needle which is pressed against the object with a spring; the other end of the needle operates the recording mechanism. Find the spring pressure on the needle which is necessary to hold it down on an object vibrating as $a_0 \sin \omega t$. The mass of the needle and the connected moving parts of the recording device is m.

Solution: If there were no spring at all, the vibrating object would lose contact with the needle point as soon as the object would have a receding acceleration. If there is no contact, the acceleration of the needle toward the object is P/m, where P is the spring pressure. This acceleration must be at least equal to the maximum receding acceleration of the vibrating object, so that

$$\frac{P}{m} = a_0\omega^2$$

or

$$P = ma_0\omega^2$$

For recording *torsional* vibrations, a seismic instrument is used which is a modification of a vibrograph. Instead of a mass on linear springs the *torsiograph* contains a flywheel on torsional springs. A very light aluminum pulley a (Fig. 2.29) is keyed to the shaft b. The heavy flywheel c can turn freely on the shaft but is coupled to it by a soft torsional spring *d.* When the pulley is held, the flywheel can perform free torsional vibrations about the shaft with a low natural frequency. When an alternating angular motion is given to the pulley, the relative motion between flywheel and pulley is again governed by the diagram of Fig. 2.20 (on account of the equivalence of the Figs. 2.3 and 2.4). Torsiographs of this type are widely used for measuring the torsional vibrations of crank shafts of slow- and medium-speed internal-combustion engines. Besides the vibration to be measured, such a shaft has also a uniform rotation. In use, the pulley *a* is driven from the crank shaft by means of a small canvas belt. When the crank shaft rotates uniformly, the flywheel follows and no relative motion between *a* and *c* occurs. When the shaft rotates non-uniformly (*i.e.*, has a torsional vibration superimposed on its rotation), the light pulley *a* will follow the shaft motion faithfully. The flywheel *c*, however, has so much inertia that it can rotate only at uniform speed. Thus the vibration appears as a relative motion between *a* and *c*, which is transmitted through a system of small bell cranks and a thin rod located along the center line of the hollow shaft *b*. The rod in turn operates a pen which scribes the record on a strip of paper, moved under the pen by clockwork. This instrument, known as the *Geiger* vibro- and torsiograph dating back to 1916, is marketed by the Commercial Engineering Laboratories, Detroit, Mich. It is still suitable for slow-speed machines, such as ship drives. However, for modern high-speed Diesel engines the recording-pen system comes to local resonance and, moreover, the magnification of the record obtainable (up to 24) is not sufficient. Then the *Summers* mechanical torsiograph, made by the General Motors Research Laboratories, Detroit, Mich., can be used to advantage. It is good up to 10,000 cycles per minute and gives a record in the form of a polar diagram.

Example 2: Let the flywheel *c* of the torsiograph of Fig. 2.29 be represented approximately by a solid steel disk of $4\frac{1}{2}$ in. diameter and 2 in. thickness. The outside diameter of the pulley is 5 in. If the flywheel *c* is held clamped, a string is wrapped round the pulley, and a $\frac{3}{4}$-lb. weight is suspended from one end of the string, the pulley circumference turns $\frac{1}{2}$ in. (*i.e.*, the weight descends $\frac{1}{2}$ in.).

If, with this instrument, a record is taken of a torsional vibration of 3 cycles per second, what is the error in the reading? What is the error in the recorded amplitude of the third harmonic of this curve?

Solution: First we have to find the natural frequency of the instrument. The torsional stiffness *k* in inch-pounds per radian follows from the fact that a torque of $\frac{3}{4}$ lb. $\times 2\frac{1}{2}$ in. causes an angular deflection of $\frac{\frac{1}{2} \text{ in.}}{2\frac{1}{2} \text{ in.}} \times 1$ radian. Thus

$$k = \frac{\frac{3}{4} \times 2\frac{1}{2}}{\frac{1}{5}} = 9.37 \text{ in.-lb./rad.}$$

The weight of the flywheel is

$$\frac{\pi}{4} \cdot \left(4\frac{1}{2}\right)^2 \times 2 \times 0.28 \text{ lb.} = 8.9 \text{ lb.}$$

Its moment of inertia is

$$I = \frac{1}{2} mr^2 = \frac{1}{2} \cdot \frac{8.9}{386} \left(2\frac{1}{4}\right)^2 = 0.059 \text{ lb. in sec.}^2$$

The natural frequency thus is

$$\omega_n = \sqrt{\frac{k}{I}} = \sqrt{\frac{9.37}{0.059}} = \sqrt{159} = 12.6 \text{ radians per second}$$

$$f_n = \frac{\omega_n}{2\pi} = \frac{12.6}{2\pi} = 2.0 \text{ cycles per second}$$

The frequency to be recorded is 50 per cent higher. Thus by Eq. (2.26) the ratio of the recorded to the actual amplitudes is

$$\frac{(1.5)^2}{1 - (1.5)^2} = \frac{2.25}{1.25} = 1.80$$

The third harmonic is $4\frac{1}{2}$ times as fast as the natural vibration of the instrument, so that its magnification factor is

$$\frac{(4\frac{1}{2})^2}{1 - (4\frac{1}{2})^2} = \frac{20.25}{19.25} = 1.05$$

2.11. Electrical Measuring Instruments. The rapid development in radio technique during the last decades has made possible a number of instruments that are generally much smaller and more sensitive than the older mechanical types discussed in the previous section. Most of these electrical "pickups" are still *seismic* instruments, for either linear or torsional vibrations, which operate on the same principle as the devices described in the previous section but have electrical windings in them that convert the mechanical vibration into an electrical voltage which can then be amplified and recorded by means of an oscillograph. Figure 2.30 shows schematically a pickup for linear vibrations, developed by Draper and Bentley, made and marketed under the name "Sperry-M.I.T." by the Sperry Gyroscope Company, Brooklyn, N.Y., and the

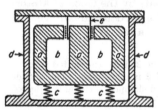

FIG. 2.30. Seismo-electric pickup, being essentially a loud-speaker element.

Consolidated Engineering Corporation, Pasadena, Calif. The electrical apparatus inside this unit, which has over-all dimensions of about 1 in. and a weight not exceeding 2 oz., is practically the same as that found in the usual radio loud-speaker. The instrument is a body of revolution which can be conceived of as generated by a rotation about its vertical center line. The part a is a piece of steel which is seismically supported on springs c. An important item, not shown in the figure, is the guiding of the mass a, the motion of which is restricted to the vertical direction entirely. No lateral motion of a can be allowed. In the hollow interior

of a, a coil b is mounted around the central cylindrical core.   This coil is
energized by direct current so as to make a magnet out of a.   Sometimes,
for simplicity, the coil b is omitted and the part a is fashioned as a permanent magnet of some special alloy steel.   The magnet a, being a body
of revolution, has a ring-shaped air gap with a radial magnetic field, into
which is inserted a thin paper cylinder e carrying a coil around it of
extremely thin wire.   The paper cylinder e is attached to the cover of
the housing d and the entire apparatus is supposed to be attached to
the machine of which the vibration is to be measured.   Any motion of
the magnet a in a vertical direction will cause a relative motion between the
magnet and the "voice coil" e and will set up an electrical alternating
voltage in e.   This voltage, which is proportional to the *velocity* of relative
motion, is now fed into an amplifier and after
sufficient magnification is recorded on an oscillograph film.   Oscillographs suitable for this work
have been developed in the last decades primarily in connection with applications of oil pros-
pecting and are now readily available on the
market.

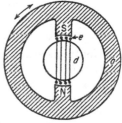

A torsiograph pickup of a similar type is
illustrated in Fig. 2.31 where a is the torsionally seismic element comparable to the part c
in Fig. 2.29.   This seismic element is made to

FIG. 2.31. Torsiograph-seis-
mo-electric pickup.

be a permanent magnet with a north and a south pole as indicated.   It
can revolve freely on a soft torsional spring around the core d which is
rigidly attached to the shaft of which the torsional vibration is to be
measured.   The core d carries a voice coil e.   The magnetic field travels
from the north pole to the south pole across the core d, and any relative
torsional motion between a and d will cause voltage variations in the
voice coil e, the intensity of which is proportional to the angular *velocity* of
the relative motion.

The records obtained on the oscillograph from either of these two
instruments therefore indicate veloc-
ity rather than amplitude.   This in
itself is no particular disadvantage,
but for certain applications it is more
convenient to have a direct record of
the amplitude instead of performing
the necessary integration numerically
or graphically on the record.   This

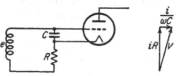

FIG. 2.32. Integrating circuit to transform a velocity record into an amplitude
record.

can be done electrically by means of the so-called "integrating circuit"
illustrated in Fig. 2.32.   In this figure, e is again the "voice coil," carrying a voltage proportional to the velocity.   This voltage is fed into a

C-R-series circuit so proportioned that the voltage across the resistance is many times, say ten times, greater than the voltage across the condenser. The voltage across the resistance is iR and the voltage across the condenser is $\dfrac{1}{C}\displaystyle\int i\,dt$, and if the first voltage is very much greater than the second, it is permissible to say that the voltage iR is practically equal to the total voltage V of the voice coil. Since, therefore, V is directly proportional to i (or to the velocity), the voltage across the condenser is directly proportional to $\int i\,dt$ (or to the integral of the velocity) which is exactly the quantity we are looking for. These relations are illustrated for harmonic variations in the vector diagram of Fig. 2.32. The integrated voltage is then put on the grid of the first tube of the amplifier. Since the voltage across the condenser is about one-tenth part of the total voltage, the sensitivity of the scheme is cut down by a factor 10, which means that an additional stage of amplification is necessary.

Amplifiers of a sensitivity independent of the frequency can be easily built for frequencies higher than 10 cyles per second and have been made even down to ¾ cycle per second, and up to 15,000 cycles per second, thus covering the entire practical frequency range for mechanical work.

For vibrations of very slow frequency another electrical principle known as the "variation of reluctance" has been employed, which is illustrated in Figs. 2.33 to 2.35. In Fig. 2.33, the two pieces a are rigidly attached to each other and they carry coils c which are energized by a constant voltage of a frequency that is high with respect to the frequencies that are to be measured. Usually, ordinary 60-cycle current will suffice for vibrations slower than 15 cycles per second; however, if vibrations considerably faster than this are to be recorded, a special alter-

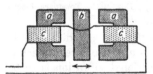

Fig. 2.33. Instrument operating on the principle of variation of reluctance, employing a carrier current of a frequency substantially higher than that of the vibration to be measured.

nator of say 500 cycles per second is used to energize the coils c. The voltage of the alternator is fed through the two coils c in series. A core b, made of laminated steel sheets like the U-pieces a, is mounted between these U-pieces so that the air gaps between them are as narrow as practicable. The central piece b vibrates back and forth between the two pieces a, thus varying the air gaps with the frequency of the vibration. If the two air gaps on the two sides of b are exactly alike, the voltage of the alternator is equally divided between the coils c; but if the air gaps of one of the pieces a are wider than those of the other piece a, then the voltages of the two coils c differ. The instrument is connected in a Wheatstone-bridge circuit as shown in Fig. 2.34 in which the coils are balanced by two equal impedances d. For equal air gaps and consequently equal voltages across c, the instrument in the Wheatstone bridge will show a zero reading, and the reading of that instrument will be proportional to the difference between the two air gaps. Naturally, the meter is affected by a current of a frequency equal to that of the exciting source; and if the instrument is replaced by an oscillograph, a record such as the upper one in Fig. 2.35 results. The fast variations in this record are those of the exciting alternator and the slow variation of the envelope is the effect we are looking for. For greater ease of reading, sometimes an electrical rectifier is inserted in the instrument branch of a Wheatstone bridge which transforms the upper record of Fig. 2.35 into the lower one. The apparatus of Fig. 2.33 can be used as a seismic instrument where the two pieces a are

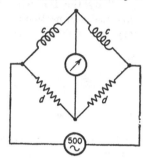

Fig. 2.34. Wheatstone-bridge circuit for the instrument of Fig. 2.33.

mounted seismically, whereas b is directly attached to the object to be measured. It has also been used as a strain meter where the two pieces a are attached to one part of the structure to be measured, while the central piece b is attached to some other part of that structure.

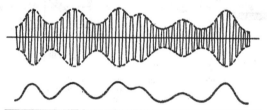

Fig. 2.35. Record obtained from the Wheatstone bridge.

The device under the name "Siemens-McNab Electric Torsion Meter" has been used for measuring the horse power of ships' shafts while under way. The part a of Fig. 2.33 is attached to a sleeve clamped on one section of the propeller shaft. The part b is attached to another sleeve, clamped to a section of the shaft some 3 ft. away from the first. If this length of 3 ft. of shafting twists with the strain, the parts b and a change position relative to each other, while rotating with the shaft. Turning to Fig. 2.34, the parts c, c rotate with the shaft, and the current is supplied the shaft through three slip rings. But the non-rotating instrument contains not just dead resistances d, but again a complete set-up like Fig. 2.33. The relative position of the (non-rotating) pieces b and a is varied with an accurate micrometer screw until the ammeter reads zero. Then the rotating and non-rotating air gaps must be alike; their position, and hence the shaft torque, is read off the non-rotating micrometer screw.

A device which has become very important is the resistance-strain-sensitive wire gage, first used by Simmons and Datwyler, further developed by Ruge and De Forest, marketed under the trade name "SR-4 gage" by the Baldwin Southwark Co., Philadelphia, Pa., and now in universal use, particularly in the aircraft industry. The gage is made of very thin (0.001 in.) wire of high electric resistance (nichrome) arranged as shown in Fig. 2.36 and mounted between two thin sheets of paper. The total length is about an inch; the total electric resistance is about 500 ohms. The gage is glued to the metal object under test, and if the metal (and consequently the nichrome wire) is strained, its electric resistance changes. The strain-sensitivity factor, which is the percentage change in resistance divided by the percentage change in length, is about 3. This means that for a stress of 30,000 lb./in.^{2} in steel, where the strain is 0.001, the resistance changes by 0.003, so that in a gage of 500 ohms resistance the change in resistance is 1.5 ohms. Figure 2.37 shows how the gage may be connected in a circuit. The battery voltage is divided

FIG. 2.36. Wire strain gage.

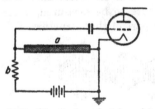 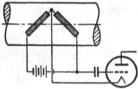

FIG. 2.37. Circuit for electric-resistance strain gage.

FIG. 2.38. Two strain gages mounted at 45 deg. on a shaft to form a torsion-sensitive unit.

between the gage a and a steady resistance b. If the strain and hence the resistance of a varies with time, so will the voltage across its terminals, and this varying voltage is put on the grid of the first vacuum tube in an amplifier, and from there passed on to an oscillograph.

Figure 2.38 shows the adaptation of this method to the measurement of twist in a shaft. It is well known that in a shaft in torsion the maximum strains have directions of 45 deg. with respect to the longitudinal axis of the shaft. Therefore, if two strain gages are glued on as shown, and the shaft is twisted, one of the gages will be elongated and the other one will be shortened. The voltage of the direct-current battery, therefore, will be unequally divided between the two strain gages and the variations in voltage will follow the strain and consequently the torque in the shaft.

The particular advantage of the strain gages just described lies in their extreme lightness. For the measurement of stresses in airplane propellers or turbine blades, where the centrifugal field is as high as 9,000g, only a pickup of practically no weight is at all feasible. The introduction of electric-resistance-strain gages has made possible for the first time the reliable measurement of vibrational phenomena in airplane propellers.

For variations of very slow frequency, the ordinary amplifier does not work, and the gages are energized by a high-frequency current, much as in Fig. 2.34. The Foxboro Company, Foxboro, Mass., is marketing an instrument under the trade name "Dynalog" with a 1,000-cycle carrier current generated by a vacuum-tube oscillator built in the instrument. The wiring diagram is somewhat like Fig. 2.34, where c, c are the two gages, one in tension, the other in compression, and d, d are condensers, one fixed, the other variable. The unbalance current of the bridge, instead of passing through the ammeter of Fig. 2.34 passes through a small motor which turns the shaft changing the capacity of the variable condenser d, until new balance is obtained and the motor current is zero. The position of the condenser shaft indicates the strain, which can be read easily to 1 per cent of full scale, the full scale commonly being a strain of 0.001 in./in.

In conjunction with this Dynalog there are available a number of "pickups" for the measurement of various quantities, such as strain, stress, and pressure. The pressure pickups have the appearance and size of spark plugs and can be screwed into the pipe line. They contain a member which is strained proportionally to the fluid or gas pressure, and to which an SR-4 gage is attached. They come in various sensitivities, the most sensitive being 0 to 600 lb./in.^{2} full scale, while the least sensitive ranges from 0 to 20,000 lb./in.^{2} full scale. Also there are spark-pluglike differential pressure gages, the most sensitive of which registers from 0 to 100 in. of water head full scale, superposed on a basic pressure of 500 lb./in.^{2} or higher.

The *stroboscope* is a device for producing intermittent flashes of light by means of which rapid vibratory motions can be made to appear to stand still or to move very slowly. In a good stroboscope the flashes of light are of extremely short duration. Imagine a vibrating object illuminated with this kind of light which is adjusted to the same frequency as the vibration. The object will be seen in a certain position; then it will be dark, and consequently the object is invisible while traveling through its cycle. When it returns to the first position after one cycle, another flash of light occurs. Thus the object appears to stand still. If the frequency of the flashes differs slightly from the frequency of the motion, the vibration will apparently take place very slowly. There have to be at least 15 flashes per second in order to create a good, non-flickering illusion of standstill, just as in a moving-picture projector. The sharpness of the picture obtained depends on the fact that during the time of the flash the object moves very little. A flash of long duration will blur the picture. The modern developments in vacuum and gas-filled tubes have made it possible to construct stroboscopes giving flashes of great intensity and of very short duration. The frequency of the flashes can be read on a calibrated dial as in a radio receiver. Thus for rather large amplitudes the instrument can be used as frequency and amplitude meter combined.

For smaller amplitudes, the stroboscope in conjunction with a seismically mounted microscope is useful. Take a seismic mass of very low frequency, carrying a microscope. Paste a very small piece of emery cloth to the vibrating object and focus the microscope on the emery, which is illuminated by stroboscopic light. The individual emery particles will appear as sharp points, which, on account of the stroboscope, run through closed curves. Thus the frequency and the amplitude can be determined.

Some stroboscopes have two or more lamps available which are operated from the same circuit and thus flash simultaneously. This is very useful for finding phase relations. Suppose that two parts of a machine are vibrating at the same frequency and that it is desired to know whether the vibrations are in phase or in opposition. Each of two observers takes a lamp, the flash frequency being regulated so that the vibration appears very slow. They now observe the two spots and the first observer signals each time his vibration is in one of the two extreme positions. The other observer can then easily check whether his motion is in phase or in opposition. A very convenient instrument, developed by Edgerton, is marketed by the General Radio Company, Cambridge, Mass., under the trade name "Strobotac."

Example: We wish to observe stroboscopically a point located 4 in. from the axis of a machine rotating at 10,000 r.p.m. If we desire a blurring of less than $\frac{1}{32}$ in., what should be the duration of the light flashes?

Solution: The point in question travels per second

$$\frac{10,000}{60} \cdot 2\pi \cdot 4 = 4,200 \text{ in.} = 135,000 \times \frac{1}{32} \text{ in.}$$

Thus the flash should last 1/135,000 sec. or less.

An interesting torsiograph, based on an entirely different principle, was developed by the General Motors Research Laboratories. It is called the "phase-shift torsiograph" and consists of a thin (say $\frac{1}{16}$ in.) wheel with a large number of equally spaced teeth (say 300) mounted on the rotating shaft. Two small electromagnets with windings are brought close to the toothed wheel, which operates somewhat like an inverted electric clock. The teeth passing by set up an alternating voltage of tooth-passing frequency in the two coils. This frequency is constant only if the shaft rotates uniformly; if the shaft executes a torsional vibration the record of the current shows alternate sine waves bunched close together and further apart.   This variable frequency output current is fed into a box and mixed with a constant frequency current of average frequency generated by a vacuum tube oscillator.   Thus the two currents will have a constantly varying phase angle between them, and by a clever trick it is possible to take an oscillograph record in which the torsional vibration amplitude shows directly against time.   The advantages of this method are the absence of slip rings, the possibility of installing it on engines so compactly built that there is no space for any other instrument, and a record which is independent of the amplification ratio of the electronic apparatus, since it depends on phase angles only.   It is interesting to note that the "seismic" element in this method is no longer a mechanical flywheel running at constant speed, but rather the vacuum tube oscillator producing a current of constant frequency.

Finally, for electric wave analyzers, see page 22 in the section on Fourier series.

2.12. Theory of Vibration Isolation.   An unbalanced machine has to be installed in a structure where vibration is undesirable.   Such a situation is not uncommon.   An alternating-current elevator motor in a hospital or hotel and the engine in an automobile are examples.   The problem consists in mounting the machine in such a manner that no vibrations will appear in the structure to which it is attached.

Its universal solution consists in properly mounting the machine on springs, and again Figs. 2.18 and 2.20 contain the information for the correct design of such mountings.   In Fig. 2.39 the machine is represented as a mass m with a force $P_0 \sin \omega t$ acting on it.   In Fig. 2.39a it is attached solidly to its substructure, while in 2.39b it is mounted on springs with a combined vertical flexibility k (the k of Fig. 2.39a is infinitely large).

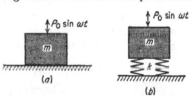

For simplicity the substructure is assumed to be rigid; the more complicated case of a movable foundation will be discusssed later on page 117.   If now P_0 is held constant and the frequency is varied, the amplitude of motion of m varies according to the diagram of Fig. 2.18.

Fig. 2.39.   A support of very flexible springs prevents vibrations from being transmitted to the foundation.

Our problem consists in finding the magnitude of the *force* transmitted to the substructure by the machine.   Since only the springs k are in contact with the foundation, the only transmitted force can be the spring force, which has the amplitude kx (damping being considered absent).   The ordinates of Fig. 2.18 represent the ratio of the maximum displacement x_0 of the mass to the static displacement $x_{st} = P_0/k$. Thus

$$\text{Ordinate} = \frac{x_0}{x_{st}} = \frac{x_0}{P_0/k} = \frac{kx_0}{P_0} = \frac{\text{spring force}}{\text{impressed force}}$$
$$= \frac{\text{transmitted force}}{\text{impressed force}} = \text{"transmissibility"}$$

The ideal is to have this ratio zero; the practical aim is to make it rather small. In Fig. 2.39a the spring constant $k = \infty$ and hence the natural or resonant frequency is infinite. Therefore, the operating frequency ω of the force is very slow with respect to the natural frequency; i.e., we are at the point A of Fig. 2.18, so that the transmitted force equals the impressed force. Physically this is obvious, since a rigid foundation was assumed and thus the mass m cannot move: the whole force P_0 must be transmitted to the foundation. The diagram of Fig. 2.18 shows immediately that *it is necessary to design the supporting springs so as to make the natural frequency of the whole machine very slow compared with the frequency of the disturbance;* in other words, the springs should be very soft.

An inspection of this diagram and its formula (2.24) reveals that, if ω is smaller than $\omega_n \sqrt{2} = \sqrt{2k/m}$, the springs actually make matters *worse:* the transmissibility is greater than one. If the natural frequency is one-fifth of the disturbing frequency, the transmissibility is 1 part in 24. This is fairly good, but in many cases it is better to make the springs softer yet.

Thus far, the support has been considered to be entirely without damping, which is practically the condition existing in steel springs. Sometimes, however, rubber or cork padding is used for this purpose, and then the damping is not negligible. The system can then be symbolized by Fig. 2.40; the amplitude of the motion of m being shown by one of the

FIG. 2.40. A spring support with damping.

curves of Fig. 2.22. In this case the displacement curve is *not* directly proportional to the amplitude of the transmissibility curve, as was the case with no damping. Now the transmitted force is made up not only of the spring force kx_0 but of the damping force $c\omega x_0$ as well. It was shown on page 48 that these two forces (being in phase with the displacement and the velocity respectively) have a 90-deg. phase angle between them. Consequently their sum, being the total transmitted force, is [Eq. (1.6), page 5]

$$x_0 \sqrt{k^2 + (c\omega)^2} \tag{2.32}$$

The amplitude x_0 is given by formula (2.28a) on page 49 so that (2.32) becomes

$$\text{Transmitted force} = P_0 \frac{\sqrt{1 + \left(\dfrac{c\omega}{k}\right)^2}}{\sqrt{\left(1 - \dfrac{\omega^2}{\omega_n^2}\right)^2 + \left(2\dfrac{c}{c_c}\dfrac{\omega}{\omega_n}\right)^2}}$$

or, since P_0 is the impressed force,

$$\text{Transmissibility} = \sqrt{\frac{1 + \left(2\dfrac{c}{c_c}\dfrac{\omega}{\omega_n}\right)^2}{\left(1 - \dfrac{\omega^2}{\omega_n^2}\right)^2 + \left(2\dfrac{c}{c_c}\dfrac{\omega}{\omega_n}\right)^2}} \qquad (2.33)$$

which actually reduces to formula (2.24) on page 43 for the case of zero damping, $c/c_c = 0$. This relation is shown graphically in Fig. 2.41.

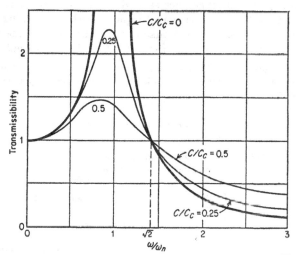

FIG. 2.41. Damping in the spring support is advantageous for $\omega < \omega_n \sqrt{2}$ but is detrimental for $\omega > \omega_n \sqrt{2}$.

Damping is seen to be advantageous only in the region $\omega/\omega_n < 1.41$ (where spring mounting makes matters worse); for all values of ω/ω_n where spring mounting helps, the presence of damping makes the transmissibility worse.

This rather paradoxical statement is not quite so important as it sounds. In the first place, the bad effect of damping is not great and can be easily offset by making the springs somewhat weaker, *i.e.*, by moving somewhat more to the right in Fig. 2.41. On the other hand, though it is not our *intention* to run at the resonance point $\omega/\omega_n = 1$, this unfortunately may sometimes occur, and *then* the presence of damping is highly desirable. Thus in spite of the dictum of Fig. 2.41, some damping in the springs generally is of advantage.

2.13. Application to Single-phase Electrical Machinery. Practical cases of isolation by means of springs occur in many machines. The main field of application, however, lies in apparatus which is inherently unbalanced or inherently has a non-uniform torque. Among the latter, single-phase electric generators or motors and internal-combustion engines are the most important.

First, single-phase machines are to be discussed. As is well known, the torque in any electric machine is caused by the pull of the magnetic field on current-carrying conductors. The magnetic field itself is caused by a current flowing through the field coils. If the machine is operated by single-phase alternating current of say 60 cycles per second, it is clear that the current flowing into the machine (and through the field coils) must become zero 120 times per second. But at zero current there is zero magnetic field and hence zero torque. Without knowing anything about the mechanism of such a machine we may suspect the torque to be some alternating periodic function of 120 cycles per second.

A more exact analysis is as follows: In any electric machine the instantaneous power in watts (which is of the dimension of work per second) equals the product of voltage and current, or

$$\text{Watts} = ei$$

If the voltage on the machine is $e = e_{max} \sin \omega t$ (where $\omega = 60 \times 2\pi$ radians per second), and $i = i_{max} \sin (\omega t - \varphi)$,

$$
\begin{aligned}
\text{Watts} &= e_{max}i_{max} \sin \omega t \sin (\omega t - \varphi) \\
&= e_{max}i_{max} \sin \omega t(\sin \omega t \cos \varphi - \cos \omega t \sin \varphi) \\
&= e_{max}i_{max}(\sin^2 \omega t \cos \varphi - \sin \omega t \cos \omega t \sin \varphi) \\
&= \frac{e_{max}i_{max}}{2} [\cos \varphi(1 - \cos 2\omega t) - \sin \varphi \sin 2\omega t] \\
&= \frac{e_{max}i_{max}}{2} [\cos \varphi - \cos (2\omega t - \varphi)]
\end{aligned}
$$

This is seen to consist of two terms, one independent of the time, representing a steady flow of power (which is the purpose for which the machine is built), and another harmonically alternating with frequency 2ω. This latter term does not deliver power during a long period of time, because its positive parts are neutralized by corresponding negative parts. The *torque* is found from the power as follows:

$$\text{Power} = \frac{\text{work}}{\text{second}} = \frac{\text{torque} \times \text{angle}}{\text{second}} = \text{torque} \times \text{angular velocity}$$

Thus all conclusions drawn for the power hold also for the torque when the angular velocity is constant, which is practically the case for a running machine.

The torque-time relation is given in Fig. 2.42, showing in this particular case that the amplitude of torque variation a is 30 per cent larger than the steady rated torque b of the machine. Though this represents a bad condition, the best that can possibly occur is that $a = b$. Then the torque merely becomes zero 120 times per second but does *not* become negative.

The machine consists of two parts, a *rotor* and a *stator*. Though it is the object of the machine to deliver torque to the rotor, Newton's law that action equals reaction requires that an equal and opposite torque act on the stator. If this stator is solidly bolted to its foundation, we have the torsional equivalent of the case of Fig. 2.39*a*. The torque reaction is fully transmitted to the foundation and from there can travel far and wide. Though the vibratory motion thus broadcast is usually very small, it may be that at quite a distance from the source there is a beam or other structure having for its natural frequency the same 120 cycles. That structure will pick up the motion and magnify it by resonance. A

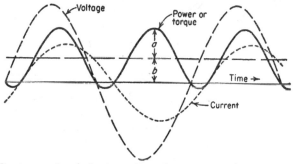

FIG. 2.42. The torque of a single-phase alternating-current motor is a periodic function having twice the frequency of the line voltage.

case is on record concerning a number of large single-phase generators installed in a basement in New York City. Complaints of a bad humming noise came from the occupants of an apartment house several blocks from where the generators were located, while the neighbors much closer to the source did *not* complain. The obvious explanation was that the complainers were unfortunate enough to have a floor or ceiling just tuned to 120 cycles per second. The cure for the trouble was found in mounting the generators on springs, as shown in Fig. 2.43.

Since the disturbance is a pure *torque* and not an up-and-down *force*, the springs have to be arranged in such a fashion that the stator can twist (*i.e.*, yield to the torque). The stiffness of the springs has to be so chosen that the torsional natural frequency of the stator on the springs is about one-seventh of 120 cycles per second.

In an actual construction for a large machine the springs of Fig. 2.43 are usually not coil springs as shown but rather beams of spring-steel loaded in bending, arranged with their length direction parallel to the axis of rotation of the generator. Figure 2.44 is a sketch of such a construction (cross section AA of Fig. 2.43); a denotes the stator, b the supporting foot, and c the beam spring, which carries its load on four points.

Small single-phase motors are used extensively in domestic appliances like refrigerators, washing machines, etc. Sometimes such motors have

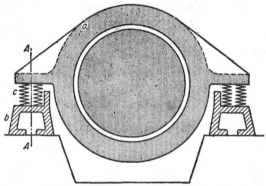

Fɪɢ. 2.43. Spring support for large single-phase generators to take the torque reaction.

a pinion on the shaft, driving a gear, and then it becomes imperative to support the rotor bearings so that they are very stiff against either vertical or lateral displacements in order to secure good operation of the gears. On the other hand, the stator should be mounted very flexibly in the rotational mode of motion.

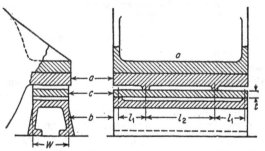

Fɪɢ. 2.44. Detail of beam spring for machine of Fig. 2.43.

There are several constructions on the market whereby both these requirements are satisfied. Two of them will be described here. Their common feature is that the rotor bearings are built solidly into the stator (which constitutes a difference from Fig. 2.43 where the bearings are mounted solidly on the floor so that the springs are between the rotor bearings and the stator). This solid rotor-stator unit is mounted on springs to the base or floor. The manner in which this is done, however, varies considerably. In the first construction (Fig. 2.45) each end of the stator is mounted in a heavy rubber ring a which is held in the foot b bolted to the floor. Rubber is a material which can be stretched enormously within the elastic limit, but at the same time it is extremely resistant to changes in volume: if a band of rubber is stretched to twice its length, its average cross section becomes half as small. (Another way of stating this is that rubber has a Poisson's ratio of one-half.) Owing to this property, the bearing inside the rubber ring can hardly

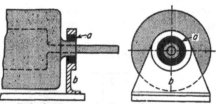

FIG. 2.45. Support of small single-phase motor in a rubber ring a, which is flexible in torsion and stiff against vertical or lateral displacements.

move sidewise with respect to the foot, because that would mean thinning of the ring on one side, which can occur only if rubber escapes vertically. This, however, is prevented by friction, so that the ring forms a stiff link between the bearing and the foot as far as lateral (or vertical) motions are concerned. Against rotation of the bearing in the foot, however, the rubber opposes only a shearing reaction, which can take place without a change in volume, making the ring flexible with respect to that motion.

The second method of accomplishing the same result is equally ingen-

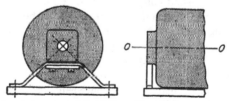

FIG. 2.46. Small-motor spring support consisting of two 45-deg. sections passing through the center of the machine.

ious and is shown in Fig. 2.46. The bearing is supported on a strip of steel, bent so as to have two 45-deg. sections and three horizontal sections (being the spring and supporting foot in one). This amounts to having two 45-deg. beams between the floor and the bearing, *built in* at each end. The design is such that the center lines of the beams pass through the bearing center. Any vertical or horizontal displacement of the bearing is associated with either *tension* or *compression* in the beams, whereas a turning of the bearing only *bends* the beams. Since thin strips are flexible in bending but very much stiffer in direct tension or compression, the desired result is obtained.

2.14. Application to Automobiles; "Floating Power." Internal-combustion engines have a torque-time diagram which does not differ appreciably from that of Fig. 2.42. For a four-cycle engine its frequency is $\frac{n}{2} \times$ (r.p.m.) cycles per minute where n is the number of cylinders. This will be explained in detail on page 197; here it is of interest only to know that the non-uniformity in torque exists. With the engine mounted rigidly on the frame, these torque variations have reactions on the car which may make themselves felt very uncomfortably. The obvious remedy is to mount the engine so that the free rotary vibration about the torque axis takes place very slowly, or, more precisely, so that the natural

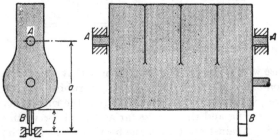

FIG. 2.47. Scheme of "floating-power" automobile engine.

frequency of such a vibration is appreciably lower than $n/2$ times the running speed.

This can be accomplished conveniently by mounting the whole engine block on two journals, fore and aft, supported in bearings attached to the chassis, enabling the block to rotate about an axis practically parallel to the torque axis and passing through the center of gravity (shown as AA in Fig. 2.47). Without anything other than the construction just described, the block would be free to rotate about the A-axis. This is prevented by a cantilever leaf spring B between the block and the frame, of which the stiffness is so chosen as to make the natural frequency sufficiently low.

Besides having an unbalanced *torque*, a four-cylinder engine also experiences some horizontal and vertical inertia *forces* (see page 177), which naturally have reactions at A and B. For this reason the bearings A as well as the end of the spring B are embedded in rubber.

In the actual construction, the axis AA is *not* quite parallel to the torque axis. This is correct procedure, for generally the torque axis is not a principal axis of inertia and consequently does not coincide with the corresponding axis of rotation.

Any rigid body has three "principal axes of inertia." Consider, for instance, an elongated solid piece of rectangular steel (Fig. 2.48) and attach to it a (weightless) shaft passing through the center of gravity but *not* coinciding with one of the principal axes (here axes of symmetry). The bar and shaft lie in the plane of the drawing. Apply a sudden torque to the shaft, and consider the acceleration caused by it. The upper part of the bar is accelerated *into* the paper, the lower part comes *out of* the paper (as indicated by dots and crosses in the figure). Multiplied by the mass of the respective elements these accelerations become "inertia forces." It is clear from the figure that these inertia forces multiplied by their distances from the shaft form a torque, which is equal and opposite to the impressed torque. *Moreover,* these forces multiplied by their distances to the vertical dotted line have a torque about that line as an axis. This will have its reaction in the bearings; the right-hand bearing will feel a force pushing it toward the reader out of the paper, and the left-hand bearing is pushed into the paper. Now if the bearings were absent, it is clear that under the influence of the torque the body would *not* rotate about the torque axis (since forces at the bearings are required in order to make it do so). Thus, in general, a body under the influence of a torque will rotate about an axis *not* coinciding with the torque axis (if the torque axis is not a principal axis).

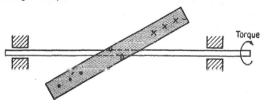

FIG. 2.48. Rotation about an axis different from a principal axis of inertia results in rotating reaction forces at the bearings.

The axis about which the "floating-power" engine has to be suspended therefore should not be the torque axis itself but rather the axis of rotation belonging to the torque axis. Only when the torque axis is a principal axis do the two coincide.

There are several other constructions of spring-supported automobiles on the market, most of which are similar in principle to the scheme of Fig. 2.47. Some have one rubber support at the rear of the engine and two rubber supports close together at the same height in the front. These two are virtually a combination of the single bearing A and the restoring spring B of Fig. 2.47.

Example: A four-cylinder automobile engine weighing 400 lb. is supported as indicated in Fig. 2.47. The radius of gyration of the engine about the axis AA is 6 in., the distance a is 18 in., and the length l of the cantilever is 4 in. The diameter of the rear wheels is 30 in. and in high gear the engine makes three revolutions per revolution of the rear wheels. It is desired that the engine be in resonance at a speed corresponding to $3\frac{1}{2}$ m.p.h. in high gear.

a. What should be the spring constant of the cantilever?

b. If one of the four cylinders does not spark properly, at what other speed is trouble to be expected?

Solution: a. $3\frac{1}{2}$ m.p.h. = 61 in. per second. The circumference of the wheel is $30\pi = 94.2$. At the critical speed the wheel makes $61/94.2 = 0.65$ r.p.s. and the engine therefore runs at $3 \times 0.65 = 1.95$ r.p.s. The torque curve of the engine goes through a full cycle for every firing. Since there are two firings per revolution in a four-cylinder, four-cycle engine, there are 3.9 firings per second. The natural frequency of the engine is desired to be $f_n = 3.9$ cycles per second or $\omega_n^2 = 4\pi^2(3.9)^2 = 600$ rad.^{2}/sec.$^2 = k/I$. Here k is the torque caused by the cantilever per radian twist. The deflection at the end of the cantilever for a twist of φ radians is 18φ in. If k_1 be the linear stiffness of the cantilever in lb./in., the spring force is $18k_1\varphi$ lb., acting on a moment arm of 18 in., so that the torque is $18 \times 18k_1\varphi$. Thus

$$k = 324k_1$$

Further
$$I = \frac{400}{386} \cdot (6)^2 = 37 \text{ lb. in. sec.}^2$$

so that
$$\omega_n^2 = 600 = \frac{324k_1}{37}$$

and
$$k_1 = \frac{37 \times 600}{324} = 69 \text{ lb. per inch}$$

b. If one cylinder fires inadequately, there is another periodicity in the torque curve for each two revolutions of the engine. Since this disturbance is four times as slow as the one discussed, it comes to resonance with the natural frequency of the engine at a speed of $4 \times 3.5 = 14$ m.p.h.

Problems 12 *to* 63.

CHAPTER 3

TWO DEGREES OF FREEDOM

3.1. Free Vibrations; Natural Modes. In the preceding chapter we discussed the theory of the vibrations of a system with a single degree of freedom with viscous damping. Though the exact idealized system with which the theory dealt occurs rarely, it was seen that a number of actual cases are sufficiently close to the ideal to permit conclusions of practical importance. The theory of the single-degree-of-freedom system enabled us to explain the resonance phenomenon in many machines, to calculate natural frequencies of a number of structures, to explain the action of most vibration-measuring instruments, and to discuss spring suspension and vibration isolation.

This exhausts the possibilities of application pretty thoroughly, and in order to explain additional phenomena it is necessary to develop the theory of more complicated systems. As a first step consider two degrees of freedom, which will yield the explanation of most "vibration dampers," of the action of a number of contrivances for stabilizing ships against rolling motions in a rough sea, and of the operation of automobile shock absorbers.

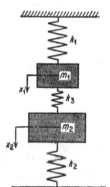

The most general undamped two-degree-of-freedom system can be reduced to that of Fig. 3.1 and consists of two masses m_1 and m_2 suspended from springs k_1 and k_2 and tied together by a "coupling spring"

Fig. 3.1. Undamped two-degree-of-freedom system with spring coupling.

k_3. Assuming that the masses are guided so as to be capable of purely vertical motions only, there are evidently two degrees of freedom, since the two masses can move independently of each other. By specifying their vertical positions x_1 and x_2 the configuration is entirely determined.

As in the single-degree-of-freedom case, there are a number of torsional, electrical, etc., two-degree-of-freedom systems which are completely equivalent to Fig. 3.1.

Proceeding now to a calculation of the *free* vibrations, we notice that there are two distinct forces acting on the mass m_1, namely the force of the main spring k_1 and that of the coupling spring k_3. The main force is $-k_1x_1$ acting downward (in the $+x_1$-direction). The shortening of the coupling spring is $x_1 - x_2$, so that its compressive force is $k_3(x_1 - x_2)$. A compressed coupling spring pushes m_1 upward, so that the force has to be taken with the negative sign. These two are the only tangible forces acting on m_1, so that its equation of motion is

$$m_1\ddot{x}_1 = -k_1x_1 - k_3(x_1 - x_2)$$

or
$$m_1\ddot{x}_1 + (k_1 + k_3)x_1 - k_3x_2 = 0 \tag{3.1}$$

The equation of motion for the second mass can be derived in the same manner. But by turning Fig. 3.1 upside down and reversing the directions of x_1 and x_2, m_2 and k_2 assume the positions of m_1 and k_1 and

$$m_2\ddot{x}_2 + (k_2 + k_3)x_2 - k_3x_1 = 0 \tag{3.2}$$

Assume now that the masses m_1 and m_2 execute harmonic motions with the same frequency ω (as yet unknown) and different amplitudes a_1 and a_2 (also unknown).

$$\left.\begin{array}{l} x_1 = a_1 \sin \omega t \\ x_2 = a_2 \sin \omega t \end{array}\right\} \tag{3.3}$$

This is a mere guess; we do not know whether such a motion is possible. By substituting in the differential equations we shall soon find out if it is possible.

$$[-m_1a_1\omega^2 + (k_1 + k_3)a_1 - k_3a_2] \sin \omega t = 0$$
$$[-m_2a_2\omega^2 + (k_2 + k_3)a_2 - k_3a_1] \sin \omega t = 0$$

These equations must be satisfied at any instant of time. They represent sine waves, so that in order to make them zero *at all times* the amplitudes in the brackets have to be zero.

$$\left.\begin{array}{l} a_1(-m_1\omega^2 + k_1 + k_3) - k_3a_2 = 0 \\ -k_3a_1 + a_2(-m_2\omega^2 + k_2 + k_3) = 0 \end{array}\right\} \tag{3.4a, b}$$

If the assumption (3.3) is correct, it is necessary that Eqs. (3.4) be satisfied. In general this is not true, but we must remember that in (3.3) nothing was specified about the amplitudes a_1 and a_2 or about the frequency ω. It will be possible to choose a_1/a_2 and ω so that (3.4) is satisfied, and with these values of a_1/a_2 and ω Eq. (3.3) becomes a solution. In order to find the correct values we have only to solve them from (3.4). Thus from (3.4a)

$$\frac{a_1}{a_2} = \frac{-k_3}{m_1\omega^2 - k_1 - k_3} \tag{3.5}$$

From (3.4b), also, the amplitude ratio can be solved:

$$\frac{a_1}{a_2} = \frac{m_2\omega^2 - k_2 - k_3}{-k_3} \tag{3.6}$$

In order to have agreement, it is necessary that

$$\frac{-k_3}{m_1\omega^2 - k_1 - k_3} = \frac{m_2\omega^2 - k_2 - k_3}{-k_3}$$

or $\qquad \omega^4 - \omega^2\left\{\dfrac{k_1 + k_3}{m_1} + \dfrac{k_2 + k_3}{m_2}\right\} + \dfrac{k_1k_2 + k_2k_3 + k_1k_3}{m_1m_2} = 0 \qquad (3.7)$

This equation, known as the "frequency equation," leads to two values for ω^2. Each one of these, when substituted in either (3.5) or (3.6), gives a definite value for a_1/a_2. This means that (3.3) can be a solution of the problem and that there are *two* such solutions.

For readers familar with *Mohr's* circle diagram in two-dimensional elasticity, the following construction is of interest. Let in Fig. 3.1

$$\omega_a^2 = \frac{k_1 + k_3}{m_1}, \qquad \omega_b^2 = \frac{k_2 + k_3}{m_2}, \qquad \omega_{ab}^2 = \frac{k_3}{\sqrt{m_1m_2}}$$

The quantities ω_a and ω_b are the frequencies of the system in which one of the masses is held clamped, while ω_{ab} expresses the strength of the coupling. With this notation, Eq. (3.7) can be written as

$$\omega^4 - \omega^2(\omega_a^2 + \omega_b^2) + (\omega_a^2\omega_b^2 - \omega_{ab}^4) = 0$$

Lay off in the diagram of Fig. 3.2 the following distances:

$$OA = \omega_a^2 \qquad OB = \omega_b^2 \qquad BC = \omega_{ab}^2$$

Then draw a circle through C about the mid-point between A and B as center. The

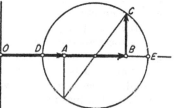

FIG. 3.2. Mohr's circle for determining the natural frequencies of Fig. 3.1.

new points D and E thus found determine the natural frequencies of the system:

$$\omega_1^2 = OD \qquad \text{and} \qquad \omega_2^2 = OE$$

which can be verified from the equation. In particular, when there is no coupling ($BC = 0$), the points D and E coincide with A and B, so that then ω_a and ω_b are the natural frequencies.

For further discussion, let us simplify the system somewhat by making it symmetrical. Let $k_1 = k_2 = k$ and $m_1 = m_2 = m$. The frequency equation then reduces to

$$\omega^4 - 2\omega^2\frac{k + k_3}{m} + \frac{k(k + 2k_3)}{m^2} = 0 \tag{3.8}$$

with the solutions

$$\omega^2 = \frac{k + k_3}{m} \pm \sqrt{\left(\frac{k + k_3}{m}\right)^2 - \frac{k(k + 2k_3)}{m^2}}$$

or

$$\omega_1^2 = \frac{k}{m} \quad \text{and} \quad \omega_2^2 = \frac{k + 2k_3}{m}$$

which are the *two natural frequencies* of the system. Substituting these frequencies in (3.5) or (3.6),

$$\frac{a_1}{a_2} = +1 \quad \text{and} \quad \frac{a_1}{a_2} = -1$$

The physical significance of these results is obvious. The fact that $a_1/a_2 = +1$ means (Eq. 3.3) that the two masses move in the same direction through the same distance. The coupling spring is not stretched or compressed in this process. Naturally the frequency of this motion is $\omega^2 = k/m$, since the system reduces to two independent single-degree-of-freedom systems. The fact that $a_1/a_2 = -1$ means that the two masses move through the same distance but in opposition to each other. This motion is wholly symmetrical, so that the mid-point of the coupling spring k_3 does not move. If this mid-point were held clamped, no change in the motion would take place. Thus the system is again split up into two independent single-degree-of-freedom systems. This time, however, the mass is connected to ground by *two* springs, one of stiffness k and another of stiffness $2k_3$ (see page 36), so the frequency is $\omega^2 = (k + 2k_3)/m$.

Thus there are *two "natural modes of motion,"* each with its corresponding natural frequency. The solution shows that if the system is given an initial disturbance of $x_1 = +1$ and $x_2 = +1$ (Fig. 3.1) and then released, the ensuing motion will be purely sinusoidal with the frequency $\omega_1^2 = k/m$; it swings in the first natural mode. On the other hand, if the initial displacement is $x_1 = +1$ and $x_2 = -1$, again a purely sinusoidal motion follows with the frequency $\omega_2^2 = (k + 2k_3)/m$, the second mode.

Assume next that the initial displacement is $x_1 = 1$ and $x_2 = 0$, from which position the system is released. As yet we have no solution for this case. But this initial displacement can be considered as the sum of two parts: first $x_1 = \frac{1}{2}$, $x_2 = \frac{1}{2}$ and second $x_1 = \frac{1}{2}$, $x_2 = -\frac{1}{2}$, for each of which a solution is known.

Assume now that the ensuing motion is the "superposition" of these two partial motions as follows:

$$\left. \begin{array}{l} x_1 = \frac{1}{2} \cos \omega_1 t + \frac{1}{2} \cos \omega_2 t \\ x_2 = \frac{1}{2} \cos \omega_1 t - \frac{1}{2} \cos \omega_2 t \end{array} \right\} \tag{3.9}$$

That this is the correct solution can be concluded from the fact that on substitution in (3.1) and (3.2) the differential equations are satisfied. Moreover at $t = 0$, the initial conditions are satisfied.

Equation (3.9) shows that the ensuing motion will be one in the first mode with amplitude $\frac{1}{2}$ and frequency ω_1, superposed on a motion with amplitude $\frac{1}{2}$ and frequency ω_2. As long as there is a coupling spring k_3, it is seen that ω_1 and ω_2 are different. Thus the combined motion of either mass can *not* be sinusoidal but must be composed of two frequencies. Naturally "beats" will occur if the two frequencies are close together (Fig. 1.8). This happens if $k_3 \ll k$, or, in words, if the coupling spring is very soft in comparison to the main springs. With an initial displacement $x_1 = 1$, $x_2 = 0$, first m_1 will vibrate with amplitude 1 and m_2 will stand practically still. After a time, however, the difference in the two frequencies will have changed the phase between the two vibrations by 180 deg. (see Fig. 1.7). Then instead of

$$x_1 = \tfrac{1}{2}, x_2 = \tfrac{1}{2} \text{ (first mode)} \quad \text{and} \quad x_1 = \tfrac{1}{2}, x_2 = -\tfrac{1}{2} \text{ (second mode)}$$

we have

$$x_1 = \tfrac{1}{2}, x_2 = \tfrac{1}{2} \text{(first mode)} \quad \text{and}$$
$$x_1 = -\tfrac{1}{2}, x_2 = +\tfrac{1}{2} \text{ (second mode)}$$

Thus the first mass stands still and the second one executes vibrations of amplitude 1. The phenomenon is periodic so that all motion travels from one mass to the other continuously.

This very interesting experiment can be shown in a number of variations, of which Fig. 3.3 gives five possibilities. The first case consists of two pendulums capable of swinging in the plane of the paper. The main springs have been replaced here by gravity, but the coupling spring exists in the form of a very soft coil spring. For "small" vibrations (say below 30-deg. amplitude) a gravity pendulum behaves like the fundamental mass-spring system. The spring constant k, which is the restoring force for unit displacement, is mg/l, so that for a simple pendulum $\omega^2 = k/m = g/l$. In further reducing Fig. 3.3a to Fig. 3.1, it is seen that the coupling-spring constant k_3 in Fig. 3.1 is the force at the masses caused by the coupling spring if the masses are pulled one unit apart. Applying this experimental definition to Fig. 3.3a, we find that, in the absence of gravity, a force of $k \dfrac{a^2}{l^2}$ at one of the masses pulls those masses 1 in. apart (see also page 37). Thus the equivalent of k_3 is ka^2/l^2.

The two natural modes of motion are easily recognized. The pendulums swing either with each other or against each other, the frequencies being $\omega_1 = \sqrt{\dfrac{g}{l}}$ and $\omega_2 = \sqrt{\dfrac{g}{l} + 2\dfrac{k}{m} \cdot \dfrac{a^2}{l^2}}$.

Pulling the left pendulum 1 in. to the left and keeping the right pendulum in its place is equivalent to the sum of the two displacements shown in Fig. 3.4b and 3.4c. Upon releasing the left pendulum, it will perform vibrations as indicated by Fig. 3.4a (the right-hand pendulum

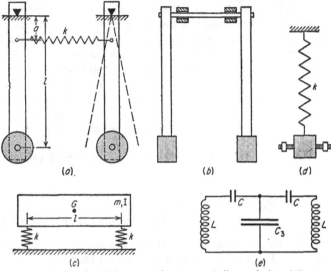

FIG. 3.3. Five experiments in which we can observe a periodic wandering of the energy from one part to another.

stands still). This motion can be regarded as the sum of two others with frequencies ω_1 and ω_2 as shown in the diagram. For the first few cycles this motion of one pendulum only will persist, because the two natural frequencies are sufficiently close together to keep in step for a short time. However, the second mode actually goes somewhat faster

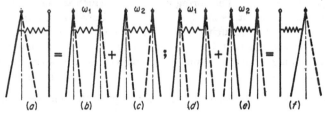

FIG. 3.4. Any motion can be broken up into the sum of two natural motions having the two different natural frequencies ω_1 and ω_2.

than the first one and gains on it since $\omega_2 > \omega_1$. After a sufficient time interval (say 20 cycles), it will be 180 deg. in advance of the first mode, which is indicated in Fig. 3.4d and e. Performing the addition shown in the figure, it is seen that the left pendulum now stands still, while the right pendulum swings with the full amplitude. Then the phenomenon repeats itself; the amplitude wanders from one pendulum to the other continuously, until the inevitable damping brings everything to rest.

In Fig. 3.3b the pendulums swing perpendicular to the plane of the paper. Two natural motions are possible: (1) the pendulums swing together, or (2) they swing against each other, thereby twisting the very slender connecting shaft, which causes some increase in the frequency. Pulling out *one* of the pendulums while keeping the other in place (thereby slightly twisting the coupling rod) and then releasing leads to the same phenomenon of continuous transfer of all motion from one pendulum to the other.

Figure 3.3c shows a system resembling in some respects an automobile chassis on its springs. Two natural motions of the mass are possible: (1) a bobbing up and down parallel to itself with the frequency $\omega_1^2 = 2k/m$ and (2) a rocking about the center of gravity G in the plane of the drawing with a frequency $\omega^2 = kl^2/2I$. The derivation of these frequency formulas is left to the reader. Now suppose the left-hand end of the chassis

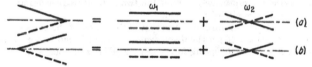

FIG. 3.5. The energy transfer of the experiment of Fig. 3.3c.

is pulled up 1 in. while the right-hand end is kept in place. From this position the system is released. Again the motion is split up into two parts (Fig. 3.5a reading from left to right).

If the quantities m, I, k, and l are such that ω_1 and ω_2 are nearly the same, the motion of Fig. 3.5a will keep on for the first few cycles without marked change. But after a larger number of cycles one of the motions (say the rocking one) gains 180 deg. on the other. Read now Fig. 3.5b from right to left and it is seen that the body vibrates with the *left*-hand end stationary. Of course, after an equal interval of time the first motion occurs again and so on until everything dies out on account of damping.

While in Fig. 3.3a and b the coupling spring could be easily seen as a separate part of the system, this is not the case in Fig. 3.3c. But the essential requirement for the experiment is that the system have two degrees of freedom with slightly different natural frequencies, and it does not matter whether the "coupling spring" can be recognized or not.

A striking experiment is shown in Fig. 3.3d known as *Wilberforce's spring*. A mass, suspended from a coil spring, has two protruding screws with adjustable nuts. The two degrees of freedom consist of an up-and-down motion and of a twisting motion. The "coupling" exists due to the fact that a coil spring when pulled out causes a slight torque and when twisted gives a slight pull. By changing the position of the nuts the moment of inertia I is changed while the mass m remains constant. Thus by a proper adjustment of the nuts the two natural frequencies can be brought to nearly the same value. Then by pulling down and releasing, an up-and-down motion of the mass without twist is initiated. After a while only twisting occurs without vertical motion, and so on.

The last case, illustrated in Fig. 3.3e, is the electrical analogue of this phenomenon (see pages 27, 28). Two equal masses (inductances) L connected to equal main springs (condensers) C are coupled with a *weak* coupling spring (*large* coupling condenser C_3 since k is equivalent to $1/C$). A current initiated in one mesh will after a time be completely transferred to the other mesh, and so on. Electrically minded readers may reason out how the currents flow in each of the two "natural modes" and what the frequencies are, and may also construct a figure similar to 3.4 or 3.5 for this case.

Example: A uniform bar of mass m and length $2l$ is supported by two springs, one on each end (Fig. 3.3c). The springs are *not* equally stiff, their constants being k (left) and $2k$ (right), respectively. Find the two natural frequencies and the shapes of the corresponding modes of vibration.

Solution: Let x be the upward displacement of the center of the bar and φ its (clockwise) angle of rotation. Then the displacement of the left end is $x + l\varphi$ and that of the right end $x - l\varphi$. The spring forces are $k(x + l\varphi)$ and $2k(x - l\varphi)$, respectively. Thus

$$m\ddot{x} + k(x + l\varphi) + 2k(x - l\varphi) = 0$$

and

$$(\tfrac{1}{3}ml^2)\ddot{\varphi} + kl(x + l\varphi) - 2kl(x - l\varphi) = 0$$

are the differential equations. With the assumption of Eq. (3.3) we obtain

$$(-m\omega^2 + 3k)x_0 - kl\varphi_0 = 0$$
$$-klx_0 + (-\tfrac{1}{3}m\omega^2 l^2 + 3kl^2)\varphi_0 = 0$$

from which follows the frequency equation

$$(-m\omega^2 + 3k)(-\tfrac{1}{3}m\omega^2 l^2 + 3kl^2) - k^2 l^2 = 0$$

or

$$\omega^4 - 12\frac{k}{m}\omega^2 + 24\left(\frac{k}{m}\right)^2 = 0$$

with the solutions

$$\omega_1^2 = 2.54\frac{k}{m} \qquad \text{and} \qquad \omega_2^2 = 9.46\frac{k}{m}$$

The shapes of the motion corresponding to these frequencies are found from the second differential equation, which can be written as

$$\frac{x_0}{l\varphi_0} = -\frac{1}{3}\frac{m}{k}\omega^2 + 3$$

Substituting the values for ω^2 just found, this becomes

$$\left(\frac{x_0}{l\varphi_0}\right)_1 = +2.16 \qquad \left(\frac{x_0}{l\varphi_0}\right)_2 = -0.15$$

This means a rotary vibration of the bar about a point which lies at a distance of $2.16l$ to the right of the center of the bar for the first natural frequency and about a point at $0.15l$ to the left of the center for the second natural frequency.

3.2. The Undamped Dynamic Vibration Absorber. A machine or machine part on which a steady alternating force of *constant* frequency is acting may take up obnoxious vibrations, especially when it is close to resonance. In order to improve such a situation, we might first attempt to eliminate the force. Quite often this is not practical or even possible. Then we may change the mass or the spring constant of the system in an attempt to get away from the resonance condition, but in some cases this also is impractical. A third possibility lies in the application of the *dynamic vibration absorber*, invented by Frahm in 1909.

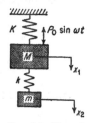

In Fig. 3.6 let the combination K, M be the schematic representation of the machine under consideration, with the force $P_0 \sin \omega t$ acting on it. The vibration absorber consists of a comparatively small vibratory system k, m attached to the main mass M. *The natural frequency $\sqrt{k/m}$ of the attached absorber is chosen to be equal to the frequency ω of the disturbing force. It will be shown that* then the main mass M does not vibrate at all, and that the small system k, m vibrates in such a way that its spring force is at all instants equal and opposite to $P_0 \sin \omega t$. Thus there is no net force acting on M and therefore that mass does not vibrate.

FIG. 3.6. The addition of a small k-m system to a large machine K-M prevents vibration of that machine in spite of the alternating force $P_0 \sin \omega t$.

To prove this statement, write down the equations of motion. This is a simple matter since Fig. 3.6 is a special case of Fig. 3.1 in which k_2 is made zero. Moreover, there is the external force $P_0 \sin \omega t$ on the first mass M. Equations (3.1) and (3.2) are thus modified to

$$\left. \begin{array}{c} M\ddot{x}_1 + (K + k)x_1 - kx_2 = P_0 \sin \omega t \\ m\ddot{x}_2 + k(x_2 - x_1) = 0 \end{array} \right\} \qquad (3.10)$$

The *forced* vibration of this system will be of the form

$$\left. \begin{array}{c} x_1 = a_1 \sin \omega t \\ x_2 = a_2 \sin \omega t \end{array} \right\} \qquad (3.11)$$

This is evident since (3.10) contains only x_1, $\ddot{x}_1$, and x_2, $\ddot{x}_2$, but *not* the first derivatives $\dot{x}_1$ and $\dot{x}_2$. A sine function remains a sine function after two differentiations, and consequently, with the assumption (3.11), all terms in (3.10) will be proportional to $\sin \omega t$. Division by $\sin \omega t$ transforms the *differential* equations into *algebraic* equations as was seen before with Eqs. (3.1) to (3.4). The result is that

$$a_1(-M\omega^2 + K + k) - ka_2 = P_0 \atop -ka_1 + a_2(-m\omega^2 + k) = 0 \Big\} \tag{3.12}$$

For simplification we want to bring these into a dimensionless form and for that purpose we introduce the following symbols:

$x_{st} = P_0/K$ = static deflection of main system
$\omega_a^2 = k/m$ = natural frequency of absorber
$\Omega_n^2 = K/M$ = natural frequency of main system
$\mu = m/M$ = mass ratio = absorber mass/main mass

Then Eq. (3.12) becomes

$$a_1\left(1 + \frac{k}{K} - \frac{\omega^2}{\Omega_n^2}\right) - a_2 \frac{k}{K} = x_{st} \atop a_1 = a_2\left(1 - \frac{\omega^2}{\omega_n^2}\right) \Bigg\} \tag{3.13}$$

or, solving for a_1 and a_2,

$$\frac{a_1}{x_{st}} = \frac{1 - \dfrac{\omega^2}{\omega_a^2}}{\left(1 - \dfrac{\omega^2}{\omega_a^2}\right)\left(1 + \dfrac{k}{K} - \dfrac{\omega^2}{\Omega_n^2}\right) - \dfrac{k}{K}} \atop \frac{a_2}{x_{st}} = \frac{1}{\left(1 - \dfrac{\omega^2}{\omega_a^2}\right)\left(1 + \dfrac{k}{K} - \dfrac{\omega^2}{\Omega_n^2}\right) - \dfrac{k}{K}} \Bigg\} \tag{3.14}$$

From the first of these equations the truth of our contention can be seen immediately. The amplitude a_1 of the main mass is zero when the numerator $1 - \dfrac{\omega^2}{\omega_a^2}$ is zero, and this occurs when the frequency of the force is the same as the natural frequency of the absorber.

Let us now examine the second equation (3.14) for the case that $\omega = \omega_a$. The first factor of the denominator is then zero, so that this equation reduces to

$$a_2 = -\frac{K}{k} x_{st} = -\frac{P_0}{k}$$

With the main mass standing still and the damper mass having a motion $-P_0/k \cdot \sin \omega t$ the force in the damper spring varies as $-P_0 \sin \omega t$, which is actually equal and opposite to the external force.

These relations are true for any value of the ratio ω/Ω_n. It was seen, however, that the addition of an absorber has not much reason unless the original system is in resonance or at least near it. We therefore consider, in what follows, the case for which

$$\omega_a = \Omega_n \quad \text{or} \quad \frac{k}{m} = \frac{K}{M} \quad \text{or} \quad \frac{k}{K} = \frac{m}{M}$$

The ratio $\qquad\qquad\qquad \mu = \frac{m}{M}$

then defines the size of the damper as compared to the size of the main system. For this special case, (3.14) becomes

$$\left.\begin{aligned}
\frac{x_1}{x_{st}} &= \frac{1 - \dfrac{\omega^2}{\omega_a^2}}{\left(1 - \dfrac{\omega^2}{\omega_a^2}\right)\left(1 + \mu - \dfrac{\omega^2}{\omega_a^2}\right) - \mu} \sin \omega t \\[2em]
\frac{x_2}{x_{st}} &= \frac{1}{\left(1 - \dfrac{\omega^2}{\omega_a^2}\right)\left(1 + \mu - \dfrac{\omega^2}{\omega_a^2}\right) - \mu} \sin \omega t
\end{aligned}\right\} \qquad (3.15a, b)$$

A striking peculiarity of this result and of Eq. (3.14) is that the two denominators are equal. This is no coincidence but has a definite physical reason. When multiplied out, it is seen that the denominator contains a term proportional to $(\omega^2/\omega_a^2)^2$, a term proportional to $(\omega^2/\omega_a^2)^1$ and a term independent of this ratio. When equated to zero, the denominator is a quadratic equation in ω^2/ω_a^2 which necessarily has two roots. Thus for two values of the external frequency ω both denominators of (3.15) become zero, and consequently x_1 as well as x_2 becomes infinitely large. These two frequencies are the *resonant* or *natural* frequencies of the system. If the two denominators of (3.15) were *not* equal to each other, it could occur that one of them was zero at a certain ω and the other one not zero. This would mean that x_1 would be infinite and x_2 would not. But, if x_1 is infinite, the extensions and compressions of the damper spring k become infinite and necessarily the force in that spring also. Thus we have the impossible case that the amplitude x_2 of the damper mass m is finite while an infinite force $k(x_1 - x_2)$ is acting on it. Clearly, therefore, if one of the amplitudes becomes infinite, so must the other, and consequently the two denominators in (3.15) must be the same.

The natural frequencies are determined by setting the denominators equal to zero:

$$\left(1 - \frac{\omega^2}{\omega_a^2}\right)\left(1 + \mu - \frac{\omega^2}{\omega_a^2}\right) - \mu = 0$$

or

$$\left(\frac{\omega}{\omega_a}\right)^4 - \left(\frac{\omega}{\omega_a}\right)^2 (2 + \mu) + 1 = 0$$

with the solutions

$$\left(\frac{\omega}{\omega_a}\right)^2 = \left(1 + \frac{\mu}{2}\right) \pm \sqrt{\mu + \frac{\mu^2}{4}} \qquad (3.16)$$

This relation is shown graphically in Fig. 3.7, from which we find, for example, that an absorber of one-tenth the mass of the main system causes two natural frequencies of the combined system at 1.17 and 0.85 times the natural frequency of the original system.

The main result (3.15) is shown in Fig. 3.8a and b for $\mu = \frac{1}{5}$, i.e., for an absorber of one-fifth the mass of the main system.

Follow the diagram 3.8a for an increasing frequency ratio $\omega/\Omega_n = \omega/\omega_a$. It is seen that $x_1/x_{st} = 1$ for $\omega = 0$, while for values somewhat larger than zero x_1 is necessarily positive, since both the numerator and the

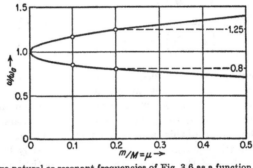

FIG. 3.7. The two natural or resonant frequencies of Fig. 3.6 as a function of the mass ratio m/M, expressed by Eq. 3.16.

denominator of Eq. (3.15a) are positive. At the first resonance the denominator passes through zero from positive to negative, hence x_1/x_{st} becomes negative. Still later, at $\omega = \Omega_n = \omega_a$, the numerator becomes negative and x_1/x_{st} becomes positive again, since both numerator and denominator are negative. At the second resonance the denominator changes sign once more with negative x_1 as a result.

The x_2/x_{st} diagram passes through similar changes, only here the numerator remains positive throughout, so that changes in sign occur only at the resonance points. It was seen in the discussion of Fig. 2.18 that such changes in sign merely mean a change of 180 deg. in the phase angle, which is of no particular importance to us. Therefore we draw the dotted lines in Fig. 3.8a and b and consider these lines as determining the amplitude, eliminating from further consideration the parts of the diagrams below the horizontal axes.

The results obtained thus far may be interpreted in another manner, which is useful in certain applications. In Fig. 3.6 let the Frahm absorber

k, m be replaced by a mass m_{equiv} attached solidly to the main mass M, and let this equivalent mass be so chosen that the motion x_1 is the same as with the absorber. Since the absorber is more complicated than just a mass, it is clear that m_{equiv} cannot be constant but must be different for each disturbing frequency ω. The downward force transmitted by the absorber to the main system M is the spring force $k(x_2 - x_1)$, which, by Eq. (3.10), is equal to $-m\ddot{x}_2$. If a mass m_{equiv} were solidly attached to M, its downward reaction force on M would be the pure inertia force

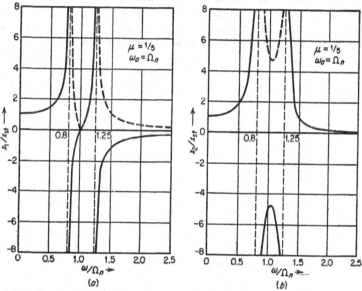

Fig. 3.8a and b. Amplitudes of the main mass x_1 and of the absorber mass x_2 of Fig. 3.6 for various disturbing frequencies ω. The absorber mass is one-fifth of the main mass.

$-m_{\text{equiv}}\ddot{x}_1$. For equivalence these two reactions must be equal, so that, by Eq. (3.11) and the second Eq. (3.13), we have

$$\frac{m_{\text{equiv}}}{m} = \frac{\ddot{x}_2}{\ddot{x}_1} = \frac{x_2}{x_1} = \frac{a_2}{a_1} = \frac{1}{1 - \dfrac{\omega^2}{\omega_a^2}}$$

which is the well-known resonance relation, shown in Fig. 2.18, page 44. Thus it is seen that the Frahm dynamic-absorber system can be replaced by an equivalent mass attached to the main system, so that the equivalent mass is positive for slow disturbing frequencies, is infinitely large for excitation at the absorber resonant frequency, and is *negative* for high frequency excitation. This way of looking at the operation of the absorber will be found useful on page 220.

From an inspection of Fig. 3.8a, which represents the vibrations of the main mass, it is clear that the undamped dynamic absorber is useful only in cases where the frequency of the disturbing force is nearly constant. Then we can operate at $\omega/\omega_a = \omega/\Omega_n = 1$ with a very small (zero) amplitude. This is the case with all machinery directly coupled to synchronous electric motors or generators. In variable-speed machines, however, such as internal-combustion engines for automotive or aeronautical applications, the device is entirely useless, since we merely replace the original system of one resonant speed (at $\omega/\Omega_n = 1$) by another system with two resonant speeds. But even then the absorber can be made to work to advantage by the introduction of a certain amount of damping in the absorber spring, as will be discussed in the next section.

An interesting application of the absorber is made in an electric hair clipper which was recently put on the market. It is shown in Fig. 3.9 and consists of a 60-cycle alternating-current magnet a which exerts a

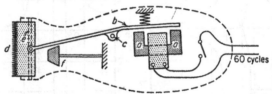

Fig. 3.9. Electric hair clipper with vibration absorber: a, magnet; b, armature tongue; c, pivot; d, cutter; e, guide for cutter; f, vibration absorber.

120-cycle alternating force on a vibrating system b. System b is tuned to a frequency near 120 cycles but sufficiently far removed from it (20 per cent) to insure an amplitude of the cutter d, which is not dependent too much on damping. Thus the cutter blade d will vibrate at about the same amplitude independent of whether it is cutting much hair or no hair at all.

The whole mechanism, being a free body in space without external forces, must have its center of gravity, as well as its principal axes of inertia, at rest. Since the parts b, d are in motion, the housing must move in the opposite direction to satisfy these two conditions. The housing vibration is unpleasant for the barber's hands and creates a new kind of resistance, known as sales resistance. This is overcome to a great extent by the dynamic vibration absorber f, tuned exactly to 120 cycles per second, since it prevents all motion of the housing at the location of the mass f. With stroboscopic illumination the masses d and f are clearly seen to vibrate in phase opposition.

The device as sketched is not perfect, for the mass f is not located correctly. At a certain instant during the vibration, the cutter d will have a large inertia force upward, while the overhung end b will have a small inertia force downward. The resultant of the inertia forces of the moving parts b, d therefore is an alternating force located to the left of the cutter d in Fig. 3.9.

The effect of the absorber is to completely eliminate 120-cycle motion of a point of the housing right under the absorber mass f, but it does not prevent the housing from rotating about that motionless point. Complete elimination of all 120-cycle motion of the housing can be accomplished by mounting two absorbers f in the device with a certain distance (perpendicular to the direction of the cutter motion) between their two masses. The two masses will then automatically assume such amplitudes as to cause two inertia forces which will counteract the force as well as the moment of the inertia action of the cutter assembly d, b, or in different words the two masses will enforce two motionless points of the housing.

For a *torsional* system, such as the crank shaft of an internal-combustion engine, the Frahm dynamic vibration absorber takes the shape of a flywheel A that can rotate freely on the shaft on bearings B and is held to it by mechanical springs k only (Fig. 3.10a). Since the torsional impulses on such an engine are harmonics of the firing frequency, *i.e.*, have a frequency proportional to the engine speed, the device will work for one engine speed only, while there are two neighboring speeds at which the shaft goes to resonance (Fig. 3.8a). In order to overcome this, the

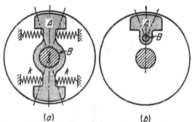

(a)   (b)

Fig. 3.10. Torsional dynamic vibration absorber (a) with mechanical springs and (b) with centrifugal springs.

system is modified by replacing the mechanical springs of Fig. 3.10a by the "centrifugal spring" of Fig. 3.10b. The pendulum in the centrifugal field of that figure acts in the same manner as an ordinary gravity pendulum in which the field g is replaced by the centrifugal field $r\omega^2$. Since the frequency of a gravity pendulum is $\sqrt{g/l}$, the frequency of a centrifugal pendulum becomes $\omega\sqrt{r/l}$, that is, proportional to engine speed. Thus a centrifugal pendulum will act as a Frahm dynamic absorber that is tuned correctly at all engine speeds. Further details of this device are discussed on page 219.

3.3. The Damped Vibration Absorber. Consider the system of Fig. 3.6 in which a dashpot is arranged parallel to the damper spring k, between the masses M and m. The main spring K remains without dashpot across itself. Newton's law applied to the mass M gives

$$M\ddot{x}_1 + Kx_1 + k(x_1 - x_2) + c(\dot{x}_1 - \dot{x}_2) = P_0 \sin \omega t \qquad (3.17)$$

and applied to the small mass m

$$m\ddot{x}_2 + k(x_2 - x_1) + c(\dot{x}_2 - \dot{x}_1) = 0 \qquad (3.18)$$

The reader should derive these equations and be perfectly clear on the various algebraic signs. The argument followed is analogous to that of page 25 and of page 80. The four terms on the left-hand side of (3.17) signify the "inertia force" of M, the main-spring force, the damper-spring force, and the dashpot force. We are interested in a solution for the *forced* vibrations only and do not consider the transient free vibration. Then both x_1 and x_2 are harmonic motions of the frequency ω and can be represented by vectors. Any term in either (3.17) or (3.18) is representable by such a vector rotating with velocity ω. The easiest manner of solving these equations is by writing the vectors as complex numbers. The equations then are

$$-M\omega^2 x_1 + Kx_1 + k(x_1 - x_2) + j\omega c(x_1 - x_2) = P_0$$
$$-m\omega^2 x_2 + k(x_2 - x_1) + j\omega c(x_2 - x_1) = 0$$

where x_1 and x_2 are (unknown) complex numbers, the other quantities being real.

Bringing the terms with x_1 and x_2 together:

$$\left.\begin{array}{l} [-M\omega^2 + K + k + j\omega c]x_1 - [k + j\omega c]x_2 = P_0 \\ -[k + j\omega c]x_1 + [-m\omega^2 + k + j\omega c]x_2 = 0 \end{array}\right\} \quad (3.19)$$

These can be solved for x_1 and x_2. We are primarily interested in the motion of the main mass x_1, and, in order to solve for it, we express x_2 in terms of x_1 by means of the second equation of (3.19) and then substitute in the first one. This gives

$$x_1 =$$
$$P_0 \frac{(k - m\omega^2) + j\omega c}{\{(-M\omega^2 + K)(-m\omega^2 + k) - m\omega^2 k\} + j\omega c\{-M\omega^2 + K - m\omega^2\}} \quad (3.20)$$

For readers somewhat familiar with alternating electric currents this result will also be derived by means of the equivalent electric circuit shown in Fig. 3.11. The equivalence can be established by setting up the voltage equations and comparing them with (3.17) and (3.18) or directly by inspection as follows. The extension (or velocity) of the spring K, the displacement (or velocity) of M, and the displacement (or velocity) of the force P_0 are all equal to x_1 (or $\dot{x}_1$). Consequently the corresponding electrical elements $1/C$, L, and E_0 must carry the same current (i_1) and thus must be connected in series. The velocities across k or across the dashpot ($\dot{x}_1 - \dot{x}_2$) are also equal among themselves, so that $1/c$ and r electrically must be in series but must carry a different current from that in the main elements L, C, and E_0. The velocity of

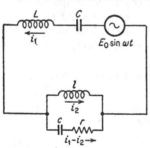

FIG. 3.11. Equivalent electric circuit. The small l-c-r "wave trap" corresponds to the absorber.

m is $\dot{x}_2$, equal to the difference of the velocity of $M(\dot{x}_1)$ and the velocity across the damper spring $(\dot{x}_1 - \dot{x}_2)$. Hence the current i_2 through l must be equal to the difference of i_1 and $(i_1 - i_2)$. The equivalence of the electrical circuit and the mechanical system is thus established.

We are interested in the main current i_1. The impedance of a coil is $j\omega L$, that of a condenser is $1/j\omega C$, that of a resistance simply R. Impedances in series, when expressed in complex, add directly, and impedances in parallel add reciprocally. Thus the impedance of the c, r branch is $r + \dfrac{1}{j\omega c}$ and that of the l branch is $j\omega l$. The two branches in parallel have an impedance

$$\cfrac{1}{\cfrac{1}{r + 1/j\omega c} + \cfrac{1}{j\omega l}}$$

To this has to be added the impedance of the other elements in series, giving

$$Z = j\omega L + \frac{1}{j\omega C} + \cfrac{1}{\cfrac{1}{r + 1/j\omega c} + \cfrac{1}{j\omega l}} = \frac{E}{i_1}$$

By performing some algebra on this expression and translating back into mechanics, the result (3.20) follows.

The complex expression (3.20) can be reduced to the form

$$x_1 = P_0(A_1 + jB_1) \tag{3.21}$$

where A_1 and B_1 are real and do not contain j. The meaning which has to be attached to (3.20) is then that in vector representation the displacement x_1 consists of two components, one in phase with the force P_0 and another a quarter turn ahead of it (compare Fig. 2.21 on page 48). Adding these two vectors geometrically, the magnitude of x_1 is expressed by

$$x_1 = P_0 \sqrt{A_1^2 + B_1^2}$$

But (3.20) is not yet in the form (3.21); it is rather of the form

$$x_1 = P_0 \frac{A + jB}{C + jD}$$

which can be transformed as follows:

$$x_1 = P_0 \cdot \frac{(A + jB)(C - jD)}{(C + jD)(C - jD)} = P_0 \cdot \frac{(AC + BD) + j(BC - AD)}{C^2 + D^2}$$

Hence the length of the x_1 vector is

$$\begin{aligned}
\frac{x_1}{P_0} &= \sqrt{\left(\frac{AC + BD}{C^2 + D^2}\right)^2 + \left(\frac{BC - AD}{C^2 + D^2}\right)^2} \\
&= \sqrt{\frac{A^2C^2 + B^2D^2 + B^2C^2 + A^2D^2}{(C^2 + D^2)^2}} = \sqrt{\frac{(A^2 + B^2)(C^2 + D^2)}{(C^2 + D^2)^2}} \\
&= \sqrt{\frac{A^2 + B^2}{C^2 + D^2}}
\end{aligned}$$

Applying this to (3.20), we may write

$$\frac{x_1^2}{P_0^2} = \frac{(k - m\omega^2)^2 + \omega^2 c^2}{[(-M\omega^2 + K)(-m\omega^2 + k) - m\omega^2 k]^2 + \omega^2 c^2 (-M\omega^2 + K - m\omega^2)^2} \tag{3.22}$$

which is the *amplitude of the motion of the main mass M.*

It is instructive to verify this result for several particular cases and see that it reduces to known results as previously obtained. The reader is advised to do this for some of the following cases:

1. $k = \infty$
2. $k = 0; c = 0$
3. $c = \infty$
4. $c = 0; \omega = \Omega_n = \sqrt{K/M} = \sqrt{k/m}$
5. $m = 0$

Thus we are in a position to calculate the amplitude in all cases. In Eq. (3.22) x_1 is a function of seven variables: P_0, ω, c, K, k, M, and m. However, the number of variables can be reduced, as the following consideration shows. For example, if P_0 is doubled and everything else is kept the same, we should expect to see x_1 doubled, and there are several relations of this same character. In order to reveal them, it is useful to write Eq. (3.22) in a dimensionless form, for which purpose the following symbols are introduced:

$$\left.\begin{array}{l} \mu = m/M = \text{mass ratio} = \text{absorber mass/main mass} \\ \omega_a^2 = k/m = \text{natural frequency of absorber} \\ \Omega_n^2 = K/M = \text{natural frequency of main system} \\ \mathbf{f} = \omega_a/\Omega_n = \text{frequency ratio (natural frequencies)} \\ \mathbf{g} = \omega/\Omega_n = \text{forced frequency ratio} \\ x_{st} = P_0/K = \text{static deflection of system} \\ c_c = 2m\Omega_n = \text{``critical'' damping (see page 38)} \end{array}\right\} \tag{3.23}$$

After performing some algebra Eq. (3.22) is transformed into

$$\frac{x_1}{x_{st}} = \sqrt{\frac{\left(2\frac{c}{c_c}\mathbf{g}\,\mathbf{f}\right)^2 + (\mathbf{g}^2 - \mathbf{f}^2)^2}{\left(2\frac{c}{c_c}\mathbf{g}\,\mathbf{f}\right)^2 (\mathbf{g}^2 - 1 + \mu\mathbf{g}^2)^2 + [\mu\mathbf{f}^2\mathbf{g}^2 - (\mathbf{g}^2 - 1)(\mathbf{g}^2 - \mathbf{f}^2)]^2}} \tag{3.24}$$

This is the amplitude ratio x_1/x_{st} of the main mass as a function of the four essential variables μ, c/c_c, $\mathbf{f}$, and $\mathbf{g}$. Figure 3.12 shows a plot of x_1/x_{st} as a function of the frequency ratio $\mathbf{g}$ for the definite system: $\mathbf{f} = 1$, $\mu = \frac{1}{20}$, and for various values of the damping c/c_c. In other words, the figure describes the behavior of a system in which the main mass is 20 times as great as the damper mass, while the frequency of the damper is equal to the frequency of the main system ($f = 1$).

It is interesting to follow what happens for increasing damping. For $c = 0$ we have the same case as Fig. 3.8a, a known result. When the damping becomes infinite, the two masses are virtually clamped together and we have a single-degree-of-freedom system with a mass $2\frac{1}{20}M$. Two other curves are drawn in Fig. 3.12, for $c/c_c = 0.10$ and 0.32.

In adding the absorber to the system, the object is to bring the resonant peak of the amplitude down to its lowest possible value. With $c = 0$ the peak is infinite; with $c = \infty$ it is again infinite. Somewhere

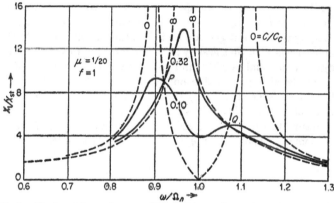

Fig. 3.12. Amplitudes of the main mass of Fig. 3.6 for various values of absorber damping. The absorber is twenty times as small as the main machine and is tuned to the same frequency. All curves pass through the fixed points P and Q.

in between there must be a value of c for which the peak becomes a minimum.

This situation also can be understood physically as follows. It was learned on page 52 that the amplitude at resonance of a single-degree-of-freedom system is limited by damping only. It was seen that damping energy is dissipated, i.e., converted into heat. When the damping force does considerable work, the amplitude remains small at resonance. This is a relation that holds for more complicated systems also. The work done by the damping force is given by the force times the displacement through which it operates. In our case the displacement is the relative motion between the two masses or also the extension of the damper spring. If $c = 0$, the damping force is zero, no work is done, and hence the resonant amplitude is infinite. But when $c = \infty$, the two masses are locked to each other so that their relative displacement is zero and again no work is done. Somewhere in between 0 and ∞ there is a damping for which the product of damping force and displacement becomes a maximum, and then the resonant amplitude will be small.

Before proceeding to a calculation of this "optimum damping," we observe a remarkable peculiarity in Fig. 3.12, *viz.*, that all four curves intersect at the two points P and Q. (See Fig. 2.41, p. 71.) This, we shall presently prove, is no accident; *all curves pass through these two points independent of the damping*. If we can calculate their location, our problem is practically solved, because *the most favorable curve is the one which passes with a horizontal tangent through the highest of the two fixed points P or Q*. The best obtainable "resonant amplitude" (at optimum damping) is the ordinate of that point.

Even this is not all that can be done. By changing the relative "tuning" $f = \omega_a/\Omega_n$ of the damper with respect to the main system, the two fixed points P and Q can be shifted up and down the curve for $c = 0$. By changing f, one point goes up and the other down. Clearly the most favorable case is such that *first* by a proper choice of f the two fixed points are adjusted to equal heights, and *second* by a proper choice of c/c_c the curve is adjusted to pass with a horizontal tangent through *one* of them. It will be seen later (Fig. 3.13) that it makes practically no difference which one of the two (P or Q) we choose.

Now return to Eq. (3.24) to see if there are any values of g for which x_1/x_{st} becomes independent of c/c_c. The formula is of the form

$$\frac{x_1}{x_{st}} = \sqrt{\frac{A\left(\dfrac{c}{c_c}\right)^2 + B}{C\left(\dfrac{c}{c_c}\right)^2 + D}}$$

This is independent of damping if $A/C = B/D$, or written out fully, if

$$\left(\frac{1}{g^2 - 1 + \mu g^2}\right)^2 = \left(\frac{g^2 - f^2}{\mu f^2 g^2 - (g^2 - 1)(g^2 - f^2)}\right)^2$$

We can obliterate the square sign on both sides but then have to add a $\pm$ in front of the right-hand side. With the minus sign, after cross-multiplication,

$$\mu f^2 g^2 - (g^2 - 1)(g^2 - f^2) = -(g^2 - f^2)(g^2 - 1 + \mu g^2) \quad (3.25)$$

It is seen that the whole of the second term on the left-hand side cancels a part of the right-hand side, so that

$$\mu f^2 g^2 = -\mu g^2(g^2 - f^2)$$
or $$\qquad f^2 = -g^2 + f^2 \qquad \text{so that} \qquad g^2 = 0$$

This is a trivial (but true) result. At $g = 0$ or $\omega = 0$ the amplitude is

x_{st}, independent of the damping, simply because things move so slowly that there is no chance for a damping force to build up (damping is proportional to velocity).

The other alternative is the plus sign before the right-hand side of (3.25). After a short calculation the equation then becomes

$$g^4 - 2g^2 \frac{1 + f^2 + \mu f^2}{2 + \mu} + \frac{2f^2}{2 + \mu} = 0 \qquad (3.26)$$

This is a quadratic equation in g^2, giving *two values*, the "fixed points" we are seeking. Let the two roots of this equation be g_1^2 and g_2^2. It is seen that g_1 and g_2 (*i.e.*, the horizontal coordinates of the fixed points P and Q) are still functions of μ and f.

Our next objective is to adjust the tuning f so that the ordinates x/x_{st} of P and Q are equal. To solve Eq. (3.26) for g_1 and g_2, to substitute these values in (3.24), and then to equate the two expressions so obtained is very time consuming. Fortunately, it is not necessary. In the first place, we remember that at P and Q the value of x/x_{st} is *independent* of the damping, so we may as well select such a value of c/c_c that (3.24) reduces to its simplest possible form. This happens for $c = \infty$, when (3.24) becomes

$$\frac{x_1}{x_{st}} = \frac{1}{1 - g^2(1 + \mu)} \qquad (3.27)$$

Substituting g_1 and g_2 in this equation gives

$$\frac{1}{1 - g_1^2(1 + \mu)} = \frac{1}{1 - g_2^2(1 + \mu)} \qquad (3.28)$$

However, this is not quite correct for the following reason: Equation (3.27) is really not represented by the curve $c = \infty$ of Fig. 3.12 but rather by a curve which is negative for values of g larger than $1/\sqrt{1 + \mu}$ (see also Fig. 2.18). Since P and Q lie on different sides of this value of g, the ordinate of P is positive and that of Q negative, so that Eq. (3.28) should be corrected by a minus sign on one side or the other. By simple algebra the equation, thus corrected, becomes

$$g_1^2 + g_2^2 = \frac{2}{1 + \mu} \qquad (3.29)$$

Now it is not even necessary to solve Eq. (3.26) for g_1 and g_2, if we remember that the negative coefficient of the middle term in a quadratic equation is equal to the sum of the roots. In Eq. (3.26) that sum is

$$g_1^2 + g_2^2 = \frac{2(1 + f^2 + \mu f^2)}{2 + \mu}$$

Substitute this in Eq. (3.29) with the result that

$$f = \frac{1}{1 + \mu} \tag{3.30}$$

This very simple formula gives the correct "tuning" for each absorber size. For a very small absorber ($\mu \approx 0$) the tuning $f \approx 1$, or the damper frequency should be the same as the main-system frequency. For a damper one-fifth as large as the main mass, $f = \frac{5}{6}$ or the damper has to be made 17 per cent slower than the main system.

Now we know how to tune, but we do not know yet what amplitude x/x_{st} we shall finally get. Figure 3.13 is a case of such tuning for $\mu = \frac{1}{4}$.

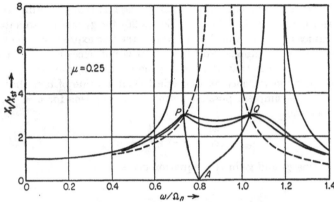

FIG. 3.13. Resonance curves for the motion of the main mass fitted with the most favorably tuned vibration-absorber system of one-fourth of the size of the main machine.

Two curves are drawn. One passes horizontally through P and then is *not* horizontal at Q; the other is horizontal at Q and not at P. It is seen that practically no error is made by taking the amplitude of either point as the maximum amplitude of the curve. This amplitude is easily calculated. Merely substitute a root of (3.26) in the expression for x_1/x_{st}, and since, at this point (P or Q), x_1/x_{st} is independent of damping, take for it form (3.27). The result is

$$\frac{x_1}{x_{st}} = \sqrt{1 + \frac{2}{\mu}} \tag{3.31}$$

This represents the most favorable possibility, if the natural frequency of the damper differs from that of the main system in the manner prescribed by (3.30).

It is interesting to compare the result (3.31) with some other cases which are sometimes encountered in actual machines (Fig. 3.14).

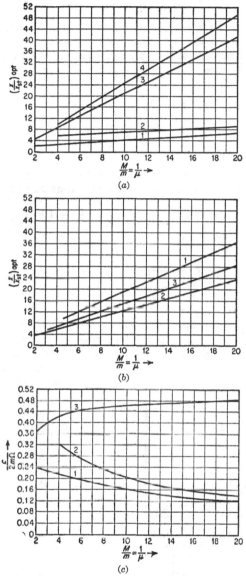

FIG. 3.14. (a) Peak amplitudes of the main mass as a function of the ratio m/M for various absorbers attached to the main mass. (b) Peak relative amplitudes between the masses M and m for various absorbers. (c) Damping constants required for most favorable operation of the absorber, i.e., for obtaining the results of (a) and (b). (See Problem 92.)

Curve 1 for the most favorably tuned and damped absorber; curve 2 for the most favorably damped absorber tuned to the frequency of the main system; curve 3 for the most favorably damped viscous Lanchester damper; curve 4 for the most favorably damped Coulomb Lanchester damper.

101

First, consider the *vibration absorber with constant tuning*, $f = 1$, where the small damper is tuned to the same frequency as the main system, independent of the size of the damper. The equation for the two fixed points (3.26) becomes

$$g^4 - 2g^2 + \frac{2}{2 + \mu} = 0$$

or

$$g^2 = 1 \pm \sqrt{\frac{\mu}{2 + \mu}}$$

For the usual damper sizes, the peak for the *smaller* g is higher than for the larger g (see Fig. 3.12; also check the location of the fixed points with the formula). Thus we substitute $g^2 = 1 - \sqrt{\dfrac{\mu}{2 + \mu}}$ in (3.27), with the result that

$$\frac{x_1}{x_{st}} = \frac{1}{-\mu + (1 + \mu) \sqrt{\dfrac{\mu}{2 + \mu}}} \tag{3.32}$$

Next, consider the apparatus known as the "Lanchester damper" (see page 210) with viscous friction, consisting of the system of Fig. 3.6, in which the damper spring has been replaced by a linear dashpot. Thus $k = 0$ and it is seen from Eq. (3.23) that ω_a and f also are zero. The fixed-point equation (3.26) becomes

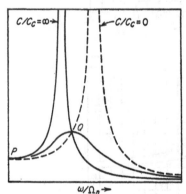

$$g^4 - 2g^2 \frac{1}{2 + \mu} = 0$$

so that one of the fixed points is permanently at $g_P = 0$, and the other is given by

FIG. 3.15. Resonance curves of a simple system equipped with a Lanchester damper with viscous friction for zero damping, infinite damping and optimum damping. All curves pass through the fixed points P and Q.

$$g_Q^2 = \frac{2}{2 + \mu} \tag{3.33}$$

The undamped and the infinitely damped constructions are single-degree-of-freedom systems, because in the first case the damper mass is completely loose and in the second case it is rigidly coupled to the main mass. This is shown clearly in Fig. 3.15, from which we also can conclude that the most favorable resonant amplitude is that of the fixed point Q.

Substitute (3.33) in (3.27) and find, for the optimum amplitude,

$$\frac{x_1}{x_{st}} = 1 + \frac{2}{\mu} \tag{3.34}$$

In some constructions of the Lanchester damper the viscous friction is replaced by dry friction or "Coulomb" friction. The analysis for that case is quite complicated and will not be given here, but the result for the most favorable resonant amplitude with such a damper is approximately

$$\frac{x_1}{x_{st}} = \frac{\pi^2}{4\mu} = \frac{2.46}{\mu} \tag{3.35}$$

The four cases already treated are shown in the curves of Fig. 3.14a. A damper of $\mu = \frac{1}{10}$ or $\frac{1}{12}$ is a practical size. It is seen that the spring-less or Lanchester dampers are much less efficient than the spring dampers or "damped dynamic absorbers." However, the design of the correct spring in the dynamic absorber is often difficult, because the small amplitudes of the main mass are obtained at the expense of large deflections and stresses in the damper spring.

Before proceeding with the calculation of the stress in the damper spring, it is necessary to find the optimum damping: $(c/c_c)_{opt}$. The optimum amplitude was found merely by stating that there must be a value of c/c_c for which the curve passes horizontally through either P or Q in Fig. 3.13. The damping at which this occurs has not been determined as yet, and now for the first time complications arise.

Start from Eq. (3.24), and substitute Eq. (3.30) into it in order to make it apply to the case of "optimum tuning." Differentiate the so-modified Eq. (3.24) with respect to g, thus finding the slope, and equate that slope to zero for the point P. From the equation thus obtained c/c_c can be calculated. This is a long and tedious job, which leads to the result

$$\left(\frac{c}{c_c}\right)^2 = \frac{\mu(3 - \sqrt{\mu/\mu + 2}}{8(1 + \mu)^3}$$

as shown by Brock.† On the other hand, if dx/dg is set equal to zero, not at point P, but rather at point Q, and the resulting equation is solved for c/c_c, we get

$$\left(\frac{c}{c_c}\right)^2 = \frac{\mu(3 + \sqrt{\mu/\mu + 2}}{8(1 + \mu)^3}$$

A useful average value between the two gives *the optimum damping for the case, Eq. (3.30), of optimum tuning,*

$$\left(\frac{c}{c_c}\right)^2 = \frac{3\mu}{8(1 + \mu)^3} \tag{3.36}$$

The same procedure applied to the case of the *constantly tuned absorber* $f = 1$, for zero slope at P, gives

$$\left(\frac{c}{c_c}\right)^2 = \frac{\mu(\mu + 3)(1 + \sqrt{\mu/\mu + 2})}{8(1 + \mu)} \tag{3.37}$$

† John E. Brock, A Note on the Damped Vibration Absorber, *Trans. A.S.M.E.*, 1946, A284.

Similarly, for the *Lanchester damper* f = 0 (Fig. 3.15), zero damping at Q is attained for

$$\left(\frac{c}{c_c}\right)^2 = \frac{1}{2(2 + \mu)(1 + \mu)} \tag{3.38}$$

These results are shown graphically in Fig. 3.14c.

Now we are ready to find the relative motion between the two masses M and m, determining the stress in the damper spring. An exact calculation of this quantity would be very laborious, because it would be necessary to go back to the original differential equations. Therefore we are satisfied with an approximation and make use of the relation found on page 50, stating that near a maximum or resonant amplitude the phase angle between force and motion is 90 deg.

Thus the work done per cycle by the force P_0 is [see Eq. (1.9), page 12]

$$W = \pi P_0 x_1 \sin 90° = \pi P_0 x_1$$

This is approximate, but the approximation is rather good because, even if φ differs considerably from 90 deg., sin φ does not differ much from unity.

On the other hand, the work dissipated per cycle by damping is, by the same formula, $\pi \times$ damping force $\times$ relative amplitude x_{rel}, since the damping force being in phase with the *velocity* has *exactly* 90-deg. phase angle with the displacement amplitude. Thus

$$W_{dissipated} = \omega(c\omega x_{rel}) \cdot x_{rel} = \pi c \omega x_{rel}^2$$

Equating the two,

$$\pi P_0 x_1 = \pi c \omega x_{rel}^2$$

or

$$x_{rel}^2 = \frac{P_0 x_1}{c\omega}$$

Written in a dimensionless form this becomes

$$\left(\frac{x_{rel}}{x_{st}}\right)^2 = \frac{x_1}{x_{st}} \cdot \frac{1}{2\mu g c/c_c} \tag{3.39}$$

This formula determines the relative motion and consequently the stress in the damper spring. Upon substitution of the proper values for μ, g, etc., this formula is applicable to the viscous Lanchester damper, as well as to the two kinds of dynamic absorbers.

The curves of Fig. 3.14b show the results of these calculations. It is seen that the relative motions or spring extensions are quite large, three or four times as large as the motion of the main system. If springs can be designed to withstand such stresses in fatigue, all is well, but this quite often will prove to be very difficult, if not impossible, within the space available for the springs. This is the reason why the Lanchester damper, though very much less effective than the spring absorber, enjoys a wide practical use.

Example: It is desired to design a damper for the system of Fig. 3.6, in which $Mg = 10$ lb.; $mg = 1$ lb.; $P_0 = 1$ lb., and $K = 102$ lb./in., which will operate for all frequencies of the disturbing force. If first the absorber spring is taken as $k = 10.2$ lb./in.:

    a. What is the best damping coefficient across the absorber?
    b. What is the maximum amplitude of the main mass?
    c. What is the maximum stress in the absorber spring?
Further, if we drop the requirement $k/K = m/M$,
    d. For what k is the best over-all effect obtained?
    e. Same question as *a* but now for the new value of k.
    f. Same question as *b* but now for the new value of k.
    g. Same question as *c* but now for the new value of k.
Solution: The answers are all contained in Fig. 3.14 *a*, *b*, and *c*.
    a. From Fig. 3.14*c* we find: $c/2m\Omega_n = 0.205$ or

$$c = 0.41 m\Omega_n = 0.41 \tfrac{1}{386} 20\pi = 0.067 \text{ lb./in./sec.}$$

    b. Figure 3.14*a* or Eq. (3.32) gives $x/x_{st} = 7.2$,

$$x_{st} = P_0/K = \tfrac{1}{102}, \text{ so that } x = 7.2/102 = 0.071 \text{ in.}$$

    c. Figure 3.14*b* gives for the relative motion across the absorber spring $x_{rel}/x_{st} = 12.8$ so that $x_{rel} = 12.8/102 = 0.126$ in. The force is $kx_{rel} = 10.2 \times 0.126 = 1.28$ lb.

    d. The most favorable tuning follows from Eq. (3.30), $\dfrac{\omega_a}{\Omega_n} = \dfrac{1}{1 + \mu} = \dfrac{10}{11}$, so that $\left(\dfrac{\omega_a}{\Omega_n}\right)^2 = \dfrac{100}{121}$. Since m, M, and K are the same now as in all previous questions, $(\omega_a/\Omega_n)^2$ is proportional to k. Thus the new absorber spring should be

$$k = \tfrac{100}{121} \times 10.2 = 8.4 \text{ lb./in.}$$

    e. Figure 3.14*c* gives $c/2m\Omega_n = 0.166$. Since $2m\Omega_n$ is the same as in question *a*, we have

$$c = \frac{0.166}{0.205} \times 0.067 = 0.054 \text{ lb. in.}^{-1} \text{ sec.}$$

    j. From Fig. 3.14*a* or Eq. (3.31) we find $x/x_{st} = 4.6$. Since from *b* we have $x_{st} = \tfrac{1}{102}$, the maximum amplitude is

$$x = \frac{4.6}{102} = 0.045 \text{ in.}$$

    f. Figure 3.14*b* gives $x_{rel}/x_{st} = 19.5$, so that $x_{rel} = 19.5/102 = 0.191$ in. With $k = 8.4$ lb./in., this leads to a maximum force in the spring of $8.4 \times 0.191 = 1.60$ lb.

The principal applications of dampers and absorbers of this type are in internal-combustion engines (page 210), in ship stabilization, which will be treated in the next section, and in electric transmission lines (page 306, Fig. 7.23). However, an "absorber" may be present in a construction without being very conspicuous.

An example of this is found in gears which, in operation, may be ringing like bells if no precautions are taken. It has been found by experience that this noise can be eliminated to a great extent (the gears "deadened") by shrinking two steel or cast-iron rings *a*, *a* (Fig. 3.16) on the inside of the rim. If the shrink fit is too loose, no deadening occurs; if it is shrunk very tight the effect is again very small, but for some intermediate shrink pressure the deadening effect is astonishingly complete. Two identical gears, one with and the other without rings, may be placed upright on the ground and their rims struck with a hammer. The first gear will sound like a piece of lead while the second one will ring for ten or more seconds. The cast-iron inserts evidently act as Lanchester dampers.

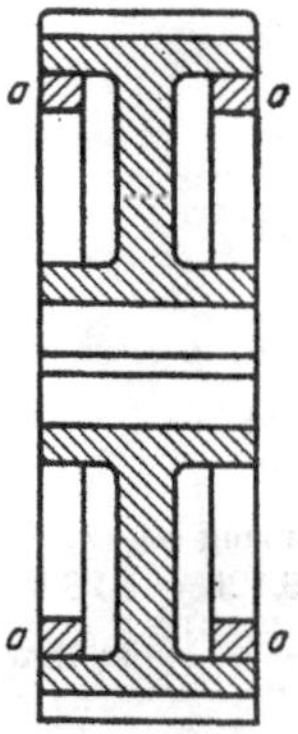

3.4. Ship Stabilization. One of the most interesting applications of the rather lengthy theory of the preceding section is the prevention of the "rolling" of ships in a rough sea by means of certain devices installed on board.

First consider the rolling of the ship itself without any damping device. Imagine the ship to be floating in still water (Fig. 3.17a), the weight **W** and the buoyancy **B** being two equal and opposite forces both passing through the center of gravity G. Now hold the ship at a *slightly* inclined position by some external couple (Fig. 3.17b). The weight **W** still acts through the point G, but the buoyancy force **B** is displaced slightly to the left. The line of action of this force intersects the center line of the ship in some point M, which is technically known as the *metacenter*. It is clear that the location of this point is determined by the geometry of the hull of the ship. The distance h between M and G is called the *metacentric height*.

The determination of this quantity from a drawing of the ship is an important task of the designer, since upon it the rolling *stability* depends. In Fig. 3.17b it is seen that the forces **W** and **B** form a couple tending to restore the ship to its vertical position. This is always the case when the metacenter is above the center of gravity or when the metacentric height h is positive. If h were negative, the **W-B** couple of Fig. 3.17b would tend to increase the inclination of the ship and the equilibrium would be unstable.

FIG. 3.16. Gear with sound-deadening rings inserted. These should be either shrunk or tack-welded in a few spots so as to allow some relative rubbing during the vibration.

Example: A ship has a rectangular cross section and the submerged part has a square section of which the sides have a length $2a$. The center of gravity lies in the vertical line of symmetry at a height x above the bottom of the ship. For small values of x the ship is stable, for large values of x it is statically unstable. Find the value of x where the equilibrium is just indifferent.

Solution: Consider a submerged piece of the ship of dimensions $2a \times 2a \times 1$ in. By taking such a slab of unit thickness we gain the advantage that the submerged volumes become numerically equal to the corresponding cross-sectional areas. By tilting through the angle φ the submerged figure changes from a square to a square from which a small triangle has been subtracted on the right side and to which a similar triangle has been added at the left side. The area of such a triangle is $a/2 \times a\varphi = a^2\varphi/2$. Since the center of gravity of these triangles is at one-third of the height from the base, the shift of the triangle from right to left shifts the center of gravity of an area $a^2\varphi/2$ through a distance of $\frac{2}{3} \cdot 2a$. The product of these quantities equals the total area of the square $4a^2$ multiplied by the horizontal shift y of the center of gravity of the whole figure. Thus

$$4a^2y = \frac{2}{3}a^3\varphi \qquad \text{or} \qquad y = \frac{a\varphi}{6}$$

The center of gravity of the submerged figure is shifted to the left over this distance from the original vertical axis of symmetry. A vertical line through this new center of gravity intersects the symmetry axis at a distance $a/6$ above the original location of the center of gravity. Since this intersection is the metacenter M, we find that M lies at a distance of $a + \frac{a}{6} = \frac{7}{6}a$ above the bottom of the ship. This is also the desired position of the center of gravity of the ship for indifferent equilibrium.

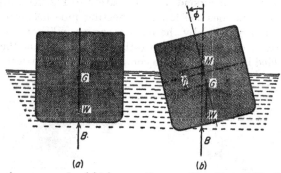

Fig. 3.17. The buoyancy and weight forces acting on a ship. For stability the metacenter M has to be located above the center of gravity G. The distance MG is the metacentric height h.

The ship is a vibratory system because when it is displaced from its equilibrium position it shows a tendency to come back. For small angles φ the location of M is independent of φ. The restoring couple is $-Wh \sin \varphi$ or $-Wh\varphi$ for sufficiently small φ. By the action of this couple the ship will roll back about some longitudinal axis. Let the moment of inertia about that axis be I_s (the subscript s stands for ship). Newton's law can be written

$$I_s\ddot{\varphi} = -Wh\varphi$$

or

$$\ddot{\varphi} + \frac{Wh}{I_s}\varphi = 0 \tag{3.40}$$

which we recognize as Eq. (2.7) of page 31 for the undamped single-degree-of-freedom system. Consequently the ship rolls with a natural frequency

$$\omega_s = \sqrt{\frac{Wh}{I_s}} \tag{3.41}$$

Imagine the ship to be in a rough sea. Waves will strike it more or less periodically and exert a variable couple on it. Though this action is not very regular, it may be regarded approximately as a harmonic disturbing torque $T_0 \sin \omega t$ to be written on the right-hand side of Eq. (3.40). In case the wave frequency ω is near to the natural frequency ω_s of the ship's roll, the oscillations may become very large. In rough seas the angle φ has been observed to reach 20 deg. Equations (3.40) and (3.41) tell us that, as far as vibrational properties go, the system of Fig. 3.17 is equivalent to Fig. 2.4 or to the upper part of Fig. 3.6. Therefore the addition of a damper of the type shown in Fig. 3.6 should help. This was done by Frahm, in 1902, who built into a ship a system of two tanks (Fig. 3.18) half filled with water, communicating through a water pipe below and through an air pipe with valve V above. The secondary or "absorber" system corresponds approximately to Fig. 2.11, page 35.

In other constructions the lower connecting pipe between the tanks was omitted and replaced by the open ocean as indicated in Fig. 3.19. These "blisters" extended along two-thirds the length of the ship and were subdivided into three or more compartments by vertical partitions.

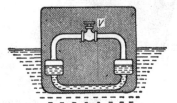 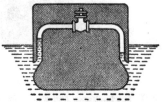

FIG. 3.18. Frahm antirolling tanks, old type.   FIG. 3.19. Modern "blister" construction of Frahm's antirolling tanks.

Both these constructions are really more complicated than Fig. 3.10, though the older construction, Fig. 3.18, comes quite close to it.

Frahm antirolling tanks were installed on the large German liners "Bremen" and "Europa."

Another method of reducing ship roll, which apparently is entirely different from Frahm's tanks but really operates on much the same principle, is the *gyroscope of Schlick* (Fig. 3.20). This device consists of a heavy gyroscope rotating at high speed about a vertical axis. The gyroscope bearings AA are mounted in a frame which is suspended in two bearings BB so that the frame is capable of rotation about an axis across the ship. The axis BB lies *above* the center of gravity of the gyroscope and its frame. A brake drum C is attached to BB, so that the swinging motion of the gyroscope frame can be damped. The weight of the gyrorotor is of the order of 1 per cent of the ship's weight. It is driven electrically to the highest possible speed compatible with its bursting strength under centrifugal stress.

For an understanding of the operation of this device, it is necessary to know the main property of a gyroscope, namely that the torque exerted on it is represented vectorially by the rate of change of the angular momentum vector.

Let the direction of rotation of the rotor be counterclockwise when viewed from above, so that the momentum vector $\overline{\mathfrak{M}}$ points upward. When the ship is rolling clockwise (viewed from the rear) with the angular velocity $\dot{\varphi}$, the rate of change of $\overline{\mathfrak{M}}$ is a vector of length $\mathfrak{M}\dot{\varphi}$ directed across the ship to the right. This vector represents the torque exerted on the rotor by its frame. The torque exerted by the rotor on its frame is directed opposite to this, so that the frame is accelerated in the

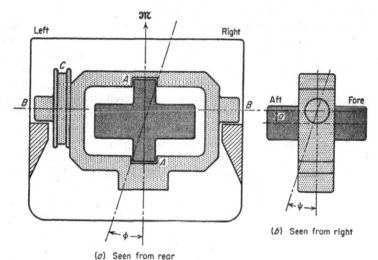

(a) Seen from rear

FIG. 3.20. Scheme of Schlick's anti-ship-rolling gyroscope. It operates by virtue of energy dissipation at the brake drum C.

direction of increasing ψ (so that the lower part of the frame tends to go to the rear of the ship).

On the other hand, if the rotor frame is swinging with a positive angular velocity ψ, the momentum vector $\overline{\mathfrak{M}}$ increases by an amount $\mathfrak{M}\psi$ each second in a direction pointing toward the front of the ship. This vector is a torque tending to rotate the rotor clockwise, and consequently the ship counterclockwise, when viewed from the rear.

Thus the ship is "coupled" to the gyroscope in much the same sense as the ship is coupled to the Frahm water tanks, though the mechanism is entirely different.

Without damping in the swing motion of the rotor frame, the presence of the gyroscope merely changes the *one* natural rolling frequency of the ship into *two* other natural rolling frequencies. A resonance with sea waves leads to infinite amplitudes φ of the ship. An infinite amount of damping clamps the rotor frame solidly in the ship. Then a roll of the ship merely creates a pitching torque on the ship's frame and conversely the clamped gyroscope will convert a pitching motion of the ship into a rolling torque on it. At resonance of the sea waves with the one natural rolling frequency again an infinite rolling amplitude results. But at some intermediate damping the two resonant peaks can be materially decreased.

Activated Ship Stabilizers. The motion of the water in the Frahm tank, as well as the precession of the Schlick gyroscope, is brought about by the rolling of the ship itself, and in both cases is impeded by a brake. This is not a perfect solution, since the best brake adjustment is different for different frequencies and other conditions. These systems are designated as "passive" systems to distinguish them from the more modern "active" systems, where the Frahm water is *pumped* from one tank to the other, and the Schlick gyro precession is *forced.* There is no longer a brake, but there is a governor or device which feels the roll of the ship and gives the proper signals controlling the Frahm pump or the Schlick precession drive, so that the phase of the counter torque is always correct.

The first of these activated devices reaching practical perfection was the Sperry gyroscopic ship's stabilizer, illustrated schematically in Fig. 3.21. It consists of a main gyroscope, which differs from Schlick's only in the fact that the axis BB passes through the center of gravity, and that the brake drum C is replaced by a gear segment meshing with a pinion on the shaft of a direct-current motor D. Besides the main gyroscope there is the *pilot gyroscope* (Fig. 3.21b, c) which has an over-all dimension of some 5 in. and is nearly an exact replica of the main one. The only difference is that there is no gear C, but instead of that there are two electrical contacts d_1 and d_2, one in front and one behind the rotor frame.

The operation is as follows. When the ship has a clockwise rolling velocity $\dot{\varphi}$ (looking from the rear) the top of the pilot rotor frame is accelerated toward the front of the ship and closes the contact d_2. This action sets certain electrical relays working which start the precession motor D so as to turn the main frame about the axis BB in the same direction as the pilot frame. In other words, the top of the main frame moves to the front of the ship. This necessitates a clockwise φ-torque on the main rotor, which has a counterclockwise reaction on the main-rotor frame and thus on the ship. Therefore the main gyroscope creates a torque on the ship which is in opposition to the *velocity* of roll and in that manner most effectively counteracts the roll. As soon as the velocity of roll of the ship becomes zero, the pilot torque disappears and the pilot rotor is pulled back to its neutral position by two springs e as shown in Fig. 3.21c. Only when the roll acquires a velocity in the opposite direction does the pilot go out of its equilibrium position again closing the contact d_1, which sets the precession motor going in the opposite direction. Thus there is always a torque acting on the ship in opposition to the instantaneous velocity of rolling. With the torque always against the angular velocity, a maximum amount of energy of the rolling motion is destroyed. (See the three rules on page 16.)

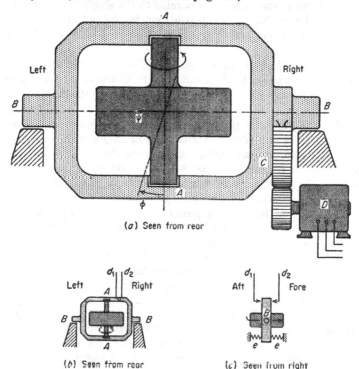

(a) Seen from rear

Left d_1 d_2 Right   Aft d_1 d_2 Fore
            A

(b) Seen from rear   (c) Seen from right

FIG. 3.21. Sperry's gyroscope for diminishing ship roll. The precession is forced by a motor D, which is controlled by a small pilot gyroscope shown in (b) and (c)

The direction of the desired ψ-precession of the main gyro was seen to be the same as that of the free pilot gyro, which means that the motor D turns the main gyro in the direction in which it would go by itself, if it were free to move in the bearings B. However, it can be easily verified that, if such freedom existed, the main gyro would precess extremely fast in an accelerated manner and would reach $\psi = 90$ deg. in a very short fraction of the roll period. At this position the roll would no longer affect the gyro. Therefore the motor D does not *push* the main gyro (except at the very beginning of the precession) but really acts like a brake, holding the speed of precession down to a proper value. Schemes have been proposed to do away altogether with the motor D, reverting to the old Schlick brake drum, with the difference, however, that the tightness of the brake would be controlled electrically by signals coming from the pilot gyro.

In actual constructions the pilot gyroscope has its axis AA horizontal and across the ship, while its frame axis BB is vertical. The line connecting the contacts d_1 and d_2 remains parallel to the ship's longitudinal axis as before. The reader should reason out for himself that with this arrangement the same action is obtained as with the one shown in Fig. 3.21.

Sperry gyro stabilizers have been installed with success on many yachts. An application to the Italian liner "Conte di Savoia" showed that a large roll was very effectively damped down by the device. However, in the roughest Atlantic storms *single* waves were found to tilt the ship 17 deg.; and since the power of the gyros was sufficient only to swing the ship 2 deg. at one time, the greatest roll angles with and without stabilizer did not differ materially. A gyroscope that would hold the ship down even in the roughest weather would become prohibitively large, of the order of 5 per cent of the weight of the ship. The same objection attaches to the activated Frahm tanks, in which the water is pumped from side to side of the ship, the pump being controlled by a pilot gyroscope. Experiments with this system on a destroyer showed that it had to be very large to be effective.

A third antiroll device utilizes the principle of lift on airplane wings. Imagine an airplane of a wing span of say 20 ft., and swell the fuselage of that plane to the size of an ocean liner, leaving the wing size unchanged. The wings are located below the water line. While the ship moves through the water, a lift will be developed on the wings. The wings can be rotated through a small angle about their longitudinal (athwartship) axis, and thus the angle of attack can be changed and consequently the hydrodynamic lift force from the water can be changed. For example, if the left (or port) hydrofoil has a large positive angle of attack and an upward lift, while the right (or starboard) hydrofoil has a negative angle and downward lift, there is a hydrodynamic rolling torque in a clockwise direction seen from behind. If now these angles of attack are continuously changed by a driving motor (controlled by a pilot gyroscope) so that the torque on the ship is opposite to that of its roll velocity, the rolling motion will be damped. This system was in operation on some British destroyers during the last war. Although its weight is small compared with that of the ship, it has the disadvantage that it increases the resistance of the ship by a small precentage so that the cost of a portion of all the fuel used during the lifetime of the ship must be charged against it. This disadvantage has been overcome recently by making the hydrofoils retractable, so that they are being used only during foul weather. This, then, is at present the best answer to the question. It should be mentioned that the water-tank and gyroscope stabilizers will work when the ship stands still; this is not the case with the hydrofoil stabilizer, which depends for its operation on the speed of the ship. The "passive," or inactivated, form of the hydrofoil system has been used for years in the form of "bilge keels," which are crude forms of hydrofoil permanently attached to the ship's sides (Fig. 3.22). When the ship

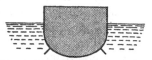

Fig. 3.22. Bilge keels, extending over more than half the length of a ship.

has forward speed, no lift force is generated on these keels, because the angle of attack is made zero. But when the ship rolls, its own rolling motion induces an apparent angle of attack which causes lift forces forming a torque opposite to the direction of the rolling velocity. Bilge keels are quite ineffective in stopping the roll when the ship lies still but become effective with an intensity roughly proportional to the square of the ship's forward speed.

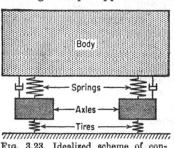

Fig. 3.23. Idealized scheme of conventional automobile with front and rear axles and shock absorbers.

3.5. Automobile Shock Absorbers. An automobile of conventional design on its springs and tires is a very complicated vibrational system. There are three distinct "masses": the body, the front axle, and the rear axle; and eight distinct "springs": the four springs proper and the four tires (Fig. 3.23). A solid body free in space has six degrees of freedom: it can bob up and down, sway back and forth, move forward and backward (*the three translations*); and, moreover, it can have three *rotations*, known under the technical names of:

1. *Rolling* about a longitudinal axis.
2. *Pitching* about a lateral axis.
3. *Yawing* or *nosing* about a vertical axis.

Since the automobile has three such bodies, it really has 18 degrees of freedom. However, a good many of these 18 are rather unimportant. The most important motions are:

1. A bobbing up and down of the body with the axles practically steady.
2. A pitching of the body with the axles nearly steady.
3. A bobbing up and down of each axle on the tire elasticity with the chassis practically undisturbed.
4. A rolling of the axles with little motion of the body.

The first two motions were discussed on page 85. For an entirely symmetrical car (which naturally does not exist) the two natural modes are a pure vertical parallel motion and a pure pitching about the center of gravity, but in the actual unsymmetrical case each mode is a mixture of the two. In practice, the natural frequencies for the first two modes are close together, being somewhat slower than 1 cycle per second in modern cars. The motions 3 and 4 have frequencies roughly equal to each other but much faster than the body motions. With older cars the axle natural frequency may be as high as 6 or 8 cycles per second; with modern cars having balloon tires and heavier axles on account of front wheel brakes, the frequency is lower. On account of the fact that the body frequency and the axle frequency are so far apart, the one motion (1 or 2) can exist practically independent of the other (3 or 4). For when the body moves up and down at the rate of 1 cycle per second, the force variation in the main spring is six times as slow as the natural frequency of the axle mass on the tire spring and thus the axle ignores the alternating force. And similarly, while the axle vibrates at the rate of 6 cycles per second, the main body springs experience an alternating force at that rate, which, however, is far too fast to make an appreciable impression on the car body (Fig. 2.18, page 44).

Resonances with either frequency occur quite often and can be observed easily on any old-model car or also on a modern car when the shock absorbers (dampers) are removed. The pitching motion of the body gets in resonance at medium speeds when running over a road with unevennesses of long wave length. For example, at some 30 m.p.h. on old concrete highways having joints spaced regularly at about 40 ft. apart, very violent pitching usually occurs in cars with insufficient shock absorbers. The other natural frequency often comes to resonance at rather low speeds when running over cobblestones. The axles then may vibrate so that the tires leave the ground at each cycle.

The worst of the evils just described have been eliminated by introducing shock absorbers across the body springs, which introduce damping in the same fashion as a dashpot would. Before starting a discussion of their action, it is well to consider first the influence of the springs and tires themselves on the "riding quality," or "riding comfort."

Assuming that the car is moving forward at a constant speed, what quantity should be considered to be a measure of comfort? It might be the vertical displacement of the chassis or any of its derivatives. It is not the displacement amplitude itself, for a ride over a mountain, being a "vibration" of amplitude 3,000 ft. at the rate of 1 cycle per hour, may be very comfortable. It is not the vertical velocity, for there are no objections to a fast ride up a steep slope. Nor is it the vertical acceleration, for a steady acceleration is felt as a steady force, which amounts only to an apparent change in g that cannot be felt. But *sudden* shocks produce uncomfortable sensations. Therefore a criterion for comfort is the rate of change of acceleration d^3y/dt^3, a quantity that has been called the "jerk."

Figure 3.24 represents a wheel or axle on its tire spring. The wheel runs over a road of which the surface is a sinusoid. If the car moves at a constant speed, the bottom of the tire experiences a motion $a_0 \sin \omega t$. Consider various wheels of the same mass m running with the same speed over the same road $a_0 \sin \omega t$ but differing among each other in the elasticity k of their tire springs. The force

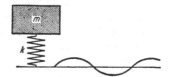

FIG. 3.24. Automobile riding over a wavy road.

F transmitted by the spring from the road to the wheel or axle is k times the relative displacement, which by Eq. (2.26) page 45, is

$$F = ky_0 = \frac{m\omega^2 a_0}{1 - \omega^2/\omega_n^2}$$

or in a dimensionless form,

$$-\frac{F}{m\omega^2 a_0} = \frac{(\sqrt{k/m\omega^2})^2}{1 - (\sqrt{k/m\omega^2})^2} \tag{3.42}$$

If the dimensionless force $F/m\omega^2 a_0$ is plotted vertically against the dimensionless square root of the tire spring constant $\sqrt{k}/\sqrt{m\omega^2}$, Eq. (3.42) shows that the diagram Fig. 2.20 (page 46) is obtained.

We see that stiff springs (large k or steel-rimmed wheels) are represented by points in the right-hand part of the diagram, which means considerable force transmission. Little force transmission occurs for weak springs (*i.e.*, balloon tires) represented by points close to the origin of Fig. 2.20.

This can be appreciated also from a somewhat different standpoint. Consider a given "sinusoidal" road or a smooth road with a single bump on it, and let the steel-tired wheel be completely rigid. The vertical accelerations of the wheel now *increase with the square of the speed*, which

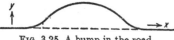

FIG. 3.25. A bump in the road.

can be seen as follows. Let the bump on the road be represented by $y = f(x)$ as in Fig. 3.25. For a car with speed v we have $x = vt$. Then the vertical velocity is

$$\frac{dy}{dt} = v\frac{dy}{d(vt)} = v\frac{dy}{dx}$$

and the vertical acceleration is

$$\frac{d}{dt}\left(\frac{dy}{dt}\right) = v\frac{d}{d(vt)}\cdot\left(\frac{dy}{dt}\right) = v\frac{d}{dx}\left(v\frac{dy}{dx}\right) = v^2\frac{d^2y}{dx^2}$$

Since d^2y/dx^2 is a property of the shape of the bump only, independent of the velocity, it is seen that the vertical acceleration increases with the square of the speed. If the wheel is rigid (no tire), the forces acting on the wheel as well as on the road are the product of the wheel mass and this acceleration. Thus the force on the road also increases with the square of the speed, making the rubber tire an absolute necessity even for moderate speeds, which is a matter of common observation.

The tires are primarily there for a protection of the road and of the wheels, whereas the main springs take care of riding comfort. With a given axle movement a_0, how do we have to design the main springs for maximum riding comfort, *i.e.*, for minimum "jerk" d^3y/dt^3? From Eq. (2.26) we have

$$\ddot{y} = \frac{\omega^2 a_0(\omega/\omega_n)^2}{1 - \omega^2/\omega_n^2}\sin\omega t$$

so that by differentiation

$$\frac{\dddot{y}}{\omega^3 a_0} = \frac{1(\omega/\omega_n)^2}{1 - \omega^2/\omega_n^2}\cdot\cos\omega t \tag{3.43}$$

Again Fig. 2.20 represents this relation, and the springs have to be made as soft as possible in the vertical direction. Then most road shocks will be faster than the natural frequency of the car and will not give it any appreciable acceleration. The introduction of damping is undesirable at these high road frequencies. But the case of resonance is not excluded, and from that standpoint damping is very desirable.

There is still another viewpoint to the question. Figure 2.20 pertains to steady-state forced vibrations, *i.e.*, to road shocks following each other with great regularity. Practically this does not occur very often as the bumps on actual roads are irregularly spaced. Thus the motion will consist of a combination of forced and free vibrations, and damping is desirable to destroy the free vibrations quickly after the road is once again smooth.

The shock absorbers on most automobiles are hydraulic and operate on the dashpot principle. Any relative motion between the axle and the car body results in a piston moving in a cylinder filled with oil. This oil has to leak through small openings, or it has to pass through a valve which has been set up by a spring so that it opens only when a certain pressure difference exists on the two sides of the piston. In this manner a considerable force opposing the relative motion across the car body springs is created, and this force is roughly proportional to the *velocity* of the relative spring motion.

The most desirable amount of damping in these shock absorbers depends on the road condition. When running over a smooth road with rolling hills and valleys which are taken at the rate of approximately one hill per second, it is clear that critical damping is wanted. On the other hand, if the road has short quick bumps, a small damping is desirable.

Some shock absorbers have one-way valves in them, so that for a spreading apart of the axle and the body a different damping occurs than for their coming together. This is accomplished by forcing the oil through different sets of openings by means of check valves. Usually the arrangement is such that when the body and axle are spreading apart the damping is great, while when they are coming together a small force is applied by the shock absorbers. The theories and arguments given by the manufacturers as a justification of this practice do not seem to be quite rational.

3.6. Isolation of Non-rigid Foundations. On page 69 we discussed the problem of protecting a foundation from the vibrations of an unbalanced rotating machine placed on it and found that the insertion of a soft spring between the machine and the foundation was a proper remedy. The spring had to be designed so that the natural frequency of the machine on it would be several times, say three times, as slow as the frequency of the disturbing vibrations. In deriving this result it was assumed that the foundation was rigid, which is a reasonable enough assumption for many cases where a machine is mounted on a foundation attached directly to the ground. However, when we deal with a large Diesel engine mounted in a ship's hull or with a powerful aircraft engine mounted on the wing of an airplane, the assumption is not at all justified, because the weight of the

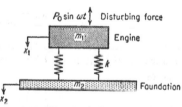

FIG. 3.26. Vibration isolation on a foundation which is not infinitely large, such as the case of an engine mounted in a ship's hull.

nearby parts of the "foundation" is considerably less than that of the engine itself. In order to make a first study of such cases, we assume the foundation to be a pure mass m_2 (Fig. 3.26), while the engine mass will be designated as m_1.

The differential equations are

$$m_1\ddot{x}_1 + k(x_1 - x_2) = P_0 \sin \omega t \atop m_2\ddot{x}_2 + k(x_2 - x_1) = 0 \Bigg\}$$

Assuming for the solution

$$x_1 = x_{1m} \sin \omega t$$
$$x_2 = x_{2m} \sin \omega t$$

and substituting, we find

$$x_1(-m_1\omega^2 + k) + x_2(-k) = P_0$$
$$x_1(-k) + x_2(-m_2\omega^2 + k) = 0$$

The engine itself is not in trouble, so that we do not care about x_1, but we are interested in the motion of the foundation x_2. From the second equation we get

$$x_1 = x_2\left(1 - \frac{m_2\omega^2}{k}\right)$$

and, substituting into the first equation,

$$x_2 = \frac{P_0 k}{m_1 m_2 \omega^4 - k(m_1 + m_2)\omega^2}$$

This expression can be written into other forms; a convenient one is

$$x_2 = \frac{P_0}{(m_1 + m_2)\omega^2\left(\dfrac{\omega^2}{\omega_n^2} - 1\right)}$$

where the natural frequency ω_n of the system is

$$\omega_n^2 = \frac{k}{m_1 m_2/(m_1 + m_2)} \tag{3.44}$$

The force transmitted to the foundation can be found most easily by remarking that it must be the inertia force of the foundation itself or $m_2\omega^2 x_2$.

$$\text{Transmitted force} = \frac{m_2}{m_1 + m_2} \cdot \frac{P_0}{\dfrac{\omega^2}{\omega_n^2} - 1} \tag{3.45}$$

Comparing this result with Eq. (2.33) and Fig. 2.41 for the old theory of a rigid foundation, we see that the old theory (for zero damping) still holds; in other words, Eq. (3.45) is still represented by the simple resonance diagram. The difference is, however, that the natural frequency ω_n of the old theory was $\omega_n^2 = k/m$, whereas now it is given by Eq. (3.44). For any kind of foundation the protecting spring has to be so designed as to make the natural frequency about one-third of the disturbing frequency. This correct statement sounds as if there is no difference between light and heavy foundations, which is misleading. Take the case of an engine 10 times as heavy as its foundation. Then Eq. (3.44) becomes

$$\omega_n^2 = \frac{k}{m_1 m_2/(m_1 + m_2)} = \frac{k}{m_1} \cdot \frac{m_1 + m_2}{m_2} = \frac{k_1}{m_1} \cdot \frac{10 + 1}{1} = 11 \frac{k}{m}$$

The natural frequency squared is 11 times as high with the light foundation as it would be with a very heavy one, and hence to get the same kind of protection we must make our protecting spring 11 times as flexible.

Suppose the engine exciting frequency to be about 1,200 v.p.m. Then for a solid foundation, a decent protecting spring would be one for a natural frequency of 400 v.p.m., which means a static sag of about $\frac{1}{4}$ in. (Fig. 2.9, page 34): a reasonable design. If this same engine is to be mounted in a ship with a light foundation $m_2/m_1 = 0.1$, the static sag in the spring must be 11 times larger, or some 2.5 in. In the first place this is very difficult to do, and moreover such an engine in a rolling and pitching ship would be entirely impossible. This shows that the protection of a ship's hull from the vibrations of the machinery inside is a very difficult proposition. The vibrations of the hull radiate noise under water, which disturbs the peace of fish and sometimes is undesirable for other reasons as well.

The idealization of the "foundation," or hull, by a mass m_2 is a crude one. The point of the hull to which the machinery is attached acts partly like a spring, partly like a mass, and partly even like a dashpot, because radiation of the vibration is a damping action. To try to determine how much spring, mass, and dashpot there is from a blueprint drawing is a hopeless proposition, but once the ship or airplane is built we can find this out experimentally without too much trouble. We attach to the location in question a vibrator which applies to it a harmonic force of which we can vary the frequency gradually. Then we measure the force as well as the motion amplitude and phase angle for each frequency. The result can best be presented in the form of a ratio Z, the *mechanical impedance*, which is a function of the frequency ω:

$$Z(\omega) = \frac{\text{force amplitude}}{\text{displacement amplitude}} \qquad (3.46)$$

As an example consider a simple spring, attached to the ground at its bottom, while the top end is actuated by the vibrator. If the top amplitude is $a \sin \omega t$, then the force is $ka \sin \omega t$, so that for a spring $Z = k$, independent of frequency. As a second example take a mass by itself. If the vibrator is operating on it, the motion is $a \sin \omega t$ and the force is $-ma\omega^2 \sin \omega t$, so that $Z = -m\omega^2$. The third simple example is a dashpot of which the cylinder is attached to ground, while the piston is actuated. If the piston motion is $a \sin \omega t$, then the force is $ca\omega \cos \omega t$, out of phase by 90 deg. with respect to the motion. In complex notation we say that the motion is a and the force $jca\omega$, so that $Z = j\omega c$, an imaginary quantity. Other cases of relatively simple impedance calculations are Probs. 84 to 90. When we know the system, we can calculate the impedance, but we do not know the properties of a point of a ship's hull. However, we can measure the impedance at this point.

Now we consider the system of a machine m_1 actuated by a force $P_0 \sin \omega t$, attached through a spring k to a foundation of impedance $Z(\omega)$, thus replacing in Fig. 3.26 the mass m_2 by the more general foundation, characterized by $Z(\omega)$, which usually is a complex quantity. The differential equations are

$$m_1\ddot{x}_1 + k(x_1 - x_2) = P_0 \sin \omega t \ \Big\}$$
$$k(x_1 - x_2) = x_2 Z \qquad\qquad$$

Assuming harmonic motion of frequency ω, the amplitudes x_1 and x_2 will be complex numbers if Z is a complex or imaginary quantity,

$$-m_1\omega^2 x_1 + k(x_1 - x_2) = P_0 \ \Big\}$$
$$k(x_1 - x_2) = x_2 Z \qquad$$

or, rearranged,

$$x_1(-m_1\omega^2 + k) + x_2(-k) = P_0$$
$$x_1(-k) + x_2(k + Z) = 0$$

Solve for x_1 from the second equation,

$$x_1 = x_2 \left(1 + \frac{Z}{k}\right)$$

and substitute into the first; then solve for x_2.

$$x_2 = \frac{P_0}{Z\left(1 - \dfrac{m_1\omega^2}{k}\right) - m_1\omega^2} \qquad (3.47)$$

The force transmitted to the ground is Zx_2, and the exciting force on the machine is P_0 so that the ratio between the two is

$$\text{Transmissibility} = \frac{Z}{Z\left(1 - \dfrac{m_1\omega^2}{k}\right) - m_1\omega^2} \qquad (3.48)$$

This general formula contains all possible foundation characteristics. As a first example take Fig. 3.26, where $Z = -m_2\omega^2$. Substituting this into (3.48) leads to the result previously found for that case. As a second example let the foundation consist of a pure spring K, so that the engine rests on the ground through k and K in series or through an equivalent spring $kK/(k + K)$.   Substituting $Z = K$ into (3.48),

$$\text{Transmissibility} = \frac{K}{K - m_1\omega^2(1 + K/k)}$$
$$= \frac{1}{1 - \omega^2 \dfrac{m_1(k + K)}{kK}} = \frac{1}{1 - \omega^2/\omega_n^2}$$

as it should be from Eq. (2.33).

For the case of a dashpot c across the protecting spring k (Fig. 3.27)

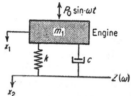

FIG. 3.27. An engine mounted on a foundation Z through a spring and dashpot.   The transmissibility is given by Eq. (3.49).

the analysis is similar and is left to the reader as Prob. 88.   The result is

$$\text{Transmissibility} = \frac{Z}{Z\left(1 - \dfrac{m_1\omega^2}{k + j\omega c}\right) - m_1\omega^2} \tag{3.49}$$

In an actual case where the foundation characteristic Z is known from experiment this formula must be used to see how the system behaves with different protecting elements k and c, and the best compromise is to be chosen.

Problems 64 to 98.

CHAPTER 4

MANY DEGREES OF FREEDOM

4.1. Free Vibrations without Damping. When the number of degrees of freedom becomes greater than two, no essential new aspects enter into the problem. We obtain as many natural frequencies and as many modes of motion as there are degrees of freedom. The general process of analysis will be discussed in the next few sections for a three-degree system; for four or more degrees of freedom the calculations are analogous.

Consider for example Fig. 4.1, representing a weightless bar on two rigid supports, carrying three masses m_1, m_2, and m_3. If the upward deflections of these masses be denoted by x_1, x_2, and x_3, the first of the equations of motion can be obtained by equating $m_1\ddot{x}_1$ to the elastic force on the first mass. This force is the difference between the lateral shear forces in the bar to the left and to the right of m_1, a quantity depending on all three deflections x_1, x_2, and x_3, complicated and difficult to calculate.

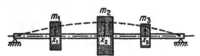

FIG. 4.1. A round shaft with three disks on stiff bearings is a system having three degrees of freedom in bending.

It is more in the nature of this particular problem to describe its elasticity by the *influence numbers*. The definition of an influence number α_{12} is "deflection of mass 1 caused by a force of 1 lb. at the location of mass 2." We have three *direct* influence numbers, α_{11}, α_{22}, and α_{33} where the unit force and the deflection are measured at the same location, and six *cross* influence numbers, α_{12}, α_{21}, α_{13}, α_{31}, α_{23}, and α_{32}, where the two locations are different. By Maxwell's theorem of reciprocity,

$$\alpha_{12} = \alpha_{21}$$

or, in words: the deflection at one location caused by a unit load at another location equals the deflection at this second location caused by a unit load at the first location. These influence numbers can be calculated for any system by the principles of strength of materials. The equations of motion can be written with them as follows. In the posi-

122

tion x_1, x_2, x_3 of maximum deflection of the bar (Fig. 4.1), the masses have accelerations $\ddot{x}_1$, $\ddot{x}_2$, $\ddot{x}_3$ and consequently experience forces $m_1\ddot{x}_1$, $m_2\ddot{x}_2$, $m_3\ddot{x}_3$. These forces are exerted *by* the bar *on* the masses. By the principle of action and reaction, the masses exert the inertia forces $-m_1\ddot{x}_1$, $-m_2\ddot{x}_2$, $-m_3\ddot{x}_3$ on the bar. The deflection at the first mass caused by these three forces is

$$
\left.
\begin{aligned}
x_1 &= -\alpha_{11}m_1\ddot{x}_1 - \alpha_{12}m_2\ddot{x}_2 - \alpha_{13}m_3\ddot{x}_3 \\
x_2 &= -\alpha_{21}m_1\ddot{x}_1 - \alpha_{22}m_2\ddot{x}_2 - \alpha_{23}m_3\ddot{x}_3 \\
x_3 &= -\alpha_{31}m_1\ddot{x}_1 - \alpha_{32}m_2\ddot{x}_2 - \alpha_{33}m_3\ddot{x}_3
\end{aligned}
\right\}
\tag{4.1}
$$

and analogously for the second and third masses,

Although these equations cannot be interpreted directly as the Newton equation for each mass, nevertheless the three together determine the three unknown motions x_1, x_2, and x_3.

As before, on page 80, in order to reduce them from differential equations to algebraic equations, we put

$$
\left.
\begin{aligned}
x_1 &= a_1 \sin \omega t \\
x_2 &= a_2 \sin \omega t \\
x_3 &= a_3 \sin \omega t
\end{aligned}
\right\}
\tag{4.2}
$$

and substitute, with the result

$$
\left.
\begin{aligned}
a_1 &= \alpha_{11}m_1\omega^2 a_1 + \alpha_{12}m_2\omega^2 a_2 + \alpha_{13}m_3\omega^2 a_3 \\
a_2 &= \alpha_{21}m_1\omega^2 a_1 + \alpha_{22}m_2\omega^2 a_2 + \alpha_{23}m_3\omega^2 a_3 \\
a_3 &= \alpha_{31}m_1\omega^2 a_1 + \alpha_{32}m_2\omega^2 a_2 + \alpha_{33}\mathrm{w}_3\omega^2 a_3
\end{aligned}
\right\}
\tag{4.3}
$$

These equations are homogeneous in a_1, a_2, and a_3, which can be seen better after rearranging and dividing by ω^2:

$$
\left.
\begin{aligned}
\left(m_1\alpha_{11} - \frac{1}{\omega^2}\right)a_1 + \quad m_2\alpha_{12}a_2 + \quad m_3\alpha_{13}a_3 &= 0 \\
m_1\alpha_{21}a_1 + \left(m_2\alpha_{22} - \frac{1}{\omega^2}\right)a_2 + \quad m_3\alpha_{23}a_3 &= 0 \\
m_1\alpha_{31}a_1 + \quad m_2\alpha_{32}a_2 + \left(m_3\alpha_{33} - \frac{1}{\omega^2}\right)a_3 &= 0
\end{aligned}
\right\}
\tag{4.4}
$$

If such homogeneous equations are divided by a_1, for example, we have *three* equations in *two* unknowns, a_2/a_1 and a_3/a_1. If we solve these unknowns from the first two equations of (4.4) and substitute the answers in the third one, we usually find that the result is *not* zero. Only if a certain relation exists among the coefficients of a_1, a_2, and a_3, can there be a solution. In the theory of determinants it is shown that this relation is

$$
\begin{vmatrix}
m_1\alpha_{11} - \dfrac{1}{\omega^2} & m_2\alpha_{12} & m_3\alpha_{13} \\[2ex]
m_1\alpha_{21} & m_2\alpha_{22} - \dfrac{1}{\omega^2} & m_3\alpha_{33} \\[2ex]
m_1\alpha_{31} & m_2\alpha_{32} & m_3\alpha_{33} - \dfrac{1}{\omega^2}
\end{vmatrix} = 0 \qquad (4.5)
$$

The argument is analogous to that given on page 80 for the two-degree-of-freedom system. The determinant expanded is a cubic equation in terms of $1/\omega^2$, known as the "frequency equation," which has three solutions and hence three natural frequencies. To each of these solutions belongs a set of values for a_2/a_1 and a_3/a_1, which determines a configuration of vibration. Thus there are three natural modes of motion.

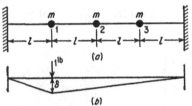

(a)

(b)

FIG. 4.2. Showing calculation of influence numbers for a string with three masses.

We shall carry out these calculations in detail for the simplest possible example, obtained by making all masses equal $m_1 = m_2 = m_3 = m$ and replacing the bar by a string of tension T and length $4l$ (Fig. 4.2).
If a load of 1 lb. is placed on location 1, the deformation will be as shown in Fig. 4.2b. The tension in the string is T and the vertical component of the tension in the part of the string to the left of m_1 is $\dfrac{\delta}{l}$ T while to the

right of m_1 it is $\dfrac{\delta}{3l}$ T. The sum of these vertical components must be

equal to the load of 1 lb. so that $\delta = \dfrac{3}{4}\dfrac{l}{T}$. This is the deflection at 1

caused by 1 lb. at 1, or $\alpha_{11} = \dfrac{3}{4}\dfrac{l}{T}$.

The deflection at the masses 2 and 3 caused by the same load can also be found from Fig. 4.2b.

$$
\alpha_{21} = \frac{2}{3} \cdot \frac{3}{4}\frac{l}{T} = \frac{1}{2}\frac{l}{T}
$$

$$
\alpha_{31} = \frac{1}{3} \cdot \frac{3}{4}\frac{l}{T} = \frac{1}{4}\frac{l}{T}
$$

The other influence numbers can be found in a similar manner:

$$\left.\begin{aligned}
\alpha_{22} &= \frac{l}{T} \\
\alpha_{11} &= \alpha_{33} = \frac{3}{4}\frac{l}{T} \\
\alpha_{12} &= \alpha_{21} = \alpha_{32} = \alpha_{23} = \frac{1}{2}\frac{l}{T} \\
\alpha_{31} &= \alpha_{13} = \frac{1}{4}\frac{l}{T}
\end{aligned}\right\} \tag{4.6}$$

and Maxwell's reciprocity relations are seen to be true. The equations of motion are obtained by substituting these values for the influence numbers in Eq. (4.1). However, since nearly every term is proportional to ml/T, we divide by this quantity and introduce the abbreviation

$$\frac{T}{ml\omega^2} = F \text{ (the frequency function)} \tag{4.7}$$

Then Eqs. (4.4) become

$$\left.\begin{aligned}
(\tfrac{3}{4} - F)a_1 + \quad \tfrac{1}{2}a_2 + \quad \tfrac{1}{4}a_3 &= 0 \\
\tfrac{1}{2}a_1 + (1 - F)a_2 + \quad \tfrac{1}{2}a_3 &= 0 \\
\tfrac{1}{4}a_1 + \quad \tfrac{1}{2}a_2 + (\tfrac{3}{4} - F)a_3 &= 0
\end{aligned}\right\} \tag{4.8}$$

Dividing the first of these by a_1, the second by $2a_1$, and subtracting them from each other leads to

$$\frac{a_2}{a_1} = 2 - \frac{1}{F} \tag{4.9}$$

Substituting this in the first equation of (4.8) and solving for a_3/a_1 gives

$$\frac{a_3}{a_1} = -7 + 4F + \frac{2}{F} \tag{4.10}$$

Substituting both these ratios in the third equation of (4.8) gives the following equation for F (the frequency equation):

$$F^3 - \tfrac{5}{2}F^2 + \tfrac{3}{2}F - \tfrac{1}{4} = 0 \tag{4.11}$$

This result could have been found also by working out the determinant (4.5). Evidently (4.11) has three roots for F. We note that none of these can be negative since for a negative F all four terms on the left become negative and then their sum cannot be zero. Since by (4.7) a negative F corresponds to an imaginary ω, we see that our three-degree-of-freedom system must have three *real* natural frequencies. This is true not only for the particular system under consideration. In general it can be shown that *an n-degree-of-freedom vibrational system without damping has n real natural frequencies, i.e.,* the roots of a frequency equation such as (3.7), (4.5), or (4.11) are always real and positive.

The cubic (4.11) is solved by trial of some values for F. $F = 0$ makes the left-hand side $-\frac{1}{4}$, while $F = 2$ makes it $+\frac{3}{4}$; evidently at least one

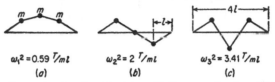

$$\omega_1{}^2 = 0.59\ T/ml \qquad \omega_2{}^2 = 2\ T/ml \qquad \omega_3{}^2 = 3.41\ T/ml$$
$$(a) \qquad\qquad\qquad (b) \qquad\qquad\qquad (c)$$

Fig. 4.3. The three natural modes of a string with three equal and equidistant masses.

root must be between 0 and 2. A few trials will show that $F = \frac{1}{2}$ is a root, so that Eq. (4.11) can be written

$$(F - \tfrac{1}{2})(F^2 - 2F + \tfrac{1}{2}) = 0$$

having the three roots

$$F_2 = \tfrac{1}{2} \qquad F_{1,3} = 1 \pm \sqrt{\tfrac{1}{2}}$$

With the relations (4.7), (4.9), and (4.10) the complete result becomes

$$F_1 = 1.707 \qquad \omega_1^2 = 0.59\,\frac{T}{ml} \qquad \frac{a_2}{a_1} = 1.41 \qquad \frac{a_3}{a_1} = 1$$

$$F_2 = 0.500 \qquad \omega_2^2 = 2\,\frac{T}{ml} \qquad \frac{a_2}{a_1} = 0 \qquad \frac{a_3}{a_1} = -1$$

$$F_3 = 0.293 \qquad \omega_3^2 = 3.41\,\frac{T}{ml} \qquad \frac{a_2}{a_1} = -1.41 \qquad \frac{a_3}{a_1} = f$$

This gives the shapes of the vibration, or the "normal modes" as shown in Fig. 4.3. These are the *only* three configurations in which the system

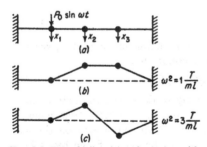

(a)

$$\omega^2 = 1\,\frac{T}{ml}$$

(b)

$$\omega^2 = 3\,\frac{T}{ml}$$

(c)

Fig. 4.4. Forced vibrations of a string with three masses. There are two frequencies at which the disturbed mass does not move; these are the frequencies of the generalized-dynamic-vibration-absorber effect.

can be in equilibrium under the influence of forces which are proportional to the displacements x (as the inertia forces are). The second mode is of particular interest because the middle mass does not move at all. If that fact had been known in advance, the frequency could have been found very easily by considering the left half of the system as one of a single degree of freedom with the spring constant $k = 2T/l$ (see Problem 28).

4.2. Forced Vibrations without Damping. Suppose an alternating force $P_0 \sin \omega t$ to be acting on the first mass of the previous example (Fig. 4.4a). The force $P_0 \sin \omega t$ *by itself* would cause "static" deflections at 1, 2, and 3 of $\alpha_{11}P_0 \sin \omega t$, $\alpha_{21}P_0 \sin \omega t$, and $\alpha_{31}P_0 \sin \omega t$. The equations of *forced* motion are obtained from (4.1) by adding these terms to the right-hand sides. With the assumption (4.2) the equations then are reduced to the algebraic form

$$\left(m_1\alpha_{11} - \frac{1}{\omega^2}\right)a_1 + \qquad m_2\alpha_{12}a_2 + \qquad m_3\alpha_{13}a_3 = -\alpha_{11}\frac{P_0}{\omega^2}$$

$$m_1\alpha_{21}a_1 + \left(m_2\alpha_{22} - \frac{1}{\omega^2}\right)a_2 + \qquad m_3\alpha_{23}a_3 = -\alpha_{21}\frac{P_0}{\omega^2}$$

$$m_1\alpha_{31}a_1 + \qquad m_2\alpha_{32}a_2 + \left(m_3\alpha_{33} - \frac{1}{\omega^2}\right)a_3 = -\alpha_{31}\frac{P_0}{\omega^2}$$

With the influence numbers (4.6) and with the definition of F given in (4.7), they become

$$\left.\begin{array}{l}\left(\dfrac{3}{4} - F\right)a_1 + \qquad \dfrac{1}{2}a_2 + \qquad \dfrac{1}{4}a_3 = -\dfrac{3}{4}\dfrac{P_0}{m\omega^2} \\[2mm] \dfrac{1}{2}a_1 + (1 - F)a_2 + \qquad \dfrac{1}{2}a_3 = -\dfrac{1}{2}\dfrac{P_0}{m\omega^2} \\[2mm] \dfrac{1}{4}a_1 + \qquad \dfrac{1}{2}a_2 + \left(\dfrac{3}{4} - F\right)a_3 = -\dfrac{1}{4}\dfrac{P_0}{m\omega^2}\end{array}\right\} \quad (4.12)$$

These equations are no longer homogeneous in a_1, a_2, a_3, as were the corresponding ones (4.8) for free vibration. They are truly a set of three equations with three unknowns and can be solved by ordinary algebra. In the calculations, the cubic (4.11) appears in the denominators and is broken up into its three linear factors, with the result that

$$\left.\begin{array}{l}a_1 = \dfrac{P_0}{m\omega^2}\dfrac{\frac{3}{4}F^2 - F + \frac{1}{4}}{(F - 1.707)(F - 0.500)(F - 0.293)} \\[3mm] a_2 = \dfrac{P_0}{m\omega^2}\dfrac{\frac{1}{2}F(F - \frac{1}{2})}{(F - 1.707)(F - 0.500)(F - 0.293)} \\[3mm] a_3 = \dfrac{P_0}{m\omega^2}\dfrac{\frac{1}{4}F^2}{(F - 1.707)(F - 0.500)(F - 0.293)}\end{array}\right\} \quad (4.13)$$

The physical meaning of these expressions is best disclosed by plotting them as resonance diagrams corresponding to Fig. 2.18 on page 44 or to Figs. 3.8a and b on page 91. For that purpose note that F, being proportional to $1/\omega^2$, is not a suitable variable. For the ordinate y of our diagrams we take the quantities

$$y_{1,2,3} = \frac{a_{1,2,3}}{\dfrac{P_0 l}{T}}$$

The denominator $P_0 l/T$ would be the "static deflection" of the middle of the string if the (constant) load P_0 were placed there $(\alpha_{22} = l/T)$, so that y is a "dimensionless amplitude." For the abscissa x we take

$$x = \frac{1}{F} = \frac{\omega^2}{T/ml}$$

The denominator T/ml can be interpreted as the ω^2 of a mass m on a spring constant T/l, so that $\sqrt{x}$ is a "dimensionless frequency." With these two new variables, Eqs. (4.13) are transformed into

$$\left. \begin{aligned} y_1 &= \frac{-x^2 + 4x - 3}{(x - 0.59)(x - 2)(x - 3.41)} \\ y_2 &= \frac{(x - 2)}{(x - 0.59)(x - 2)(x - 3.41)} \\ y_3 &= \frac{-1}{(x - 0.59)(x - 2)(x - 3.41)} \end{aligned} \right\} \tag{4.14}$$

plotted in Figs. 4.5, 4.6, and 4.7. The reader should satisfy himself that for the static case $x = 0$, all three expressions (4.14) give the proper static deflections. An interesting property of (4.14) is that the factor $(x - 2)$ can be divided out in the expression for y_2. This means physically that the middle mass does not get infinite amplitudes at the second resonance, while both the first and third masses do go to infinity. A glance at the second normal mode of Fig. 4.3 shows that this should be so.

While the numerators of y_2 and y_3 show no peculiarities, it is seen that the numerator of y_1 is a quadratic which necessarily becomes zero for two frequencies, viz., for $x = 1$ and $x = 3$ (Fig. 4.5). At these frequencies the first mass, on which the force is acting, does not move, whereas the two other masses do vibrate. We have before us a *generalization of the dynamic vibration absorber* of page 87. If the first mass does not move, we can consider it clamped and the system reduces to one of two degrees of freedom (Fig. 4.4). Such a system has two natural frequencies which can easily be calculated to be $x = 1$ and $x = 3$. The action can then be imagined as follows. At two resonant frequencies the two-dimensional system can be excited to finite amplitudes by an infinitely small excitation, in this case by an infinitely small alternating motion of mass 1. On mass 1 in Fig. 4.4b or c two alternating forces are acting, one being the vertical component of the string tension from the right and the other one being the external force $P_0 \sin \omega t$. These two forces must be always equal and opposite, because m_1 does not move.

Generalizing, we thus might be tempted to make the following statement: If an alternating force acts on a mass of an n-degree-of-freedom system, there will be $n - 1$ frequencies at which that mass will stand still while the rest of the system vibrates.   Care has to be exercised, however, in making such sweeping generalizations.   For example, an exception to the rule can be pointed out immediately by exciting our system at the middle mass.   On account of this mass being a *node* at the second resonance

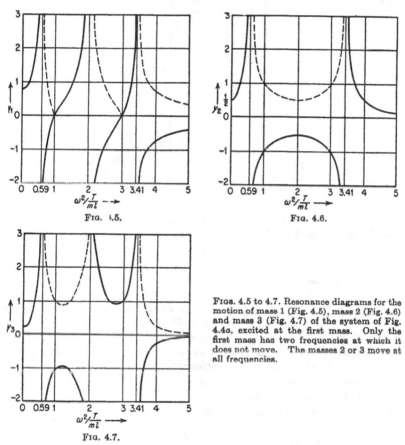

FIG. 4.5.

FIG. 4.6.

FIG. 4.7.

FIGS. 4.5 to 4.7. Resonance diagrams for the motion of mass 1 (Fig. 4.5), mass 2 (Fig. 4.6) and mass 3 (Fig. 4.7) of the system of Fig. 4.4a, excited at the first mass.   Only the first mass has two frequencies at which it does not move.   The masses 2 or 3 move at all frequencies.

(Fig. 4.3), the force can perform no work on it at that frequency so that no infinite amplitudes can be built up.   The "resonant frequency" and the "vibration absorber frequency" happen to coincide in this case.   In reasoning out the shape of the three resonance curves for excitation at the middle mass, it should be borne in mind that the system is completely symmetrical so that the y_1 and the y_3 diagrams must be alike.   Without carrying out the calculations in detail, we can conclude that the result must have the general shape shown in Fig. 4.8.   Below $x = 2$ all three masses are in phase, somewhat like Fig. 4.3a; above that frequency they are in opposite phase, somewhat like Fig. 4.3c.   At the second natural

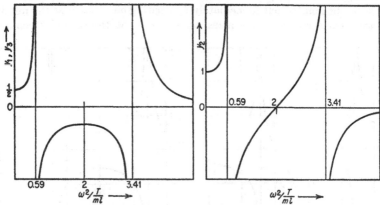

Fig. 4.8. Resonance diagrams for the symmetrical string with the three masses of which the middle mass is excited by an alternating force.

frequency, however, the configuration must, for reasons of symmetry, be as shown in Fig. 4.9.   The amplitude of motion of the masses 1 and 3 must be determined by the value of the exciting force, so that the sum of the vertical components of the tensions in the two pieces of string attached to m_2 must be equal and opposite to the exciting force.

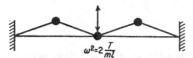

Fig. 4.9. Vibration-absorber effect in string with three masses, of which the middle one is excited.

4.3. Free and Forced Vibrations with Damping.

If there is damping in a system of many degrees of freedom, we are practically interested in two questions: (a) in the rate of decay of amplitude of the free vibration; (b) in the amplitude at resonance of the forced vibration.   The method of calculation employed in the exact classical theory will be shown in the example of the string with three equal and equidistant masses.

Let a damping force $-c\dot{x}_2$ be acting on the middle mass (Fig. 4.10). This force causes deflections of

Fig. 4.10. Damping at the central mass of the string.

$-\alpha_{12}c\dot{x}_2$, $-\alpha_{22}c\dot{x}_2$, and $-\alpha_{32}c\dot{x}_2$ at the three masses.   The differential equations (4.1) for the free vibration become

$$x_1 = -\alpha_{11}m\ddot{x}_1 - \alpha_{12}m\ddot{x}_2 - \alpha_{13}m\ddot{x}_3 - \alpha_{12}c\dot{x}_2$$
$$x_2 = -\alpha_{21}m\ddot{x}_1 - \alpha_{22}m\ddot{x}_2 - \alpha_{23}m\ddot{x}_3 - \alpha_{22}c\dot{x}_2$$
$$x_3 = -\alpha_{31}m\ddot{x}_1 - \alpha_{32}m\ddot{x}_2 - \alpha_{33}m\ddot{x}_3 - \alpha_{22}c\dot{x}_2$$

$$(4.15)$$

where the various influence numbers have the values expressed by (4.6). By algebraic manipulations these can be transformed into

$$m\ddot{x}_1 + \frac{T}{l}x_1 \qquad + \frac{T}{l}(x_1 - x_2) \qquad = 0$$
$$m\ddot{x}_2 + \frac{T}{l}(x_2 - x_1) + \frac{T}{l}(x_2 - x_3) + c\dot{x}_2 = 0 \qquad (4.16)$$
$$m\ddot{x}_3 + \frac{T}{l}(x_3 - x_2) + \frac{T}{l}x_3 \qquad = 0$$

The first equation of (4.16) is found by subtracting the second of (4.15) from twice the first of (4.15), i.e., by forming $2x_1 - x_2$. The second equation of (4.16) is obtained by calculating $x_1 + x_3 - 2x_2$ and the third one by forming $x_2 - 2x_3$. The physical significance of Eqs. (4.16) is apparent. They are the Newtonian equations for the various masses, the first term being the inertia force, the second the vertical component of the string tension to the left of the mass, the third that same component to the right, and the fourth the damping force.

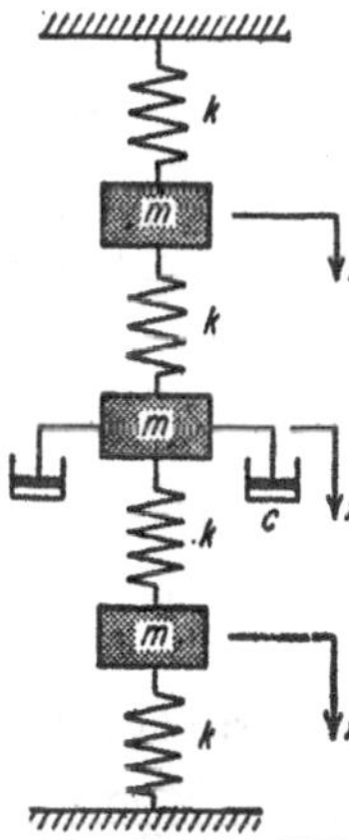

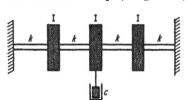

In this case it would have been possible and easier to write the equations in the last form directly without using the influence numbers. However, for the example of the beam with which this chapter started (Fig. 4.1), influence numbers afford the simplest manner of approach.

Before proceeding with the solution of (4.16), it may be well to point out that these equations may represent two other systems as well, shown in Figs. 4.11 and 4.12. In Fig. 4.11 the masses are restricted to vertical motion alone, and the spring constant k

FIG. 4.11. The longitudinal vibrations of this system are completely equivalent to the vibration of either Fig. 4.10 or Fig. 4.12.

has to be made equal to T/l to give complete analogy with Fig. 4.10. The second example, Fig. 4.12, is a torsional one. The reader will do well to interpret the results shown in Figs. 4.2 to 4.9 for these two cases.

In solving Eqs. (4.16), we know from the last two chapters that an assumption of the form $x = a \sin \omega t$, which is perfectly justifiable for the undamped case, will not lead to a result if damping is present. The

FIG. 4.12. Torsional equivalent of the system of Fig. 4.10 or Fig. 4.11.

solution is rather expected to be of the form $x = a \cdot e^{-pt} \sin qt$. This is met by assuming

$$\left.\begin{array}{l} x_1 = a_1 e^{st} \\ x_2 = a_2 e^{st} \\ x_3 = a_3 e^{st} \end{array}\right\} \qquad (4.17)$$

where s is a *complex* number, $s = p + iq$. The value $-p$ gives the exponent of decay of amplitude and q is the natural frequency (see page 38). Substituting (4.17) in (4.16),

$$\left.\begin{array}{l} \left(ms^2 + 2\dfrac{T}{l}\right)a_1 - \dfrac{T}{l}a_2 + 0 = 0 \\[2mm] -\dfrac{T}{l}a_1 + \left(ms^2 + cs + 2\dfrac{T}{l}\right)a_2 - \dfrac{T}{l}a_3 = 0 \\[2mm] 0 - \dfrac{T}{l}a_2 + \left(ms^2 + 2\dfrac{T}{l}\right)a_3 = 0 \end{array}\right\}$$

This is a homogeneous set of equations in a_1, s_2, and a_3 and can have a solution only if the determinant vanishes:

$$\begin{vmatrix} ms^2 + 2\dfrac{T}{l} & -\dfrac{T}{l} & 0 \\[2mm] -\dfrac{T}{l} & ms^2 + cs + 2\dfrac{T}{l} & -\dfrac{T}{l} \\[2mm] 0 & -\dfrac{T}{l} & ms^2 + 2\dfrac{T}{l} \end{vmatrix} = 0$$

or, written out,

$$\left(ms^2 + 2\dfrac{T}{l}\right)\left[\left(ms^2 + 2\dfrac{T}{l}\right)\left(ms^2 + cs + 2\dfrac{T}{l}\right) - 2\left(\dfrac{T}{l}\right)^2\right] = 0 \quad (4.18)$$

This equation of the sixth degree in s is known also as the "frequency equation," though s in this case is not the frequency but a complex number expressing frequency and rate of decay combined. The quantity s is called the "complex frequency."

In this particular case the equation falls into two factors of which the first one leads to

$$s^2 = -\frac{2T}{ml} \qquad \text{or} \qquad s_{1,2} = \pm j\sqrt{\frac{2T}{ml}}$$

with a solution of the form

$$Ae^{j\sqrt{\frac{2T}{ml}}t} + Be^{-j\sqrt{\frac{2T}{ml}}t}$$

which can be transformed to [see Eq. 1.8), page 11]

$$C_1 \cos\sqrt{\frac{2T}{ml}}\,t + C_2 \sin\sqrt{\frac{2T}{ml}}\,t$$

This solution therefore gives a frequency $\omega^2 = 2T/ml$, while the rate of decay of amplitude is zero, since s does *not* contain any real part. The frequency coincides with that of Fig. 4.3*b* for the undamped case, in which the middle mass is a node. Therefore the damping force can do no work, which is the reason for the absence of a rate of decay in this second mode and also the reason for the fact that the natural frequency is not affected at all by the damping.

The other factor of (4.18), after multiplying out, becomes

$$s^4 + \frac{c}{m} s^3 + 4 \frac{T}{ml} s^2 + 2 \frac{T}{ml} \frac{c}{m} s + 2 \left(\frac{T}{ml}\right)^2 = 0$$

having four roots for s, which we *do* expect to have *real* parts, since in the modes of Figs. 4.3*a* and *c* the damping does perform work. The roots of s will be of the form

$$s_3 = -p_1 + jq_1$$
$$s_4 = -p_1 - jq_1$$
$$s_5 = -p_2 + jq_2$$
$$s_6 = -p_2 - jq_2$$

because the complex roots of algebraic equations always occur in conjugate pairs.

The numerical calculation of these roots from the numerical values of m, c, T, and l is cumbersome even for the comparatively simple equation of the fourth degree.† Therefore this classical method is unsuited to a practical solution of the problem. It has been discussed here merely because in Chap. 7 we shall consider cases in which the *real* parts of s become *positive*, which means a decay function of the form e^{+pt}, which is *not* decay but actually a building up of the vibration; the motion is then called "self-excited."

In practical cases the damping is usually so small that the natural frequency and the mode of motion are very little affected by it (Fig. 2.16, page 40). Hence the rate of decay of the free vibration may be calculated by assuming the configuration and frequency which would occur if no damping existed, as follows.

If the amplitude of the middle mass be a_2 and the frequency be ω, Eq. (2.30), page 52, gives for the work dissipated per cycle by the damping force $ca_2\omega$:

$$W = \pi c \omega a_2^2$$

The kinetic energy of the system when passing through its neutral position is

$$\tfrac{1}{2}m\omega^2(a_1^2 + a_2^2 + a_3^2) = \tfrac{1}{2}m\omega^2 \mathbf{f} a_2^2 \tag{4.19}$$

† The mathematical method by which this can be done is discussed in "Mathematical Methods in Engineering" by Th. von Kármán and M. A. Biot, p. 246.

where the factor f depends on the configuration.   This energy is diminished by $\pi c \omega a_2^2$ each cycle, or

$$d(\tfrac{1}{2}m\omega^2 f a_2^2) = m\omega^2 f a_2 da_2 = \pi c \omega a_2^2$$

Hence,
$$\frac{da_2}{a_2} = \frac{\pi c}{m\omega f}$$

If in a natural mode of motion one of the masses reduces its amplitude to one-half, all other masses do the same, so that

$$\frac{da_1}{a_1} = \frac{da_2}{a_2} = \frac{da_3}{a_3} = \frac{\pi c}{m f \omega}$$

In the first mode of motion, Fig. 4.3a, the factor f, as defined by (4.19), is seen to be 2, whereas $\omega = \omega_1 = \sqrt{0.59 \frac{T}{ml}}$, so that the percentage decay per cycle is

$$\frac{da_1}{a_1} = 2.04c \sqrt{\frac{l}{Tm}}$$

In the third mode of motion f is also 2, but $\omega_3 = \sqrt{3.41 \frac{T}{ml}}$ so that

$$\frac{da_1}{a_1} = 0.85c \sqrt{\frac{l}{Tm}}$$

This method gives perfectly satisfactory results for the usual damping values.   Of course, when the damping becomes an appreciable fraction of c_c, the procedure ceases to be reliable.

For *forced* vibrations with damping, the "classical" method is even more complicated than for free vibrations.   It becomes so cumbersome as to be entirely useless for practical numerical purposes.   However, for technically important values of the damping the above energy method gives us a good approximation for the amplitude at *resonance* in which we are most interested.

As before, we assume that at resonance the damping force and exciting force are so small with respect to the inertia and elastic forces (see Fig. 2.21, page 48, for the single-degree case) that the mode of motion is practically undistorted.   Then the damping dissipation per cycle can be calculated in the same manner as has just been done for the free vibration.   In the steady-state case this dissipation must be equal to the work per cycle done on the system by the exciting force or forces.   In general, there is some phase angle between the force and the motion.   At "resonance," however, this phase angle becomes 90 deg., as explained on page 50, at which value of the phase angle the work input for a given force and motion becomes a maximum.

As an example, take the combined Figs. 4.4a and 4.10. The work input of the force per cycle is $\pi P_0 a_1$, and the resonant amplitude is calculated from

$$\pi P_0 a_1 = \pi c \omega a_2^2 \quad \text{or,} \quad \pi P_0 = \pi c \omega \left(\frac{a_2}{a_1}\right)^2 a_1$$

Hence
$$(a_1)_{res} = \frac{P_0}{c\omega}\left(\frac{a_1}{a_2}\right)^2$$

In the first mode we have $a_2/a_1 = 1.41$ and $\omega = \sqrt{0.59\,\dfrac{T}{ml}}$ (page 126), so that

$$(a_1)_{res} = 0.65\,\frac{P_0}{c}\sqrt{\frac{ml}{T}}$$

For the two other natural frequencies we find

$$(a_1)_{res} = \infty \quad \text{(second mode)}$$
$$(a_1)_{res} = 0.27\,\frac{P_0}{c}\sqrt{\frac{ml}{T}} \quad \text{(third mode)}$$

The most important technical application of this method is in connection with torsional vibration in the crank shafts of Diesel engines, as discussed in Chap. 5.

4.4. Strings and Organ Pipes; Longitudinal and Torsional Vibrations of Uniform Bars. These four types of problem will be treated together because their mathematical and physical interpretations are identical.

In the last few sections a string with *three* masses has been investigated. The "string" itself was supposed to have no weight; the masses were supposed to be concentrated at a few distinct points. By imagining the number of masses to increase without limit we arrive at the concept of a uniform string with *distributed* mass.

The equation of motion is derived by writing Newton's law for a small element dx of the string, of which again the tension T is assumed to be constant. Let the deflection curve during the vibration be $y(x, t)$, where the ordinate varies both with the location along the string and with the time. The vertical component of the tension T pulling to the left at a certain point x of the string is (Fig. 4.13)

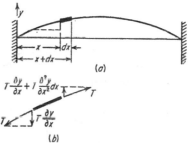

FIG. 4.13. Vertical components of the tensions acting on an element dx of a stretched string.

$$-T\frac{\partial y}{\partial x}$$

negative because it pulls downward, whereas y is positive upward. The differential coefficient is *partial*, because the string is considered at a certain instant, *i.e.*, t is a constant in the differentiation. At the right-hand end of the element dx, the vertical component of the tension is

$$+\mathbf{T}\frac{\partial y}{\partial x} + \partial\left(\mathbf{T}\frac{\partial y}{\partial x}\right) = \mathbf{T}\frac{\partial y}{\partial x} + \frac{\partial}{\partial x}\left(\mathbf{T}\frac{\partial y}{\partial x}\right)dx = \mathbf{T}\frac{\partial y}{\partial x} + \mathbf{T}\frac{\partial^2 y}{\partial x^2}dx$$

This quantity is positive because it pulls upward. The factor $\dfrac{\partial^2 y}{\partial x^2}dx$ expresses the increase in slope along dx. Since the two vertical forces on the element dx are not equal (Fig. 4.13b), there is an excess upward pull of

$$\mathbf{T}\frac{\partial^2 y}{\partial x^2}dx$$

which must accelerate the element in the upward direction. If we denote the mass per unit length of the string by μ_1, the mass of dx is $\mu_1 dx$ and Newton's law gives

$$\mu_1 dx\frac{\partial^2 y}{\partial t^2} = \mathbf{T}\frac{\partial^2 y}{\partial x^2}dx$$

Dividing by dx we obtain the partial *differential equation of the string:*

$$\mu_1\frac{\partial^2 y}{\partial t^2} = \mathbf{T}\frac{\partial^2 y}{\partial x^2} \tag{4.20}$$

The reader should compare the structure of this formula with the first of the equations (4.16) and determine the physical meaning of each term.

The problem of *longitudinal vibrations in a bar* is quite similar to that of the string and is a generalization of Fig. 4.11 (without damping)

Fig. 4.14. Longitudinal vibrations of a bar: x determines the position of any point, and ξ is the displacement during vibration of each point x.

when we take more and smaller masses and more and shorter springs. Now the masses are not numbered 1, 2, 3 as in Fig. 4.11 but designated by their position x along the bar (Fig. 4.14). Let the longitudinal displacement of each point x be indicated by the Greek equivalent of x, namely ξ. Thus the state of motion of the bar is known if we know $\xi(x, t)$, again a function of two variables.

The cross section x goes to $x + \xi$, and the section $x + dx$ goes to $(x + dx) + (\xi + d\xi)$. At some instant t the length dx becomes

$$dx + \frac{\partial \xi}{\partial x}dx$$

Thus $\partial\xi/\partial x$ is the unit elongation which causes at the section x of the bar a tensile stress of

$$E\,\frac{\partial\xi}{\partial x}$$

where E is the modulus of elasticity. If the bar were stretched with a constant stress, $E\,\frac{\partial\xi}{\partial x}$ would be constant along the length of the bar, and the element dx would be pulled to the left with the same force as to the right. But if the stress $E\,\frac{\partial\xi}{\partial x}$ varies from point to point, there will be an excess force on the element to accelerate it longitudinally.

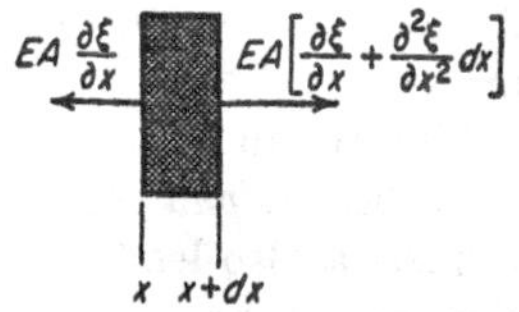

FIG. 4.15. Longitudinal elastic forces on an element of the beam of Fig. 4.14.

In Fig. 4.15 let the element dx be represented with its two forces which are the stresses multiplied by the cross-sectional area A. The force to the left is $AE\,\frac{\partial\xi}{\partial x}$, and that to the right is $AE\,\frac{\partial\xi}{\partial x}$ plus the increment due to the increase dx in the abscissa. This increment of force is $\frac{\partial}{\partial x}\left(AE\,\frac{\partial\xi}{\partial x}\right)dx$. Hence the excess force to the right is

$$AE\,\frac{\partial^2\xi}{\partial x^2}\,dx$$

Let the mass per unit *length* of the bar be μ_1, and Newton's law becomes

$$(\mu_1 dx)\,\frac{\partial^2\xi}{\partial t^2} \;=\; AE\,\frac{\partial^2\xi}{\partial x^2}\,dx$$

or

$$\mu_1\,\frac{\partial^2\xi}{\partial t^2} \;=\; AE\,\frac{\partial^2\xi}{\partial x^2} \tag{4.21}$$

where AE is the tension stiffness of the bar. This is the same differential equation as (4.20).

A variant of this case is the *organ pipe*, where an air column instead of a steel column executes longitudinal vibrations. Equation (4.21) evidently must be the same; μ_1 signifies the mass of air per unit length of pipe, and E is its modulus of elasticity. Instead of the *stress* in the above derivation, we have here the *pressure* and since the definition of E in elasticity is

$$\frac{\text{Stress}}{E} \;=\; \frac{\text{elongation}}{\text{original length}}$$

we have correspondingly for the E in gases

$$\frac{\text{Increase in pressure}}{E} = \frac{\text{decrease in volume}}{\text{original volume}}$$

or
$$E = -v\frac{dp}{dv}$$

As in elasticity, the quantity E in gases is measured in pounds per square inch.

Finally, an inspection of Figs. 4.10, 4.11, and 4.12 will make it clear that the *torsional* vibration of a uniform shaft with distributed moment of inertia also leads to the same differential equation. The variable in this case is the angle of twist $\varphi(x, t)$, and the differential equation is

$$\mu_1 \frac{\partial^2 \varphi}{\partial t^2} = GI_p \frac{\partial^2 \varphi}{\partial x^2} \tag{4.22}$$

where μ_1 is the moment of inertia per inch length of shaft and GI_p is the torsional stiffness of the shaft. It is left as an exercise to the reader to derive this result.

Proceeding to a solution of (4.20), (4.21), or (4.22), we assume that the string vibrates *harmonically* at some natural frequency and in some natural or normal configuration. It remains to be seen whether such an assumption is correct. In mathematical language this means that we assume

$$y(x, t) = y(x) \sin \omega t \tag{4.23}$$

Substitute this in (4.20), which then becomes

$$\frac{d^2y}{dx^2} + \frac{\mu_1 \omega^2}{T} y = 0 \tag{4.24}$$

which is an ordinary differential equation. Whereas in all previous problems this sort of assumption simplified the ordinary differential equations to algebraic ones, we have here the simplification of a partial differential equation to an ordinary differential equation.

It is seen that (4.24) has the same mathematical form as Eq. (2.7), page 31, or in words: the amplitude of the string as a function of *space* acts in the same manner as the amplitude of a single-degree-of-freedom system as a function of *time*.

Therefore the general solution of (4.24) is by Eq. (2.8)

$$y(x) = C_1 \sin x \sqrt{\frac{\mu_1 \omega^2}{T}} + C_2 \cos x \sqrt{\frac{\mu_1 \omega^2}{T}} \tag{4.25}$$

which determines the shape of the string at the instant of maximum deflection. The integration constants C_1 and C_2 can be determined from the condition that at the ends of the string the amplitudes must be zero, or

$$y = 0 \quad \text{for} \quad x = 0 \quad \text{and for} \quad x = l$$

Substituting $x = 0$ gives

$$y(0) = 0 = C_1 \cdot 0 + C_2 \cdot 1$$

so that $C_2 = 0$. With $x = l$, we get

$$y(l) = 0 = C_1 \sin l \sqrt{\frac{\mu_1 \omega^2}{T}} \tag{4.26}$$

This can be satisfied by making $C_1 = 0$, which gives the correct but uninteresting solution of the string remaining at rest. However, (4.26)

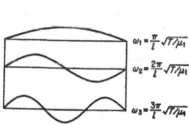

$$\omega_1 = \frac{\pi}{l}\sqrt{T/\mu_1}$$

$$\omega_2 = \frac{2\pi}{l}\sqrt{T/\mu_1}$$

$$\omega_3 = \frac{3\pi}{l}\sqrt{T/\mu_1}$$

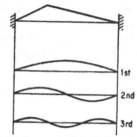

1st

2nd

3rd

FIG. 4.16. The first three natural modes of motion of the lateral vibration of a uniform string or of the longitudinal or torsional vibration of a uniform bar built in at both ends.

FIG. 4.17. Shape of a "plucked" string with the first three Fourier components of that shape.

can also be satisfied by making the argument of the sine an integer multiple of π or 180 deg.

$$l\sqrt{\frac{\mu_1 \omega^2}{T}} = 0, \pi, 2\pi, 3\pi, \ldots \tag{4.27}$$

This determines the natural frequencies, while the corresponding normal modes can be found at once by substitution of Eq. (4.27) in Eq. (4.25). The results are illustrated in Fig. 4.16.

There is an infinite number of normal elastic curves and correspondingly an infinite number of natural frequencies. The motion in each one of these modes is such that the amplitude of every point of the string varies harmonically with the time, and consequently the normal curve remains similar to itself. Therefore, if a string is deflected in one of the shapes of Fig. 4.16 and then released, it will return to its original position in an interval of time determined by the natural period of the vibration. At that frequency and shape the inertia force and spring force of each element dx of the string are in equilibrium with each other at any instant.

If the string is given an initial displacement of a shape different from any of those of Fig. 4.16, e.g., a displacement such as is shown in Fig. 4.17, the shape can be considered to be composed of a (Fourier) series of the normal shapes (see page 17).   Each Fourier component then will execute a motion conformal to itself, but each one will do this *at its own particular frequency*.   Thus after one-eighth period of the fundamental mode, the amplitude of that fundamental component will have decreased to 0.707 of its original value, the second component will have zero amplitude, while the fourth mode will have reversed its amplitude.   Thus the compound shape of Fig. 4.17 is *not* preserved during the motion.   However, after a *full* period of the fundamental motion the original shape recurs.

The shapes of Fig. 4.16 pertain also to the longitudinal (or torsional) vibrations of a bar with both ends built in or to the vibrations of an "organ pipe" with both ends closed.   The ordinates then signify displacements along the bar.   The frequencies are evidently the same, except for a substitution of the "tension stiffness" AE instead of the tension $\mathbf{T}$.

For the longitudinal (or torsional) vibrations of a *cantilever* bar or of an organ pipe with one open end, the general expression (4.25) for the shape still holds, but the end conditions for determining C_1 and C_2 are different.

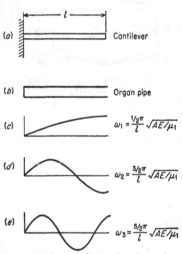

At the closed end $x = 0$, we still have $y = 0$, because the air cannot penetrate the solid wall at the closed end of the pipe. At the open end, however, there can be displacement but no stress (in the bar) or no pressure excess (in the organ pipe). In the derivation of the differential equation this stress was seen to be proportional to $\partial\xi/\partial x$ (or dy/dx in the string notation). The end conditions are therefore

$$x = 0 \qquad\qquad y = 0$$
$$x = l \qquad dy/dx = 0$$

FIG. 4.18. Longitudinal vibrations of a steel column or air column of which one end is fixed and one end free.

The first of these makes $C_2 = 0$ in (4.25), while the second one can be satisfied by equating the length of the bar to ¼, ¾, ⅝, etc., wave lengths, as shown in Fig. 4.18.

In conclusion, a number of results previously obtained are assembled in Fig. 4.19.   The first of these is half of Fig. 4.3b; the second one is Fig. 4.4b; and the third one is Fig. 4.3a.   The inscribed frequencies also have been taken from the same sources, except that M now stands for all the masses combined and L for the total length of the string.

In the right half of Fig. 4.19 two masses have been added at the points of support. These masses do not affect the frequency since they do not move. However, they do affect the value of M, which is the total mass. By increasing the number of masses from 1 to 2, 3, etc., we must finally approach the fundamental frequency of the continuous string. In the left half of the figure the frequency of the continuous string is approached from below, because the masses are concentrated too close to the center where their inertia is very effective. Conversely, in the right half of the figure the mass is too close to the supports where it contributes a very small amount of kinetic energy; hence the frequencies are too large.

It is seen that the exact factor $\pi^2 = 9.87$ is approached very slowly, and therefore that a quick approximate method for finding the natural frequency based on such shifting of masses is rather unsatisfactory.

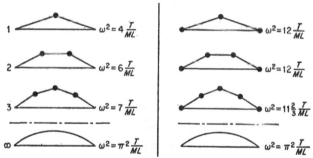

FIG. 4.19. By increasing the number of equidistant masses on the string the uniform mass distribution is approached gradually. The convergence is too slow to have practical significance.

4.5. Rayleigh's Method. The string problem is the simplest one among all those having an infinite number of degrees of freedom. Though for this problem an exact solution of the natural frequencies can be obtained, this is far from possible for the general problem of a system with distributed mass and distributed flexibility. Therefore it is of great importance to have an approximate method for finding the lowest or fundamental frequency, a method which will always work. Such a procedure has been developed by *Rayleigh;* it is a generalization of the energy method discussed on page 33.

Briefly, a shape is *assumed* for the first normal elastic curve; with this assumption the (maximum) potential and kinetic energies are calculated and are equated. Of course, if the *exact* shape had been taken as a basis for the calculation, the calculated frequency would be exactly correct also; for a shape differing somewhat from the exact curve a very useful and close approximation for the frequency is obtained. Since the exact solution for the string is known, we choose it as an example for the explanation of Rayleigh's method, which will enable us to judge the error of the approximate result.

For a calculation of the potential energy we observe that the deflected

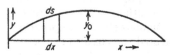

string has a greater length than the straight one. It is subjected to a tension T all the time, so that in going into the deflected shape an amount of work $T \Delta l$ has to be performed on it. This is stored in the string in the form of poten-

Fig. 4.20. Calculation of the potential energy of a string.

tial energy. For a calculation of the increase in length Δl, we observe that the length of an element ds is (Fig. 4.20)

$$ds = \sqrt{(dx)^2 + (dy)^2} = dx \sqrt{1 + \left(\frac{dy}{dx}\right)^2} \approx dx \left[1 + \frac{1}{2}\left(\frac{dy}{dx}\right)^2\right]$$

The increase in length of that element is

$$ds - dx = \frac{1}{2}\left(\frac{dy}{dx}\right)^2 dx$$

so that
$$Pot = \frac{T}{2} \int_0^l \left(\frac{dy}{dx}\right)^2 dx \qquad (4.28)$$

This result can be derived somewhat differently as follows. In the derivation of Eq. (4.20), page 136, it was seen that the right-hand side $T \frac{\partial^2 y}{\partial x^2}$ signifies the downward force per unit length of the string. Imagine the string to be brought into its deflected shape by a static loading $q(x)$ which grows proportional to the deflection $y(x)$. The work done on an element dx by $q(x)$ in bringing it to the fully deflected position $y(x)$ is $\frac{1}{2}q(x)y(x)\,dx$, and the potential energy is

$$Pot = \frac{1}{2} \int_0^l q(x)y(x)\,dx$$

Since $q(x) = -T\frac{d^2y}{dx^2}$,

$$Pot = -\frac{T}{2} \int_0^l y \cdot \frac{d^2y}{dx^2}\,dx \qquad (4.29)$$

By a process of partial integration this can be shown to be equal to (4.28):

$$\int_0^l y \frac{d^2y}{dx^2}\,dx = \int_0^l y\,d\left(\frac{dy}{dx}\right) = y\frac{dy}{dx}\Big|_0^l - \int_0^l \frac{dy}{dx}\,dy$$

The first term is zero because y is zero at 0 and l. The integral in the second term can be written

$$-\int_0^l \frac{dy}{dx}\frac{dy}{dx}\,dx = -\int_0^l \left(\frac{dy}{dx}\right)^2 dx$$

The total kinetic energy is the sum of the kinetic energies $\frac{1}{2}mv^2 =$

$\frac{1}{2}(\mu_1\,dx)(y\omega)^2$ of the various elements:

$$Kin = \frac{1}{2}\,\mu_1\omega^2\int_0^l y^2\,dx \qquad (4.30)$$

As in the case of a single degree of freedom (page 33), the expressions (4.28) and (4.30) are the *maximum* energies; the maximum potential energy occurs in the most deflected position, and the maximum kinetic energy occurs in the undeformed position where the velocity is greatest. Equating the two energies we find for the frequency:

$$\omega^2 = \frac{\mathbf{T}}{\mu_1}\frac{\displaystyle\int_0^l \left(\frac{dy}{dx}\right)^2 dx}{\displaystyle\int_0^l y^2\,dx} \qquad (4.31)$$

The value ω^2 obtained with this formula depends on the form $y(x)$ which we assume. First consider the exact shape:

$$y = y_0 \sin\frac{\pi x}{l}$$

By Eq. (4.28) the potential energy is

$$Pot = \frac{\mathbf{T}}{2}\int_0^l \left(y_0\,\frac{\pi}{l}\cos\frac{\pi x}{l}\right)^2 dx = \frac{\mathbf{T}}{2}\,y_0^2\,\frac{\pi^2}{l^2}\,\frac{l}{2} \qquad \text{(see page 14)}$$

Similarly we find for the kinetic energy: $Kin = \dfrac{\mu_1\omega^2}{2}\,y_0^2\,\dfrac{l}{2}$, so that the frequency becomes

$$\omega_1 = \frac{\pi}{l}\sqrt{\frac{\mathbf{T}}{\mu_1}} = \frac{3.142}{l}\sqrt{\frac{\mathbf{T}}{\mu_1}} \qquad (4.32)$$

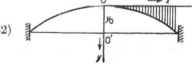

which is the exact value.

Next assume a parabolic arc for the shape of the string. The equation of

FIG. 4.21. A parabolic arc as the approximate (Rayleigh) shape of a vibrating string.

a parabola in the xy system of Fig. 4.21 is $y = px^2$. The parabola can be made to pass through the two points $y = y_0$ and $x = \pm l/2$ by giving p the value $4y_0/l^2$. The equation $y = 4y_0\dfrac{x^2}{l^2}$ describes the shaded ordinates of Fig. 4.21. The deflection of the string is y_0 minus the shaded ordinate:

$$y = y_0\left(1 - \frac{4x^2}{l^2}\right)$$

Using this value for y in (4.28) and (4.30), we have after a simple integration:

$$Pot = \frac{8}{3} T \frac{y_0^2}{l}$$

$$Kin = \frac{4}{15} \mu_1 \omega^2 l y_0^2$$

and
$$\omega_1 = \frac{\sqrt{10}}{l} \sqrt{\frac{T}{\mu_1}} = \frac{3.162}{l} \sqrt{\frac{T}{\mu_1}}$$

which is only 0.7 per cent greater than the exact value. The error is surprisingly small, since it can be seen physically that the parabola cannot be the true shape. The spring effect driving a particle dx of the string back to equilibrium lies in the *curvature*, or d^2y/dx^2, of the string. At the ends the string particles do not move, so that there they have

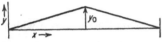

FIG. 4.22. Another Rayleigh approximation for half a sine wave.

obviously neither inertia force nor spring force. Therefore the exact shape must have no curvature at the ends, which condition is violated by the parabola. To test the power of Rayleigh's method we shall now apply it to a most improbable shape of deflection curve (Fig. 4.22):

$$y = y_0 \frac{x}{l/2}, \qquad \text{for} \qquad x = l/2$$

We find successively,

$$Pot = 2T y_0^2 / l$$
$$Kin = \mu_1 \omega^2 l y_0^2 / 6$$

and
$$\omega_1 = \frac{\sqrt{12}}{l} \sqrt{\frac{T}{\mu_1}} = \frac{3.464}{l} \sqrt{\frac{T}{\mu_1}}$$

which is 10 per cent greater than the exact value (4.32).

Rayleigh's approximation always gives for the lowest natural frequency a value which is somewhat too high. Among a number of approximate results found in this manner the smallest is always the best one. A proof for the statement will be given on page 161.

Finally, we shall solve the combination problem of a heavy string of total mass M, in the middle of which is attached a single concentrated weight of the same mass M. This problem is again equivalent to that of the longitudinal (or torsional) vibrations of a bar clamped at both ends and having a concentrated disk in the middle with a mass (or moment of inertia) equal to that of the bar itself.

Regarding the elastic curve, it can be said that, if the central mass were absent, the curve would be sinusoidal, whereas if the string mass were absent, it would be as shown in Fig. 4.22. The actual shape will lie between these two. Assuming first a sinusoid, we note that the potential energy is not affected by the presence of the central mass. The kinetic energy, however, is increased by $\frac{1}{2}M\omega^2 y_0^2$, which is twice as great as the kinetic energy of the string itself, since $M = \mu_1 l$. Thus the total kinetic energy is three times as large as without the central mass and consequently the frequency is $\sqrt{3}$ times as small:

$$\omega_1 = \frac{\pi}{\sqrt{3}}\frac{1}{l}\sqrt{\frac{T}{\mu_1}} = 1.81\sqrt{\frac{T}{Ml}}$$

With the string deformed as shown in Fig. 4.22, again the potential energy is not affected, and the kinetic energy becomes $M\omega^2 y_0^2/2$ larger, i.e., $(\frac{1}{2} + \frac{1}{6})/\frac{1}{6} = 4$ times as great as before. Thus the frequency is

$$\omega_1 = \frac{\sqrt{12}}{2l}\sqrt{\frac{T}{\mu_1}} = 1.73\sqrt{\frac{T}{Ml}}$$

Since this last value is smaller than the one found before, it is the better approximation. The exact solution for this problem is

$$\omega_1 = 1.721\sqrt{\frac{T}{Ml}}$$

This exact solution, though somewhat complicated, can be found from the theory developed on page 138. Equation (4.25) gives the general shape of a vibrating string, which we apply now to the left half of our string. The condition that the left end is at rest gives $C_2 = 0$ as before, so that the shape of the left half of the string is determined by

$$y = C \sin x \sqrt{\frac{\mu_1\omega^2}{T}} \qquad (4.33)$$

FIG. 4.23. Exact calculation of heavy string with central mass.

where C and ω are unknown. The amplitude C is of no particular importance, but the frequency ω determines the "wave length" of the sine curve. In Fig. 4.23 the shape is shown, with the right half of the string as a mirrored image of the left half. The central mass M experiences an inertia force $M\omega^2 y_0$ and an elastic force $2T \tan \alpha$ and, as these two forces must be in equilibrium,

$$2T \tan \alpha = M\omega^2 y_0 \qquad (4.34)$$

The value y_0 and $\tan \alpha$ are the ordinate and the slope of Eq. (4.33) at the point where $x = l/2$, or

$$y_0 = C \sin \frac{l}{2}\sqrt{\frac{\mu_1\omega^2}{T}}$$

$$\tan \alpha = C \sqrt{\frac{\mu_1\omega^2}{T}} \cos \frac{l}{2}\sqrt{\frac{\mu_1\omega^2}{T}}$$

Since $\mu_1 l = M$ a substitution of these expressions in (4.34) gives

$$\frac{\omega}{2} \sqrt{\frac{Ml}{T}} = \cot \frac{\omega}{2} \sqrt{\frac{Ml}{T}}$$

Thus we have to find an angle of which the magnitude in radians equals the value of the cotangent. For zero degrees the angle is zero and the cotangent infinite; for 90 deg. the angle is 1.6 radians and the cotangent is zero. Clearly the equality must occur somewhere between 0 and 90 deg. From a trigonometric table we find that it occurs at 49.3 deg. = 0.8603 radian. Thus

$$\frac{\omega}{2} \sqrt{\frac{Ml}{T}} = 0.8603$$

or

$$\omega_1 = 1.721 \sqrt{\frac{T}{Ml}}$$

Since the smallest value obtained for the frequency is always the best one, Rayleigh sometimes writes down a formula for the shape which is not entirely determined but contains an arbitrary parameter. With this formula the frequency is calculated in the regular manner, giving a result which also contains the parameter. By giving the parameter various values, the frequency also assumes different values. The best value among these is the smallest one, *i.e.*, the minimum frequency as a function of the parameter. The approximation thus obtained is very much better than with the normal Rayleigh method.

Ritz has generalized this procedure to more than one parameter. The *Ritz method* of finding natural frequencies is very accurate but unfortunately requires rather elaborate calculations.

Example: A ship drive consists of an engine, a propeller shaft of 150 ft. length and 10 in. diameter, and a propeller of which the moment of inertia is the same as that of a solid steel disk of 4 in. thickness and 4 ft. diameter. The inertia of the engine may be considered infinitely great. Find the natural frequency of torsional vibration.

Solution: On account of the great engine inertia the engine end of the shaft can be considered as built in, so that the system might be described as a "torsional cantilever." The shape of the deflection curve (*i.e.*, angle φ vs. distance x from engine) would be a quarter sine wave if there were no propeller, and it would be a straight line through the origin if the shaft inertia were negligible with respect to that of the propeller. We choose the latter straight line as our Rayleigh shape, thus: $\varphi = Cx$.

From the strength of materials we take two results:

1. The relation between torque M and angle of twist φ:

$$d\varphi = \frac{M\,dx}{GI_p}$$

2. The potential energy stored in a slice dx of the shaft:

$$d\,Pot = \frac{M^2\,dx}{2GI_p}$$

where GI_p is the torsional stiffness of the shaft.

Since our assumed Rayleigh curve has a constant slope $d\varphi/dx = C$, it follows from the first of these equations that the torque $M = CGI_p$ is constant along the length of the shaft. The second equation can thus be integrated immediately:

$$Pot = \frac{M^2 l}{2GI_p}$$

The kinetic energy of a shaft element dx is $\frac{1}{2}(I_1 dx)\dot{\varphi}^2$, where I_1 is the mass moment of inertia per unit length of the shaft. But $\dot{\varphi} = \omega\varphi = \omega C x = \omega M x/GI_p$.

The kinetic energy of the shaft becomes

$$Kin_s = \frac{I_1}{2}\left(\frac{\omega M}{GI_p}\right)^2 \int_0^l x^2\, dx = \frac{I_1}{6} \cdot \frac{\omega^2 M^2 l^3}{G^2 I_p^2}$$

The angular amplitude of the propeller (of which the inertia is I) is $\varphi_p = Cl = Ml/GI_p$, and its kinetic energy

$$Kin_p = \frac{I}{2}\omega^2 \frac{M^2 l^2}{G^2 I_p^2}$$

Equating the sum of the two kinetic energies to the potential energy and solving for ω^2, we find:

$$\omega^2 = \frac{GI_p}{l\left(I + \dfrac{I_1 l}{3}\right)}$$

from which it is seen that one-third of the shaft inertia is to be thought of as concentrated at the propeller.

With the numerical data of the problem we find:

$$I = \frac{1}{2}mr^2 = \frac{1}{2}\left(\frac{0.28}{386}\pi r^2 4\right)r^2 = 1,510 \text{ in. lb. sec.}^2$$

$$I_1 l = \frac{1}{2}mr^2 l = \frac{1}{2}\left(\frac{0.28}{386}\pi r^2 1\right)r^2 l = 1,280 \text{ in. lb. sec.}^2$$

$$\frac{GI_p}{l} = \frac{G}{l}\frac{\pi}{2}r^4 = \frac{12.10^6}{150 \times 12} \cdot \frac{\pi}{2}\cdot 5^4 = 6.55\ 10^6 \text{ in. lb.}$$

so that

$$\omega^2 = \frac{6.55\ 10^6}{1,510 + 427} = 3,380 \text{ rad.}^2/\text{sec.}^2$$

and

$$f = \frac{\omega}{2\pi} = \frac{1}{2\pi}\sqrt{3,380} = 9.3 \text{ cycles/sec.}$$

An exact solution can be found by a process very similar to that discussed on page 145. In fact, Fig. 4.23 can be suitably interpreted for this propeller shaft. The frequency equation becomes

$$\alpha \tan \alpha = \frac{I_1 l}{I} = \frac{1,280}{1,510} = 0.846$$

where α is an abbreviation for

$$\alpha = l\sqrt{\frac{I_1\omega^2}{GI_p}}$$

By trial the solution of this transcendental equation is found to be

$$\alpha = 46.3 \text{ deg.} = 0.809 \text{ radian}$$

from which

$$\omega^2 = (0.809)^2 \frac{GI_p}{l^2 I_1} = 3,350 \text{ rad.}^2/\text{sec.}^2$$

which is 1 per cent smaller than the Rayleigh result.

4.6. Bending Vibrations of Uniform Beams. In the various textbooks on strength of materials the differential equation of the static loading of a beam is usually given in the following form:

$$\left.\begin{array}{l} \mathbf{M} = EI\,\dfrac{d^2y}{dx^2} \\[2mm] \mathbf{q} = \dfrac{d^2\mathbf{M}}{dx^2} \\[2mm] \mathbf{q} = \dfrac{d^2}{dx^2}\left(EI\,\dfrac{d^2y}{dx^2}\right) \end{array}\right\} \qquad (4.35a,\ b,\ c)$$

or combined

where $\mathbf{q}$ is the load per running inch and $\mathbf{M}$ is the bending moment.

If the cross section of the beam is constant along its length, the factor EI does not depend on x and the equation simplifies to

$$\mathbf{q} = EI\,\frac{d^4y}{dx^4} \qquad (4.36)$$

The various diagrams for a beam on two supports under two stretches of uniform loading are shown in Fig. 4.24, but Eqs. (4.35) and (4.36) are generally true and hold just as well for other manners of support, *e.g.*, for cantilevers.

If a beam is in a state of sustained vibration at a certain natural frequency, the "loading" acting on it is an alternating inertia load. In order to get a physical conception of this statement, note that in the position of maximum *downward* deflection (Fig. 4.24e) each particle of the beam is subjected to a maximum *upward* acceleration. Multiplied by the mass of the particle, this gives an *upward* inertia force which the beam must exert on the particle. By the principle of action and reaction the particle in question must exert a *downward* force on the beam. All these downward forces of the various particles constituting the beam form a loading $\mathbf{q}$ which is responsible for the deflection and is related to it by (4.35)

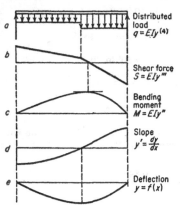

Fig. 4.24. Illustrating the differential equations of a beam in bending.

or (4.36). Naturally, while the beam is passing through its equilibrium position, the accelerations and therefore the loadings are zero, but then the deflections are also zero.

Thus the differential equation of the vibrating *bar of uniform cross section* is

$$EI\,\frac{\partial^4 y}{\partial x^4} = -\mu_1\,\frac{\partial^2 y}{\partial t^2} \qquad (4.37)$$

where μ_1 is the mass of the beam per unit length.   Assuming a sustained free vibration at a frequency ω, we have, as on page 138,

$$y(x, t) = y(x) \sin \omega t \qquad (4.23)$$

which gives to (4.37) the form

$$EI \frac{d^4y}{dx^4} = \mu_1\omega^2 y \qquad (4.38)$$

The left side of this is the elastic expression for the loading [Eq. (4.36)], while the right side is the maximum value of the inertia load.   From it we see that the physical characteristic of any "normal elastic curve" of the beam is that *the q loading diagram must have the same shape as the deflection diagram.*   Any loading that can produce a deflection curve similar to the loading curve can be regarded as an inertia loading during a vibration; the natural frequency appears merely in the numerical factor $\mu_1\omega^2$ connecting the two.

The functions which satisfy (4.38) must have the property that, when differentiated four times, they return to their original form multiplied by a positive constant $\mu_1\omega^4/EI$.   We may remember four functions that will do this, *viz.*:

$$e^{ax}, \qquad e^{-ax}, \qquad \sin ax, \qquad \text{and} \qquad \cos ax$$

where the coefficient a has to be so chosen that

$$a = \sqrt[4]{\frac{\mu_1\omega^2}{EI}} \qquad (4.39)$$

Thus the general solution of (4.38) containing four integration constants can be written

$$y(x) = C_1e^{ax} + C_2e^{-ax} + C_3 \sin ax + C_4 \cos ax \qquad (4.40)$$

This expression determines the shape of the various "normal elastic curves."   The four integration constants C have to be calculated from the end conditions.   For each end of the beam there are two such conditions, making the required four for the two ends.   They are for a

Simply supported end: $y \;\; = 0, \; y'' = 0$
                                 (zero deflection and bending moment)
Free end: $y'' = 0, \; y''' = 0$
                               (zero bending moment and shear force)
Clamped end: $y \;\; = 0, \; y' \;\; = 0$
                               (zero deflection and slope)

which will be clear from a consideration of the physical meaning of the various derivatives as shown in Fig. 4.24.   For any specific case the four end conditions substituted in (4.40) give four homogeneous algebraic equations in the four C's. The determinant of that system equated to zero is an equation in a, which by (4.39) is the frequency equation. This process has been carried out for the various kinds of beams (beam on two supports, cantilever or "clamped-free" beam, clamped-clamped beam, etc.), but we prefer here to find approximate solutions by using Rayleigh's method. Only for the beam on two simple supports can the exact solution be recognized from (4.40) in a simple manner. The end conditions are in this case

$$x = 0, \; y = y'' = 0 \qquad \text{and} \qquad x = l, \; y = y'' = 0$$

We see immediately that a sine-wave shape satisfies these conditions, and that the cosine or e-functions violate them. Thus for a beam on two supports (4.40) simplifies to

$$y(x) = C \sin ax$$

so that the normal elastic curves of a uniform beam on two supports are the same as those of the string shown in Fig. 4.16, but the frequencies are different. They are found by making the argument of the sine equal to an integer number times π or

$$al = l \sqrt[4]{\frac{\mu_1 \omega^2}{EI}} = n\pi \qquad (n = 1, 2, 3, \ldots)$$

so that

$$\omega_1 = \frac{\pi^2}{l^2} \sqrt{\frac{EI}{\mu_1}}, \qquad \omega_2 = \frac{4\pi^2}{l^2} \sqrt{\frac{EI}{\mu_1}}, \qquad \cdots, \qquad \omega_n = \frac{n^2\pi^2}{l^2} \sqrt{\frac{EI}{\mu_1}} \quad (4.41)$$

Whereas the consecutive natural frequencies of the string increase as 1, 2, 3, 4, etc. (page 139), for the beam on two supports they increase as 1, 4, 9, 16, etc.

We have seen that in a natural shape of the uniform beam the inertia loading diagram is similar to the deflection diagram, because the inertia load at each point is $\mu_1 dx\omega^2 y$, proportional to the deflection y. Thus to each natural shape there belongs a natural loading curve $\mu_1\omega^2 y$. This concept is useful for solving a group of problems, of which the following is a typical example:

A beam on two supports is in a state of rest. A load P is suddenly applied to the center and remains on it for t_0 seconds. Then it is removed. What is the ensuing state of motion?

The concentrated load, being not one of the natural loadings, will excite many of the natural motions. In order to see through the situation, the applied loading is resolved into a series of natural loadings, in this case into a Fourier series. A concentrated load P is hard to work with; we replace it by a distributed load of intensity q acting over a short length δ, such that $q\delta = P$. Then, by Eq. (1.12a), page 18, the various Fourier coefficients become

$$a_n = \frac{2}{\pi} \int_0^l F(x) \sin n\frac{\pi x}{l} \cdot d\frac{\pi x}{l} = \frac{2q}{\pi} \int_{\frac{l}{2}-\frac{\delta}{2}}^{\frac{l}{2}+\frac{\delta}{2}} \sin n\frac{\pi x}{l} d\frac{\pi x}{l}$$

$$= \pm \frac{2q}{\pi} \cdot \frac{\pi\delta}{l} = \pm \frac{2P}{l}$$

where the $+$ sign holds for $n = 1, 5, 9$ and the $-$ sign for $n = 3, 7, 11,$
. . . . Thus a concentrated force P at the center of a beam is equivalent to a series of sine loadings of the same intensity $2P/l$. The first few terms are illustrated in Fig. 4.25.

We investigate the influence on the motion of each of these natural loadings individually. Any of them will influence only the natural motion to which they belong, and under one of these loadings

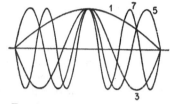

FIG. 4.25. Fourier components of a concentrated load.

the system acts as one of a single degree of freedom, to which the solution of Problem 48 may be applied. Thus for the first loading

$$y = y_{st}[\cos \omega_1(t - t_0) - \cos \omega_1 t]$$

The static deflection curve under a loading $q = \dfrac{2P}{l} \sin \dfrac{n\pi x}{l}$ is found by integrating Eq. (4.36) four times:

$$(y_{st})_n = \frac{2Pl^3}{n^4\pi^4 EI} \cdot \sin\frac{n\pi x}{l}$$

The entire motion is the superposition of the individual motions for each mode and can be written as

$$y(x, t) = \frac{2Pl^3}{\pi^4 EI} \sum_{1,3,5}^{n=\infty} (-1)^{\frac{n-1}{2}} \frac{\sin\dfrac{n\pi x}{l}}{n^4} [\cos \omega_n(t - t_0) - \cos \omega_n t]$$

where the values of ω_n are to be found from Eq. (4.41).

Suppose the load is applied during a time t_0 which is a multiple of a period of the first harmonic motion (and therefore a multiple of the period of any higher harmonic as well). Then $\cos \omega_n(t - t_0) = \cos \omega_n t$, and the whole solution $y(x, t)$ reduces to zero. No motion results after the load ceases to apply.

Next consider the case where the load stays on for $\frac{1}{2}$ period of the first harmonic (and therefore for $\frac{3}{2}$ period of the third harmonic, $\frac{25}{2}$ period of the fifth, etc.). Then $\cos \omega_n(t - t_0) = -\cos \omega_n t$, and the square bracket becomes $-2 \cos \omega_n t$, so that

$$y(x, t) = \frac{4Pl^3}{\pi^4 EI} \sum_{1,3,5} \frac{1}{n^4} (-1)^{\frac{n-1}{2}} \sin \frac{n\pi x}{l} \cos \omega_n t$$

All harmonics are present in the motion, but their amplitudes are proportional to $1/n^4$. Thus, while the first harmonic has an amplitude of $2Pl^3/n^4 EI$ at the center of the span, the third harmonic is only $\frac{1}{81}$ times as large, the fifth $\frac{1}{625}$, etc.

In applying Rayleigh's method, the expression (4.30) for the kinetic energy holds for the bar as well as for the string. But the expression (4.28) for the potential energy will be different since the spring effect in this case is due to the bending resistance EI rather than to the tension $\mathbf{T}$. From strength of materials we have the following formulas for the potential or elastic energy stored in an element of length dx of the beam:

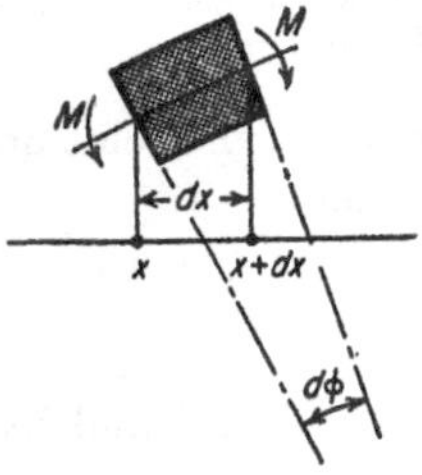

$$d\, Pot = \frac{\mathbf{M}^2}{2EI}\, dx$$

or

$$d\, Pot = \frac{EI}{2} \left(\frac{d^2y}{dx^2}\right)^2 dx$$

FIG. 4.26. Potential energy of flexure in a beam element.

These can be derived simply as follows. Consider an element dx under the influence of the bending moment $\mathbf{M}$ (Fig. 4.26). The element is originally straight and is bent through an angle $d\varphi$ by the moment $\mathbf{M}$. If the left-hand end of the element be assumed to be clamped, the moment $\mathbf{M}$ at the right-hand end turns through the angle $d\varphi$. The work done by $\mathbf{M}$ on the beam is $\frac{1}{2}\mathbf{M}\, d\varphi$, where the factor $\frac{1}{2}$ appears because both $\mathbf{M}$ and $d\varphi$ are increasing from zero together. This work is stored as potential energy in the beam element.

Now calculate the angle $d\varphi$. If the slope at the left-hand end x be dy/dx, then the slope at the right-hand end is $\frac{dy}{dx} + \left(\frac{d^2y}{dx^2}\right) \cdot dx$ and the difference in slope $d\varphi$ is

$$d\varphi = \frac{d^2y}{dx^2}\, dx$$

so that $d\, Pot = \frac{1}{2}\mathbf{M}y''\, dx$

With the differential equation of bending $\mathbf{M} = EIy''$, the two forms given above follow immediately.

Thus the total potential energy in the beam is

$$Pot = \frac{EI}{2} \int_0^l \left(\frac{d^2y}{dx^2}\right)^2 dx \tag{4.42}$$

It is left as an exercise to the reader to derive the first natural frequency of a beam on two supports by substituting in the expressions (4.30) and (4.42) half a sine wave for the shape y.

Let us now calculate the fundamental frequency of a *cantilever* or "clamped-free" beam. We have to choose a curve (Fig. 4.27) which is horizontal at $x = 0$ and has no curvature or bending moment y'' at the end l. A quarter cosine wave has these properties:

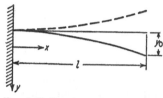

$$y = y_0 \left(1 - \cos \frac{\pi x}{2l}\right) \qquad (4.43)$$

Fig. 4.27. Quarter cosine wave as a Rayleigh shape for a cantilever.

Since this expression cannot be forced into the form (4.40) by manipulating the four C's, (4.43) is not the *exact* form of the normal curve. Substituted in (4.42) and (4.30), we find with the aid of the integral of page 14

$$Pot = \frac{\pi^4}{64} \frac{EI}{l^3} y_0^2$$

$$Kin = \mu_1 \omega^2 y_0^2 l \left(\frac{3}{4} - \frac{2}{\pi}\right)$$

Equating these two expressions, the frequency becomes

$$\omega = \frac{\pi^2}{8 \sqrt{\frac{3}{4} - \frac{2}{\pi}}} \sqrt{\frac{EI}{\mu_1 l^4}} = \frac{3.66}{l^2} \sqrt{\frac{EI}{\mu_1}}$$

$$(4.44)$$

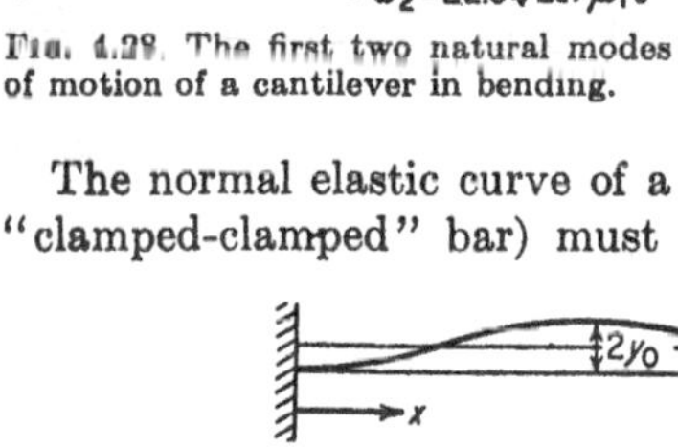

$\omega_1 = 3.52 \sqrt{EI/\mu_1 l^4}$

$\omega_2 = 22.0 \sqrt{EI/\mu_1 l^4}$

Fig. 4.28. The first two natural modes of motion of a cantilever in bending.

The *exact* solution contains the factor 3.52, which is 4 per cent smaller than 3.66. Figure 4.28 gives the exact shape together with that of the second mode.

The normal elastic curve of a beam which is built in at both ends (a "clamped-clamped" bar) must have a shape that is symmetrical and

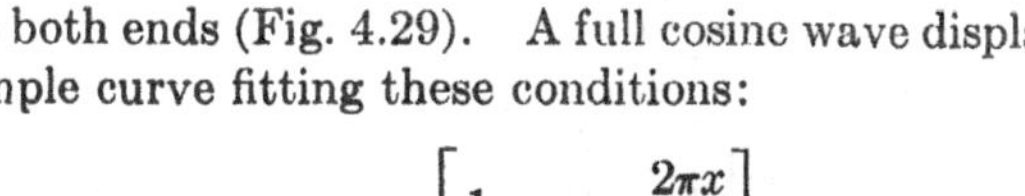

$2y_0$

$\omega_1 = 22.4 \sqrt{EI/\mu_1 l^4}$

Fig. 4.29. Normal elastic curve of a clamped-clamped bar.

horizontal at both ends (Fig. 4.29). A full cosine wave displaced upward by y_0 is a simple curve fitting these conditions:

$$y = y_0 \left[1 - \cos \frac{2\pi x}{l}\right]$$

We find successively:

$$Pot = \frac{EI}{2} y_0^2 \frac{16\pi^4}{l^4} \frac{l}{2}$$

$$Kin = \frac{\mu_1}{2} y_0^2 \omega^2 l [1 + \frac{1}{2}]$$

$$\omega = \frac{4\pi^2}{\sqrt{3}} \sqrt{\frac{EI}{\mu_1 l^4}} = \frac{22.7}{l^2} \sqrt{\frac{EI}{\mu_1}}, \tag{4.45}$$

whereas the exact solution is 22.4 or 1.3 per cent smaller than 22.7.

Finally, we consider the "free-free" bar, *i.e.*, a bar which is suspended freely from one or more strings or which is floating on a liquid. The simplest mode of vibration (Fig. 4.30) must have two nodes and no curvature y'' at either end. Such a shape can be had conveniently in the form of half a sine wave displaced vertically through a short distance a:

$$y = y_0 \sin \frac{\pi x}{l} - a$$

The amount of vertical displacement a is important, since it determines the location of the two nodes. For $a = 0$ they are at the ends of the beam; for $a = y_0$ they are both at the center. The actual value of a between 0 and y_0 can be found from the fact that since no external alternating force is acting on the beam, its total vertical momentum must be zero. While the beam is passing through its equilibrium position, the ends have downward velocities ωy and the middle has an upward velocity ωy. Since the beam is uniform, *i.e.*, since all particles dx have the same mass, these values ωy are proportional to the momentum as well. The total momentum is zero if the areas above and below the center line in Fig. 4.30 are equal or if

$$0 = \int_0^l y\, dx = y_0 \int_0^l \sin \frac{\pi x}{l}\, dx - \int_0^l a\, dx = \frac{2}{\pi} y_0 l - al$$

so that

$$a = \frac{2y_0}{\pi}$$

With that expression for the shape of the vibration we find

$$Pot = \frac{\pi^4}{4} \frac{EI y_0^2}{l^3}$$

$$Kin = \mu_1 \omega^2 y_0^2 l \left[\frac{1}{4} - \frac{2}{\pi^2} \right]$$

$$\omega = \frac{\pi^2}{2\sqrt{\left(\frac{1}{4} - \frac{2}{\pi^2}\right)}} \sqrt{\frac{EI}{\mu_1 l^4}} = \frac{22.72}{l^2} \sqrt{\frac{EI}{\mu_1}} \tag{4.46}$$

The exact result is the same as that of the clamped-clamped bar, namely 22.4, which is 1 per cent smaller than 22.72.

Example: A cantilever beam EI, of length l and of mass μ_1 per unit length (total mass $m = \mu_1 l$) carries a concentrated mass M at its end. Find the natural frequency by Rayleigh's method, and in particular find what fraction of m should be added to M in order to make the simple formula (2.10) applicable (page 33).

Solution: The shape of the deflection curve has to satisfy the same requirements that were used in deriving Eq. (4.43), so that we shall retain the expression employed there. The potential energy is not affected by the addition of a mass M at the end of the bar, but since the amplitude of that end is y_0, the kinetic energy is increased by $\frac{1}{2}M\omega^2 y_0^2$. With $m = \mu_1 l$, the total kinetic energy can be written as

$$Kin = \frac{1}{2}\omega^2 y_0^2 \left[M + m\left(\frac{3}{2} - \frac{4}{\pi}\right)\right] = \frac{1}{2}\omega^2 y_0^2 (M + 0.23m)$$

With the expression of page 147 for the potential energy the frequency becomes

$$\omega^2 = \frac{3.03EI}{l^3(M + 0.23m)}$$

Thus 23 per cent of the mass of the bar has to be added to the end mass. In case the bar is supposed weightless, $m = 0$ and the result for ω^2 found here is 1 per cent greater than the exact value, where the coefficient is 3.

$$\omega_1 = 22.4\sqrt{EI/\mu_1 l^4}$$

FIG. 4.30. Normal elastic curve of a free-free bar.

4.7. Beams of Variable Cross Section. In many practical cases the cross section of the beam is not constant over its length. The most common example of a beam on two supports is a shaft in its bearings, the shaft usually having a greater cross section in its middle portion than near its ends. A steel ship in the water sometimes executes vibrations as a free-free bar, somewhat in the form of Fig. 4.30. These vibrations become of importance if the unbalanced forces of the propelling machinery have the same frequency as the natural frequency of the ship. But the bending stiffness of a ship is by no means constant over its entire length.

The method of Rayleigh can be applied to such non-uniform beams also, since it is always possible to make some reasonable estimate regarding the shape of the deflection curve. The calculations are the same as those for the beam of constant section, with the evident exception that the expression (4.42) for the potential energy has to be modified by bringing the now variable stiffness EI under the integral sign. If the stiffness varies in a more or less complicated manner along the length x, the evaluation of the integral for the potential energy may become difficult, but, even if the exact calculation is impossible, the integral can always be evaluated graphically.

A somewhat different manner of finding the frequency has been developed by *Stodola*, primarily for application to turbine rotors. His process is capable of being repeated a number of times and of giving a better result after each repetition. Briefly it consists of drawing first some reasonable assumed deflection curve for the shaft in question. By multiplying this curve with the mass and the square of the (unknown) frequency $\mu_1(x)\omega^2$, it becomes an assumed inertia loading. Since ω^2 is not known, it is arbitrarily taken equal to unity to begin with. Then with the inertia loading $y(x)\mu_1(x)$ the deflection curve $_2y(x)$ is constructed by the regular methods of graphical statics. Of course this second deflection curve $_2y(x)$ coincides with the originally assumed one $y(x)$ only if

1. $y(x)$ is exactly the normal elastic curve.
2. The natural frequency ω^2 is exactly unity.

The first of these conditions is fulfilled approximately, but the second is generally far from the facts. The deflection $_2y(x)$ has more or less the shape of the original assumption $y(x)$, but its ordinates may be 10,000 times smaller. If that is so, we could have obtained approximately equal ordinates for $_2y(x)$ and $y(x)$ by assuming a frequency $\omega^2 = 10,000$. In that case, the original inertia load would have been 10,000 times as large and the final deflection $_2y(x)$ also 10,000 times as large, i.e., approximately equal to the original assumption. Therefore, the ratio of the ordinates of $y(x)$ and $_2y(x)$ gives a first approximation for the frequency ω^2.

With a fairly reasonable guess at a deflection curve, the accuracy obtained with this procedure is very good. If greater accuracy is desired, we can repeat the construction with $_2y(x)$ as our original estimate, finding a third curve $_3y(x)$. It will be proved on page 162 that the process for finding the fundamental mode of vibration is convergent, i.e., each successive curve is nearer to the true shape than the previous one. In fact, the convergence is so rapid that usually no difference can be detected between the shape of $_3y(x)$ and $_2y(x)$.

For the second and higher modes of vibration the process is not convergent. Nevertheless Stodola's method, properly modified, can be used for the higher modes, as explained on page 162.

The details of the construction belong to the field of graphical statics rather than to vibration dynamics. As a practical example consider a shaft of 72 in. length, on two solid bearings, shown in Fig. 4.31, I. Dividing it into six sections of equal lengths, the masses and bending stiffness EI of the various sections are shown in the table below, where the modulus of elasticity E has been taken as 30×10^6 lb. per square inch:

Section No.	Mass per inch, lb. in.$^{-2}$ sec.^{2}	Section mass, lb. in.$^{-1}$ sec.^{2}	EI, lb. in.^{2}
1	0.0142	0.17	9.13×10^8
2	0.0320	0.38	46.2×10^8
3	0.0568	0.68	146×10^8
4	0.0568	0.68	146×10^8
5	0.0320	0.38	46.2×10^8
6	0.0142	0.17	9.13×10^8

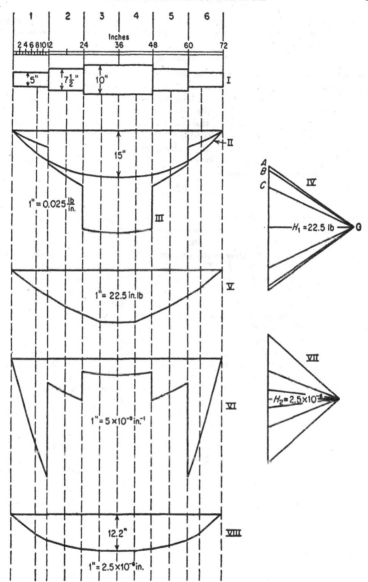

FIG. 4.31. Stodola's construction for determining the fundamental frequency of a rotor.

The assumed deflection curve is designated by II. It has been made rather flat in the center portion because that part is much stiffer than the rest of the structure. In order to obtain the inertia load

$$y\mu_1\omega^2 = y\mu_1 \cdot 1,$$

the ordinates y have to be multiplied with the mass per running inch μ_1, $i.e.$, with the second column of the table. This gives curve III, which is drawn so that each ordinate "inch" represents 0.025 lb./in. All lengths are measured in actual shaft inches indicated by the scale above I. Thus one "inch" of the shaft is roughly $\frac{1}{25}$ in. in the printed figure. The ordinate of II in the center of the shaft is 15 in. and the middle ordinate of III is 0.852 lb./in. (15 × 0.0568).

In order to find the deflection curve under this loading, four integrations have to be performed, divided into two groups of two each. In the first group we integrate twice and arrive at the bending moment M:

$$q = \frac{d^2M}{dx^2} \tag{4.35b}$$

The first integration is performed by evaluating the areas of the six sections of curve III. For instance, the area of the first section, being nearly triangular, is $\frac{1}{2}$ × 12 in. × 0.138 lb./in. = 0.83 lb. This is the combined inertia force (for $\omega = 1$ rad./second) of the whole first section and thus is the change in the shear force between the left end and the right end of section 1. The six areas of curve III are set off vertically below each other in diagram IV, such that AB is 0.83 lb.; $BC = 4.40$ lb. = the area of section 2 of curve III. Thus the vertical line on the left of IV represents the shear forces S and is the result of the first integration. Now take an arbitrary horizontal distance H_1, here taken equal to 22.5 lb. and connect its end O with A, B, C, etc. Then, in curve V, draw lines parallel to the rays of diagram IV, so that the line parallel to OB in IV (which separates section 1 from section 2) runs between the vertical dotted lines through the centers of gravity of the areas 1 and 2 of curve III. The diagram V represents the bending moments; the scale being 1 in. = H_1 = 22.5 in.-lb. Thus, for example, the bending moment in the middle of the shaft is 396 in.-lb.

In order to pass from the bending moment curve V to the deflection curve VIII, we have to perform two more integrations:

$$\frac{M}{EI} = \frac{d^2y}{dx^2}$$

This equation is built exactly like 4.35b; in fact, the deflection y can be considered to be the "bending moment curve of a beam with the loading M/EI." The values of EI for the various sections are given in the last column of the table, and curve VI shows the M/EI diagram. We can repeat the process that has led from III via IV to V, and find VIII from VI via VII. The ordinates of III were measured in lb./in. and those of VI in in.$^{-1}$; so that the dimensions of VI, VII, VIII are found from their counterparts III, IV, V by dividing through by pounds. In particular, the horizontal distance H_2 of VII has no dimension; it is a pure number.

The deflection curve VIII has more or less the appearance of the first guess II; however, its middle ordinate is

$$12.2 \times 2.5 \times 10^{-6} \text{ in.} = 30.5 \times 10^{-6} \text{ in.},$$

whereas the same ordinate in diagram II was 15 in. Thus the first approximation for the natural frequency of the shaft is

$$\omega_1 = \sqrt{\frac{15}{30.5 \times 10^{-6}}} = 700 \text{ rad./sec.}$$

For other graphical and numerical methods to solve the problem of the natural frequencies of flexural vibration of a bar of variable stiffness and inertia, see page 229.

4.8. Normal Functions and Their Applications. We now turn to the proofs of Rayleigh's minimum theorem and of the convergence of Stodola's process. Though these proofs are not essential for an understanding of the subsequent parts of the book, they may give the reader a clearer insight into the nature of "normal modes of motion."

With the *string* and the *beam on two supports*, it was seen that the various normal elastic curves are sine functions:

$$y_1 = \sin \frac{\pi x}{l}, \qquad y_2 = \sin \frac{2\pi x}{l}, \qquad \cdots \cdots, \qquad y_n = \sin \frac{n\pi x}{l}$$

In these expressions the amplitudes of the motions have been arbitrarily assumed to be such that their maximum deflections are 1 inch.

On the other hand, the normal elastic curves of a *cantilever beam* (page 153) or of a beam with non-uniform cross section are curves of much greater complication.

We know from page 17 that any arbitrary curve between 0 and l can be developed into a trigonometric or Fourier series and that one of the most important properties of such a series is

$$\int_0^l \sin \frac{m\pi x}{l} \sin \frac{n\pi x}{l} \, dx = 0, \qquad (m \neq n)$$

as explained on page 15.

Applied to the special case of string vibration, this means that any elastic curve $y(x)$ which may be given to the string by an external loading can be split up into a series of "normal" components. This is true not only for the string with its sine functions, but generally for any elastic system.

If the normal elastic curves of a system of length l are $y_1(x)$, $y_2(x)$, $\cdots$, $y_n(x)$ $\cdots$, then any arbitrary deflection curve $y(x)$ of that system can be developed into a series:

$$y(x) = a_1 y_1(x) + a_2 y_2(x) + \cdots a_n y_n(x) + \cdots \tag{4.47}$$

Moreover, the relation

$$\int_0^l \mu_1(x) y_n(x) y_m(x) \, dx = 0 \qquad (n \neq m) \tag{4.48}$$

holds, so that any coefficient a_n in (4.47) can be found by exactly the same process as that employed on page 18:

$$a_n = \frac{\int_0^l \mu_1(x) y(x) y_n(x) \, dx}{\int_0^l \mu_1(x) y_n^2(x) \, dx} \tag{4.49}$$

This gives us a wide generalization of the concept of Fourier series.

To prove (4.48), consider an elastic system (beam) of length l of which the elastic properties are determined by the "influence functions" $I(x, x_1)$, with the following definition (Fig. 4.32): the deflection at a point x of the beam caused by a load of 1 lb. at a point x_1 is $I(x, x_1)$. In this expression both x and x_1 are variables running from 0 to l (see page 122).

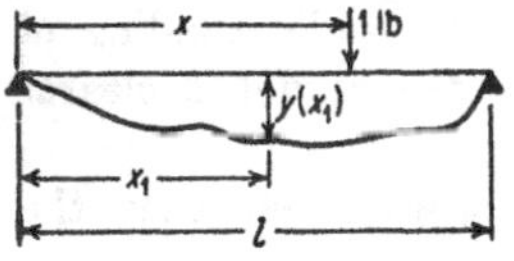

Maxwell's reciprocity theorem on the strength of materials states that the deflection at point 1 due to a unit load at point 2 equals the deflection at 2 due to a unit load at 1. Thus the influence function satisfies the relation

FIG. 4.32. Definition of influence function $I(x_1, x)$.

$$I(x, x_1) = I(x_1, x)$$

Let the beam be vibrating at one of its natural frequencies with the shape $y_n(x)$. The maximum inertia force acting on a section dx_1 of the beam with mass μ_1 per unit length is

$$\mu_1(x_1)\ dx_1\omega_n^2 y_n(x_1)$$

and the deflection caused by that load at a point x is

$$\omega_n^2 y_n(x_1)I(x, x_1)\mu_1(x_1)\ dx_1$$

There are inertia loads of this kind on every section dx_1 between 0 and l, so that the actual deflection curve is the sum of all the partial deflection curves

$$y_n(x) = \omega_n^2 \int_0^l y_n(x_1)I(x, x_1)\mu_1(x_1)\ dx_1 \tag{4.50}$$

This relation holds only when $y_n(x)$ is a natural mode, because only then can the beam be in equilibrium with loads proportional to its own displacements.

In order to prove (4.48), we multiply (4.50) by $\mu_1(x)y_m(x)\ dx$ and integrate:

$$\int_0^l \mu_1(x)y_m(x)y_n(x)\ dx = \omega_n^2 \int_0^l \int_0^l y_n(x_1)y_m(x)I(x, x_1)\mu_1(x_1)\mu_1(x)\ dx_1\ dx \tag{4.51}$$

Since (4.50) holds for any natural frequency, we may replace n by m. Then we can multiply by $\mu_1(x)y_n(x)\ dx$ and integrate, with the result:

$$\int_0^l \mu_1(x)y_m(x)y_n(x)\ dx = \omega_m^2 \int_0^l \int_0^l y_m(x_1)y_n(x)I(x, x_1)\mu_1(x_1)\mu_1(x)\ dx_1\ dx$$

In this last double integral we may reverse the order of integration, i.e., reverse x and x_1:

$$\int_0^l \mu_1(x)y_m(x)y_n(x)\ dx = \omega_m^2 \int_0^l \int_0^l y_m(x)y_n(x_1)I(x_1, x)\mu_1(x)\mu_1(x_1)\ dx\ dx_1$$

This double integral is seen to be the same as that in (4.51) on account of Maxwell's theorem that $I(x, x_1) = I(x_1, x)$. Let the value of the double integral be A; then, on subtracting the last result from (4.51), we obtain:

$$0 = (\omega_n^2 - \omega_m^2)A$$

This means that for $\omega_m \neq \omega_n$, the double integral A is zero, which makes the left-hand side of (4.51) also zero, so that the proposition (4.48) is proved.

Proof of Rayleigh's Minimum Theorem. The approximate curve $y(x)$ assumed in the Rayleigh procedure is not a normal elastic curve but can be expanded in a series of such curves:

$$y(x) = y_1(x) + a_2 y_2(x) + a_3 y_3(x) + \cdots + a_n y_n(x) + \cdots$$

In order to express the fact that $y(x)$ is an approximation of $y_1(x)$, its coefficient has been taken equal to unity, whereas the other coefficients a_2, a_3, etc., may be small numbers. A normal elastic curve $y_n(x)$ is a curve that can be caused by a static loading $\mu_1 \omega_n^2 y_n(x)$.

Thus the static loading $p(x)$ which causes the assumed curve $y(x)$ is

$$p(x) = \mu_1 [\omega_1^2 y_1(x) + a_2 \omega_2^2 y_2(x) + \cdots + a_n \omega_n^2 y_n(x)]$$

The potential energy of an element dx is $\frac{1}{2} y(x) p(x)\, dx$, and the total potential energy is

$$Pot = \frac{1}{2} \int_0^l \mu_1 [y_1(x) + a_2 y_2(x) + \cdots + a_n y_n(x) + \cdots][\cdots + a_n \omega_n^2 y_n(x) + \cdots]\, dx$$

Since by (4.48) all integrals of products with $m \neq n$ are zero, this becomes

$$Pot = \frac{1}{2} \left(\omega_1^2 \int_0^l \mu_1 y_1^2\, dx + \cdots + a_n^2 \omega_n^2 \int_0^l \mu_1 y_n^2\, dx + \cdots \right)$$

The kinetic energy of an element dx vibrating through the neutral position with a velocity $\omega y(x)$ is $\frac{1}{2} \omega^2 y^2 \mu_1\, dx$, and

$$Kin = \frac{1}{2} \omega^2 \int_0^l \mu_1 y^2\, dx = \frac{1}{2} \omega^2 \left(\int_0^l \mu_1 y_1^2\, dx + \cdots + a_n^2 \int_0^l \mu_1 y_n^2\, dx + \cdots \right)$$

since again all terms with products $y_m y_n$ drop out.

It is seen that both the potential and kinetic energies consist of the sum of the various energies of the components y_1, y_2, etc. This is so only if y_1, y_2 are normal modes; if this is not the case, the integrals of the products $y_n y_m$ have to be considered also.

By Rayleigh's procedure we equate the two energies and solve for ω^2:

$$\omega^2 = \frac{\omega_1^2 \int_0^l \mu_1 y_1^2\, dx + \cdots + a_n^2 \omega_n^2 \int_0^l \mu_1 y_n^2\, dx + \cdots}{\int_0^l \mu_1 y_1^2\, dx + \cdots + a_n^2 \int_0^l \mu_1 y_n^2\, dx + \cdots}$$

or

$$\omega^2 = \omega_1^2 \frac{1 + \dfrac{\omega_2^2}{\omega_1^2} a_2^2 \binom{2}{1} + \dfrac{\omega_3^2}{\omega_1^2} a_3^2 \binom{3}{1} + \cdots}{1 + a_2^2 \binom{2}{1} + a_3^2 \binom{3}{1} + \cdots} \tag{4.52}$$

where the symbols $\binom{n}{1}$ are abbreviations for

$$\binom{n}{1} = \frac{\displaystyle\int_0^l \mu_1 y_n^2\, dx}{\displaystyle\int_0^l \mu_1 y_1^2\, dx}$$

Since $\omega_2 > \omega_1$ and $\omega_3 > \omega_2$, etc., it is seen that in (4.52) the various entries in the numerator are larger than the ones just below them in the denominator. Thus the fraction in (4.52) is greater than 1, from which it follows that

$$\omega > \omega_1$$

or the frequency ω found by Rayleigh's procedure is greater than the first natural frequency ω_1, which was to be proved.

Moreover, an inspection of (4.52) will show that this property holds only for the first, or lowest, frequency but not for the second or higher ones.

Proof of the Convergence of Stodola's Process. Let the first assumption for the deflection curve be $y(x)$, where

$$y(x) = y_1(x) + a_2 y_2(x) + a_3 y_3(x) + \cdots + a_n y_n(x) + \cdots$$

With a mass distribution $\mu_1(x)$ and an arbitrary frequency $\omega = 1$ the inertia loading becomes

$$\mu_1 y = \mu_1 y_1 + a_2 \mu_1 y_2 + a_3 \mu_1 y_3 + \cdots + a_n \mu_1 y_n + \cdots$$

The deflection curve for the loading $\mu_1 \omega^2 y_n$ is y_n; consequently the loading $a_n \mu_1 y_n$ gives a deflection $a_n y_n / \omega_n^2$, so that the second deflection curve of the process becomes

$$_2 y(x) = \frac{y_1(x)}{\omega_1^2} + \cdots + \frac{a_n y_n(x)}{\omega_n^2} + \cdots,$$

which differs from the first curve in that each term is divided by the square of its natural frequency. Proceeding in this manner we find for the $(n + 1)$st deflection curve

$$_{n+1} y = \frac{1}{\omega_1^{2n}} \left[y_1 + \left(\frac{\omega_1}{\omega_2}\right)^{2n} a_2 y_2 + \left(\frac{\omega_1}{\omega_3}\right)^{2n} a_3 y_3 + \cdots \right]$$

Since $\omega_1 < \omega_2$ and $\omega_1 < \omega_3$, etc., it is seen that with increasing n the impurities y_2, y_3, $\cdots$ decrease, and the first mode y_1 appears more and more pure.

4.9. Stodola's Method for Higher Modes.

The above proof shows that an attempt to construct the second normal elastic curve by Stodola's method will end in failure because any impurity of the fundamental elastic curve contained in the guess for the second curve will be magnified more than the second curve itself. After a large number of repetitions it will be found that the second mode disappears altogether and that only the fundamental mode is left. Still it is possible to find the second mode if before each operation the deflection curve is purified from its first-mode content. For this it is necessary first to know the shape of the first mode with sufficient accuracy.

Let $y(x)$ be the assumed shape for the second mode which unfortunately will contain some first harmonic impurity, say $A y_1(x)$. Then we want to find

$$y(x) - A y_1(x)$$

which will be free from first harmonic contamination. In order to find

A, substitute the above expression in Eq. (4.48)

$$\int_0^l \mu_1(x)[y(x) - Ay_1(x)]\, y_1(x)\, dx = 0$$

or

$$\int_0^l \mu_1(x)y(x)y_1(x)\, dx = A \int_0^l \mu_1(x)y_1^2(x)\, dx$$

or

$$A = \frac{\int \mu_1(x)y(x)y_1(x)\, dx}{\int \mu_1(x)y_1^2(x)\, dx} \tag{4.53}$$

The integrand in the numerator, apart from the factor $\mu_1(x)$, is the product of the known first harmonic deflection curve and the assumed approximation for the second harmonic deflection curve $y(x)$. In the denominator the integrand is the product of the mass $\mu_1(x)$ and the square of the first harmonic curve. Both integrals can be evaluated graphically; thus A is determined, and the assumed shape for the second mode can be purified from its first-mode contamination. Then the Stodola process is applied to this curve.

For the third or higher modes the procedure is similar, but the assumed curve for the third harmonic has to be purified from the first as well as from the second harmonic by Eq. (4.53). Thus the Stodola process cannot be applied to a higher mode of vibration until after all lower modes have been determined with sufficient accuracy.

The method is not necessarily restricted to the graphical form of page 157. It is sometimes applied arithmetically, as will now be shown for the simple example of the string with three equal masses of Fig. 4.2. In the equations (4.3) the terms on the right are the deflections caused by the individual inertia forces. With the influence numbers of Eq. (4.6), the elastic deflection equations (4.3) are rewritten ($m_1 = m_2 = m_3 = m$).

$$\left.\begin{aligned}
\frac{T}{m\omega^2 l}\, a_1 &= \frac{3}{4}\, a_1 + \frac{1}{2}\, a_2 + \frac{1}{4}\, a_3 \\[4pt]
\frac{T}{m\omega^2 l}\, a_2 &= \frac{1}{2}\, a_1 + a_2 + \frac{1}{2}\, a_3 \\[4pt]
\frac{T}{m\omega^2 l}\, a_3 &= \frac{1}{4}\, a_1 + \frac{1}{2}\, a_2 + \frac{3}{4}\, a_3
\end{aligned}\right\} \tag{4.54}$$

With Stodola, we now assume a shape for the deformation in the first mode, and for the purpose of illustrating the convergence of the method we intentionally make a stupid choice: $a_1 = a_2 = a_3 = 1$. Substitute that into the right-hand sides of Eq. (4.54), and calculate their sums.

$$Ca_1 = 1\tfrac{1}{2} \qquad Ca_2 = 2 \qquad Ca_3 = 1\tfrac{1}{2}$$

where $C = T/m\omega^2 l$. By reducing the middle amplitude to unity (the same value as assumed first), we thus find for the second approximation of the deflection curve

$$a_1 = \tfrac{3}{4} \qquad a_2 = 1 \qquad a_3 = \tfrac{3}{4}$$

Put this into the right sides of Eq. (4.54), and find

$$Ca_1 = 1\tfrac{1}{4}, \qquad Ca_2 = 1\tfrac{3}{4}, \qquad Ca_3 = 1\tfrac{1}{4}$$

or again reduced to unity at the center, the third approximation becomes

$$a_1 = \tfrac{5}{7}, \qquad a_2 = 1, \qquad a_3 = \tfrac{5}{7} = 0.714$$

Another substitution leads to the fourth approximation

$$a_1 = 1\tfrac{7}{24}; \qquad a_2 = 1; \qquad a_3 = 1\tfrac{7}{24} = 0.707$$

The fifth approximation is

$$a_1 = \tfrac{29}{41}; \qquad a_2 = 1; \qquad a_3 = \tfrac{29}{41} = 0.707$$

which is identical with the previous one within slide rule accuracy. Substituting this into the first of the equations (4.54), we have

$$0.707 \, \frac{T}{m\omega^2 l} = 1.207 \qquad \text{or} \qquad \omega_1^2 = 0.586 \, \frac{T}{ml}$$

as found before on page 126.

Proceeding to the second mode, its shape is obvious (page 126) from the symmetry of the case. However, for the purpose of illustrating the method we start with a very bad assumption:

$$a_1 = 1.000, \qquad a_2 = 0.500, \qquad a_3 = -0.750 \qquad \text{(first)}$$

First this expression is to be purified from its fundamental harmonic content by means of Eq. (4.53). All masses are equal and divide out' from (4.53). The expression thus is

$$A = \frac{1.000 \times 0.707 + 0.500 \times 1.000 - 0.750 \times 0.707}{0.707 \times 0.707 + 1.000 \times 1.000 + 0.707 \times 0.707} = 0.338$$

The first harmonic amount to be subtracted from the above assumption then is

$$a_1 = 0.338 \times 0.707 = 0.240, \qquad a_2 = 0.338, \qquad a_3 = 0.240$$

which leads to

$$a_1 = 0.760, \qquad a_2 = 0.162, \qquad a_3 = -0.990$$

or, multiplying by a constant so as to make a_1 equal to unity, for purposes of comparison,

$$a_1 = 1.000, \qquad a_2 = 0.213, \qquad a_3 = -1.302 \quad \text{(first, purified)}$$

Substituting this into Eq. (4.54) and multiplying by a constant so as to make $a_1 = 1.000$ leads to

$$
\begin{aligned}
a_1 &= 1.000, & a_2 &= 0.116, & a_3 &= -1.181 & \text{(second)} \\
a_1 &= 1.000, & a_2 &= 0.051, & a_3 &= -1.125 & \text{(third)} \\
a_1 &= 1.000, & a_2 &= -0.024, & a_3 &= -1.148 & \text{(fourth)}
\end{aligned}
$$

By this time considerable first harmonic error has crept into the solution, so that it is necessary to purify again by means of Eq. (4.53).

$$a_1 = 1.000, \qquad a_2 = +0.038, \qquad a_3 = -1.058 \qquad \text{(fourth, purified)}$$

Continuing

$$a_1 = 1.000, \qquad a_2 = +0.018, \qquad a_3 = -1.035 \qquad \text{(fifth)}$$
$$a_1 = 1.000, \qquad a_2 = \ \ 0.000, \qquad a_3 = -1.034 \qquad \text{(sixth)}$$

Again it becomes necessary to throw out the first harmonic, which has crept in,

$$a_1 = 1.000, \qquad a_2 = +0.012, \qquad a_3 = -1.018 \qquad \text{(sixth, purified)}$$
$$a_1 = 1.000, \qquad a_2 = +0.006, \qquad a_3 = -1.012 \qquad \text{(seventh)}$$
$$a_1 = 1.000, \qquad a_2 = \ \ 0.000, \qquad a_3 = -1.012 \qquad \text{(eighth)}$$
$$a_1 = 1.000, \qquad a_2 = +0.004, \qquad a_3 = -1.006 \qquad \text{(eighth, purified)}$$
$$a_1 = 1.000, \qquad a_2 = +0.002, \qquad a_3 = -1.004 \qquad \text{(ninth)}$$
$$a_1 = 1.000, \qquad a_2 = \ \ 0.000, \qquad a_3 = -1.004 \qquad \text{(tenth)}$$

It is seen that the convergence is very slow, and that the first harmonic creeps in continually and has to be thrown out about every other step.

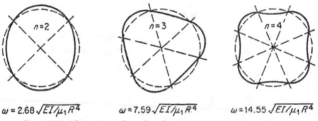

$$\omega = 2.68\sqrt{EI/\mu_1 R^4} \qquad \omega = 7.59\sqrt{EI/\mu_1 R^4} \qquad \omega = 14.55\sqrt{EI/\mu_1 R^4}$$

FIG. 4.33. Normal modes of a ring bending in its own plane.

4.10. Rings, Membranes, and Plates. The strings and beams thus far discussed suffice in many cases to give a tolerably accurate idealization of the actual constructions or machines with which we are dealing. Where this is no longer possible, an idealization in terms of rings (curved beams), membranes, or plates may be helpful. But the calculation of the natural frequencies of these elements is much more complicated than anything we have thus far considered. Therefore, in this section the results only will be given, while for the detailed derivations the reader is referred to the literature, especially to the book of S. Timoshenko, "Vibration Problems in Engineering."

Full Ring. Of the many possible motions of a full ring, the bending vibrations are the most important. If the ring has uniform mass and stiffness, it can be shown that the *exact* shape of the mode of vibration consists of a curve which is a sinusoid on the developed circumference of the ring. In Fig. 4.33 these shapes are shown for the four, six, and eight noded modes or for two, three, and four full waves along the circumference of the ring.

The exact formula for the natural frequencies is

$$\omega_n = \frac{n(n^2 - 1)}{\sqrt{n^2 + 1}} \sqrt{\frac{EI}{\mu_1 R^4}} \qquad (4.55)$$

where n is the number of full waves, μ_1 is the mass per unit length of the ring, EI the bending stiffness, and R the radius.

One of the most important applications of this result is to the frames of electric machines. As these machines often carry salient poles, which act as concentrated masses (Fig. 6.37, page 266), the exact shapes of vibration are no longer developed sinusoids, although in the spirit of Rayleigh's procedure the sinusoid may be considered as an approximate shape. The potential energy of deformation is not altered by the addition of the poles, but the kinetic energy changes from Kin_r to $Kin_r + Kin_p$, where the subscripts pertain to the ring and poles, respectively. Therefore, the result (4.55) for the frequency has to be corrected by the factor

$$\sqrt{\frac{Kin_r}{Kin_r + Kin_p}} \qquad (4.56)$$

In case the number of poles is $2n$, i.e., equal to the number of half waves along the ring, and in case these poles are located in the antinodes so that they move parallel to themselves (Fig. 6.38b), the correction (4.56) becomes specifically

$$\sqrt{\frac{M_r}{M_r + M_p \dfrac{2n^2}{n^2 + 1}}} = \sqrt{\frac{1}{1 + \dfrac{2n^2}{n^2 + 1} \dfrac{M_p}{M_r}}} \qquad (4.57)$$

where M_r is the mass of the complete ring and M_p is the mass of all poles combined, so that M_p/M_r is the ratio of one pole mass to the ring mass per pole.

Another important case occurs when the $2n$ poles are located at the nodes of the radial vibration and there execute rocking motions about the node axis. The correction factor for this case (Fig. 6.38c) is

$$\sqrt{\frac{1}{1 + \dfrac{n^5}{n^2 + 1} \cdot \dfrac{4I_p}{M_r R^2}}} \qquad (4.58)$$

in which I_p is the moment of inertia of a single pole for the axis about which it rotates during the vibration. The actual location of that axis is somewhat doubtful (on account of the fact that the "node" of the ring is a node only in the radial motion but moves back and forth tangentially), but no great error is made by taking the axis on the center line of the ring at the node.

Partial Ring. Quite often the stators of electric motors or generators are bolted on a foundation in the manner shown in Fig. 4.34a. If the foundation or bedplate is very stiff, the stator may be regarded as a partial ring of angle α built in (clamped) at both ends. The fundamental mode of vibration of such a ring in its own plane will be approximately as sketched in Fig. 4.34b. Its natural period, calculated by the procedure of Rayleigh, leads to a result which dimensionally is the same as (4.55) but the numerical factor depends on the central angle α and has to be written $f(\alpha)$:

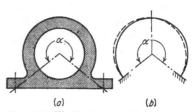

FIG. 4.34. The fundamental flexural mode in its own plane of a partial ring.

$$\omega = f(\alpha) \sqrt{\frac{EI}{\mu_1 R^4}} \qquad (4.59)$$

The values of the constant $f(\alpha)$ for the various angles between $\alpha = 180$ deg. (half circle) and $\alpha = 360$ deg. (full circle clamped at one point) are shown in Fig. 4.35.

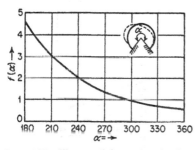

FIG. 4.35. The coefficient $f(\alpha)$ in Eq. (4.59) for the frequency of Fig. 4.34.

In case the stator carries salient poles, the correction (4.56) has to be applied. No greater error is committed by distributing the pole masses uniformly along the ring, since the various pole-carrying points of Fig. 4.34b move through roughly the same amplitudes (which is totally different from some of the cases of Fig. 4.33). The natural frequency calculated from Eq. (4.59) and Fig. 4.35 is usually somewhat (of the order of 10 per cent) high on account of the fact that the feet of the stator do not constitute a complete "clamping" but admit some angular motion.

If the ring of Fig. 4.34 has a small dimension in the direction perpendicular to the paper (*i.e.*, in the direction of the axis of the cylinder), another motion has caused trouble in some cases. It is a vibration perpendicular to the plane of the paper. If Fig. 4.34 were viewed from the side, it would be seen as a cantilever beam of height h. The lateral vibration would then appear in a form similar to that shown in Fig. 4.28a. In this case the elastic resistance of the ring consists of a combination of bending and twist determined by the quantities

EI_2 = bending stiffness (now in a plane perpendicular to the paper, 90 deg. from the EI in Eqs. (4.55) and (4.59), and

C = torsional stiffness, which has the form GI_p for a bar of circular cross section

The frequency can be written in the form

$$\omega = f\left(\alpha, \frac{EI_2}{C}\right)\sqrt{\frac{EI_2}{\mu_1 R^4}} \qquad (4.60)$$

where the numerical constant is shown in Fig. 4.36. This figure was found by a modified Rayleigh method and subsequently verified by laboratory tests, showing the results to be substantially correct.

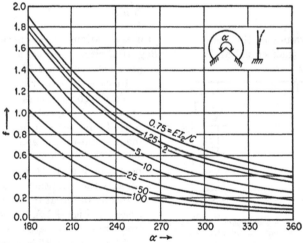

FIG. 4.36. Coefficients $f(\alpha, EI_2/C)$ of Eq. (4.60) for the frequency of a partial ring vibrating perpendicularly to its own plane.

A *membrane* is a skin which is stretched with a great tension and which has no bending stiffness whatever. It is therefore to be considered as a two-dimensional generalization of a string. A circular membrane or drumhead has an infinite number of natural modes of motion whereby the nodes appear as diameters and also as smaller concentric circles. However, we shall discuss here the fundamental mode only, having no nodes except the boundary. The shape of the vibration is practically that of a hill formed by the revolution of a sine curve (Fig. 4.37). The

frequency of this motion is

$$\omega = 2.40 \sqrt{\frac{T}{\mu_1 R^2}} = 4.26 \sqrt{\frac{T}{\mu_1 A}} \qquad (4.31)$$

where T is the tension per running inch across any section of the membrane, μ_1 is the mass per unit area, and A is the total area πR^2.

The formula in its second form is useful also when the membrane is no longer circular but has some other boundary which roughly resembles the circle (square, triangle, half or quarter circle, etc.). Even then (4.61) is approximately correct if the area A of the non-circular membrane is substituted. In such a case the numerical factor is somewhat greater than 4.26. An idea of the error involved can be had from the fact that for a square membrane the factor 4.26 in Eq. (4.61) becomes 4.44, for a 2×1 rectangular membrane 4.97, and for a 3×1 rectangle 5.74.

Fig. 4.37. Fundamental mode of a drumhead with the frequency $\omega = 2.40 \sqrt{T/\mu_1 R^2}$.

Just as a membrane is a two-dimensional string, so a plate may be considered as a two-dimensional "beam." The theory of the vibrations of plates even in the approximate form of Rayleigh-Ritz is extremely complicated. The results are known for circular and rectangular plates with either free, clamped, or simply supported edges, and the reader who may have occasion to use these formulas should refer to the more elaborate books on the subject by Rayleigh, Prescott, or Timoshenko.

Problems 99 *to* 138.

CHAPTER 5

MULTICYLINDER ENGINES

5.1. Troubles Peculiar to Reciprocating Engines. There are two groups of vibration phenomena of practical importance in reciprocating machines, namely:

1. Vibrations transmitted to the foundation by the engine as a whole.

2. Torsional oscillations in the crank shaft and in the shafting of the driven machinery.

Each one of these two effects is caused by a combination of the periodic accelerations of the moving parts (pistons, rods, and cranks) and the periodic variations in cylinder steam or gas pressure.

Consider a vertical single-cylinder engine. The piston executes an alternating motion, *i.e.*, it experiences alternating vertical accelerations. While the piston is accelerated downward there must be a downward force acting on it, and this force must have a reaction pushing upward against the stationary parts of the engine. Thus an alternating acceleration of the piston is coupled with an alternating force on the cylinder frame, which makes itself felt as a vibration in the engine and in its supports. In the lateral direction, *i.e.*, perpendicular to both the crank shaft and the piston rod, moving parts are also being accelerated, namely the crank pin and part of the connecting rod. The forces that cause these accelerations must have equal and opposite reactions on the frame of the engine. This last effect is known as "horizontal unbalance. In the longitudinal direction, *i.e.*, in the crank-shaft direction, no inertia forces appear, since all moving parts remain in planes perpendicular to the crank shaft.

The mathematical relation describing these effects is Newton's law, stating that in a mechanical system the rate of change of momentum equals the resultant $\bar{F}$ of all external forces:

$$\frac{d}{dt} (\Sigma m \bar{v}) = \bar{F} \qquad (5.1)$$

This is a vector equation and is equivalent to three ordinary equations. Two of these equations are of importance, while the third (in the longitudinal direction) is automatically satisfied because v is always zero in that direction.

Equation (5.1) can be interpreted in a number of ways. First, consider the "mechanical system" as consisting of the *whole* engine, and assume it is mounted on extremely flexible springs so as to be floating freely in space. No external forces $\bar{F}$ are acting, and Eq. (5.1) states that, while the piston is accelerated downward (*i.e.*, acquires downward momentum), the cylinder must be accelerated upward. If the cylinder mass is 50 times the piston mass, the cylinder acceleration must be 50 times as small as the piston acceleration.

Next consider only the moving parts, *i.e.*, piston, rod, and crank shaft, as the mechanical system. During rotation these parts have a definite acceleration, or $\frac{d}{dt}(m\bar{v})$, in the vertical and lateral directions. Equation (5.1) determines the value of the force $\bar{F}$ acting on these parts, and consequently the value of the reaction $-\bar{F}$ on the stationary parts.

Equation (5.1) is sometimes written with the differentiation carried out:

$$\Sigma\left(m\,\frac{d\bar{v}}{dt}\right) = \bar{F} \tag{5.2}$$

The expression $-m\,d\bar{v}/dt$ is called the "inertia force," and the theorem states that the external force acting on the system plus the sum of all the inertia forces of the moving parts equals zero.

These various inertia forces can form moments. Consider a two-cylinder vertical engine with the two cranks set 180 deg. apart. While one piston is accelerated downward the other one is accelerated upward, and the two inertia forces form a couple tending to rock the engine about a lateral axis. Similarly, the horizontal or lateral inertia forces of the two cranks are equal and opposite forming a couple tending to rock the engine about a vertical axis.

A rocking about the crank-shaft axis can occur even in a *single*-cylinder engine. If the piston be accelerated downward by a pull in the connecting rod, it is clear that this pull exercises a torque about the crank-shaft axis. Since the piston acceleration is alternating, this inertia torque is also alternating.

Newton's law for moments states that in a mechanical system on which an external torque or moment $\bar{M}$ is acting

$$\frac{d}{dt}(\Sigma\overline{mva}) = \bar{M} \tag{5.3}$$

where a is the moment arm of the momentum $m\bar{v}$. In words: the external torque equals the rate of change of moment of momentum. With the differentiation performed the relation reads

$$\Sigma a \left(m \frac{d\bar{v}}{dt} \right) = \overline{\mathbf{M}} \tag{5.4}$$

or the sum of the moments of the inertia forces of the various moving parts equals the external moment.

As before, we can take for our mechanical system either the whole engine mounted on very weak springs, or we can take merely the moving parts. In the first case the external torque is zero, and therefore any increase in the clockwise angular momentum of the moving parts must be neutralized by an increase in counterclockwise angular momentum of the stationary parts of the engine. In the second case the increase in clockwise angular momentum of the moving parts must be caused by a clockwise torque or moment on these parts, which has a counterclockwise reaction torque on the frame. If this frame is mounted solidly on its foundation, this countertorque is communicated to the foundation and may cause trouble. On the other hand, if the engine is mounted on soft springs, no reaction to the foundation can penetrate through these springs and the countertorque is absorbed as an inertia torque of the frame and cylinder block. Hence that block must vibrate, but no appreciable torque gets into the foundation: we have "floating power" (page 76).

The formulas (5.1) and (5.3) suffice for a derivation of the *inertia* properties of the engine which will be carried out in the next two sections. We shall turn our attention now to the effect of alternating steam or gas pressure in the cylinders.

In Fig. 5.1, let any inertia effect be excluded by assuming either that the moving parts have a negligible mass or that the engine is turning over very slowly at a constant speed ω. Let the pressure force on the piston be P, which is variable with the time (or with the crank angle ωt). The gas pressure not only pushes the piston downward, but it also presses upward against the cylinder head. The piston force P is transmitted through the piston rod (force 1) to the crosshead. Neglecting friction,

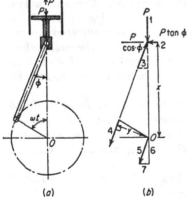

(a)   (b)

FIG. 5.1. Gas pressure forces on a single-cylinder engine.

it is there held in equilibrium by the forces 2 and 3. The forces 1, 2, and 3 of Fig. 5.1b are those acting on the crosshead; 3 is a compression in the connecting rod and 2 has a reaction pressure on the guide or frame to the *right* and of magnitude P tan φ. The force 3 of magnitude $P/\cos \varphi$ is transmitted through the connecting rod to the crank pin (force 4). By shifting this force parallel to itself to O we add a torque $yP/\cos \varphi$, which is the driving torque of the gas pressure. The force 5 is taken up by the main bearings at O and can be resolved into a vertical component 6, and a horizontal component 7. From the similarity of the triangles 1, 2, 3 and 5, 6, 7 it can be seen immediately that the magnitude of 6 is P and that of 7 is P tan φ.

The forces transmitted to the stationary parts of the engine are:

First, P upward on the cylinder head.

Second, P tan φ to the right on the crosshead guide.

Third, P downward on the main bearings at O.

Fourth, P tan φ to the left on the main bearings at O.

The total *resultant* force on the frame is zero, but there is a resultant torque Px tan φ. By Newton's law of action and reaction this torque must be equal and opposite to the driving torque on the crank shaft, $yP/\cos \varphi$. The truth of this statement can easily be verified because it can be seen in Fig. 5.1b that $y = x \sin \varphi$. Thus the gas pressures in the cylinder do not cause any resultant forces on the engine frame but produce only a torque about the longitudinal axis.

Summarizing, we note that no forces occur along the longitudinal axis of an engine, while in the lateral and vertical directions only inertia forces appear. About the vertical and lateral axes only inertia torques are found, whereas about a longitudinal axis both an inertia torque and a cylinder-pressure torque occur.

If we assume the engine to be built up of rigid bodies, *i.e.*, elastically non-deformable bodies, the problem is one of "balance" only. The frame or stationary parts usually fulfill this condition of rigidity, but as a rule the crank shaft can be twisted comparatively easily, which makes torsional vibrations possible. The subject is usually divided into three parts:

a. Inertia Balance: By this is meant the balance of the engine against vertical and lateral forces and against moments about vertical and lateral axes.

b. Torque Reaction: Under this heading we study the effect of the torque (due to inertia and cylinder-pressure effects) acting on the stationary parts about the longitudinal axis (floating power).

c. Torsional Vibrations of the Crank Shaft: Here we deal with the consequences of this same longitudinal torque on the moving parts of the engine.

The effect c is of particular importance since many crank shafts have been broken on account of it. Now that the theory is understood such failures are unnecessary.

The first step in the discussion of the subject is the derivation of the expressions for the vertical and lateral inertia forces of a single-crank mechanism as well as a formula for its inertia torque.

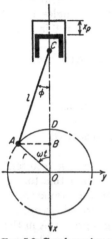

5.2. Dynamics of the Crank Mechanism. Let Fig. 5.2 represent a simple piston and crank, and let

x_p = downward displacement of piston from top.

ωt = crank angle from top dead center.

r = crank radius.

l = length of connecting rod.

Assume the crank shaft to be rotating at uniform speed, *i.e.*, ω is constant. Our first object is the calculation of the position of the piston in terms of the angle ωt. The distance x_p would be equal to the length DB in the figure, were it not that the connecting rod has assumed a slanting position in the meantime. The distance DB, which is a first approximation of x_p can be written

FIG. 5.2. Crank mechanism.

$$r(1 - \cos \omega t)$$

In order to calculate x_p exactly, we must add to this as a correction term the difference between AC and BC or

$$l(1 - \cos \varphi)$$

The auxiliary angle φ can be expressed in terms of ωt by noting that

$$AB = l \sin \varphi = r \sin \omega t$$

or

$$\sin \varphi = \frac{r}{l} \sin \omega t \qquad (5.5)$$

and consequently

$$\cos \varphi = \sqrt{1 - \frac{r^2}{l^2} \sin^2 \omega t}$$

Hence the exact expression for the piston displacement x_p in terms of the crank angle ωt is

$$x_p = r(1 - \cos \omega t) + l\left(1 - \sqrt{1 - \frac{r^2}{l^2} \sin^2 \omega t}\right) \qquad (5.6)$$

On account of the square root this formula is not very convenient for further calculation. It can be simplified by noting that the second term under the square root is small in comparison to unity. In the usual engine, r/l differs little from $\frac{1}{4}$, so that the second term is less than $\frac{1}{16}$. Therefore, the square root is of the form $\sqrt{1 - \delta}$, where $\delta \ll 1$. Expanding this into a power series and retaining only the first term gives the approximation

$$\sqrt{1 - \delta} \approx 1 - \frac{\delta}{2}$$

With $\delta = \frac{1}{16}$, the error made is less than one part in 2,000. Equation (5.6) becomes

$$x_p \approx r(1 - \cos \omega t) + \frac{r^2}{2l} \sin^2 \omega t$$

A further simplification is obtained by converting the square of the sine into the cosine of the double angle by means of

$$\cos 2\omega t = 1 - 2 \sin^2 \omega t$$

or

$$\sin^2 \omega t = \frac{1 - \cos 2\omega t}{2}$$

Thus the piston displacement is

$$x_p = \left(r + \frac{r^2}{4l} \right) - r \left[\cos \omega t + \frac{r}{4l} \cos 2\omega t \right] \tag{5.7}$$

The velocity and acceleration follow from the displacement by differentiation:

$$\dot{x}_p = r\omega \left[\sin \omega t + \frac{r}{2l} \sin 2\omega t \right] \tag{5.8}$$

$$\ddot{x}_p = r\omega^2 \left[\cos \omega t + \frac{r}{l} \cos 2\omega t \right] \tag{5.9}$$

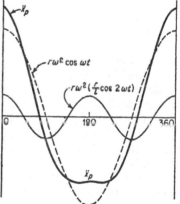

After multiplication by the mass of the piston, these expressions become the vertical momentum and the vertical inertia force. They are seen to consist of two terms, one varying with the same frequency as the rotation and known as the "primary" term, and the other varying at double frequency and known as the "secondary" term. If the connecting rod is

Fig. 5.3. The piston acceleration as a function of the crank angle for $r/l = \frac{1}{4}$.

infinitely long, the secondary term diappears and the piston executes a harmonic motion. With a short connecting rod the motion, and especially the acceleration, deviates considerably from a sinusoid. As an example, Fig. 5.3 gives the piston acceleration (or inertia force) of an engine in which $l/r = 4$.

Having found the dynamic properties of the piston, we proceed to the rotating parts of the crank. The problem is first simplified by concentrating the entire rotating crank mass in its center of gravity. (The inertia force of this mass is the same as the resultant of all the small inertia forces on the various small parts of the crank.) Next the mass is shifted from the center of gravity to the crank pin A, but in this process it is diminished inversely proportional to the distance from the center of the shaft, so that the inertia force (which is here a centripetal force) remains unchanged.

In this manner the whole crank structure is replaced by a single mass m_c at the crank pin, and the vertical displacement can be found immediately from Fig. 5.2:

$$x_c = r(1 - \cos \omega t) \tag{5.10}$$

so that the vertical components of velocity and acceleration become

$$\left.\begin{aligned} \dot{x}_c &= r\omega \sin \omega t \\ \ddot{x}_c &= r\omega^2 \cos \omega t \end{aligned}\right\} \tag{5.11}$$

The horizontal components are

$$\left.\begin{aligned} y_c &= -r \sin \omega t \\ \dot{y}_c &= -r\omega \cos \omega t \\ \ddot{y}_c &= r\omega^2 \sin \omega t \end{aligned}\right\} \tag{5.12}$$

The momentum (or inertia force) is obtained from the velocity (or acceleration) by multiplying by the rotating crank mass m_c.

Returning to Fig 5.2 we note that the inertia forces of the piston and the crank have been successfully put into formulas so that only the characteristics of the connecting rod remain to be determined. This seems to be the most difficult part of the problem, since the motion of the rod is rather complicated. The top point of the rod describes a straight line, while the bottom point moves on a circle. All other points describe ellipses, so that the determination and subsequent integration of the inertia forces of all these points require considerable algebra. Fortunately, however, this is not necessary. If the connecting rod is replaced by another structure, having the same mass and the same center of gravity, so that the path of the center of gravity is not changed, then the total inertia force of the rod is equal to that of the new structure. This follows directly from Newton's law which states that the component of the inertia force of a body in a certain direction equals the product of its mass and the acceleration of the center of gravity in that direction.

With the aid of this relationship the problem can be easily solved by replacing the rod by two concentrated masses, one at each end, so that the center of gravity stays where it is and so that the sum of the two concentrated masses equals the total mass of the original connecting rod. This division of mass is the same as the division of the weight into two parts by placing the rod horizontally on two scales as shown in Fig. 5.4.

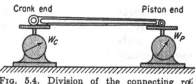

Crank end   Piston end

W_c W_p

FIG. 5.4. Division of the connecting rod, weight into its reciprocating and rotating parts.

Although the division of the connecting rod into two distinct masses leaves the center of gravity in its place and also leaves the total mass constant, the moment of inertia of the two distinct masses is different from the moment of inertia of the original connecting rod. Therefore the division of Fig. 5.4 is correct procedure for finding the inertia *forces*, but it is not exact for determining the moments of these forces, *i.e.*, the inertia *couple*.

Having thus divided the connecting-rod mass into a part moving with the piston (reciprocating) and another part moving with the crank pin (rotating), we can denote the total reciprocating and rotating masses by m_{rec} and m_{rot}. In other words, m_{rec} is the sum of the mass of the piston and of a part of the connecting rod and m_{rot} is the sum of the equivalent mass of the crank and the other part of the connecting rod.

With this notation the total vertical inertia force X (for *all* moving parts) of one cylinder is

$$X = m_{rec}\ddot{x}_p + m_{rot}\ddot{x}_c$$
$$= (m_{rec} + m_{rot})r\omega^2 \cos \omega t + m_{rec}\frac{r^2}{l}\omega^2 \cos 2\omega t \qquad (5.13)$$

and the horizontal inertia force Y is

$$Y = m_{rot}\ddot{y}_c = m_{rot}r\omega^2 \sin \omega t \qquad (5.14)$$

In words: the vertical component of the inertia force consists of two parts, a "primary part" equal to the inertia action of the combined reciprocating and rotating masses as if they were moving up and down harmonically with crank-shaft frequency and amplitude r, and a "secondary part" equal to the inertia action of a mass $\frac{r}{4l} m_{rec}$ moving up and down with twice the crank-shaft frequency and with the same amplitude r.

The horizontal or lateral component has a primary part only, *viz.*, that due to the rotating mass.

Finally we have to determine the torque of the inertia forces about the longitudinal axis O. For the purpose of finding the vertical and horizontal inertia *forces*, the connecting rod was replaced by two masses at the piston and crank pin in the manner of Fig. 5.4, and this procedure was shown to give exact results. For the inertia *torque* the result so obtained is no longer exact, but it will be correct to an acceptable degree of approximation. Thus again the complicated piston-rod-crank system is replaced by a mass m_{rec}, reciprocating according to (5.7), and a mass m_{rot} rotating uniformly around O so that it has no torque about O. The inertia torque is caused wholly by the reciprocating mass m_{rec}, and its magnitude can be deduced from Fig. 5.1b, where it was seen that the torque equals the downward piston force multiplied by $x \tan \varphi$. That the downward force in the present argument is an inertia force expressed by $-m_{rec}\ddot{x}_p$ instead of being a gas-pressure force as assumed in Fig. 5.1 does not make any difference. The distance x is

$$x = l \cos \varphi + r \cos \omega t \approx \left(l - \frac{r^2}{4l}\right) + r \cos \omega t + \frac{r^2}{4l} \cos 2\omega t$$

Further, $\tan \varphi = \dfrac{\sin \varphi}{\sqrt{1 - \sin^2 \varphi}} \approx \sin \varphi \left(1 + \frac{1}{2} \sin^2 \varphi\right)$

$$= \frac{r}{l} \sin \omega t \left(1 + \frac{r^2}{2l^2} \sin^2 \omega t\right),$$

so that the torque becomes

$$\mathbf{M} = -m_{rec}\ddot{x}_p \cdot x \tan \varphi$$

$$= -m_{rec}r\omega^2 \left(\cos \omega t + \frac{r}{l} \cos 2\omega t\right) \times \frac{r}{l} \sin \omega t \left(1 + \frac{r^2}{2l^2} \sin^2 \omega t\right)$$

$$\times \left\{\left(l - \frac{r^2}{4l}\right) + r \cos \omega t + \frac{r^2}{4l} \cos 2\omega t\right\}$$

Upon multiplying this out we disregard all terms proportional to the second or higher powers of r/l. This involves an error of the same order as that committed in passing from (5.6) to (5.7). Thus

$$\mathbf{M} = -m_{rec}\omega^2 r^2 \sin \omega t \left\{\frac{r}{2l} + \cos \omega t + \frac{3r}{2l} \cos 2\omega t\right\}$$

With the trigonometric relation

$$\sin \omega t \cos 2\omega t = \tfrac{1}{2} \sin 3\omega t - \tfrac{1}{2} \sin \omega t$$

the torque becomes finally

$$\mathbf{M} = \frac{1}{2} m_{rec}\omega^2 r^2 \left(\frac{r}{2l} \sin \omega t - \sin 2\omega t - \frac{3r}{2l} \sin 3\omega t\right) \qquad (5.15)$$

This important formula for the inertia torque (acting on the shaft in the direction of its rotation, or also on the frame about O in the opposite direction) is quite accurate for the usual type of engine where the connecting rod consists of two substantial bearings at the ends, joined by a relatively light stem. On the other hand, in a radial aircraft engine, the "master connecting rod" has a crank end carrying not only the crank-pin bearing, but $n-1$ other bearings to which the other $n-1$ connecting rods are attached. It does not seem reasonable to replace this structure by two concentrated masses at the ends, and for this case the exact connecting rod analysis, given below, is of interest.

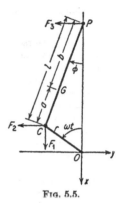

FIG. 5.5.

In Fig. 5.5 let the crank rotate in a counterclockwise direction at the uniform speed ω, and let it drag with it the connecting rod. The piston is supposed to be massless, since its inertia force is given by Eq. (5.9). The piston moreover is supposed to have no friction, so that the reaction force of the cylinder wall on the rod must be F_3. Let further F_1 and F_2 be the forces exerted by the crank pin on the rod, which moves in its prescribed manner under the influence of the three forces F. This is a case of plane motion, governed by the three equations of Newton:

In the x-direction, $\qquad F_1 = m\ddot{x}_G$
In the y-direction $\qquad F_2 + F_3 = m\ddot{y}_G$
Moments about c.g., $\quad -F_1 a \sin \varphi + F_2 a \cos \varphi - F_3 b \cos \varphi = I_G \ddot{\varphi}$

The geometry of the motion is prescribed; in particular the center of gravity moves thus:

$$x_G = x_p + (x_c - x_p)(b/l) = (x_p a/l) + (x_c b/l) \qquad \text{and} \qquad y_G = y_c b/l$$

where the subscripts c and p denote crank pin and piston, while a and b are the distances to the center of gravity G as shown in Fig. 5.5. The accelerations $\ddot{x}_p$, $\ddot{x}_c$, and $\ddot{y}_c$ are given by Eqs. (5.9), (5.11), and (5.12). The angle φ and its functions, including $\ddot{\varphi}$, are determined by Eq. (5.5). Thus the Newton equations can be solved for their three unknowns F_1, F_2, and F_3. It is noted that the first Newton equation becomes

$$F_1 = (m\ddot{x}_p a/l) + (m\ddot{x}_c b/l) = m_{rec}\ddot{x}_p + m_{rot}\ddot{x}_c$$

which leads to the result Eq. (5.13), known before. Similarly the combination $F_2 + F_3$ was designated before as Y in Eq. (5.14). Thus, for the inertia *forces* it is seen once more that the statement at the bottom of page 177 is correct. Now we wish to calculate the *torque* in the clockwise direction exerted on the shaft by the inertia of the rod. It is

$$M = -F_1 r \sin \omega t - F_2 r \cos \omega t$$

so that it is necessary to find F_2 separately by eliminating F_3 from between the last two Newton equations. This gives

$$F_2 = -m_{rot} \frac{b}{l} r\omega^2 \sin \omega t - \frac{I_G \ddot{\varphi}}{\cos \varphi} + \frac{F_1 a \sin \varphi}{l \cos \varphi}$$

In working this out by means of Eq. (5.5) we neglect all terms containing powers of r/l higher than 2. This leads to

$$F_2 = -m_{rot} \frac{b}{l} r\omega^2 \sin \omega t - \frac{I_G}{l^2} r\omega^2 \sin \omega t + \frac{1}{2} m_{rec} \frac{r}{l} r\omega^2 \sin 2\omega t$$

With this expression the inertia torque, after some trigonometry, becomes

$$\mathbf{M}_{\text{shaft}} = \frac{1}{2} m_{\text{rec}}\omega^2 r^2 \left[\frac{r}{2l} \sin \omega t - \left(1 + \frac{ab - k^2}{al} \right) \sin 2\omega t - \frac{3r}{2l} \sin 3\omega t \right] \quad (5.16)$$

in which k is the radius of gyration of the rod, defined by $mk^2 = I_G$. This result is approximate only in the sense that higher powers of r/l have been neglected; otherwise, it is exact. It differs from (5.15) only in the double-frequency term, which now depends on the moment of inertia mk^2.

Equation (5.15) is the expression for the inertia torque on the shaft of a connecting rod consisting of two concentrated masses ma/l and mb/l at distances b and a from the center of gravity. Such a rod has a radius of gyration $k^2 = ab$, and it is seen that Eq. (5.16) reduces to Eq. (5.15) if this substitution is made.

It is interesting to consider two cases of rods that have no end concentrations in order to see how (5.16) differs numerically from (5.15). First take the uniform rod, $a = b = l/2$ and $k^2 = l^2/12$. In this case the double-frequency term of (5.16) is 33 per cent greater than the term in the approximate formula (5.15). Next consider a rod with $m_{\text{rec}} = 0$, $(b = l)$, having its center of gravity at the crank pin and a certain dimension around it, which is a rough picture of the master rod of a radial aircraft engine. Assuming $k^2 = l^2/10$, we find a middle term in (5.16) which is the same as that in (5.15) if only m_{rec} is replaced by $m/10$. But, moreover, *the sign is reversed.*

The aircraft master rod of actual practice is a combination of the two cases just discussed, and the increase in moment due to the "uniform rod effect" more or less balances the decrease in moment due to the large moment of inertia of the crank end. Thus, even for so unusual a rod as that of a radial aircraft engine, the approximate result (5.15) is fairly accurate.

The torque acting on the *frame* of the engine about the shaft center O (Fig. 5.5) is found by multiplying the force F_3 by its moment arm.

$$\mathbf{M}_{\text{frame}} = F_3(l \cos \varphi + r \cos \omega t)$$

Solving for F_3 from the Newton equation, substituting it into the above expression, and working it out, neglecting higher powers of r/l, involves more algebra than it is expedient to reproduce here. The answer becomes

$$\mathbf{M}_{\text{frame}} = \frac{1}{2} m_{\text{rec}}\omega^2 r^2 \left\{ \left[\frac{(r^2 + 8l^2)(k^2 - ab)}{4ral^2} + \frac{r}{2l} \right] \sin \omega t \right.$$
$$\left. - \left[\frac{ab - k^2}{al} + 1 \right] \sin 2\omega t - \left[\frac{3r(k^2 - ab)}{4al^2} + \frac{3r}{2l} \right] \sin 3\omega t \right\} \quad (5.17)$$

Again, for the connecting rod with two concentrated ends ($k^2 = ab$) this result reduces to Eq. (5.15). Thus for the general connecting rod the inertia torques on the shaft and the frame are not equal but differ by the moment of the inertia forces of the various rod points about O. Only when the rod degenerates into two concentrated masses is this moment zero, since the two inertia forces are along the center line and along a radius, both passing through O.

5.3. Inertia Balance of Multicylinder Engines.

The unbalance or inertia forces on a single-cylinder engine are given by Eqs. (5.13) and (5.14). In these formulas the reciprocating mass m_{rec} is always positive, but the rotating mass m_{rot} can be made zero or even negative by "counterbalancing" the crank (Fig. 5.6). It is therefore possible to reduce the horizontal inertia force Y to zero, but the vertical unbalance force X always exists.† Thus a single-cylinder engine is inherently unbalanced.

Consider a two-cylinder engine with 180-deg. crank angle. Since the two cranks are opposed to each other, the two horizontal inertia forces are also in opposition and cancel each other (except for a moment about the vertical axis). Since the two pistons move against each other, the same is true for the primary vertical forces. However, the secondary vertical forces are in the same direction and add. To

FIG. 5.6. Counterbalanced crank.

understand this, it is convenient to visualize the various forces as (the horizontal projections of) rotating vectors (page 3). We shall now explain this vector method for the general case of a multicylinder engine.

In such an engine let the distance between the nth crank and the first crank be l_n and the angle between the nth crank and the first crank be α_n (the nth crank angle). In Fig. 5.7 the first crank is shown in a vertical

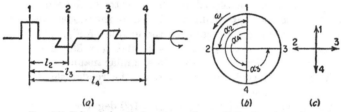

(a)   (b)   (c)

FIG. 5.7. Primary inertia forces on a four-cylinder 2-cycle engine.

position, corresponding to a maximum value of the primary vertical inertia force. The second crank is α_2 radians ahead of the first one, and consequently its vertical primary inertia force has passed through its maximum value α_2/ω sec. earlier. If the rotating vector representing the primary vertical force of the first cylinder is in its vertical position, the vector representing the second cylinder is in the position α_2, and generally the vector representing the nth cylinder is in the position α_n. The same statement is true for the primary *horizontal* inertia force.

Therefore, the crank diagram of Fig. 5.7b, regarded as a vector diagram (Fig. 5.7c), represents the primary force conditions in the engine. For example a four-cylinder engine of this type has balanced primary forces.

† A patent has been issued on a scheme whereby the connecting rod is extended beyond the crank pin so as to make W_p in Fig. 5.4 negative. In this manner M_{rec} may be made zero also. No such engine has ever been constructed on account of the large crank case required.

The secondary force vectors, however, rotate twice as fast as the crank shaft. Referring again to Fig. 5.7a, if the secondary force of crank 1 be a vertical vector, the vector of crank 2 *was* vertical at the time that crank 2 was vertical. Crank 2 has traveled α_2 radians from the vertical, and the vector of crank 2 consequently is $2\alpha_2$ radians from the vertical. The secondary-force diagram therefore is a star with the angles $2\alpha_2$, $2\alpha_3$, . . . , $2\alpha_n$ between the various vectors.

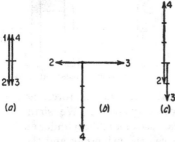

Figure 5.8a shows this diagram for the engine of Fig. 5.7.

A similar reasoning holds for the *moments* of these forces about a lateral axis. The moment of the nth inertia force about the center of the first crank shaft is that force multiplied by the moment arm l_n (Fig. 5.7a). The plane in which such a moment operates is defined by the direction of the force and the longitudinal center line of the crank shaft. Therefore,

Fig. 5.8. (a) Secondary forces, (b) primary moments, and (c) secondary moments for the four-cylinder engine of Fig. 5.7.

the moment can be represented also by a vector in the same direction as the inertia force, its length being multiplied by the proper moment arm l_n.

The primary-moment diagram of the engine of Fig. 5.7a is given in Fig. 5.8b, where $l_1 = 0$, $l_2 = l$, $l_3 = 2l$ and $l_4 = 3l$. The secondary-moment diagram (Fig. 5.8c) follows in a similar manner.

With the aid of such vector diagrams the reader should prove the following propositions:

1. A four-cylinder engine of 0, 90, 270, 180 deg. crank shaft (two-cycle engine) has balanced primary and secondary forces and also has balanced secondary moments, but the primary moments are unbalanced.

2. A four-cylinder engine of 0, 180, 180, 0 deg. crank shaft (four-cycle engine) has balanced primary forces and moments, while the secondary forces and moments are unbalanced.

3. A six-cylinder four-cycle engine (0, 120, 240, 240, 120, 0 deg.) has all forces balanced and all moments balanced.

4. An eight-cylinder in-line engine (0, 180, 90, 270, 270, 90, 180, 0 deg.) is completely balanced.

In these examples it has been tacitly assumed that all pistons are alike and are spaced at equal distances, which is the case in modern internal-combustion engines. However, the method will work just as well for unequal piston masses and unequal spacings. In fact it was for the application to large triple and quadruple expansion steam engines for ship propulsion that the theory was originally developed (Schlick's theory of balancing, about 1900).

Example: A triple expansion steam engine has pistons of which the weights are to each other as $1 : \tfrac{1}{2} : 2$. If it is desired to balance this engine for primary forces, how should the crank angles be made?

Solution: The vectors in the diagram have lengths in the required ratios. Drawing the vector of two units length vertically, as in Fig. 5.9, the equilibrium requires that the two other vectors be arranged so that their horizontal components balance and that the sum of their vertical components be two units. With the angles α and β of Fig. 5.9, we have

$$1 \cdot \sin \alpha = 1\tfrac{1}{2} \sin \beta$$
$$1 \cdot \cos \alpha + 1\tfrac{1}{2} \cos \beta = 2$$

To solve these, calculate $\cos \alpha$ from the first equation:

$$\cos \alpha = \sqrt{1 - \sin^2 \alpha} = \sqrt{1 - 2\tfrac{1}{4} \sin^2 \beta}$$

and substitute in the second one:

$$\sqrt{1 - 2\tfrac{1}{4} \sin^2 \beta} = 2 - 1\tfrac{1}{2} \cos \beta \qquad \text{FIG. 5.9.}$$

Square and simplify:

$$6 \cos \beta = 5\tfrac{1}{4}$$

from which $\cos \beta = 0.88$ and $\beta = 28$ deg

Further, $\cos \alpha = 2 - \tfrac{3}{2} \times 0.88 = 0.68$ and $\alpha = 47$ deg.

It is possible to express the results of these vector diagrams in simple mathematical language. The requirement for balanced primary forces is that the geometrical sum of all the vectors of Fig. 5.7c be zero. If this be so, the sum of their horizontal projections as well as the sum of their vertical projections must be zero or

$$\sum_n \sin \alpha_n = 0 \qquad \text{and} \qquad \sum_n \cos \alpha_n = 0 \tag{5.18}$$

Similarly, the conditions for balanced secondary forces are

$$\sum_n \sin 2\alpha_n = 0 \qquad \text{and} \qquad \sum_n \cos 2\alpha_n = 0 \tag{5.19}$$

For the primary moments

$$\sum_n l_n \sin \alpha_n = 0 \qquad \text{and} \qquad \sum_n l_n \cos \alpha_n = 0 \tag{5.20}$$

For the secondary moments

$$\sum_n l_n \sin 2\alpha_n = 0 \qquad \text{and} \qquad \sum_n l_n \cos 2\alpha_n = 0 \tag{5.21}$$

All these formulas are true only for *equal* piston masses.

For the four-cylinder engine of Fig. 5.7 we have $\alpha_1 = 0$, $\alpha_2 = 90$,

$\alpha_3 = 270$, $\alpha_4 = 0$ deg., and consequently Eqs. (5.18) become

$$0 + 1 - 1 + 0 = 0 \quad \text{and} \quad 1 + 0 + 0 - 1 = 0$$

so that the primary forces are balanced.

But Eqs. (5.20) become

$$0 \cdot 0 + 1 \cdot 1 - 2 \cdot 1 + 3 \cdot 0 = 1 - 2 \neq 0$$
$$0 \cdot 1 + 1 \cdot 0 + 2 \cdot 0 + 3 \cdot 1 - 3 \neq 0$$

so that the primary moments are unbalanced.

Thus we are able to test the inertia balance of any engine design by using either the formulas (5.18) to (5.21) or the vector diagrams.

It may be well to recall that in this analysis the engine has been considered to be a "rigid body." This is usually the case in automobile and aircraft engines where all cylinders are cast in a single block, but in marine engines the cylinders sometimes are mounted separately. Then the forces or moments of two cylinders may be in opposition to each other and not move the engine as a whole, but they may move the two cylinders against each other elastically. The problem becomes extremely complicated, and is not of sufficient practical importance to merit much time for its solution. In this connection the reader is referred to the analogous problem in rotating machinery discussed in Sec. 6.5.

5.4. Natural Frequencies of Torsional Vibration. The shafting of an internal-combustion engine with all its cranks, pistons, flywheel, and driven machinery is too complicated a structure to attempt an exact determination of its torsional natural frequency. It is necessary first to simplify or "idealize" the machine to some extent by replacing the pistons, etc., by equivalent disks of the same moment of inertia and by replacing the crank throws by equivalent pieces of straight shaft of the same torsional flexibility. In other words, the machine has to be reduced to the shape of Fig. 5.12a. This process is at best approximate.

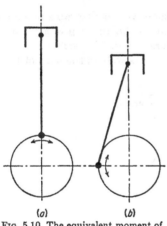

(a)   *(b)*

FIG. 5.10. The equivalent moment of inertia of a piston varies with its position.

First consider the equivalent moment of inertia of each crank mechanism. The moment of inertia I_{rot} of the purely rotating parts offers no difficulty, but it is not quite evident what should be done with the reciprocating weight. In Fig. 5.10a and b the piston is shown in two positions. Imagine the crank shaft to be non-rotating but to be executing small torsional oscillations. In Fig. 5.10a this takes place without any motion of the piston, but in Fig. 5.10b the motion (and acceleration) of the piston practically equals that of the crank pin. The equivalent inertia in position a is zero whereas in position b it is $m_{rec}r^2$. Thus while the crank shaft is rotating, the total equivalent moment of inertia of the crank mechanism varies between I_{rot} and $I_{rot} + m_{rec}r^2$, with an average value of $I_{rot} + \frac{1}{2}m_{rec}r^2$. The system with variable inertia (page 350) is now replaced by one of constant inertia I, where

$$I = I_{rot} + \frac{1}{2}m_{rec}r^2 \qquad (5.22)$$

Next consider the idealization of a crank throw into a piece of ordinary shafting of the same torsional flexibility. This is physically quite permissible, but the calculation of the flexibility is a very difficult matter.

In Fig. 5.11a it is seen that, if the main shaft is subjected to twist, the crank webs W are subjected to bending moments and the crank pin P is in twist. It is possible to calculate the angle of twist produced by a certain torque by applying to the webs and pin the usual "beam" formulas for bending and twist. However, that will give very inaccurate results because these formulas are true only for long and slender beams and will lead to serious errors if applied to short stubs of a width and thickness [nearly as great as the length. Moreover, it can be seen that

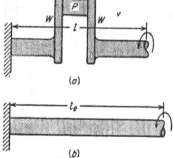

FIG. 5.11. A crank of length l is replaced by a piece of uniform shaft of length l_e having the same torsional flexibility.

the torque in Fig. 5.11a will cause not only a twisting rotation of the free end but also a sidewise displacement of it on account of the bending in the webs. In an actual machine the sidewise motion is impeded by the main bearings and the torsional stiffness of the crank shaft is increased by these bearings, especially if their clearance is small.

Experiments have been carried out on a number of crank shafts of large, slow-speed engines showing that the "equivalent length" l_e of Fig. 5.11b (i.e., the length of ordinary shaft having the same torsional stiffness) is nearly equal to the actual length l. The variation is between

$$0.95l < l_e < 1.10l$$

the lower value being for small throws and stiff webs and the higher value for large throws and thin flexible webs. In all tests the diameter of the main shaft was equal to that of the crank pin.

In cases where the crank pin has a different diameter (usually smaller) from that of the main bearing journal, the throw is replaced by a straight shaft of two different diameters; the point where the diameter jumps from one value to the other being located at the center of the crank web. For high-speed, light-weight engines, particularly aircraft engines, where the webs are no longer rectangular blocks but have shaved-off corners to save weight, the equivalent stiffness is very much smaller than would follow from the above simple calculation. In extreme cases the stiffness may be as low as 50 per cent of the value so calculated. The best guide is then a comparison of calculation and experiment of a number of previous crank shafts of similar characteristics.

In case one part of the system is connected to the other part through gears, it is convenient to reduce everything to one speed. As was explained on page 30, this is accomplished by eliminating the gears and multiplying the moments of inertia and the spring constants of the *fast* rotating parts by n^2 where $n > 1$ is the gear ratio.

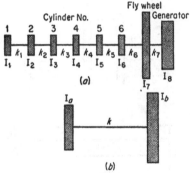

FIG. 5.12. The engine is replaced by a two-mass system for the purpose of an approximate calculation of the lowest natural frequency.

Let Fig. 5.12a represent the idealized machine, in this case a six-cylinder Diesel engine driving a flywheel and an electric generator. There are eight degrees of freedom. It is theoretically possible to find the eight natural frequencies by the method of Chap. 4, using a determinant with eight rows and eight columns and an eighth degree equation in ω_n^2. This is obviously undesirable from the standpoint of time consumption.

Instead, we use a method of successive approximations starting with a rough first guess at the frequency. Such a guess for the lowest natural frequency can be made by replacing Fig. 5.12a by Fig. 5.12b, where I_a is the inertia of all six cylinders combined and I_b that of the flywheel and generator rotor combined. The frequency of the latter system is [Eqs. (2.4) and (2.10)]

$$\omega = \sqrt{k \frac{I_a + I_b}{I_a I_b}}$$

and is an approximation to the lowest frequency of Fig. 5.12a. In the reduction of Fig. 5.12a to 5.12b the judgment of the calculator enters. With some experience the frequency can be estimated to within 10 per cent.

The rough value ω_1, thus obtained, serves as the basis for the following method of calculation due to *Holzer*. Assume the whole system to be in a torsional oscillation with the frequency ω_1. If ω_1 were a natural frequency this could occur without any external torque on the system (a *free* vibration). If ω_1 is not a natural frequency, this can occur only if at some point of the system an external torque of frequency ω_1 is acting. We have then a *forced* vibration. Assume arbitrarily that the angular amplitude of the first disk in Fig. 5.12a is 1 radian. The torque necessary to make that disk vibrate is

$$I_1\omega_1^2 \sin \omega_1 t$$

This torque can come only from the shaft to the right of I_1. If that shaft has a torsional spring constant k_1, its angle of twist is $\dfrac{I_1\omega_1^2}{k_1} \sin \omega_1 t$ with a maximum value $\dfrac{I_1\omega_1^2}{k_1}$. Since the amplitude of disk I_1 is 1 radian and the shaft twists $\dfrac{I_1\omega_1^2}{k_1}$ radians, disk I_2 must vibrate with an amplitude of $1 - \dfrac{I_1\omega_1^2}{k_1}$ radians. This requires a torque of amplitude

$$I_2\omega_1^2 \left(1 - \frac{I_1\omega_1^2}{k_1}\right)$$

This torque is furnished by the difference in the shaft torques left and right, and, since the torque in k_1 is known, the torque in k_2 can be calculated. From this we find the angle of twist of k_2, the angle of I_3, etc., finally arriving at the last disk I_8. But there is no shaft to the right of I_8 to furnish the necessary torque. In order to make the system vibrate as described, it is necessary to apply to I_8 an external torque T_{ext} of the value found by the calculation. Only when ω_1 happened to be a natural frequency would this T_{ext} be found equal to zero. The magnitude and sign of T_{ext} therefore are a measure of how far ω_1 is removed from the natural frequency. A number of such calculations with different values of ω_1 must be made, until finally the remainder torque T_{ext} is practically zero. The advantage of this method is that it gives not only the natural frequency but also the complete shape of the natural mode of vibration, and this will be needed for the calculation of the work input by the non-uniformities of the cylinder torques (page 200).

The actual course of the calculations can best be illustrated by a definite example, as follows.

5.5. Numerical Example. We take as an example a recent light-weight, high-speed Diesel engine driving an electric generator (Fig. 5.13).

Some of the characteristic constants, other than those shown in Fig. 5.13, are:

Engine: 4-cycle V-8 Diesel
V-angle: 60 deg.
Crank shaft: 0, 180, 180, 0 deg with firing order 1 3 4 2
Rated power: 50 hp. per cylinder at 2,000 r.p.m.
                       100 hp. for the engine
Full-load torque: 1,580 in.-lb. per cylinder
Generator: 250 kw; normal speed 2,000 r.p.m.

The inertias and flexibilities of Fig. 5.13 were calculated from the drawings in the manner described before. The only new thing here is the viscous damper, consisting of a housing in which a loose flywheel can turn freely. The only coupling between the damper flywheel and its surrounding housing is through the viscous friction of oil or silicone fluid filling the entire space inside the housing. This type of damper will be described in detail on page 211, where it will be shown that the equivalent inertia of the damper assembly is the sum of the entire housing inertia plus half the inertia of the damper flywheel.

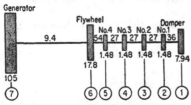

FIG. 5.13. V-8 type, 4-cycle Diesel driving a generator, with a viscous Houde-type damper. All inertias are shown as lb. in. sec.²; all flexibilities k are shown as millions of in. lb. per radian.

The value 7.94 lb. in. sec.² shown in Fig. 5.13 was so determined. Each individual crank throw carries two connecting rods and two pistons in a V-type engine. The value 1.48 lb. in. sec.² of Fig. 5.13 is made up of the inertia of the crank throw proper with its attached counterweight, the rotating part of the weight of two connecting rods plus *half* the reciprocating weights of two rods and two pistons, all at crank radius.

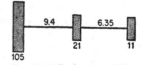

FIG. 5.14. Reduction of Fig. 5.13 to a three-mass system for an estimate of the first frequency.

First we must guess at the first frequency of Fig. 5.13. Noticing that the crank masses are much smaller than the rest, we throw two cranks with the flywheel, two others with the damper, and lump the flexibilities of the crank shaft into one:

$$\frac{1}{k} = \frac{1}{54} + \frac{1}{27} + \frac{1}{27} + \frac{1}{27} + \frac{1}{36} = \frac{7}{54} + \frac{1}{36} = 0.158 = \frac{1}{6.35}$$

This leads to the system of Fig. 5.14 with three disks. Applying the appropriate formula of page 430, we find

$$\omega_1^2 = 300,000 \qquad \omega_2^2 = 1,120,000$$

and we proceed to set up a Holzer table with the first value.

FIRST MODE, FIRST TRY, $\omega^2 = 300,000$

No.	I (1)	$I\omega^2/10^6$ (2)	β (3)	$I\omega^2\beta/10^6$ (4)	$\Sigma/10^6$ (5)	$k/10^6$ (6)	Σ/k (7)
1	7.94	2.38	1.000	2.38	2.38	36	0.066
2	1.48	0.444	0.934	0.42	2.80	27	0.104
3	1.48	0.444	0.830	0.37	3.17	27	0.118
4	1.48	0.444	0.712	0.32	3.49	27	0.129
5	1.48	0.444	0.583	0.26	3.75	54	0.069
6	17.8	5.34	0.514	2.75	6.50	9.4	0.691
7	105.0	31.5	−0.177	−5.59	+0.91		

The physical meaning of the various columns in this table is as follows: in column 2 is the inertia torque of each disk for an amplitude of 1 radian at the frequency shown at the head of the table; in column 3 is the angular amplitude β of each disk; in column 4 is the inertia torque of each disk at the amplitude β; the fifth or Σ-column gives the value of the shaft torque beyond the disk in question; the sixth shows the flexibilities; and column 7 is the windup angle in each shaft.

We start by filling in all the numbers in the first two columns and in the sixth or k-column. Then we start on the first line and proceed to the right. The shaft torque to the left of the first, or damper, disk is the damper torque itself, so that we just copy 2.38 in the Σ-column. The windup angle is found by $2.38/36 = 0.066$. Now we subtract 0.066 from $\beta = 1.000$, finding $\beta = 0.934$ for disk 2, which is the first cylinder. Proceeding farther to the right in line 2, we have no trouble until we come to the Σ-column. There we add 0.42 to 2.38 to give 2.80 in the Σ-column, or physically the shaft torque between disks 2 and 3 is the sum of the inertia torques of disks 1 and 2. The reader should now follow the calculations step by step in the entire table, never losing sight of the physical meaning of each number. The last value we find is in the Σ-column, and 0.91×10^6 is the sum of all the inertia torques of all seven disks, and it also is the torque in the (non-existing) shaft to the left of the flywheel, necessary to maintain a forced vibration at $\omega^2 = 300,000$ and $\beta_1 = 1.000$.

With a remainder torque 0.91, and not equal to zero, we have not hit on the natural frequency. Before blindly trying to construct another Holzer table with some other value of ω^2, we reason a bit and see what would happen to the Holzer table for a small value of ω^2, close to zero. The values in the second column are small and positive. The figures in the first line become small, and hence β_2 will be almost 1.000, slightly below it. Reasoning on, we conclude that the remainder torque (the last Σ-value) must be positive and small. Of course, for $\omega^2 = 0$ all values in columns 2, 4, 5, and 7 are zero. Now we sketch Fig. 5.15, in which the remainder torque is plotted against the frequency ω^2. All we know about the figure is that the curve passes through the origin, that it is

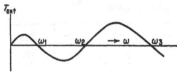

FIG. 5.15. Behavior of the remainder torque of column 5 of the Holzer calculation.

positive for small ω^2, and that it should be zero for natural frequencies. But this scanty knowledge is enough to be very useful. In this case we are aiming at the first, or lowest, natural frequency, and in our first try we ended up with a positive remainder torque. From Fig. 5.15 we conclude that our guess for ω^2 was too small (if it had been too large, the torque would have been negative). So our second try is made somewhat larger than the first one.

FIRST MODE, SECOND TRY, $\omega^2 = 320,000$

No.	I	$I\omega^2/10^6$	β	$I\omega^2\beta/10^6$	$\Sigma/10^6$	$k/10^6$	Σ/k
	(1)	(2)	(3)	(4)	(5)	(6)	(7)
1	7.94	2.54	1.000	2.54	2.54	36	0.071
2	1.48	0.473	0.929	0.44	2.98	27	0.110
3	1.48	0.473	0.819	0.39	3.37	27	0.125
4	1.48	0.473	0.694	0.33	3.70	27	0.137
5	1.48	0.473	0.557	0.26	3.96	54	0.073
6	17.8	5.70	0.484	2.75	6.71	9.4	0.715
7	105.0	33.6	−0.231	−7.75	−1.04		

Now we come out with a negative remainder torque, and from Fig. 5.15 we conclude that ω^2 is too large for the first natural frequency. We have two points on the curve, reasonably close together, and since any short piece of curve is nearly straight, we interpolate linearly and take for our third guess

$$\omega^2 = 300,000 + \frac{0.91}{0.91 + 1.04}(320,000 - 300,000) \approx 310,000$$

This time the remainder torque is insignificantly small, so that we call it good.

FIRST MODE, THIRD AND FINAL TRY, $\omega^2 = 310,000$, $\omega = 552$, V.P.M. $= 5,300$

No.	I	$I\omega^2/10^6$	β	$I\omega^2\beta/10^6$	$\Sigma I\omega^2\beta$	$k/10^6$	Σ/k
	(1)	(2)	(3)	(4)	(5)	(6)	(7)
1	7.94	2.46	1.000	2.46	2.46	36	0.068
2	1.48	0.459	0.932	0.43	2.89	27	0.106
3	1.48	0.459	0.826	0.38	3.27	27	0.121
4	1.48	0.459	0.705	0.32	3.59	27	0.133
5	1.48	0.459	0.572	0.26	3.85	54	0.071
6	17.8	5.51	0.501	2.76	6.61	9.4	0.703
7	105.0	32.6	−0.202	−6.60	+0.01		

Proceeding to the second mode, we start with the higher of the two frequencies of Fig. 5.14:

SECOND MODE, FIRST TRY, $\omega^2 = 1,120,000$

No.	I	$I\omega^2/10^6$	β	$I\omega^2\beta/10^6$	$\Sigma/10^6$	$k/10^6$	Σ/k
	(1)	(2)	(3)	(4)	(5)	(6)	(7)
1	7.94	8.90	1.000	8.90	8.90	36	0.246
2	1.48	1.65	0.754	1.24	10.14	27	0.375
3	1.48	1.65	0.379	0.63	10.77	27	0.399
4	1.48	1.65	−0.020	− 0.03	10.74	27	0.398
5	1.48	1.65	−0.418	− 0.69	10.05	54	0.188
6	17.8	19.9	−0.606	−12.05	−2.00	9.4	−0.214
7	105.0	117.0	−0.392	−44.6	−46.6		

The remainder torque is quite large and negative. Looking at Fig. 5.15 and remembering that we are aiming at the second frequency, we conclude that our choice for ω^2 was too small. Two more tries are made, with the following result:

$$\omega^2 = 1,200,000 \ldots \text{ remainder torque } -30.4$$
$$\omega^2 = 1,300,000 \ldots \text{ remainder torque } - 2.09$$

These are still too small. Plotting the three points so far available, we see that they are not on a straight line. Passing a smooth curve through them and extrapolating it (a procedure much less satisfactory than interpolating), we come to $\omega^2 = 1,310,000$.

Figure 5.16 shows the shapes of the first two natural modes of motion, where the ordinate indicates the angle of swing β of each mass. It is seen that the nth mode of motion has n nodes. Another property of the curves, useful for checking or for rough visualization before calculating, is that the total angular momentum of each curve must be zero or $\Sigma I\beta = 0$. In Fig. 5.16 the flywheel is very large; hence in the first

SECOND MODE, FOURTH AND FINAL TRY, $\omega^2 = 1,310,000$; $\omega = 1,145$;
V.P.M. $= 10,950$

No.	I	$I\omega^2/10^6$	β	$I\omega^2\beta/10^6$	$\Sigma/10^6$	$k/10^6$	Σ/k
	(1)	(2)	(3)	(4)	(5)	(6)	(7)
1	7.94	10.40	1.000	10.40	10.40	36	0.288
2	1.48	1.94	0.712	1.38	11.78	27	0.435
3	1.48	1.94	0.277	0.54	12.32	27	0.455
4	1.48	1.94	−0.178	−0.35	11.97	27	0.444
5	1.48	1.94	−0.622	−1.21	10.76	54	0.199
6	17.8	23.3	−0.821	−19.10	−8.34	9.4	−0.885
7	105.0	138	+0.064	+8.85	+0.51		

mode its inertia 105 times its amplitude 0.20 is equal and opposite to the sum of the $I\beta$ of all other masses.

The Holzer method does not take advantage of the fact that the various cylinders of the usual engine are identical. For our particular example of four cranks out of 7 masses total, it does not make too much difference, but there are engines with eight cranks out of a total of 9 or 10 masses, and then very considerable work can be saved by treating the engine as a whole, a method first used by F. M. Lewis. With this procedure the flexibilities and inertias of the various cranks are uniformly distributed along the entire length of the engine, which then becomes a "shaft," or "beam," in torsional vibration, as discussed on page 138.

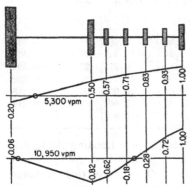

FIG. 5.16. The shapes or "modes" of the two lowest natural frequencies of the system in Fig. 5.13.

The four cyclinder masses together are $4 \times 1.48 = 5.92$, and the flexibilities were combined in passing from Fig. 5.13 to Fig. 5.17 to give 6.35×10^6 in. lb./rad. This leads to the system shown in Fig. 5.17. The engine, or "shaft" is governed by the differential equation (4.22) (page 138) and its solution [Eq. (4.25)], which can be written in the form

$$\beta(x) = A \cos\left(x \sqrt{\frac{\mu_1\omega^2}{GI_p}} + \alpha\right)$$

where A is an amplitude constant and α a phase angle. These two constants A and α must be found for each individual case to fit the boundary conditions. With the notation $I = \mu_1 l$ for the total moment of inertia of the entire engine and $K = GI_p/l$ for the stiffness of the entire length of the engine, this last result can be written as

$$\beta(x) = A \cos\left(\omega\sqrt{\frac{I}{K}} \cdot \frac{x}{l} + \alpha\right) \qquad (5.23)$$

The combination

$$\omega\sqrt{\frac{I}{K}} = \theta \qquad (5.24)$$

can be recognized as the number of radians of cosine wave along the engine (from $x = 0$ to $x = l$), and this quantity can be visualized and estimated before calculations are started.

The torque at any point along the engine shaft is

$$\mathbf{M} = GI_p\frac{d\beta}{dx} = A\omega\sqrt{IK}\sin\left(\theta\frac{x}{l} + \alpha\right) \qquad (5.25)$$

or in particular, at the two ends of the shaft,

$$\begin{array}{lll}
\text{Beginning } x = 0 & \mathbf{M}_{x=0} = A\omega\sqrt{IK}\sin\alpha \\
\text{End} \quad\quad x = l & \mathbf{M}_{x=l} = A\omega\sqrt{IK}\sin(\theta + \alpha)
\end{array} \Biggr\} \quad (5.25a, b)$$

With the three formulas (5.23), (5.24), and (5.25) the calculation can be carried out. For example, take the first mode with $\omega^2 = 310{,}000$ and $\omega = 552$. The engine starts at the left with a concentrated inertia, and we handle this as in the Holzer method. $\beta_1 = 1.000$. The inertia torque

$$\mathbf{M}_1 = I\omega^2\beta = 7.94 \times 310{,}000$$
$$\times 1.000 = 2.46 \times 10^6$$

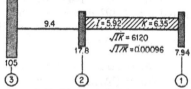

Fig. 5.17. The system of Fig. 5.13 in which the engine masses are uniformly distributed along the engine length, for a frequency calculation by Lewis's method

This torque must be carried by the engine shaft; hence it is equal to $(\mathbf{M}_{12})_1$, which we read, "$\mathbf{M}$ one-two at one." Now, by Eq. (5.23),

$$(\beta_{12})_1 = A\cos\alpha = 1.000 \qquad (5.23a)$$

and, by Eq. (5.25a),

$$(\mathbf{M}_{12})_1 = A\omega\sqrt{IK}\sin\alpha = 2.46 \times 10^6$$

Dividing the last two equations,

$$\tan\alpha = \frac{2.46 \times 10^6}{\omega\sqrt{IK}} = \frac{2.46 \times 10^6}{552 \times 6{,}120} = 0.730$$
$$\therefore \alpha = 36.1 \text{ deg.}$$

Now, by Eq. (5.24),

$$\Theta = \omega \sqrt{\frac{I}{K}} = 552 \times 0.00096 = 0.530 \text{ radian} = 30.4 \text{ deg.}$$

and $\Theta + \alpha = 30.4 + 36.1 = 66.5$ deg. This means that the β-curve along the engine is a cosine wave starting at 36.1 deg. and continuing till 66.5 deg. Now, from Eq. (5.23a),

$$A = \frac{1.000}{\cos \alpha} = \frac{1.000}{0.808} = 1.24$$

which means that the maximum height of our cosine wave is 1.24 (to the right of mass 1 in Fig. 5.17), that it has reduced to $1.24 \cos \alpha = 1.000$ at mass 1, and that it is at mass 2, by Eq. (5.23),

$$\beta_2 = A \cos (\Theta + \alpha) = 1.24 \cos 66.5° = 0.497$$

The torque in the engine at the 2-end is

$$(M_{12})_2 = A\omega \sqrt{IK} \sin (\Theta + \alpha) = 1.24 \times 552 \times 6{,}120 \times 0.917$$
$$= 3.84 \times 10^6$$

Now we are through with the engine; the rest of the calculation follows the Holzer pattern. Note that the last two numerical results check those in the Holzer table of page 191; β_6 at the flywheel there was 0.501, as compared with 0.497 just found here. The torque in the shaft just right of flywheel in the Holzer table was 3.85, while here we just found 3.84. We continue (see Fig. 5.17).

$$M_2 = I\omega^2\beta = 17.8 \times 0.310 \times 0.497 \times 10^6 = 2.76 \times 10^6$$
$$M_{23} = (M_{12})_2 + M_2 = 3.84 + 2.76 = 6.60 \times 10^6$$
$$\beta_{23} = \frac{M_{23}}{k_{23}} = 6.60/9.4 = 0.703$$
$$\beta_3 = \beta_2 - \beta_{23} = 0.497 - 0.703 = -0.206$$
$$M_3 = I\omega^2\beta = 105 \times 0.310 \times -0.206 \times 10^6 = -6.70 \times 10^6$$
$$M_{\text{remainder}} = M_{23} + M_3 = -0.10 \times 10^6, \text{ small}$$

For the purpose of showing the method more clearly, the calculation for the second mode is now given, without running comment. The data are those of Fig. 5.17, and we use the three equations (5.23), (5.24), and (5.25).

An advantage of the Lewis method is that it often enables us to make a quick and surprisingly close guess of the frequency. This is done by sketching in the curves of Fig. 5.16, before any calculation, just looking at the relative values of the flexibilities and bearing in mind that the angular moment must be balanced, $\Sigma I\beta = 0$. From the curves so sketched, we deduce the approximate value of Θ, that is, the number of radians worth of cosine wave along the engine length, and from it, by Eq. (5.24), find the estimated frequency.

LEWIS CALCULATION, SECOND MODE, $\omega^2 = 1.30 \times 10^6$, $\omega = 1,140$

$\beta_1 = 1.000 \qquad M_1 = 7.94 \times 1.30 \times 1.000 \times 10^6 = 10.30 \times 10^6 = (M_{12})_1$

$\tan \alpha = \dfrac{10.30 \times 10^6}{1,140 \times 6,120} = 1.478 \qquad \alpha = 55.9 \text{ deg.}$

$\Theta = 1,140 \times 0.00096 = 1.093 = 62.6 \text{ deg.}$

$\alpha + \Theta = 118.5 \text{ deg.} = 90 + 28.5 \text{ deg.}$

$A = \dfrac{1}{\cos 55.9} = \dfrac{1}{0.561} = 1.78$

$\beta_2 = 1.78 \cos 118.5 = -1.78 \sin 28.5 = -0.850$

$(M_{12})_2 = 1.78 \times 1,140 \times 6,120 \times 0.879 = 10.92 \times 10^6$

$M_2 = 17.8 \times 1.30 \times 0.850 \times 10^6 = -19.70 \times 10^6$

$M_{23} = 10.92 - 19.70 = -8.78 \times 10^6$

$\beta_{23} = -\dfrac{8.78}{9.4} = -0.935 \qquad \beta_3 = +0.085$

$M_3 = 105 \times 1.30 \times 0.085 \times 10^6 = +11.6 \times 10^6$

$M_{\text{remainder}} = -8.78 + 11.6 = 2.8 \times 10^6$, small

Still another method, due to F. P. Porter, is used by several engine manufacturers and consists in replacing the entire engine by an "equivalent flywheel" I_{equiv}. The torque exerted by the entire engine on the rest of the system is expressed by formula (5.25), above. If the engine were to consist of a single flywheel I_{equiv}, oscillating at the amplitude of the end of the engine, the torque would be

$$I_{\text{equiv}}\omega^2\beta(l) = I_{\text{equiv}}\omega^2 \cos \Theta$$

Equating this to the torque, Eq. (5.25), of the actual engine and considering Eq. (5.24) we get

$$I_{\text{equiv}} = I \frac{\tan \Theta}{\Theta} \qquad (5.26)$$

Thus the engine of actual inertia I with a flexible crank shaft acts as a solid flywheel (without flexibility) of inertia I_{equiv} at the assumed frequency determined by Θ. The rest of the calculation follows essentially Holzer's pattern.

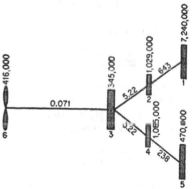

FIG. 5.18. Ship drive consisting of high-speed turbines 1 and 5, double reduction-gear drives 2, 3, 4, and propeller 6.

The Holzer method can be applied conveniently to the calculation of the frequencies of *branched* systems, such as that in Fig. 5.18, which shows the main drive of a ship built in 1940 for the U.S. Maritime Commission.   The disks 1 and 5 represent the inertia of a low-pressure and a high-pressure steam turbine, running at 7,980 r.p.m.   The disks 2 and 4 are intermediate gears running at 730 r.p.m., while 3 is the main gear, running at 85 r.p.m. and coupled to the propeller 6.   The inertias shown are in lb. in. sec.[2] and are already multiplied with the squares of their speed ratios (page 30).   The flexibilities shown must be multiplied by 10^9 to measure in in. lb./radian and likewise have been corrected for speed. To find the lowest natural frequency we notice that the engine shafting is stiff in comparison with the drive shaft.   Thus for a first estimate all turbine masses are lumped at the main gear and

$$\omega^2 \approx \frac{k}{I} = \frac{0.071 \times 10^9}{416,000} = 170$$

A Holzer trial has shown this value to be low; and, with a final value $\omega^2 = 176$, the last calculation proceeds as follows:

$$\omega^2 = 176, \qquad \beta_1 = 1.000, \qquad M_1 = I_1\omega^2\beta_1 = 1.275 \times 10^9$$
$$\beta_{12} = \frac{M_1}{k_1} = \frac{1.275}{643} = 0.002, \qquad \beta_2 = 0.998$$
$$M_2 = I_2\omega^2\beta_2 = 0.181, \qquad M_{23} = M_1 + M_2 = 1.456$$
$$\beta_{23} = \frac{M_{23}}{k_{23}} = 0.279, \qquad \beta_3 = \beta_2 - 0.279 = 0.719$$

Now we do not know the amplitudes in the 3-4-5 branch, having once assumed $\beta_1 = 1.000$.   Nevertheless we start fresh with the assumption $\beta_5 = 1.000$ and work back.

$$\beta_5 = 1.000, \qquad M_5 = 0.083, \qquad \beta_{45} = 0.000, \qquad \beta_4 = 1.000$$
$$M_4 = 0.191, \qquad M_{34} = 0.274, \qquad \beta_{34} = 0.053, \qquad \beta_3 = 0.947$$

It is clear that β_3 cannot at the same time have an amplitude of 0.719 and of 0.947.   It is possible to make the last value come out equal to 0.719 simply by multiplying all figures in the last two lines by the ratio 0.719/ 0.947 = 0.760.   Then these lines become

$$\beta_5 = 0.760, \qquad M_5 = 0.063, \qquad \beta_{45} = 0.000, \qquad \beta_4 = 0.760$$
$$M_4 = 0.145, \qquad M_{34} = 0.205, \qquad \beta_{34} = 0.04, \qquad \beta_3 = 0.719$$

Proceeding with the main gear 3, it is seen that not only its own inertia torque $M_3 = I_3\omega^2\beta_3 = 0.044$ is acting on it, but the torques M_{23} and M_{34} from the two branches as well.   Thus the torque entering the propeller shaft is

$$M_{36} = 0.044 + 1.456 + 0.205 = 1.705$$

Further,

$$\beta_{36} = 24.01, \qquad \beta_6 = -23.29, \qquad M_6 = -1.705$$

Remainder, 0

In a similar way the reader should find the second mode of motion of this system, which consists primarily of one turbine swinging against the other one.   This leads to a frequency $\omega^2 = 1,929$, and an elastic curve

$$\beta_1 = 1.000, \qquad \beta_2 = 0.978, \qquad \beta_3 = -2.064, \qquad \beta_4 = -4.87$$
$$\beta_5 = -4.89, \qquad \beta_6 = +0.200$$

In carrying out this calculation it will be found that the last result comes out to be the small difference between two large numbers, which is very inaccurate.   Therefore β_6 is calculated better by means of Eq. (2.26),

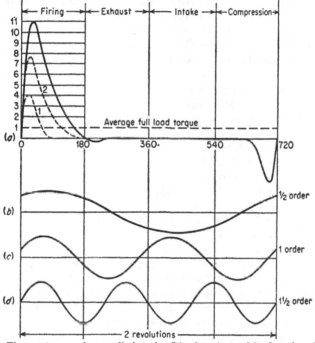

Fig. 5.19. The gas torque of one cylinder of a Diesel engine and its first three harmonic components.

page 45, considering the propeller and its shaft to be excited at $\omega^2 = 1.929$ by a motion $\beta_3 = -2.064$, which is known accurately.

5.6. Torque Analysis.   Since the torsional vibrations in the crank shaft are excited by the non-uniformities in the driving torque we proceed to an examination of the properties of this torque.   We have seen in Sec. 5.2 that it is made up of two parts, one due to cylinder pressure and the other due to inertia.

In Fig. 5.19a the cylinder-pressure torque of a four-cycle Diesel engine is shown as a function of the crank angle.   At the four dead-center positions during the two revolutions of a firing cycle the torque is zero.

When the engine is operated at partial load by a reduced injection of fuel, the curve is changed only in the firing quarter cycle. The dotted lines 1 and 2 indicate the shapes for zero and half load. At zero load the pressure during the firing period is equal to that during the compression period, so that even when there is no average torque at all there are alternating torques of considerable amplitude.

It is seen that the average torque delivered by the cylinder is only a small fraction of the maximum torque which occurs during the firing period. The fact that the torque is so irregular constitutes one of the inherent disadvantages of the reciprocating engine as compared with the turbine where the torque curve is a straight horizontal line.

It is possible to break up Fig. 5.19a into its harmonic components as explained on page 17, and as an illustration the first three harmonics are shown. They are known as the harmonic components of the order $\frac{1}{2}$, 1, and $1\frac{1}{2}$ because they show as many full sine waves *per revolution* of the engine. In two-cycle engines and in steam engines, only harmonics of integer orders occur. It is only in the four-cycle internal-combustion engine that we have half-order harmonics due to the fact that the torque curve is periodic with a firing cycle, *i.e.*, with two revolutions.

It is seen that the 1- and $1\frac{1}{2}$-order curves add up to a positive result near $\omega t = 45$ deg. and to a negative result near $720 - 45$ deg., while in a broad range near $\omega t = 360$ deg. they cancel each other approximately. Thus the three harmonic curves added together give a rough approximation of the actual torque curve, but many more harmonics are required to show the torque curve in all its detail.

The most complete and useful harmonic analyses were made by F. P. Porter in a paper entitled Harmonic Coefficients of Engine Torque Curves, published in the *Transactions of the A.S.M.E.*, 1943, page $A33$. In that paper, which is too long to reproduce here, analyses are given for eight widely different types of engine (slow and fast, Diesel and spark plug, two- and four-cycle) so that one of the eight prototype engines of Porter is always sufficiently close for practical purposes to any engine that may come up.

The set of harmonics most closely resembling those of our lightweight high-speed medium-sized Diesel engine (page 187) is designated by Porter as set P2. The most significant figures from his table are reproduced in the table at the top of page 199, trimmed down to three decimal places.

The last column $\sqrt{a^2 + b^2}$ continues on as follows:

Harmonic	4	$4\frac{1}{2}$	5	$5\frac{1}{2}$	6	$6\frac{1}{2}$	7	$7\frac{1}{2}$	8	$8\frac{1}{2}$	9
$\sqrt{a_n^2 + b_n^2}$	18.4	14.0	10.8	8.4	6.3	4.5	3.2	2.3	1.7	1.0	0.6

The original table gives all the a_n and b_n separately, up to the eighteenth harmonic for eight different values of mean indicated pressure.

M.I.P. (MEAN INDICATED PRESSURE) = 140 LB./IN.²

Harmonic	a_n	b_n	$\sqrt{a_n^2 + b_n^2}$
0	....	22.2	22.2
½	....	....	47.7
1	47.5	(20.1)	51.6
1½	....	....	49.6
2	42.3	(−4.1)	42.5
2½	....	....	34.9
3	27.5	(−8.3)	28.7
3½	....	....	23.4

all to five decimal places accuracy, which is far more than needed for practice. The a_n-coefficient is the nth component starting as a sine wave from top dead center; b_n is the nth component starting as a cosine wave; and $\sqrt{a_n^2 + b_n^2}$ is the total harmonic starting at some phase angle. All figures are pressures in lb./in.², and, in order to get the corresponding torque harmonic per cylinder, we must multiply by the piston area and by the crank radius. This, however, we can avoid doing because the zero-th coefficient b_0 is listed, which means the average or steady component of the torque, and we can express each harmonic torque as a fraction (sometimes larger than 1) of the steady average torque. All we really need is the total torque $\sqrt{a_n^2 + b_n^2}$, and we mostly can ignore its breakdown into a sine and cosine contribution. Only in the harmonics 1, 2, and 3 do we want a breakdown, because Eq. (5.15) (page 178) shows an inertia torque having these harmonics. In order to compound the gas torque with the inertia torque, we must know its phase, and from Eq. (5.15) we see that we want the *sine* component of the gas torque added to the inertia torque and then that combination added by $\sqrt{a^2 + b^2}$ to the cosine component of gas torque. Incidentally, it is important to note that the principal middle term of Eq. (5.15) is negative and the gas torque positive. The second-harmonic gas torque always opposes the second-harmonic inertia torque, and when an engine gets into a second-harmonic resonance, it is far worse off at light loads than at full load.

Now in the above table we divide the $\sqrt{a_n^2 + b_n^2}$-values by $b_0 = 22.2$, giving a set of p_n-values which denote the nth-harmonic torque as a fraction of the rated full-load steady torque.

n	½	1	1½	2	2½	3	3½	4	4½
p	2.16	2.32	2.23	1.91	1.57	1.28	1.05	0.82	0.63

n	5	5½	6	6½	7	7½	8	8½	9
p	0.48	0.38	0.28	0.20	0.14	0.10	0.08	0.05	0.04

This is in a nutshell what we really want; it shows that the peak value of the first three harmonic gas torques is more than twice the rated steady torque, and it shows that the intensity of the disturbing torque diminishes with higher orders and that when we come to the eighteenth harmonic ($n = 9$) the intensity is almost negligibly small.

The p-values of the above table strictly speaking are good only for the full-load operating condition, and for lighter loads we should use the corresponding figures for the lower-M.I.P. tables of Porter. However, it appears that the p-values as shown here are roughly independent of the load; in other words, when an engine operates at full power the disturbance of order 6 is 28 per cent of the full-power rated torque, and when the engine idles, the disturbance of order 6 still is roughly 28 per cent of the full-power rated torque, or maybe slightly less. A calculation with the p-values shown is necessary for full power and is about right for other powers as well, so that we can avoid the complication. The only exceptions to this rule are the orders 1, 2, and 3, which have to be compounded with the inertia torque.

In ship drives it is not only the Diesel engine that can excite torsional vibrations in the installation. The propeller itself, usually having three or four blades, does not experience a uniform reaction torque from the water. Each time a blade passes the rudder stem or some other near-by obstacle, the pressure field about the blade is influenced and the torque modified. Thus there will be torque fluctuations with propeller-blade frequency. Though little detailed information about the intensity of these variations is available at the present time, it has been found that an assumed torque variation of 7.5 per cent of the mean propeller torque leads to calculated torsional amplitudes that are in decent agreement with measured amplitudes on a considerable number of ships.

5.7. Work Done by Torque on Crank-shaft Oscillation. Assume the crank shaft to be in a state of torsional oscillation, superposed on its main rotating motion. If one of the harmonics of the torque of a cylinder has the same frequency as the vibratory motion, that torque performs work upon the motion. The work so done may be either positive or negative (or zero), depending on the phase relation.

Generally speaking each torque harmonic will induce in the system a forced torsional vibration of its own frequency, so that the motion of the shaft is made up of as many harmonics as are present in the torque. However, nearly all of these harmonics have frequencies so far removed from the natural frequency that the corresponding vibrational amplitude is negligibly small. Only when one of the torque harmonics coincides with one of the natural frequencies is the response appreciable, and the amplitude of vibration then may become great. The "critical speeds" of the engine at which such resonance may occur are very numerous.

For example, the Diesel-generator set of page 188 has a lowest natural frequency of $\omega = 552$, or 5,300 v.p.m. (vibrations per minute). Suppose this machine ran at 10,600 r.p.m. Being a four-cycle engine, it would then have 5,300 firings per minute, just in resonance, and there would be half a vibration cycle per revolution, so that we call this 10,600 r.p.m. the critical speed of order $\frac{1}{2}$ in the first mode. Similarly, at a speed of 2,650 r.p.m. there may be vibrations of 5,300 v.p.m., which are excited by the second harmonic of the torque curve. This speed of 2,650 r.p.m. is called the "critical" of order 2 in the first mode, or $5,300/n$ is the critical of order n. In the second mode with 10,950 v.p.m. the critical of order n is at $10,950/n$ r.p.m. The two spectrums of critical speeds are shown in Fig. 5.20.

Most of the critical speeds thus found are not dangerous as very little work is put into them by the torque. The amplitude builds up until the work done by the torque equals the work dissipated in damping in the

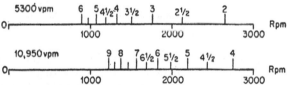

FIG. 5.20. The various critical speeds in mode 1 and mode 2 of the Diesel generator of page 188.

manner indicated in Fig. 2.23, page 52. It is now our object to calculate the work input at the various critical speeds in order to find their comparative danger, while a discussion of the dissipation by damping will be postponed to the next section.

The work done per cycle by one cylinder (the nth one) is $\pi M_n \beta_n \sin \varphi_n$, where M_n is the torque harmonic, β_n the torsional amplitude, and φ_n the phase angle between the two (see page 12). Let us investigate how these three quantities vary from cylinder to cylinder. The torque harmonic M_n has the same magnitude but a different phase at the various cylinders, because we assume that they all fire with the same intensity but naturally not all at the same time. On the other hand the angular displacement β_n varies in magnitude from cylinder to cylinder according to Fig. 5.16 but it has the same phase everywhere because all disks reach their maximum amplitude (or go through zero) simultaneously. The phase angle φ_n therefore varies from cylinder to cylinder. This is shown in Fig. 5.21, where the (horizontal projection of the rotating) doubly lined vector represents the torque harmonic and the (h.p.o.t.r.) single vector represents the angular vibration amplitude for the various cylinders. The velocity of rotation of all the diagrams is ω, the natural cir-

202   MECHANICAL VIBRATIONS

cular frequency of the vibration. This is *not* the angular velocity of the crank shaft which is m times as slow as ω for the mth-order critical speed.

Since the work done by the nth cylinder is $\pi M_n \beta_n \sin \varphi_n$, it is not changed if, as in Fig. 5.22, the *directions* of the torque and displacement vectors in each individual diagram are interchanged, so that we now consider the fictitious case of torques in phase at the various cylinders and torsional amplitudes out of phase. This is convenient for adding the work done by the individual cylinders. Since $\beta_n \sin \varphi_n$ is the horizontal

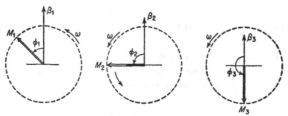

Fig. 5.21. Torque harmonic **M** and vibration amplitude β for the cylinders 1, 2, and 3. The subscripts 1, 2, 3 under **M** denote the cylinder and not the order of the harmonic. The diagram holds for *any* order of harmonic.

projection of the single-lined vector β_n in Fig. 5.22a, the work by one cylinder is πM_n times the vector obtained by projecting β_n horizontally. Hence the work done by all cylinders combined is πM_n times the vector obtained by projecting the *resultant* of all β-vectors horizontally as indicated in Fig. 5.22b. There will be some phase angle ψ in this result which will depend on the original φ_1 at the first crank.

The φ_1 or ψ is unknown, and its exact determination for each frequency ω is out of the question. However we can state that *at resonance* ψ must

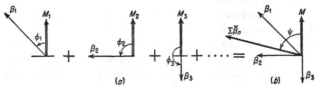

Fig. 5.22. The work input by all cylinders is found by adding the work of the various cylinders individually.

be 90 deg., which can be understood as follows. At "resonance" the amplitude (considered as a function of the frequency) is a maximum, and consequently the work dissipated by damping is a maximum. But this work is equal to the input of Fig. 5.22b. Thus the phase angle ψ is such as to make that work a maximum, *i.e.*, ψ must be 90 deg. Hence we do not need the doubly lined arrows of Fig. 5.22 for the determination of the work input. It is necessary merely to draw a star of vectors with the phases of the torques M_n and the magnitudes of the angular displacements

β_n. The vector sum of this star, numerically multiplied by π times the torque amplitude $\mathbf{M}_n$, is the work done by all the cylinders *of one bank* per cycle of oscillation.

For an in-line engine this is the whole story, but, for a V-engine with two banks or for W- or X-engines with three and four banks, something more comes in. All the banks operate on the same crank shaft and usually have the same firing order within each bank. Consider a V-engine with V-angle α_V between the two banks. A certain cylinder, say No. 1, fires when its crank is in top dead center. Then after the crank physically has turned through the angle α_V, it is in top dead center for cylinder 1 in the other bank, which then fires. The time between firings of two cylinders with the same number then is $\alpha_V/2\pi$ times the time of one revolution. For the nth-order vibration ($n = \frac{1}{2}$, 1, $1\frac{1}{2}$, etc.) there are n cycles of vibration for one revolution. Thus the time between firings of two cylinders of the same number (on the same crank) is the time of $\alpha_V/2\pi$ revolutions or the time of $n\alpha_V/2\pi$ vibration cycles. Suppose the angle $n\alpha_V$ to be 360 deg. or a multiple thereof; then the firing in the two banks takes place on the same phase point of the vibration cycle and the two banks reinforce each other: the work input for two banks is double the work input for one bank, or, in the language of the trade: the V-factor is 2. On the other hand, if $n\alpha_V$ comes out 180 deg., then the first bank fires when the crank shaft twists forward and the second bank fires when the crank shaft twists backward (in the nth harmonic, of course), and the work input of one bank is equal and opposite to the work input of the other: the V-factor is zero. Generalizing this reasoning, we recognize that the V-factor is the vector sum of two vectors of unit length (one unit is the total work input of *one* bank in the nth order) with angle $n\alpha_V$ between them (Fig. 5.23). Thus the factor is

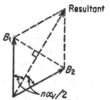

FIG. 5.23. Vector diagram to determine the V-factor for an engine. B_1 and B_2 are vectors of equal length, each representing the total nth-order torque of one bank, with angle $n\alpha_V$ between them. The resultant vector is the combined nth-order torque, and the V-factor is: $2 \cos (n\alpha_V/2)$.

$$V\text{-factor} = 2 \cos \left(\frac{n\alpha_V}{2}\right) \tag{5.27}$$

Consider the specific example of the Diesel generator of page 188, with a 0, 180, 180, 0 crank shaft and firing order 1 3 4 2. We proceed to construct the star diagram of Fig. 5.22b for the various orders of vibration, first considering the phase angles only and paying no attention to the *length* of the vectors.

At the critical speed of order $\frac{1}{2}$, *i.e.*, when half a vibration occurs during one revolution, the crank shaft makes a full turn, while the vibration vector turns through only 180 deg. Or while the crank shaft turns 180 deg., between two consecutive firings the vibration vector turns only through 90 deg. This gives Fig. 5.24a. First cylinder 1 fires; after one-quarter revolution of the diagram (*i.e.*, half a revolution of the crank shaft) cylinder 3 is on top and fires, etc.

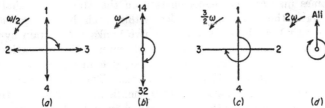

FIG. 5.24. Shapes of vector diagrams for various orders of vibration. For the time being the lengths of the vectors are not considered.

Next consider a vibration of order 1, *i.e.*, 1 v.p.m. The motion of the vector diagram is the same as that of the crank shaft, so that Fig. 5.24b is the same as the crank diagram. The 1½-order vibration gives a vector diagram turning 1½ times as fast as the crank shaft, with angles of 1½ × 180 = 270 deg. between consecutive arrows (Fig. 5.24c). Finally order 2 gives an angle 2 × 180 = 360 deg. between arrows: all arrows are in the same direction. Order 2½ gives angles of 2½ × 180 = 360

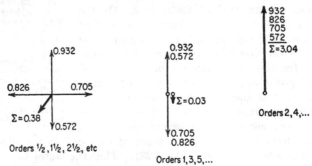

FIG. 5.25. Complete star diagrams (direction and magnitude) for one bank of the first mode of motion of the engine of page 188. The lengths of the vectors are taken from the final Holzer table on page 191.

+ 90 deg. between arrows, which is the same as 90 deg., or order ½, etc. Thus the diagrams of Fig. 5.24 are the only ones for all orders of vibration, and, even of these four, two are virtually alike: (a) and (c) are mirrored images of each other, giving the same vector resultant, although in a different direction. But since we are interested only in the value of the vector sum and not in its phase, Figs. 5.24a and c are identical for our purpose. Thus there are three different diagrams:

4-pronged for orders $\frac{1}{2}$, $1\frac{1}{2}$, $2\frac{1}{2}$, $3\frac{1}{2}$, etc.

2-pronged for orders 1, 3, 5, 7, etc.

1-pronged for orders 2, 4, 6, etc.

Now we are ready to construct the diagrams completely including the proper *lengths* of the vectors. Figure 5.25 gives them for the first mode of motion and Fig. 5.26 for the second or two-noded mode.

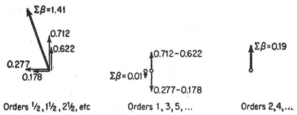

Orders $\frac{1}{2}$, $1\frac{1}{2}$, $2\frac{1}{2}$, etc   Orders 1, 3, 5, ...   Orders 2, 4, ...

FIG. 5.26. Complete star diagrams for one bank in the second mode of motion.

The critical speeds of order 2, 4, etc., are known as *major* critical speeds, all others being *minor* critical speeds. The characteristic property of a major critical speed is that all the vectors in the diagram have the same phase. The physical significance is that with a *rigid* engine (in which the crank shaft cannot twist) the major critical speeds are the only speeds at which work can be done on the vibration, because, as all magnitudes of β_n are then equal, the resultants of the star diagrams of all minor critical speeds are zero.

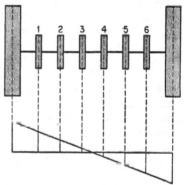

FIG. 5.27. First normal elastic curve for symmetrical engine with two very heavy flywheels.

The distinction between major and minor critical speed does not imply that a major speed is always more dangerous than a minor. In fact, for engines with a more or less symmetrical normal elastic curve, as shown in Fig. 5.27, the resultant of the major critical speeds is zero, whereas for the minor speeds of order $1\frac{1}{2}$, $4\frac{1}{2}$, etc., the resultant becomes very large (Fig. 5.24).

Returning now to our Diesel generator, we are ready to put everything together toward a calculation of the work input per cycle, using Fig. 5.20 for the critical speeds; Figs. 5.25 and 5.26 for the vector sums $\Sigma\beta$; Eq. (5.27) for the V-factor; and the table of page 199 for the harmonic intensity, or p-factor. We have the values shown in the tables for the first and second modes of motion.

FIRST MODE OF MOTION

Order	2½	3	3½	4	4½	5
R.p.m.	2,120	1,770	1,520	1,330	1,180	1,060
V-factor	0.52	0	0.52	1.00	1.41	1.73
p-factor	1.57	1.28	1.05	0.82	0.63	0.48
Σβ	0.38	0.03	0.38	3.04	0.38	0.03
pVΣβ	0.31	0	0.21	2.50	0.34	0.03

SECOND MODE OF MOTION

Order	5	5½	6	6½	7	7½	8
R.p.m.	2,190	2,000	1,830	1,690	1,570	1,460	1,370
V-factor	1.73	1.93	2.00	1.93	1.73	1.41	1.00
p-factor	0.48	0.37	0.28	0.20	0.14	0.10	0.08
Σβ	0.01	1.41	0.19	1.41	0.01	1.41	0.19
pVΣβ	0.01	1.00	0.11	0.55	0.00	0.20	0.02

In these tables the last line $pV\Sigma\beta$ is a measure for the work input per cycle. We only have to multiply the $pV\Sigma\beta$ figures by $\pi M_{mean} = \pi \times$ 1,580 in.-lb. to get the work input per cycle for amplitude $\beta = 1$ radian at the damper. This input work then has to be equated to the work dissipated in the system, which will be calculated in the next two sections. However, the $pV\Sigma\beta$ figures are a good measure for the relative severity of the various orders of vibration. We see that among all the first-mode criticals the order 4 at 1,330 r.p.m. is the most serious one. Among the second-mode criticals, that of order 5½ at 2,000 r.p.m. is the worst.

5.8. Damping of Torsional Vibration; Propeller Damping. The work done by the gas and inertia couples on the motion in a steady-state constant-amplitude vibration is exactly compensated by the work dissipated in the system. Sometimes it is reasonably easy to find where this dissipation takes place, when, for example, the system contains a friction damper, drives a hydraulic pump or a propeller, or has a slipping clutch or other joint. But in cases where no such obvious places of friction dissipation are apparent in the engine installation, it is somewhat of a mystery where the energy goes to. The literature on this subject extends back to 1920. Attempts were made to explain the energy dissipation on the basis of internal friction in the shafting, sometimes called "internal hysteresis." Actual measurements of hysteresis in steel give extremely low values, by far not sufficient to explain the damping in actual installations. But just the same, until a few years ago, it was usual to make elaborate calculations of shaft hysteresis dissipation and to multiply the answer so found by a large, empirically found multiplier. This led to reasonable estimates of the actually observed amplitude, which never was more than twice as large or smaller than half as large as the "calculated" value. The folly or unreasonableness of this procedure was impressed on the author one day when he was aboard a Diesel-driven oil tanker, in which the torsional amplitude, as calculated, appeared dangerously large but, when actually measured, proved to be only half as large as calculated. Upon completion of the measurements and walking through the ship, a point in the anchor-chain storage space was reached, which was practically the entire ship length removed from the engine. One of the links of the anchor chain was lying on the steel deck and vibrating visibly and audibly. A check with the second hand of a watch proved beyond doubt that the frequency was that of the engine torsional. In this case the engine-torsional frequency happened to coincide with a natural bending frequency of the ship beam as a whole and was exciting it. After observing that part of the energy input of the gas torque was dissipated in an anchor chain 500 ft. away, the author never again calculated shaft hysteresis losses and multiplied them by an empirical constant.

The status of the art now is that in the absence of definite dampers, propellers, pumps, or slipping torsion clutches the damping of the system is due to a number of causes which are elusive, complicated, and different from installation to installation. Since we have no way of knowing the details, we are forced to rely on statistics only. F. M. Lewis studied a large number of such engines, calculating for each the quantity $pV\Sigma\beta M_{mean}$ (page 206) for the mode and order of vibration in question. Here M_{mean} is the mean full-power torque for one cylinder of the engine ($= 1,580$ in.-lb. for our example), pM_{mean} is the amplitude of the harmonic torque under consideration, and $V\Sigma\beta$ is a pure number at most equal to the total number of cylinders of the engine and sometimes very small or zero, expressing the number of cylinders pulling together toward a vibration of the order considered. Thus the quantity $pVM_{mean}\Sigma\beta$ is called the "exciting torque" of the particular order and mode of vibration. This quantity, when operating at resonance, causes a shaft torque (in the location of maximum torque along the shafting, i.e., at the location of the maximum slope of the diagrams, Fig. 5.16) many times as large as the exciting torque. Comparing the actually measured maximum resonant shaft torque M_{shaft} to the calculated exciting torque $pV\Sigma\beta M_{mean}$ for a large number of engines, Lewis found that the resonant magnification varied from 25 to 100 times, usually being about 50 times. Thus we use the rule

$$M_{shaft\,max} = 50pV\Sigma\beta M_{mean\,1\,cyl} \qquad (5.28)$$

remembering that the factor 50 in exceptional cases may be as high as 100, when the damping is exceptionally small. Also, the rule only holds for installations without obvious dampers. In our Diesel-generator example there is a viscous-type damper at the free end of the engine, and we shall see later, on page 215, that the maximum resonant shaft torque is only 5 times the exciting torque instead of 50 times. The example would be inoperable without the damper, or rather, when operated without the damper at the critical of order 4 in the first mode, the crank shaft would fail by fatigue in a short time.

In ship-engine installations the damping provided by the action of the water on the propeller is particularly effective. The moment of inertia of a ship's propulsion engine usually is much greater than the propeller inertia, so that in the first mode of motion the propeller amplitude is large compared with the engine amplitude. A damping torque is one which is opposite in phase to the angular velocity. In the free vibration of the first mode, superposed on the steady rotation of the entire installation the propeller speed will alternately be faster and slower than normal. Since the resisting torque of the surrounding water increases with the speed there is positive damping action, which can be explained as follows. During the half cycle that the propeller speed is greater than average $(\Omega_p + d\Omega_p)$, the retarding torque is also greater than average $(M_p + dM_p)$, so that the excess torque dM_p tends to retard the motion, *i.e.*, dM_p is directed opposite to the excess velocity $d\Omega_p$. Conversely, during the half cycle that the propeller speed is smaller than average $(\Omega_p - d\Omega_p)$, the retarding torque is $M_p - dM_p$, so that the excess $-dM_p$ is accelerating. The excess velocity $d\Omega_p$ is directed against the rotation Ω_p, which also is against the direction of the excess torque.

If for these small variations in torque and speed the torque-speed characteristic is straight, the damping constant c, being the retarding torque per unit angular velocity, is $c = dM_p/d\Omega_p$.

By (Eq. 2.30), page 52, the work dissipation per cycle is

$$W = \pi c \omega \beta_p^2 = \pi \omega \beta_p^2 \frac{dM_p}{d\Omega_p}, \qquad (5.29)$$

where β_p is the amplitude of vibration at the propeller. The work input per cycle by the cylinder torques doubles if all amplitudes of vibration are doubled, since the torques are not affected by a change in amplitude. However, by (5.29) the propeller dissipation quadruples if the amplitudes are doubled. Thus there will be a definite amplitude at which input and output of energy balance each other (Fig. 2.23, page 52). It is necessary merely to find the value of $dM_p/d\Omega_p$.

In Fig. 5.28 the *steady-state* relation between the torque and the propeller speed of a typical ship is shown. The curve is a parabola or a somewhat steeper curve expressed by $M_p = \Omega_p^n$ with an exponent n between 2 and 3. This curve can be easily obtained for a given ship by the actual measurement of the torque (indicator diagrams), and the shaft revolutions per minute for a number of speeds. But the slope of *this* curve is not the damping constant we are seeking, because in it the ship's forward speed grows with the revolutions per minute, whereas during the rapid Ω_p-variations of the torsional vibration, the ship's speed is constant. It is shown below in small print that at a definite torque and speed (point P in Fig. 5.28) the slope $dM_p/d\Omega_p$ for a constant ship speed is considerably greater than the slope of the steady-state curve. The dotted line through P indicates the curve for constant ship speed, and it is usually assumed that its slope at P is twice as large as the slope of the fully drawn steady-state characteristic.

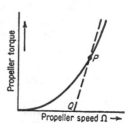

Fig. 5.28. Marine propeller characteristic.

Consider a propeller-blade element cut out by two cylinders concentric with the shaft and with radii r and $r + dr$. The section of the propeller blade so obtained has the appearance of an airplane-wing section. Let this blade be moving forward (Fig. 5.29a) with the ship's velocity V and tangentially with the velocity $\Omega_p r$. The water will flow against it from the upper left corner of the drawing with the relative velocity V_{rel}. The propeller is so designed that this direction includes a small angle α (the angle of attack) with the main direction of the blade. This causes a hydrodynamic lift force L on the blade perpendicular to the direction of flow (Fig. 5.29b). There will

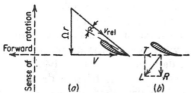

Fig. 5.29. Direction of water flow (a) and the forces (b) acting on a propeller-blade element.

be also a small drag or resistance force in the direction of flow which we may disregard in this argument. The lift L can be resolved into two components T and R: T being the thrust, and R the reaction, thus causing a torque Rr about the shaft axis. The sum of all T's for the various blade elements of the propeller add up to the thrust on the ship, and the sum of the various R's is equal and opposite to the engine torque in the steady-state case.

Imagine a periodic variation in the propeller speed Ω_p, while the ship's speed V is constant. In Fig. 5.29a the length $\Omega_p r$ varies, and consequently the angle of attack α varies. This varies the lift L and the torque Rr. Let Ω_p diminish to such an extent that the angle α and with it the lift and Rr become zero. Then the propeller torque is zero, because the propeller freely screws through the water, which in this case acts as a stationary nut. The forward speed and the rotation are adjusted so that this screwing takes place without any effort. In the usual designs the blade angle $\tan^{-1}\dfrac{\Omega_p r}{V}$ varies between 20 and 80 deg. along the blade, whereas the angle of attack α is of the order of 5 deg. Thus we see that a diminishing of Ω_p by 10 or 20 per cent is sufficient to make the torque zero. This condition is indicated by the point Q in Fig. 5.28.

In this argument it has been tacitly assumed that the rate of change $d\Omega_p/dt$ is of no influence on the phenomenon, *i.e.*, we have assumed that the flow in Fig. 5.29a is a steady-state flow for each ratio $\Omega_p r/V$. In case the variation in Ω_p is *slow*, such a succession of steady-state flows is practically the same as the actual flow, but for *rapid* variations ($d\Omega_p/dt$ = large), this analysis is inapplicable. A completely satisfactory theory of propeller damping does not exist as yet, and for important cases where the frequency is high, only an experiment on a model can give reliable information.

In engine installations without active propeller damping, without slipping couplings that absorb energy, or without other visible sources of energy dissipation, a critical speed with a comparatively large entry in the last row of the table on page 206 will inevitably cause such large amplitudes that the crank shaft or driving shaft breaks in fatigue. To prevent this we can apply one of the following procedures:

1. If the engine is to operate always at the same speed, e.g., a synchronous-generator drive, changes in the elasticity of the shaft or in the inertia of the masses can be made such as will remove the running speed sufficiently far from any important critical speed.

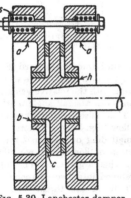

2. If the engine has to operate over a narrow speed range, course 1 usually suffices. If it does not suffice, the relative severity of the minor critical speeds may be influenced by changing the firing order. This is explained on page 223.

3. If operation over a very wide speed range is required, as for instance in Diesel locomotives or in ship drives, it may become very difficult, if not impossible, to avoid all danger of torsional vibration by the means 1 and 2. An artificial damper should then be applied.

FIG. 5.30. Lanchester damper.

Three such devices will now be discussed, i.e., the friction damper of Lanchester, the tuned centrifugal pendulum described on page 93, and the hydraulic coupling or "fluid flywheel."

5.9. Dampers and Other Means of Mitigating Torsional Vibration. One of the earliest effective dampers invented was the one of *Lanchester* (Fig. 5.30), consisting of two disks a, which can rotate freely on the shaft bearings b. Between them is a hub disk h solidly keyed to the shaft. This hub h carries brake lining c on its faces against which the disks a can be pressed by screwing down the springs s.

If the engine, i.e., the hub h, is in uniform rotation, the friction carries the disks a with the shaft, so that the disks then merely increase the inertia of the engine by a small percentage. If, however, the hub executes a torsional vibration, the motion of the disks depends on the amount of friction between them and the hub. If the friction torque is extremely small, the disks rotate uniformly and there is a relative slip between the hub and the disks with the amplitude of the hub motion. Since the friction torque is nearly zero, very little work is converted into heat. On the other hand, if the friction torque is very large, the disks lock on the hub and follow its motion.   There is then no relative slip and hence no energy dissipation.   Between these two extremes there is both slip and a friction torque, so that energy is destroyed.   There must be some optimum value of the friction torque at which the energy dissipated is a maximum, as indicated in Fig. 5.31.

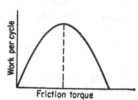

FIG. 5.31. Energy dissipation in the Lanchester damper as a function of the friction torque.

The old type of Lanchester damper depends on Coulomb, or dry, friction between steel and some sort of asbestos-type brake lining.   The coefficient of friction of such an assembly is greatly dependent on the degree of contamination with oil, and since the damper is located on the end of the crank shaft, this is difficult to avoid.   In order to be more certain about the friction in the device, the "dry" friction can be replaced by viscous friction.   A series of dampers of various sizes is now readily available on the market, for example, the Houde damper, made by the Houdaille-Hersey Corporation and shown in Fig. 5.32. H is a hub, splined inside so that it can be attached easily to a shaft end. A is the flywheel, which, through a bronze ring B, can rotate freely about the hub.

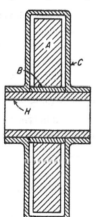

FIG. 5.32. Houde type of viscous Lanchester damper.

C is a housing, welded to the hub H.   The clearance between the flywheel A and the housing C is very small, accurately held to dimension, and completely filled with fluid, which is permanently sealed in that space. The fluid is not oil, but a silicone fluid which has a viscosity like that of oil, but which has the advantage that its viscosity varies less drastically with temperature than that of oil.   A rotating motion of A relative to C and H is associated with a viscous-friction torque, so that this damper substantially acts like the dry-friction type of Fig. 5.31.   The principal characteristics of this damper are shown in Figs. 5.34 and 5.35, the first of which gives the energy dissipation per cycle, and the second of which gives the fraction of the actual flywheel inertia of A in Fig. 5.32 which is effective on an engine, both as functions of the viscosity c of the damping fluid.

We shall now derive these relations.   Let φ_h be the instantaneous angle of the hub and φ_d the instantaneous angle of the damper flywheel.   Then the instantaneous value of the relative angle (or slip angle across the viscous film) is $\varphi_{rel} = \varphi_h - \varphi_d$.   The torque acting on the damper flywheel I_d, by Newton's law, is $I_d\ddot{\varphi}_d$, and this torque is furnished by viscous friction across the film, which is $c\dot{\varphi}_{rel}$.

Thus

$$I_d \ddot{\varphi}_d = c\dot{\varphi}_{rel} \qquad (5.30)$$

When we prescribe the hub motion as $\varphi_h = \varphi_{h_0} \sin \omega t$, this is a differential equation with the unknown φ_d and we can solve and interpret it. We shall do this by vector representation, rather than by algebra. For steady-state motion at a certain frequency ω the angles φ_h, φ_d and $\varphi_r = \varphi_h - \varphi_d$ are vectors. From (5.30) it is seen that the torque vector is 180 deg. out of phase with φ_d and 90 deg. out of phase with $\varphi_r = \varphi_h - \varphi_d$. Hence the φ_r-vector is perpendicular to the φ_d-vector, which is shown in

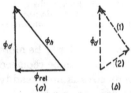

the diagram (Fig. 5.33). Rewriting Eq. (5.30) for the instant when the quantities reach their maximum value,

$$I_d \omega^2 \varphi_d = c\omega \varphi_{rel}$$

and, from Fig. 5.33,

$$I_d \omega^2 \varphi_d = c\omega \sqrt{\varphi_h^2 - \varphi_d^2}$$

Solving this for the damper-flywheel motion φ_d, we find

$$\varphi_d = \frac{1}{\sqrt{1 + (I_d\omega/c)^2}} \varphi_h \qquad (5.31)$$

and we check the physical facts that for $c = 0$ we have $\varphi_d = 0$, while for $c = \infty$ we have $\varphi_d = \varphi_h$.

FIG. 5.33. (a) Phase relationship between the damper flywheel motion, the hub motion, and the relative motion of the viscous Lanchester damper. (b) The damper flywheel motion (or torque) resolved into components in phase and 90 deg. out of phase with the hub motion.

Now the maximum torque transmitted between the hub and the damper flywheel is $I_d\omega^2\varphi_d$. In order to find the work per cycle done by this torque, we must find the motion which is 90 deg. out of phase with it. Since the torque, by Eq. (5.30), can be looked upon as a viscous-friction torque, the motion 90 deg. out of phase with it is the relative motion. Thus the work per cycle is

$$W = \pi \cdot (I_d\omega^2\varphi_d) \cdot \varphi_{rel} = \pi I_d\omega^2\varphi_d \sqrt{\varphi_h^2 - \varphi_d^2}$$

Working this out and eliminating φ_d by means of Eq. (5.31), we find

$$W \text{ per cycle} = \frac{\pi}{2} I_d\omega^2\varphi_h^2 \cdot \frac{2c/I_d\omega}{1 + (c/I_d\omega)^2} \qquad (5.32)$$

This is plotted in Fig. 5.34. The fraction at the end of the equation reaches a maximum value of unity for $c/I_d\omega = 1$, as can be found by differentiation. Thus we conclude that the optimum damping (for maximum energy dissipation) is

$$c_{opt} = I_d\omega \qquad (5.33a)$$

$$W_{opt} = \frac{\pi}{2} I_d\omega^2\varphi_h^2 \qquad (5.33b)$$

Now we shall calculate the apparent moment of inertia of the damper flywheel, *i.e.*, the amount of inertia which is felt by the hub. Physically it is clear that for $c = 0$ (no damping) the hub feels nothing at all, while for $c = \infty$ the flywheel is frozen to the hub and the hub feels the full inertia I_d. In Fig. 5.33 the torque transmitted to the hub is in line with φ_d. We now resolve that torque into components, (1) parallel to the hub motion and (2) perpendicular to the hub motion. The first of these is a torque felt by the hub in phase with its own motion, and the hub feels this as inertia. The second component is 90 deg. out of phase with the hub motion, and the hub feels it as viscous friction. We see that

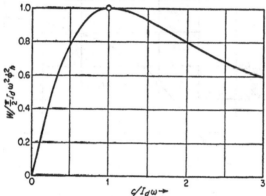

FIG. 5.34. Work per cycle dissipated by a Houde or viscous Lanchester damper as a function of the damping torque.

component 1 is φ_d/φ_h as large as the total torque. Thus the equivalent inertia torque is

$$I_{\text{equiv}}\omega^2\varphi_h = \frac{\varphi_d}{\varphi_h}(I_d\omega^2\varphi_d)$$

or

$$I_{\text{equiv}} = I_d \cdot \left(\frac{\varphi_d}{\varphi_h}\right)^2$$

and by Eq. (5.31)

$$I_{\text{equiv}} = \frac{I_d}{1 + (I_d\omega/c)^2} \tag{5.34}$$

For the optimum damping $c = I_d\omega$ this becomes

$$I_{\text{equiv}} = \tfrac{1}{2}I_d \quad \text{(optimum damping)} \tag{5.35}$$

The last two results are plotted in Fig. 5.35. Thus, when starting the Holzer calculations for a system containing a Lanchester or Houde damper, we replace the damper by a moment of inertia equal to that of the hub and housing plus half that of the loose flywheel. This was done in Fig. 5.13. Strictly speaking this is correct only when the damping is optimum [Eq. (5.33a)], which for a given damper is true only for one particular frequency of vibration. But we start by assuming optimum damping, calculate the system, find the dangerous frequency, and then specify the damping c for the damper, so as to hit the optimum. This can always be done not only by choosing the damping fluid but also by specifying the clearance between the flywheel and the housing, because both affect the value of c.

It is clear from Fig. 5.34 and Eq. (5.33b) that the damper should be placed at a point of the shaft where the torsional amplitude is great and

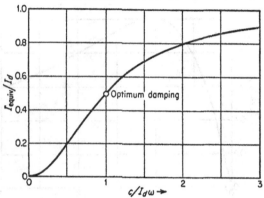

FIG. 5.35. The equivalent flywheel effect on the hub of the floating flywheel in a Lanchester or Houde damper as a function of the damping torque.

that the device becomes entirely useless when placed at a node of the vibration. This is a property a Lanchester damper has in common with a ship's propeller.

Now we are ready to apply this damper theory to our Diesel-generator example. The energy input per cycle of vibration is (page 206)

$$\pi \cdot p V \Sigma \beta \cdot M_{mean} \cdot \varphi_{hub}$$

Assuming optimum damping, the energy dissipation per cycle is

$$\frac{\pi}{2} I_d \omega^2 \varphi_{hub}^2$$

Equating the two gives

$$\varphi_{hub} = \frac{2}{I_d \omega^2} (p V \Sigma \beta) M_{mean}$$

In our engine M_{mean} = 1,580 in.-lb., and the damper flywheel is I_d = 8.72 lb. in. sec.^{2}, so that

$$\varphi_{hub} = 360 \frac{p V \Sigma \beta}{\omega^2}$$

In the first mode of motion the most dangerous critical is that of order 4 at 1,330 r.p.m. There we find $\omega_1^2 = 310,000$ (page 191) and $pV\Sigma\beta = 2.50$ (page 206) so that

$$\varphi_{\text{hub}} = \frac{360 \times 2.50}{310,000} = 0.0029 \text{ radian}$$

Looking back to the Holzer table of page 191, where $\beta_1 = 1.000$, we now interpret that 1.000 as 0.0029 radian. All other quantities in the entire table also must be multiplied by 0.0029. In particular the maximum shaft torque occurring in the engine is the largest number in the Σ-column, in the shaft between the flywheel and the generator. It is

$$\mathbf{M}_{\text{max}} = 6.61 \times 10^6 \times 0.0029 = 19,000 \text{ in.-lb.}$$

To get a conception of what this means, we compare it with two other torques:

1. The full-load mean rated torque, which for eight cylinders is $8 \times 1,580 = 12,600$ in.-lb.
2. The "exciting torque" $pV\Sigma\beta\mathbf{M}_{\text{mean}} = 2.50 \times 1,580 = 3,950$ in.-lb.

We see that the resonant torque at the fourth-order critical at 1,330 r.p.m. is about 5 times the exciting torque. In the case of no damper we know from page 207 that the torque would be 50 times as large, instead of 5 times as large. But even with this damper the (useless) alternating torque at 1,330 r.p.m. still is 50 per cent larger than the useful full rated torque. The shafts of Diesel installations always are designed quite conservatively (for a good reason!), and a torque of 19,000 in.-lb. leads to an acceptable stress.

For the second mode we have the worst speed at 2,000 r.p.m., of order $5\frac{1}{2}$ with $pV\Sigma\beta = 1.00$ and $\omega_2^2 = 1,310,000$. This gives

$$\varphi_{\text{hub}} = \frac{360 \times 1.00}{1,310,000} = 0.000275 \text{ radian}$$

The maximum torque (see Holzer table on page 192) occurs in the middle of the engine crank shaft between masses 3 and 4 and is

$$\mathbf{M}_{\text{max}} = 12.32 \times 10^6 \times 0.000275 = 3,400 \text{ in.-lb.}$$

This, being less than 20 per cent of the first-mode torque at 1,330 r.p.m., leads to stresses of very small magnitude.

The foregoing calculations are subject to criticism in that they are based on "optimum damping." The first mode being the worst one, we find for the optimum damping in that first mode

$$c = I_d\omega_1 = 8.72 \times 555 = 4,800 \text{ in.-lb. torque}$$

for a slipping velocity of 1 rad./sec. We prescribe to the damper manufacturer this value. Then, by Fig. 5.35, the effective damper inertia is half the flywheel, 8.72/2, plus that of the damper housing, 3.58 = 7.94 lb. in. sec.², which is the value appearing in Fig. 5.13 and in all the Holzer tables. Having prescribed the damping to be optimum for the first mode of motion, it unfortunately is not optimum for the second mode. There it would have to be

$$c = I_d\omega_2 = 8.72 \times 1{,}145 = 10{,}000 \text{ in.-lb./rad./sec.}$$

while we just made it to be 4,800 in.-lb./rad./sec. The damping then is slightly more than twice optimum, and, by Fig. 5.34, the effectiveness of the damper is diminished to about 80 per cent, so that the stress is 1.25 times as large as just calculated, still negligibly small. Also, the equivalent inertia, by Fig. 5.35, is not half the damper flywheel, but 80 per cent of it, or

$$I_1 = 3.58 \text{ (housing)} + 0.80 \times 8.72 = 10.50 \text{ lb. in. sec.}^2$$

instead of 7.94. Thus, strictly speaking, we should recalculate the Holzer table of page 192 with this modified figure for I_1. However, this is done in practice only if the stress involved is considerable, which is not the case here.

In order to make the Lanchester damper dissipate more energy for a given inertia of the flywheel a of Fig. 5.30, the relative motion between the flywheel a and the hub h may be increased by mounting the flywheel on tuned springs. This produces the "damped tuned vibration absorber" of which the theory for viscous damping is discussed on pages 93 to 106. That theory gives the complete behavior of such a damper when applied to a simple K-M system. In order to apply it to a multimass system the theory would become hopelessly complicated. However the results of pages 93 to 106 can be applied with decent accuracy to a multimass system as well, by replacing the multimass system by an equivalent K-M system as follows:

1. The mass M of the one-mass system is so chosen that for equal amplitudes at M and at the point of the multimass system where the damper is attached, the kinetic energy of M equals the kinetic energy of the multimass system in the mode of motion considered.

2. The spring K of the one-mass system is then so chosen that K/M is equal to the ω^2 of the multimass system in the mode of motion under consideration.

3. The exciting force P on the single mass M is so chosen that its work $\pi P x_1$ at resonance is equal to the total work input by all the exciting forces of the multimass system adjusted to the same amplitude x_1 at the point of attachment of the damper.

Another device useful for avoiding or damping torsional vibration is *Foettinger's* hydraulic coupling, also known as "fluid flywheel" (Fig. 5.36a). It consists of a piece A in the shape of half a doughnut keyed to the driver shaft. A similar piece B is keyed to the follower shaft. A cover C is attached solidly to A on the driver shaft and can turn with respect to the follower shaft. At D there is a hydraulic seal with little friction. The entire interior of the doughnut is filled with a fluid, thin oil or water, and the sole purpose of the cover C is to hold that fluid in place. The doughnut-shaped space is subdivided into a large number

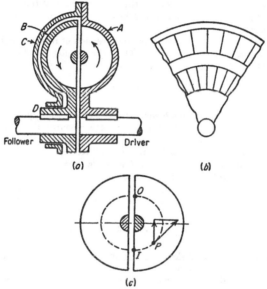

Fig. 5.36. The hydraulic coupling or "fluid flywheel" transmits torque primarily by the action of Coriolis forces.

of open compartments by many thin vanes, each having the form of a semicircle, and arranged in purely radial planes (Fig. 5.36b). By Newton's law of action and reaction the torques on driver and follower must be equal and opposite. Since the device does not operate at ideal efficiency, the speed of the follower must be somewhat less than that of the driver, the speed ratio being the same as the efficiency, which is between 97 and 99 per cent. The fluid in the coupling is under the influence of centrifugal force, which is greater in the driver than in the follower on account of the speed difference. Thus a circulation is set up, moving the fluid outward in the driver and inward in the follower. This circulation, for the existence of which a speed difference is essential, is the cause of torque transmission between the two shafts.

Consider a particle of fluid dm at point P in Fig. 5.36c. Its velocity will have a radial component v_r, and the Coriolis acceleration is $2\Omega v_r$, directed tangentially. The Coriolis force is $2\Omega v_r\,dm$ and its moment is $2\Omega v_r r\,dm$ in a direction such as to retard the rotation Ω of the driver. For all the particles in the stream tube of P the torque integrates to

$$\int 2\Omega v_r r\,dm = 2\Omega \int \frac{dr}{dt} \cdot r\,dm = 2\Omega \int r\,dr \cdot \frac{dm}{dt}$$

$$= 2\Omega \frac{dm}{dt} \int_I^O r\,dr = \Omega \frac{\Delta m}{T} (r_O^2 - r_I^2)$$

The factor dm/dt appearing in this integration is the mass flowing by P per second, which is constant and equal to $\Delta m/T$, i.e., the total mass Δm of the entire steam tube from I to O in the driver and from O to I in the follower divided by the period of circulation T in seconds.

The Coriolis torque on the follower is in the direction of rotation and is calculated similarly with the same form of answer. Only the angular speed of the follower is less, say $\Omega - \Delta\Omega$, so that the Coriolis torque is

$$(\Omega - \Delta\Omega) \frac{\Delta m}{T} (r_O^2 - r_I^2)$$

which is different from the Coriolis torque of the driver. This apparent discrepancy is removed by considering that there are some other contributions to the torques. At O, fluid of tangential velocity Ωr_O from the driver is received by the follower of which the tangential speed is less by an amount $\Delta\Omega \cdot r_O$. The loss in tangential momentum per second thus is $\Delta\Omega \cdot r_O \cdot \Delta m/T$ which is equal to the force exerted by the stream tube of P on the follower, in the direction of rotation. The moment arm of this force is r_O giving a moment $\Delta\Omega r_O^2 \Delta m/T$. Consequently the total amount on the follower is the sum of the Coriolis torque and the torque caused by the change in momentum:

$$\mathbf{M} = \frac{\Delta m}{T} [\Omega(r_O^2 - r_I^2) + \Delta\Omega \cdot r_I^2] \qquad (5.36)$$

Similarly at the inlet point I slow water from the follower enters the driver which rotates faster, thus causing a retarding torque on the driver of $\Delta\Omega r_I^2 \Delta m/T$, which in conjunction with the driver's Coriolis torque gives the same expression (5.36) for the total retarding torque on the driver.

The torque (5.36) is that due to a single stream tube only. The torque of the complete device is found by still another integration, in which r_O, r_I, and T are variables, since the period of circulation will be different for different stream lines. However, Eq. (5.36) can be interpreted as the total torque if we consider Δm to be the mass of water in the entire doughnut, r_O and r_I the radii belonging to the center stream line, and T some average period of circulation.

So far we have considered only uniform or steady-state operation of the coupling. To investigate its damping characteristics both halves of it are now given non-uniform motions. Let the driver speed be $\Omega + \varphi_d$, and let the follower speed be $\Omega - \Delta\Omega + \varphi_f$, where the φ are variable with time. If these variations are sufficiently rapid, the consequent changes in centrifugal force are so fast that the velocity of the fluid circulation is not affected. Then we can apply the above steady-state analysis, merely substituting the variable angular speeds for the constant ones. Thus the torque on the follower (in the direction of rotation) is

$$(\Omega - \Delta\Omega + \varphi_f) \frac{\Delta m}{T} (r_O^2 - r_I^2) + (\Delta\Omega - \varphi_f + \varphi_d) r_O^2 \frac{\Delta m}{T}$$

which is seen to be the sum of the steady-state torque (5.36) plus the variable part:

$$M_{var} = \frac{\Delta m}{T} [\phi_d r_0^2 - \phi_f r_f^2] \tag{5.37}$$

In the same manner the torque on the driver, in a direction opposite to that of the rotation, is written as the sum of the Coriolis and momentum transfer components. The answer again is (5.36) plus the variable part (5.37). It is noted that the torque (5.37) is proportional to the angular *speeds* and thus acts as a damping torque. It may be a positive or a negative damping torque, on account of the second term in the bracket of (5.37), but in all actual installations it is found to be a positive damping torque. See further Problems 204 and 205 on page 411.

Another means of correcting a troublesome condition of torsional vibration is the tuned centrifugal pendulum mentioned on page 93. Since there is no energy loss in the device, it cannot be considered a "damper," but, just as the Frahm absorber of page 87, it acts like an infinite mass for the frequency to which it is tuned, thus enforcing a node at the point of its application. For other frequencies it acts like a mass which is *not* infinitely large and thus does not affect the situation particularly. The proof of this statement is as follows:

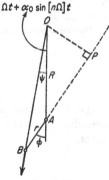

FIG. 5.37.

Let the shaft in Fig. 5.37 rotate about its center O with an average angular speed Ω on which is superposed a rotational oscillation $\alpha = \alpha_0 \sin \omega t = \alpha_0 \sin n\Omega t$, the number n being the "order" of the vibration. The (mathematical) pendulum of length r and mass m swings about A, with the *small* angle $\varphi = \varphi_0 \sin n\Omega t$ relative to the shaft. The angle AOB denoted by ψ satisfies

$$\psi = \varphi r / (R + r)$$

Considering the relative motion of the system with respect to a uniformly rotating coordinate system Ω, the Coriolis forces can be neglected for small oscillations. The tangential component of the centrifugal force (*i.e.*, normal to AB) is

$$-m\Omega^2(R + r) \sin (\varphi - \psi) \approx -m\Omega^2(R + r) (\varphi - \psi) = -m\Omega^2 R\varphi$$

and the tangential displacement of B with respect to the coordinate system is $\alpha(R + r) + \varphi r$. Thus the equation of motion is

$$(R + r)\ddot{\alpha} + r\ddot{\varphi} = -\Omega^2 R\varphi$$

which, after substitution of the harmonic values for α and φ, yields

$$\frac{\varphi_0}{\alpha_0} = \frac{n^2(R + r)}{R - n^2 r} \tag{5.38}$$

The tension in the pendulum string $m\Omega^2(R + r)$ furnishes the only reaction from the pendulum on the shaft and with the moment arm $\overline{OP} = R\varphi$ gives the reaction torque

$$\mathbf{M} = m\Omega^2(R + r)R\varphi = m\Omega^2(R + r)R\left(\frac{\varphi_0}{\alpha_0}\right) \cdot \alpha_0 \sin n\Omega t$$

After substitution of (5.38), this becomes

$$\mathbf{M} = \frac{m(R + r)^2}{1 - \dfrac{r}{R}n^2} \cdot \omega^2\alpha_0 \sin \omega t = -\frac{m(R + r)^2}{1 - \dfrac{r}{R}n^2} \cdot \ddot{\alpha}$$

If instead of the pendulum a moment of inertia I_{equiv} had been attached to the shaft, its reaction torque to an acceleration $\ddot{\alpha}$ would have been $-\ddot{\alpha}I_{\text{equiv}}$, from which follows that the undamped pendulum for small oscillations is completely equivalent to a flywheel of inertia:

$$I_{\text{equiv}} = \frac{m(R + r)^2}{1 - \dfrac{r}{R}n^2} \tag{5.39}$$

The numerator of this expression is the moment of inertia of the pendulum when clamped to the shaft; the denominator is a multiplication factor. Thus a "tuned" pendulum

$$n^2 = \frac{R}{r} \tag{5.40}$$

is equivalent to an infinite moment of inertia; an "overtuned" pendulum $(R/r > n^2)$ represents a (large) positive inertia, while an "undertuned" pendulum behaves like a large negative moment of inertia (see Fig. 2.18, page 44).

The tuning formula (5.40) carries within itself a difficult problem of design. The order of an objectionable harmonic vibration in a multi-cylinder engine is at least $n = 3$, usually higher. The radial distance R is limited by space considerations; in a radial aircraft engine, for instance, where the pendulum is conveniently located in the crank counterweight, the maximum distance R is of the order of 5 in. Thus, by (5.40), the pendulum length r is about $\frac{1}{2}$ in. for $n = 3$, and considerably shorter for higher orders of vibration. Since the pendulum must have appreciable mass, the construction indicated in Fig. 3.10b is impossible. Two solutions of the problem have been found; they are shown in Figs. 5.38 and 5.39, both located in the counterweight of a crank shaft.

The first one, known as the "bifilar" type, was invented by *Sarazin* in France and, independently, by *Chilton* in the United States. The pendulum is a large U-shaped weight, fitting loosely around the crank-shaft overhang. That overhang carries two circular holes of diameter d_1. The U-shaped loose counterweight has holes of the same diameter. The two pieces are joined by two pins of a diameter d_2, smaller than that of the holes. It is now possible for the pendulum to roll without slipping on the pins, and in doing this the center of the hole in the pendulum describes a small circle about the center of the crank-shaft hole as a

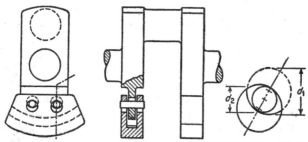

FIG. 5.38. The "bifilar" or Sarazin-Chilton type of tuned centrifugal pendulum.

center. Thus the radius of this circular path is $(d_1 - d_2)$, and it is seen that all points of the U-pendulum describe similar paths. The pendulum swings parallel to itself in a circular path of radius $d_1 - d_2$. Thus in Eq. (5.40), $R + r$ is the distance from the shaft center to the center of gravity of the pendulum, while $r = d_1 - d_2$. Thus it is possible to make r very small and still retain a large mass.

The other construction is due to *Salomon* in France and consists of a cylinder of radius r_2, rolling or sliding in a cylindrical cavity of radius r_1 (Fig. 5.39).

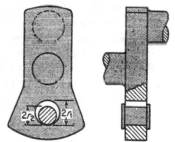

FIG. 5.39. The roller or Salomon type of centrifugal pendulum.

In case the cylinder slides without rotation, all its points describe similar paths of radius $r_1 - r_2$; this quantity thus is the equivalent pendulum length r. For a rolling cylinder the swing is slower, so that r is greater than $r_1 - r_2$. Since the mass involved in this construction is much smaller than that of Fig. 5.38, the amplitude through which it must swing for correct operation is much greater, which creates some additional difficulties.

A single pendulum arranged as a loose counterweight exerts a torque on the crank shaft by virtue of the fact that the force exerted on its guide does not pass through the center O (Fig. 5.37) but is directed along BA. Thus the tangential component of the force along BA by its moment arm

R furnishes the desired reaction torque, but in addition to that torque the pendulum exerts a *force* on the crank shaft. This alternating force is entirely unbalanced and can cause a linear vibration of the center *O*. If two pendulums were installed, one in the counterweight and another one diametrically opposite, *i.e.*, at the crank pin, these two pendulums would enforce nodes at the two points of their application. In case the shaft excitation were purely torsional, the two pendulums would acquire equal and opposite amplitudes, their reactions forming a pure torque. If, however, the shaft excitation were a purely lateral force, the two pendulums would swing in phase, furnishing a pure force as a reaction. In the case of mixed excitation, the two pendulums would assume different amplitudes such that the sum of their reactions would be a force and a torque, equal and opposite to the excitation. The argument in connection with Fig. 5.37 makes it clear that the pendulum can furnish a reaction force only in a direction perpendicular to the radius *OA*, while along that radial direction it simply acts like a dead body. Thus the two pendulums just discussed cannot prevent motion along the line *OA*. In order to prevent all motion in the plane of the crank throw when the excitation consists of a torque, a lateral force, and a radial force, three pendulums are necessary, located 120 deg. apart for convenience. They will respond with three different amplitudes causing three reactions of which the sum is equal and opposite to the sum of the excitations.

In applying a centrifugal pendulum to a multicylinder engine installation it is important to keep in mind that *the pendulum is not a damper*, so that for any given critical of a certain order and mode the Lewis equation (5.28) applies, stating that there is a dynamic resonant-stress multiplication factor of 50. The complication introduced by the centrifugal pendulum into the calculations is due to the fact that for each critical the equivalent inertia [Eq. (5.39)] is different, because it depends on the order *n*. Hence we must now compute as many Holzer tables as there are suspicious criticals in the running range. This is bewildering in the beginning and requires patience. After a number of Holzer tables are available, one sees more or less how the land lies and calculates with better guesses of ω^2.

In general the practical importance of the centrifugal pendulum has declined somewhat in recent years, and it has been forced out to a large extent by the Houde damper, which is commercially available in a convenient form and operates beneficially over the entire speed range for a large number of critical speeds. (See page 216.)

The last topic to be discussed in this chapter is the influence of firing order on the severity of minor critical speeds, which was mentioned on page 210. First let us investigate how many different possible firing orders there exist in an engine. Start with a four-crank two-cycle engine, in which the cranks are 90 deg. apart. Writing all possible firing sequences systematically we have:

$$1\ 2\ 3\ 4 \qquad 1\ 2\ 4\ 3 \qquad 1\ 3\ 2\ 4 \qquad 1\ 3\ 4\ 2 \qquad 1\ 4\ 2\ 3 \qquad 1\ 4\ 3\ 2$$

which are six possibilities. Starting with number 1, there are three different choices for the second number; and, having chosen the first two numbers, there are two choices for the third number. Thus the number of sequences we can write for a 4-cylinder engine is $3 \cdot 2 \cdot 1 = 3!$, and, in general, for an n-cylinder in-line engine we can write $(n-1)!$ such sequences. Inspecting the six sequences above, we note that the last one is the same as the first one when read backward, from right to left, and the same is true of the other combinations. The firing orders come in mirrored pairs. For the purpose of getting different vector sums like Fig. 5.25, the number of possible different firing orders in an n-cylinder engine is

$$\frac{(n-1)!}{2}$$

This runs into surprisingly large numbers; for example, for a 10-cylinder engine (they have been built) the number is 181,440. When designing such an engine from a fresh start, we should look over these 181,440 possibilities and choose the best one! Each different firing order means a different crank shaft, in which the crank angles are arranged differently along the length of the engine. Many of these crank shafts are undesirable from the standpoint of engine inertia balance (page 180), but after these have been eliminated a large number still remain. The problem then consists in constructing vector diagrams like Fig. 5.25 or 5.26, in which the vector lengths are interchanged with each other in all possible ways, constructing the vector sum, and taking the one with the smallest resultant. This is something we can do only in a new design, starting on paper. Once the engine is built and in trouble, a change of firing order is not a feasible way of correcting the trouble, because it means a new crank shaft and new camshafts.

As an example, consider a four-cylinder engine (two-cycle) with heavy flywheels at both ends in the manner of Fig. 5.27, and with Holzer-amplitudes as follows:

Cylinder number	1	2	3	4
Amplitude	+3	+1	−1	−3

Consider the three possible firing orders:

$$(A)\ 1\ 2\ 3\ 4 \qquad (B)\ 1\ 3\ 2\ 4 \qquad (C)\ 1\ 2\ 4\ 3$$

Construct the vector diagrams, and show that for orders 1, 3, 5, etc., the combination *B* is preferable, giving a vector sum of 2.82, whereas *A* gives 5.64 and *C* gives 6.30. On the other hand, for orders 2, 4, 6, etc., the combination *C* is preferable, giving a vector sum zero, while the other two cases give 4 and 8. This illustrates the fact that by changing the firing order we only shift the bad spot in the spectrum from one critical to another; we cannot eliminate trouble in all speeds by a single firing order. However, the ability to choose a firing order is a powerful tool in the satisfactory torsional design of an engine, particularly if it is required to run in a narrow speed range only.

Problems 139 *to* 172.

CHAPTER 6

ROTATING MACHINERY

6.1. Critical Speeds. Consider a disk of mass m on a shaft running at constant angular speed ω in two bearings, as shown in Fig. 6.1. Let the center of gravity of the disk be at a radial distance e (= eccentricity) from the center of the shaft. If the disk were revolving about the shaft center line, there would be a rotating centrifugal force $m\omega^2 e$ acting on the disk. Such a rotating force can be resolved into its horizontal and vertical components and thus is seen to be equivalent to the sum of a vertical and a horizontal vibratory force of the same amplitude $m\omega^2 e$. Hence we expect the disk to execute simultaneous vertical and horizontal vibrations, and in particular we expect the disk to vibrate violently when

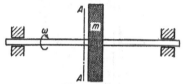

FIG. 6.1. Unbalanced rotating disk.

these impulses are in resonance with the natural frequency, *i.e.*, when the angular speed ω of the shaft coincides with the natural frequency ω_n of the non-rotating disk on its shaft elasticity.

This conclusion is not restricted to a single disk symmetrically mounted on rigid bearings but holds for more complicated systems as well. The speeds at which such violent vibrations occur are known as "critical speeds." In general the critical speeds ω of any circular shaft with several disks running in two or more rigid bearings coincide with the natural frequencies of vibration of the non-rotating shaft on its bearings. The critical speeds can be calculated from the influence numbers in the manner discussed in

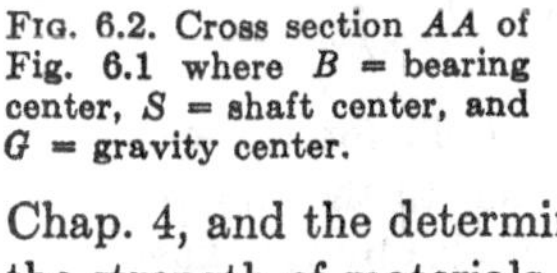

FIG. 6.2. Cross section AA of Fig. 6.1 where B = bearing center, S = shaft center, and G = gravity center.

Chap. 4, and the determination of the influence numbers is a problem in the strength of materials.

The same result can be obtained also in a slightly different manner as follows. Figure 6.2 is drawn in the plane AA of Fig. 6.1 perpendicular to the shaft. The origin of the x-y coordinate system is taken in the point

225

B which is the intersection with the plane AA of the center line connecting the two bearings. In the whirling unbalanced shaft there are three points of importance:

B = the center of the Bearings
S = the center of the Shaft (at the disk)
G = the center of Gravity (of the disk)

In Fig. 6.2 these three points have been drawn in a straight line BSG, which is supposed to rotate about B with the angular velocity ω of the disk. It will be seen that this apparently arbitrary assumption is the only one for which all forces are in equilibrium.

Further let

e = constant distance between S and G (eccentricity).
$r = BS$ = the deflection of the shaft at the disk.

If the effect of gravity be omitted, there are two forces acting on the disk: first, the elastic pull of the shaft which tends to straighten the shaft or

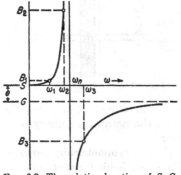

FIG. 6.3. The relative location of S, G, and B for various speeds.

to pull S toward B, and, second, the centrifugal force on the center of gravity G, which point is traveling in a circle of radius $(r + e)$. The first force depends on the bending stiffness of the shaft and is proportional to its deflection; thus we write for it kr (toward the center). The centrifugal force is $m\omega^2(r + e)$ directed from the center outward. For a steady whirling motion these two forces must be in equilibrium:

$$kr = m\omega^2 r + m\omega^2 e \qquad (6.1)$$

and solving for the shaft deflection r,

$$r = e\,\frac{\omega^2}{\dfrac{k}{m} - \omega^2} = e\,\frac{\omega^2}{\omega_n^2 - \omega^2} = e\,\frac{\left(\dfrac{\omega}{\omega_n}\right)^2}{1 - \left(\dfrac{\omega}{\omega_n}\right)^2} \qquad (6.2)$$

This formula coincides with Eq. (2.26) on page 45 for the case of a simple k-m system excited by a force proportional to the square of the frequency. Hence Eq. (6.2) may be represented also by the diagram of Fig. 2.20, which is shown again in Fig. 6.3. Taking the points S and G at the fixed distance e apart, the location of B with respect to these two points at each frequency is the projection of the ordinate of the curve on the vertical axis. It is seen immediately that for very slow rotations ($\omega \approx 0$) the radius of whirl BS is practically zero; at the critical speed, $r = BS$ becomes infinite, while for very large frequencies B coincides with G. Thus at very high speeds the center of gravity remains at rest, which can be easily understood physically, since, if it were not so, the inertia force would become very (infinitely) great.

Equation (6.1) shows that for a perfectly balanced shaft ($e = 0$), the spring force kr and the centrifugal force $m\omega^2 r$ are in equilibrium. Since both are proportional to the deflection, the shaft is in a state of indifferent equilibrium at resonance. It can rotate permanently with any arbitrary amount of bend in it. Whereas below the critical speed the shaft offered some elastic resistance to a sidewise force, this is no longer true at the critical speed. The smallest possible sidewise force causes the deflection to increase indefinitely.

Another interesting conclusion that can be drawn from Fig. 6.3 is that, for speeds below the critical, G lies farther away from the center B than S does, whereas, for speeds above the critical, S lies farther outside. The points S and G are on the same side of B at all speeds. Thus below the critical speed the "heavy side flies out," whereas above the critical speed the "light side flies out."

The inertia force or centrifugal force is proportional to the eccentricity of G, which is $r + e$; and the elastic force is proportional to the eccentricity of S, which is r. The proportionality constants are $m\omega^2$ and k, respectively. For speeds below the critical, $m\omega^2$ is smaller than k, so that $r + e$ must be larger than r since the two forces are in equilibrium. At the critical speed, $r + e$ is equal to r, which necessitates that r be infinitely large. Above the critical speed, $r + e$ is smaller than r, which makes r negative.

It is difficult to understand why the shaft, when it is accelerated gradually, should suddenly reverse the relative positions of the three points B, S, and G at the critical speed. In fact the above analysis states merely that at a *given constant* speed the configuration of the three points, as determined by Fig. 6.3, is the only one at which equilibrium exists between the two forces. Whether that equilibrium is stable or unstable, we do not know as yet. It can be shown that for certain types of friction the equilibrium is stable below as well as above the critical speed.

The stability above the critical speed is due to the *Coriolis* acceleration which is set up as soon as the center of gravity of the disk moves radially away from the center B. Then G is accelerated sidewise and ultimately driven to the other side of B, destroying the collinearity of B, S, and G during the process. If this sidewise escape is prevented, *i.e.*, if the collinearity of the three points is enforced, the equilibrium above the critical speed is indeed *unstable*.

The theory leading to Fig. 6.3 applies also to the system of Fig. 6.4 where the mass m is constrained to move without friction along a straight wire which in turn rotates with speed ω. When $\omega = 0$, the spring is not stretched and the equilibrium position of the mass is at a distance e from the vertical-shaft center. With increasing ω the mass will move more and more toward P, and just below the critical speed it will rest

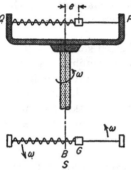

against P. Above the critical speed the equilibrium position of the mass is on the other side (the Q-side) of the vertical shaft, so that the centrifugal force toward Q is in equilibrium with the spring force toward P caused by the compression in the spring. This equilibrium, however, is unstable, as can be easily verified by displacing the mass by a small amount from the equilibrium position. Then the centrifugal force either increases or decreases at a faster rate than the spring force, with the result that the mass flies either to Q or to P, depending on the direction of the small initial displacement. In this experiment the collinearity of the three points B, S, and G is enforced by the wire, and sidewise escape is impossible. While the mass is moving along the wire, the Coriolis effect is felt only as a sidewise pressure on the wire and this does not influence the motion. In case the wire were absent, as in our original set-up of Fig. 6.1, a radial

Fig. 6.4. Rotating wire PQ along which the mass m can slide. This system is *unstable* above the critical speed.

velocity of the mass would be associated with a sidewise acceleration (Coriolis) so that the above argument would be no longer valid.

In order to prove the stability of the system of Fig. 6.1 we have to write Newton's equations for the disk in the general case, *i.e.*, dropping the assumption of collinearity. The only assumption we retain is that the disk rotates at a uniform speed ω about its center S, which is permissible if its moment of inertia is sufficiently large. In Fig. 6.5 the distance SG is constant and equal to e, whereas BS is variable and is denoted by r.

Let the coordinates of S be x and y, then the coordinates x_G and y_G of the center of gravity are $x + e \cos \omega t$ and $y + e \sin \omega t$. The only tangible force acting on the disk is the elastic force kr toward B and this force

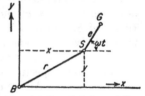

Fig. 6.5. Proof of the stability of the system of Fig. 6.1 above the critical speed.

has the components $-kx$ and $-ky$ along the axes. Newton's equations for the center of gravity G are therefore

$$m\ddot{x}_G = -kx \quad \text{and} \quad m\ddot{y}_G = -ky$$

or written out

$$\begin{aligned} m\ddot{x} + kx &= m\omega^2 e \cdot \cos \omega t \\ m\ddot{y} + ky &= m\omega^2 e \cdot \sin \omega t \end{aligned} \tag{6.3}$$

From Chap. 2 we know that the solution of these equations states that the motion of S in the x-direction as well as in the y-direction is made up of two parts, a free vibration of frequency $\omega_n^2 = k/m$ and a forced vibration of frequency ω. The two forced vibrations in the x- and y-direction being 90 deg. out of phase in time as well as in space make up the steady rotation of Fig. 6.2 (see Problem 37, page 384). If the usual type of friction exists, the free vibrations will be damped out after a time, so that indeed the circular motion with amplitude (6.2) is reached ultimately. The "free vibration" which gradually dies down expresses the sidewise escape from collinearity as before discussed. However, there are types of friction for which the whirl above the critical speed is unstable, as discussed on page 295.

Until now the bearings of the machine have been assumed rigid. By making them flexible the argument already given needs no change whatever, provided the flexibility of the bearings is the same in all directions. The meaning of k, as before, is the number of pounds to be applied at the disk in order to deflect it 1 in. With flexible bearings, k is numerically smaller than with rigid bearings, but that makes no difference in the behavior of the shaft other than somewhat lowering its critical speed.

This situation is slightly altered if the bearings have different flexibility in the horizontal and vertical directions. Usually with pedestal bearings the horizontal flexibility is greater (k is smaller) than the vertical flexibility. We merely split the centrifugal force $m\omega^2e$ into its horizontal and vertical components $m\omega^2e \cos \omega t$ and $m\omega^2e \sin \omega t$ and then investigate the vertical and horizontal motions separately. In Eqs. (6.3) this procedure introduces the difference that k in the x-equation is not the same as the k in the y-equation. At the frequency ω_1, the horizontal motion gets into resonance whereas the amplitude of the vertical motion is still small (Fig. 6.6). The path of the disk center S is an elongated horizontal ellipse. At a greater speed ω_2, there is vertical resonance and the path is an elongated vertical ellipse. Thus there are two

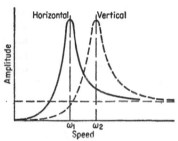

FIG. 6.6. Resonance diagram for a shaft on bearings which are stiffer vertically than they are horizontally.

critical speeds and the shaft can hardly be said to "whirl" at either of them. Rather the shaft center vibrates almost in a straight line at either critical speed.

The generalization of this theory to shafts with many disks on more than two bearings with different flexibilities in the two principal directions is obvious. In general, there will be twice as many critical speeds as there are disks.

6.2. Holzer's Method for Flexural Critical Speeds. The usual method for determining the natural frequencies or critical speeds of shafts or beams in bending is the "iteration" method of Stodola, either in its graphical form (page 157) or its numerical form (page 164). Recently another manner of arriving at the result was suggested by several authors; this method can be properly called an extension of the Holzer method, familiar in torsional calculations, to flexural vibration. The beam in question is first divided into a convenient number of sections 1, 2, 3, etc., just as in Fig. 4.31 (page 157). The mass of each section is calculated, divided into halves, and these halves concentrated at the two ends of each section. Thus the beam is weightless between cuts and at each cut there is a concentrated mass equal to half the sum of the masses of the two adjacent sections. As in the Holzer method, we assume a frequency and proceed from section to section along the beam. In the torsional problem (governed by a second-order differential equation) there are *two* quantities of importance at each cut: the angle φ and the twisting moment, proportional to $d\varphi/dx$ (page 138). In the flexural problem (governed by a fourth-order equation) there are *four* quantities of importance at each cut: the deflection y, the slope $\theta = y' = dy/dx$, the bending moment $M = EIy''$, and the shear force $S = dM/dx = EIy'''$; and it is necessary to find the relations between these quantities from one cut to

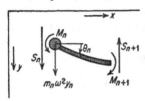

FIG. 6.7.

the next. Figure 6.7 shows the section between the nth cut and the $n + 1st$ cut, together with the various quantities. The sign of these quantities is defined as positive as shown in Fig. 6.7. It is noted that the cut is made at and immediately to the left of the concentrated mass. The mass m_n shown in the figure thus equals half the mass of the section between

cut $n - 1$ and n plus half the mass of the section between n and $n + 1$. Then we can write the following four equations for the section of length l:

$$S_{n+1} = S_n + m_n\omega^2 y_n \tag{a}$$

$$M_{n+1} = M_n + S_{n+1}l \tag{b}$$

$$\theta_{n+1} = \theta_n + \frac{M_{n+1}l}{EI} - \frac{S_{n+1}l^2}{2EI} \tag{c}$$

$$y_{n+1} = y_n + \theta_n l + \frac{M_{n+1}l^2}{2EI} - \frac{S_{n+1}l^3}{3EI} \tag{d}$$

of which (a) and (b) are the equilibrium equations of the section, subject to the inertia force or centrifugal force $m_n\omega^2 y_n$ at the chosen frequency ω^2. The equations (c) and (d) are the deformation equations of the section, considered to be a cantilever built in at the left at the proper angle θ_n, and deformed by the force S_{n+1} and the moment M_{n+1} at its right-hand end.

The equations (a) to (d) allow us to calculate y, θ, M, and S at the right-hand end of a section where they are known at the left-hand end. This can be done with a Holzer table, similar to the familiar one in torsional vibration, but much more elaborate, containing 17 columns instead of 7.

If we start from a simply supported end, where $y = 0$ and $M = EIy'' = 0$, the slope θ and the shear force S are unknown. In the torsional case *only* the amplitude was unknown, which was arbitrarily assumed to be 1.000. Here we assume $\theta = 1.000$ and $S = S_0$. If we have a single span, by the Holzer table we find values for y, θ, M, and S at the other end bearing, all in terms of the symbol S_0 and the assumed numerical value of ω^2 and the assumed slope 1.000. At the end bearing we must have $y = 0$, and from this condition S_0 is calculated numerically and substituted in. Then we find a definite numerical value for the bending moment **M** at the end bearing, which is the counterpart of Holzer's "remainder torque" in the torsional case. Repeating the calculation a number of times for different values of ω^2 and plotting the end moment against ω^2 leads to a curve like Fig. 5.15 on page 190, and the natural frequencies are the zero points of that curve.

The case of a multispan beam is essentially the same. The start is as usual, and upon arriving at the first intermediate bearing we set $y = 0$ and solve for S_0. But there is a new unknown reaction and consequently a new shear force S_1 at the intermediate bearing. Thus, between the first and second intermediate bearings the calculation proceeds as before; only with the unknown symbol S_1 instead of S_0 in the previous span.

Suppose the beam starts with an overhang instead of a bearing-supported end. Then $M = S = 0$ at that end, while y and θ are unknown. We start with $y = 1.000$ and $\theta = \theta_0$ and the calculation is the same as before. For a built-in end $y = \theta = 0$, and we start with $M = 1.000$ and $S = S_0$.

Whereas the calculation for the torsional problem can be carried out with three decimal places on the slide rule, this is no longer feasible for the more complicated case of flexure. Eight or more decimal places are necessary to arrive at a final result accurate to three places, so that calculating machines become essential. In fact the large digital computer has now become an accepted design tool in many large firms, and this iterative Holzer method of finding critical speeds is particularly adapted to computer operations. Longhand calculations by this method would take a month or more for an ordinary system, which is a prohibitive time. Before the day of computing machines the Holzer method was impractical and the Rayleigh or Stodola methods (page 141 and page 155) were used, giving approximate results only. Now the Holzer method is extensively used and the computer yields the answer in a few minutes. If the designer does not like the answer, he makes some appropriate changes in the design and feeds it back to the computer, until a satisfactory design has been evolved. In this process many man-years of longhand calculations are performed by the computer without tiring the designer.

The method can be extended to more complicated cases, for example, to a twisted turbine blade in which the principal axes of bending stiffness turn through an angle along the blade length. Then a vibration in one plane is no longer possible: the motion in the axial direction is coupled to that in the tangential direction. Let the axial direction be x, the tangential direction be y, and the radial direction along the blade be z. First calculate the mass per unit length μ_1, the bending stiffnesses I_z and I_y in the axial and tangential directions, and the product of inertia I_{zy}, or stiffness coupling. All of these are variable with the length z. Then cut the blade into a number of sections; take the average value of μ_1, I_z, I_y, and I_{zy} of each section and consider these values constant along each section, as in Fig. 6.7. The vibration takes place in the x and y coordinates simultaneously so that there will be eight equations instead of the four of page 230. Equations (a) and (b) still hold, but there is one of each for the x and for the y directions. Equations (c) and (d) become more complicated since there is cross coupling on account of the product of inertia I_{zy} term. The eight equations are given below without derivation.

$$S_{x,n+1} = S_{x,n} + m_n\omega^2 x_n$$
$$S_{y,n+1} = S_{y,n} + m_n\omega^2 y_n$$
$$M_{x,n+1} = M_{x,n} + S_{x,n+1}l$$
$$M_{y,n+1} = M_{y,n} + S_{y,n+1}l$$

$$x'_{n+1} = x'_n + \frac{l}{E(I_{zn}I_{yn} - I_{zyn}^2)}\left[I_{yn}\left(M_{x,n+1} - \frac{S_{x,n+1}}{2}l\right) + I_{zyn}\left(M_{y,n+1} - \frac{S_{y,n+1}}{2}l\right)\right]$$

$$y'_{n+1} = y'_n + \frac{l}{E(I_{zn}I_{yn} - I_{zyn}^2)}\left[I_{zn}\left(M_{y,n+1} - \frac{S_{y,n+1}}{2}l\right) + I_{zyn}\left(M_{x,n+1} - \frac{S_{x,n+1}}{2}l\right)\right]$$

$$x_{n+1} = x_n + x'_n l$$
$$+ \frac{l^2}{E(I_{zn}I_{yn} - I_{zyn}^2)}\left[I_{yn}\left(M_{x,n+1} - \frac{2}{3}S_{x,n+1}l\right) + I_{zyn}\left(M_{y,n+1} - \frac{2}{3}S_{y,n+1}l\right)\right]$$

$$y_{n+1} = y_n + y'_n l$$
$$+ \frac{l^2}{E(I_{zn}I_{yn} - I_{zyn}^2)}\left[I_{zn}\left(M_{y,n+1} - \frac{2}{3}S_{y,n+1}l\right) + I_{zyn}\left(M_{x,n+1} - \frac{2}{3}S_{x,n+1}l\right)\right]$$

Now start at the built-in bottom of the blade, at station zero with $x = y = x' = y' = 0$. The bending moments and shear forces at the root are unknown. Take $M_x = 1.000$, $M_y = M_{y0}$, $S_x = S_{x0}$, and $S_y = S_{y0}$ at the root. Start the calculation with an assumed frequency ω^2 and proceed from station to station to the free end of the blade. All quantities in the calculation will carry the three unknown letters S_{x0}, S_{y0}, and M_{y0}. At the free end we have $M_x = M_y = S_x = S_y = 0$, which are four equations. From three out of these four solve for S_{x0}, S_{y0}, and M_{y0} and substitute these values into the fourth one, which then will *not* come out zero, as it should. Take, for example, M_x, which will give $M_{x,\text{end}}$ instead of zero. Plot one point on the diagram $M_{x,\text{end}} = f(\omega^2)$. Repeat the process many times for other frequencies and plot the curve completely. The zeros of this curve are the natural frequencies. This method is quite impossible without a computer: it means months of tedious work. Once it is "programmed" for the computer, however, we receive the entire $M_{x,\text{end}} = f(\omega^2)$ curve in a matter of minutes.

6.3. Balancing of Solid Rotors.

The disk of Fig. 6.1, of which the center of gravity lies at a distance of e in. from the shaft center, will vibrate and also will cause rotating forces to be transmitted to the bearings. The vibration and the bearing forces can be made to disappear by attaching a small weight to the "light side" of the disk so as to bring its center of gravity G in coincidence with the shaft center S. If the original eccentricity is e, the disk mass M, and the correction mass m, applied at a radial distance r from S, then

$$mr = eM \qquad \text{or} \qquad m = \frac{e}{r} M$$

The "unbalance" mr of the disk is usually measured in "inch ounces." It is, of course, correct to double the balance weight for a given disk if the double weight is applied at half the original radius, since the centrifugal force is proportional to the product mr.

The determination of the location of the correction is a problem of statics. The shaft can be placed on two parallel horizontal rails, for example, then the heavy spot will roll down, and a correction weight can be attached tentatively to the top side of the disk. The amount of this weight is then varied until the disk is in indifferent equilibrium, *i.e.*, shows no tendency to roll when placed in any position. In order to minimize the errors of such a procedure (or as is sometimes said, in order to increase the sensitivity of the balancing machine), the rails must be made of hard steel and must be firmly embedded in heavy concrete, so that their deformation under the load is as small as possible.

The set of horizontal rails is the simplest *static balancing machine* in existence. For machines in which the rotating mass is of disk form, *i.e.*, has no great dimensions along the axis, static balance is the only balance required to insure quiet operation at all speeds.

In case the rotor is an elongated body, static balance alone is not sufficient. Figure 6.8 shows a rotor which is supposed to be "ideal," *i.e.*, of perfect rotational symmetry, except that two equal masses, m_1 and m_2, are attached to two symmetrically opposite points. The rotor is evidently still in static balance, since the two masses do not remove the center of gravity from the shaft center line. When in rotation, the centrifugal forces on m_1 and m_2 form a moment which causes rotating reactions R on the bearings as indicated. This rotor is said to be statically balanced but *dynamically unbalanced*, because this type of unbalance can

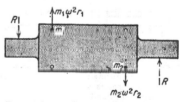

Fig. 6.8. A dynamically unbalanced rotor causes equal and opposite rotating reactions on its bearings.

be detected by a dynamic test only, while on a static balancing machine the rotor appears to be perfect.

We shall now prove that any unbalance whatever in a rigid rotor (static, dynamic, or combined) can be corrected by placing appropriate correction weights in two planes, the end planes I and II of the rotor usually being chosen on account of their easy accessibility (Fig. 6.9). Let the existing unbalance mr consist of 4 in. oz. at one-quarter of the length of the rotor and of 3 in. oz. in the middle between the planes I and II but turned 90 deg. with respect to the first unbalance. In determining the corrective masses to be placed in the planes I and II, we shall first find the corrections for the 4-unit unbalance, then find them for the 3-unit unbalance, and finally add the individual corrections together. The 4-unit unbalance will cause a 4-unit rotating centrifugal force, which can be held in static equilibrium by a 3-unit force at I and by a one-unit force at II. Thus we have to place a 3-unit correction mass in plane I, 180 deg. away from the original unbalance, and similarly a single-unit correction mass in plane II, also 180 deg. away from the original unbalance.

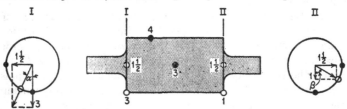

FIG. 6.9. The most general unbalance in a rigid rotor can be corrected by placing one weight in each of two planes I and II.

The 3-unit unbalance is corrected by 1½-unit masses in each of the two planes. Thus in total we have to place in plane I a 3-unit mass and a 1½-unit mass, 90 deg apart. The two centrifugal forces due to these can be added together by the parallelogram of forces so that instead of placing two correction masses in plane I we insert a single mass of $\sqrt{(3)^2 + (1\frac{1}{2})^2} = 3.36$ units at an angle $\alpha = \tan^{-1} 0.5$ from the diameter of the 4-unit unbalance. Similarly, the total correction in plane II consists of a correction mass of $\sqrt{1 + (1\frac{1}{2})^2} = 1.80$ units at an angle $\beta = \tan^{-1} 1.5$ from the same diameter.

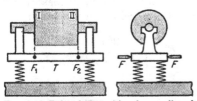

FIG. 6.10. Balancing machine for small and medium size rotors with two interchangeable fulcrums F_1 and F_2.

The process can be extended to a larger number of unbalanced masses, so that any unbalance in a rigid rotor can be corrected by a single mass in each of the two balancing planes.

In any given rotor the size and location of the existing unbalance are unknown. They can be determined in a dynamic balancing machine. A type of construction of such a machine, used for small and medium-sized rotors, is shown in Fig. 6.10. The rotor is put in two bearings which are rigidly attached to a light table T. This table in turn is supported on springs and can be made rotatable about either one of two fulcrum axes F_1 or F_2, located in the two balancing planes I and II. The rotor is driven either by a belt or by a flexible shaft, in which cases the driving motor is separate from the table T, or sometimes is driven by direct coupling to a small motor rigidly mounted on T. The latter scheme increases the weight of the table, which is undesirable. The drive is not shown in the figure.

The balancing process is as follows. Make F_1 a fulcrum by releasing F_2 and run the rotor until it, together with the table, comes to resonance on the springs. The maximum oscillating motion takes place at the right-hand end of T, and its amplitude is read on a dial indicator. By a series of operations to be described presently, the location and magnitude of the correction weight in the plane II are determined. With this weight inserted, the rotor and table do not vibrate at all. Any unbalance which still may exist in the rotor cannot have a moment about the fulcrum F_1, so that such unbalance must have a resultant located in plane I.

Next, fulcrum F_1 is released and fulcrum F_2 is tightened, and the correction weight in plane I is determined by the same process, to be described. After this correction has been applied, the moments of all centrifugal forces are zero about the axes through F_1 and F_2. But, then, by the rules of statics, there can be no moment about any other axis, and the rotor is balanced completely.

Now we proceed to discuss how the correction weights can be determined. Apparently the simplest method is by means of the phase-angle relation shown in Fig. 2.22b, page 51. If a pencil or a piece of chalk is held very close to the rotating and whirling shaft, it will "scribe the heavy spot" when the shaft runs below its critical speed; it will "scribe the light spot" when above resonance, while exactly *at* the critical speed it will scribe at a point which is 90 deg behind the heavy spot. Thus the *location* of the unbalance can be found by scribing, and the *magnitude* of the correction is then determined by a few trials.

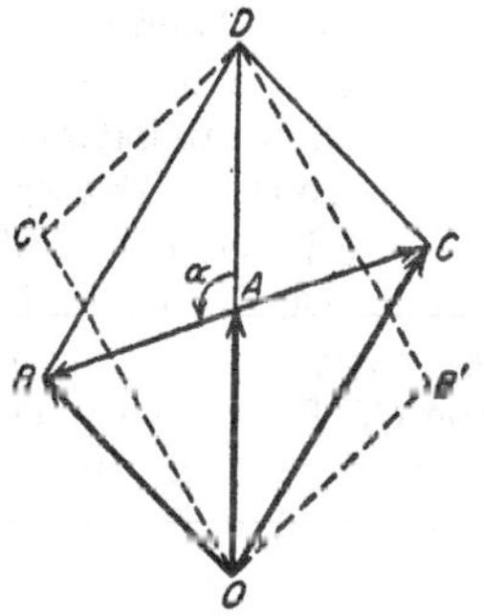

In practice this phase-angle method is very inaccurate, since near resonance the phase angle varies rapidly with small variations in speed, whereas at speeds markedly different from the critical the amplitudes of the vibration are so small that no satisfactory scribing can be obtained.

FIG. 6.11. Vector diagram for determining the unbalance in a plane by three or four observations of amplitude.

A more reliable method is based on observations of the amplitude only. It consists of conducting three test runs with the rotor in three different conditions: (1) without any additions to the rotor, (2) with a unit unbalance weight placed in an arbitrary hole of the rotor, and (3) with the same unbalance weight placed in the diametrically opposite hole. In Fig. 6.11

let OA represent to a certain scale the original unbalance in the rotor and also, to another scale, the vibrational amplitude observed as a result of this unbalance at a certain speed. Similarly let OB represent vectorially the total unbalance of the rotor after the unit addition has been placed in the first hole. It is seen that the vector OB may be considered as the sum of the vectors OA and AB, where AB now represents the extra unbalance introduced. If now this unbalance is removed and replaced in the diametrically opposite hole, necessarily the new additional unbalance is represented by the vector AC equal and opposite to AB, and consequently the vector OC, being the sum of the original unbalance OA and the addition AC, represents the complete unbalance in the third run.

As a result of the amplitude observations in these three runs, we know the relative lengths of the vectors OB, OA, and OC, but we do not as yet know their absolute lengths or their angular relationships. However, we do know that OA must be the median of the triangle OBC and the problem therefore consists in constructing a triangle OBC, of which are known the ratios of two sides and a median. Its construction by Euclid's geometry is carried out by doubling the length OA to OD and then observing that in the triangle ODC the side DC is equal to OB, so that in triangle OCD all three sides are known. Thus the triangle can be constructed, and as soon as this has been done we know the relative lengths of AB and OA. Since AB represents a known unbalance weight artificially introduced, we can deduce from it the magnitude of the original unknown unbalance OA. Also the angular location α of the original unbalance OA with respect to the known angular location AB is known.

There is one ambiguity in this construction. In finding the original triangle OCD, we might have obtained the triangle $OC'D$ instead. Consequently we would have obtained the direction $C'B'$ instead of the direction CB for our artificially introduced unbalances. This ambiguity can be removed by a fourth run which also will act as a check on the accuracy of the previous observations. It is noted that in the construction of Fig. 6.11 no assumptions have been made other than that the system is linear, i.e., that all vibration amplitudes are proportional to the unbalance masses. This relation is not entirely true for actual rotors but it is a good approximation to the truth. If after going through the motions shown in Fig. 6.11 and if after inserting the correction weight so found there still is vibration present in the machinery, that vibration will be very much less than the original one and the process of Fig. 6.11 may be repeated once more.

In factories where great numbers of small- or moderate-sized motors have to be balanced as a routine operation, the process of Fig. 6.11 takes too much time. For such applications the movable fulcrum machine of Fig. 6.10 was developed into an intricate precision apparatus in which the balancing is done by means of a so-called "balancing head."

A *balancing head* is an apparatus which is solidly coupled to the rotor to be tested and which contains two arms with weights (Fig. 6.12). These arms rotate with the rotor and keep the same relative position with respect to it, at least as long as the operator does not interfere. The possibility of rotating these arms *relative* to the rotor exists in the form of an intricate system of gears, clutches, and magnets or motors. The power for its operation is introduced necessarily through slip rings, since the whole head is rotating. The operator has before him two buttons. If he presses the first one, the two arms rotate in the same direction; if he presses the second one, the arms rotate in opposite directions at the rate of about one revolution per 5 sec. relative to the rotor in each case.

Fig. 6.12. A balancing head with two unbalanced arms.

Since the two arms form the only unbalance in the head, this makes it possible for the operator to change the magnitude as well as the direction of the added unbalance. By letting the two masses rotate in the same direction (button 1) and watching the vibration indicator, a maximum and a minimum amplitude appear every 5 sec. After taking his finger off button 1 at the minimum amplitude, the operator makes the two arms rotate against each other by pressing button 2. Since during this operation the bisecting line of arms remains at rest with respect to the rotor, the direction of the additional unbalance does not change, but the magnitude varies from two masses (when the arms coincide) to zero (when they are 180 deg apart). After the vibration has been reduced to zero, the rotor is stopped and from the position of the arms in the head the desired correction is determined immediately. As before, the process has to be performed twice for different locations of the fulcrum.

Another entirely different balancing head is the one invented by *Thearle* (1930). The machine is of the type of Fig. 6.10 with two fulcrums and with a head like Fig. 6.12 but with the important difference that the two arms are entirely free to rotate with respect to the rotor, except for the possibility of clamping them. There are no gears or magnets, merely a clutch which either clamps or releases the arms. In operation the arms are first clamped and the machine brought to above its critical speed. Upon releasing the arms, *they will automatically seek the position of complete balance* where all vibration ceases. They are clamped again in that position and the rotor is brought to rest.

The theory of operation of this device is very interesting. Suppose that the two arms are clamped in a 180-deg. position so that the head with the arms included is in perfect balance. The only unbalance in the system is in the rotor.

In Fig. 6.13, let B (center line of bearings), S (center of shaft, *i.e.*, balancing head), and G (center of gravity) have the usual meanings. We know from Fig. 6.3 that these three points appear in different sequences for speeds below and above the critical speed. The whirl of the whole assembly is about the bearing center line B, so that the centrifugal forces acting on the clamped arms must be directed away from B.

If at some speed below the critical (Fig. 6.13a) the arms are released, then the centrifugal forces will turn them toward each other to the top of the figure. Having arrived there they find themselves on the side of G, *i.e.*, on the *heavy* side. On the other hand, if they are released above the critical speed, Fig. 6.13b shows that the centrifugal forces tend to drive the arms again to the top of the figure, which is now the *light* side. In coming closer together, the arms bring the location of G up and, after they have gone a certain distance, G coincides with S (and also with B) and all vibration ceases.

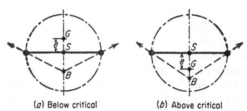

(a) Below critical   (b) Above critical

FIG. 6.13. Explains the Thearle balancing machine.

In practice the two arms are replaced by two balls which can roll in a circular concentric track. It is interesting to note that an early invention (about 1900) by Leblanc called for such a concentric track or tube filled partially with mercury. This was supposed to ensure balance automatically, and the theory given was that of Thearle and Fig. 6.13 above. However, the two arms or balls will work, whereas mercury in the same track will not work. This can be recognized by remarking that, when self-balance is achieved, the track center coincides with point S and the centrifugal forces then are everywhere perpendicular to the circular track. Any one ball is then in indifferent equilibrium in any position along the track, but the mercury will distribute itself evenly all around. Whereas two balls or arms can furnish a counter unbalance with a track centric with S, the mercury cannot do so. Hence Leblanc's invention was in error.

Recently (1947) this principle was applied to a domestic washing machine (Thor washing machine, U.S. Patent 2,420,592). The bowl of this machine, spinning at about 600 r.p.m. with a damp wad of clothes in it (centrifugal drying process), is fitted with two loose rings. They are held centrally clamped at low speeds, and above the critical speed they are released and automatically assume positions counteracting the considerable unbalance of the clothes.

With the modern developments in radio technique, it is now no longer necessary to run balancing machines at resonance. They can be operated at speeds well removed from the resonant one, the very small vibrations at the bearing being picked up by electronic devices, of which the output can be amplified to any desired degree. Machines utilizing such amplifiers are discussed in the next section.

6.4. Simultaneous Balancing in Two Planes. It is possible to simplify the methods of balancing described in the previous section if means are available to measure the phase angle between the location of the unbalance in the rotor and the "high spot" of the vibration. Let the rotor be supported in two bearings a and b which are flexible in, say, the horizontal direction and stiffly supported vertically. The balancing planes I and II do not coincide with the locations of the bearings a and b. Now imagine the rotor to be ideally balanced so that while it is rotating in the bearings no bearing vibration occurs. Then unbalance the ideal rotor by placing a unit weight in the angular location 0° of balancing plane I. This will cause a vibration in both bearings and these vibrations are denoted as α_{aI} and α_{bI}, where the first subscript denotes the bearing at which the vibration occurs and the second subscript denotes the balancing plane in which the unit unbalance at zero angular location has been placed. When there is no damping in the system, these numbers α are real numbers, by which we mean that the maximum displacement in the horizontal direction of the bearings occurs at the same instant that the unbalance weight finds itself at the end of a horizontal radius. If there is damping in the system, there will be some phase angle between the unbalance radius and the horizontal at the moment that the bearing has its maximum displacement, and this condition can be taken care of by assigning complex values to the α numbers.

In a similar manner the ideal rotor may be unbalanced with a unit weight in the zero angular location of plane II, which then causes the bearing vibrations α_{aII} and α_{bII}. The four numbers α so found are known as the complex *dynamic influence numbers* of the set-up. The rotor is run well above its critical speed so that the phase angles are 180 deg. and the influence numbers are real. These four influence numbers completely determine the elastic and inertia properties of the system for the r.p.m. at which they are determined, but they are entirely independent of the amount of unbalance present.

Next suppose that the unbalance in plane I is not a unit unbalance at zero angular location but an unbalance which numerically as well as angularly differs from the unit unbalance, and is represented by the complex number $\bar{U}_{\text{I}}$. Then this unbalance $\bar{U}_{\text{I}}$ will cause a vibration at the bearing a, expressed by the product $\alpha_{a\text{I}}\bar{U}_{\text{I}}$.

With these notations, it is now possible to write the vibration vectors $\bar{V}$ at the two bearings in terms of a general unbalance $\bar{U}_{\text{I}}$ and $\bar{U}_{\text{II}}$ as follows:

$$\left. \begin{array}{l} \bar{V}_a = \alpha_{a\text{I}}\bar{U}_{\text{I}} + \alpha_{a\text{II}}\bar{U}_{\text{II}} \\ \bar{V}_b = \alpha_{b\text{I}}\bar{U}_{\text{I}} + \alpha_{b\text{II}}\bar{U}_{\text{II}} \end{array} \right\} \qquad (6.4)$$

It is possible to measure the vibration vectors $\bar{V}_a$ and $\bar{V}_b$ and calculate from them by means of the set (6.4) the unknown unbalance vectors $\bar{U}_{\text{I}}$ and $\bar{U}_{\text{II}}$, with the following result:

$$\left. \begin{array}{l} \bar{U}_1 = \dfrac{\alpha_{b\text{II}}}{\Delta}\,\bar{V}_a - \dfrac{\alpha_{a\text{II}}}{\Delta}\,\bar{V}_b \\[2mm] \bar{U}_2 = \dfrac{\alpha_{a\text{I}}}{\Delta}\,\bar{V}_b - \dfrac{\alpha_{b\text{I}}}{\Delta}\,\bar{V}_a \end{array} \right\} \qquad (6.4a)$$

In these equations $\Delta = \alpha_{a\text{I}} \cdot \alpha_{b\text{II}} - \alpha_{b\text{I}}\alpha_{a\text{II}}$ is the determinant of the coefficients of Eq. (6.4). The set of equations (6.4a) enables us to calculate the unknown unbalance vectors if we can measure the vibration vectors at the two bearings and if we know the four dynamic influence numbers.

These $\bar{V}$-vectors can be measured in various ways. A very convenient method consists of inverted loud-speaker elements such as are described on page 62. These elements are attached to the two bearing shells a and b of the balancing machine, and their output is an electric alternating voltage which in magnitude and phase determines the vibration vector. The *Gisholt-Westinghouse* balancing machine uses such elements and also has an electric circuit by which Eqs. (6.4a) are automatically solved. In order to understand the operation of this circuit, shown in Fig. 6.14, we rewrite the first of Eqs. (6.4a) as follows:

$$\bar{U}_{\text{I}} \cdot \frac{\Delta}{\alpha_{b\text{II}}} = \bar{V}_a - \frac{\alpha_{a\text{II}}}{\alpha_{b\text{II}}} \cdot \bar{V}_b \qquad (6.5)$$

In this equation we notice that the ratio $\alpha_{a\text{II}}/\alpha_{b\text{II}}$ is smaller than 1 because the numerator is the response of a bearing to a unit unbalance far away from it, while the denominator is the response to an unbalance close to it. In all ordinary systems this ratio is smaller than unity. Thus we see from Eq. (6.5) that the unbalance in plane I is found by taking the vibration vector of bearing a, subtracting from it a *fraction* of the vibration vector of bearing b, and multiplying the result by $\alpha_{b\text{II}}/\Delta$. The fraction of $\bar{V}_b$ in general is a complex fraction but it is made real by running the machine at a speed far above its resonance. The subtraction of these two quantities is accomplished in Fig. 6.14 by connecting in series the full output of the loud-speaker coil on bearing a with a fraction of the voltage output of loud-speaker coil $\bar{V}_b$. This fraction is picked off by a potentiometer knob 1. In this way it is possible to adjust that fraction to any real number smaller than one. The fact that there is a minus sign on the right-hand side of Eq. (6.5) instead of a plus sign has no further importance than that the terminals of one of the coils have to be reversed. The voltage representing the right-hand side of Eq. (6.5) is then fed into an amplifier, and the amplified voltage is multiplied by the number α_{bII}/Δ, by picking off a fraction of it through the potentiometer knob 2. The output of the circuit is then read on a milliammeter and is simultaneously used to actuate a stroboscopic lamp which flashes once per revolution of the rotor. If it is only possible to set knob 1 so as to represent the ratio $\alpha_{aII}/\alpha_{bII}$ and to set knob 2 so as to represent the ratio α_{bII}/Δ, then the milliammeter to a certain scale will read directly the amount of the unbalance, while the stroboscopic lamp will apparently

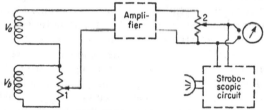

FIG. 6.14. Circuit diagram of the Gisholt-Westinghouse balancing machine. (*J. G. Baker.*)

freeze the rotor at such an angular position that a fixed needle points at the angular location of the unbalance.

The circuit thus described solves the first of Eqs. (6.5). For the solution of the second equation (6.5) it is necessary to combine the full output of $\bar{V}_b$ with a fraction of $\bar{V}_a$ and multiply the amplified output by a different number. This is done by a new circuit with knobs 3 and 4 instead of 1 and 2 in a similar manner.

The interesting feature of this circuit is that the proper setting of these knobs is not calculated but found by a series of very simple experiments. Suppose that a large number of identical rotors have to be balanced in a mass-production process. We start with balancing one rotor in any convenient manner until it is perfect and this may take us a considerable time. This *perfect* rotor placed in the two bearings a and b will cause no vibration in them, therefore no voltage $\bar{V}_a$ or $\bar{V}_b$, and hence no reading in the milliammeter. Then a unit unbalance at zero angular location is deliberately placed in plane I. This ought to cause a unit reading on the milliammeter and a zero angular reading on the stroboscope in the case where the circuit of Fig. 6.14 with knobs 1, 2 is switched in, *i.e.*, in the case where a left-right switch is set on the position I. If this switch is set to the position II, the other circuit with knobs 3 and 4 is in force and the milliammeter ought to give a zero reading. Naturally, these readings will not be as they should, because the four adjustments have not been made as yet. It follows from Eq. (6.5) that with the switch in the position II the zero reading on the milliammeter (due to unit unbalance in plane I) is not affected by the knob 4 but can be accomplished entirely by 3. We therefore turn knob 3 until the milliammeter reading becomes zero.

Now the unit unbalance in plane I is removed and brought to plane II, while the selector switch is thrown to position I. Again the milliammeter should read zero, which is accomplished quickly by adjusting knob 1. Now, leaving the unit unbalance in plane II, the selector switch is thrown to the position II and the knob 4 is adjusted until the milliammeter reads a unit unbalance and the stroboscope a zero angular position. Finally, the unit unbalance is brought back to plane I, the selector switch is set on plane I, and knob 2 adjusted to get unit reading on the milliammeter and zero angular reading on the stroboscope. This process of making the four adjustments takes only a few minutes for an experienced operator, and thereafter these adjustments are correct for every other rotor in the series to be balanced. The balancing process then consists of placing an unbalanced rotor in the bearings, starting the rotor by a foot-operated switch (belt drive), reading the milliammeter and the angular position, throwing the selector switch to the other side, and again reading the unbalance numerically as well as angularly. This process takes only a few seconds and is extremely accurate.

In cases where a *single* rotor has to be balanced instead of a whole series in mass production, such as, for example, a turbine or a generator rotor in its own bearings in a powerhouse, the problem is to produce one "ideal rotor." The procedure outlined above does not solve the difficulty, but it is still possible to use the apparatus of Fig. 6.14 by a clever expedient, due to J. G. Baker, which consists of fooling the circuit of Fig. 6.14 into believing that it deals with an ideal rotor, whereas in reality it deals with an ordinary unbalanced rotor. For this purpose two small alternating-current generators are made to be driven by the turbine to be balanced. These generators produce currents of a frequency equal to that of the r.p.m., and their voltage output can be regulated in magnitude as well as in phase. Now the circuit of Fig. 6.14 is opened in two spots at the two coils V_a and V_b. The output of the generators, suitably modified, is now fed into these openings and regulated so that the voltage induced by the vibration in each pickup coil is bucked by an equal and opposite voltage artificially introduced by the generators. With this set-up the circuit of Fig. 6.14 gets no impulses and therefore reacts as if an ideal rotor were run. Now with the bucking voltages in force the three runs of the existing rotor are made: (1) "as is," (2) with a unit unbalance in plane I, and (3) with a unit unbalance in plane II. In this manner the adjustment on all four knobs is carried through as outlined above. After this, the artificially introduced bucking voltages are removed and now the circuit responds to the actual rotor with the existing unbalance in it.

A still simpler method of balancing without fulcrums is suggested by Eqs. (6.4). It is clear that the vibration readings can be made at such a position along the rotor that the influence numbers α_{aII} and α_{bI} become zero. This means that the measurement $\bar{V}_a$ (or $\bar{V}_b$) has to be made at a position along the rotor which will not experience any vibration if an unbalanced weight is placed in plane II (or I). This position is known as the "center of percussion," belonging to the "center of shock" II (or I). In that case each loud-speaker element or other type of electrical indicator reads only the vibration caused by one of the balancing planes alone and, instead of solving a set of four algebraic simultaneous equations (6.4) with four unknowns, the problem is reduced to finding a solution to two sets of two unknowns each. This method has been used for some time in a machine developed by the General Motors Research Laboratory.

6.5. Balancing of Flexible Rotors; Field Balancing. In discussing the effects of unbalanced masses in the last two sections, we have assumed that the rotor was not deformed by them. When running at speeds far below the first critical, this assumption is perfectly justified, but for speeds higher than about half of the first critical the rotor assumes deformations which can no longer be neglected since they set up new centrifugal forces in addition to the ones caused by the original unbalance. If, for example, a unit unbalance is located in the center of a symmetrical rigid rotor, the unit centrifugal force due to this unbalance will have reactions of half a unit at each of the bearings. On the other hand, if the rotor is flexible, the unit centrifugal force will put a bend in the structure and bring its center line off the original position. Consequently, the bent center line whirls around and additional centrifugal forces are set up which will alter the bearing reactions.

The machine can evidently be balanced by adding a corrective mass in the middle directly opposite the original unbalance. But we prefer to balance it in two definite planes near the ends. Assume that the rotor consists of a straight uniform shaft and that the balancing planes are at one-sixth of the total length from each end. Evidently the rigid rotor will be balanced by putting in corrections of magnitude $\frac{1}{2}$ in each plane (Fig. 6.15a).

When the unbalanced rotor is running at its first critical speed, its deflection curve is a sinusoid (page 150) of which the amplitude is so large that the newly "induced" unbalance is far greater than the original unit unbalance. Thus the original unbalance does not influence the shape of the deflection curve, which at the balancing planes has half the amplitude of the middle. The proper corrections have to be of the same amount as the original unbalance. This can be understood by bending the shaft a little more. The centrifugal forces of the shape itself (exclusive of the original unbalance) are in equilibrium with the elastic forces at any position of the shaft, since there is resonance. When increasing the deflection at the center by δ, the work done by the unbalance is $\delta \times 1$ and the work done by each of the two correction weights $\frac{1}{2}\delta \times 1$. It is seen that the equilibrium remains indifferent (characteristic of a balanced rotor at a critical speed) when the correction weights are made a full unit each (Fig. 6.15b).

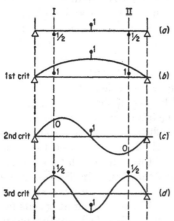

FIG. 6.15. The proper correction weights to be inserted in the planes I and II vary with the speed in a flexible rotor.

At the second critical speed the central unbalance is not displaced in position so that no correction weights are necessary. At the third critical speed the correction weights have to be made half a unit on the side opposite to where they were at slow speeds (Fig. 6.15c, d).

We thus draw the conclusion that *a flexible rotor can be balanced in two planes for a single speed only;* as a rule the machine will become unbalanced again at any other speed. Large turbine spindles or turbogenerator-rotors in modern applications usually run between their first and second critical speeds. When such units are balanced at a rather low speed in the machine sketched in Fig. 6.10, they quite often become rough when run at full speed in their permanent bearings. This is one of the reasons why shop balancing is not sufficient, and why such machines have to be balanced again in the field under service conditions. In the field no movable fulcrums are available and the process of balancing takes a considerable time. As a rule, the amplitude method discussed on page 235 is applied, but in order to secure good balance it is necessary to repeat the operation a number of times, shuttling back and forth from one balancing plane to the other.

Since the year 1950 the rotors of large turbogenerators have become so long and slender that they run between the second and third critical speeds. Balancing in two planes is no longer sufficient and it is now practice in some of the larger firms to balance these spindles in the shop in three or more balancing planes. While, theoretically, two-plane balancing at one particular running speed is possible, in practice it happens sometimes that the correction weights required become impossibly large and then balancing in three or sometimes four planes is indicated. In principle we can generalize the procedure of page 234, where vibration measurements in two planes determine the necessary correction weight in two other planes by means of two simultaneous vector equations (6.4). Then when we make n measurements in n locations we can find n vector equations determining the necessary corrections in n planes.

There are cases on record where even several weeks of systematic field balancing did not produce a smooth machine. In such cases the trouble is evidently caused by something other than unbalance. In one particular machine it was found that a careless workman had dropped a balancing weight in the hollow interior of a turbine spindle and had failed to report the fact. Consequently a loose weight of 1 lb. was flying around freely in that space, and it was impossible to balance the machine.

A remarkable series of cases of steam-turbine vibration, observed off and on during the last fifteen years, has now been explained. The turbine would vibrate with the frequency of its rotation, obviously caused by unbalance, but the intensity of the vibration would vary periodically and extremely slowly. On some turbines the period of time between two consecutive maxima of vibration intensity was as low as 15 min.; on others this period was as much as 5 hr. The seriousness of the trouble consisted in the fact that each maximum was worse than its predecessor, so that after half a dozen of these cycles the machine had to be shut down.

Observations were made of the phase angle of the vibration, *i.e.*, the angle between the vertical and the radial direction of a definite point of the rotor at the instant that, say, the horizontal vibrational displacement of a bearing was maximum to the right. This angle was observed by watching the needle of a vibrometer placed on the bearing by a stroboscopic light, flashing once per revolution and operated by a contactor driven off the rotor. The phase angle was found to increase indefinitely, growing by 360 deg. each time the vibration reached a maximum. This was explained as a "rotating unbalance" which would creep through the rotor and which would be additive to the original steady unbalance when the vibration was a maximum and in phase opposition

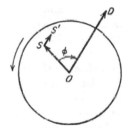

FIG. 6.16. This illustrates a spiral wandering of the unbalance within the rotor caused by unsymmetrical heating or cooling.

to the steady unbalance at times of minimum vibration. A detailed explanation of how an unbalance can creep so slowly through a rotor has been given by R. P. Kroon, as follows.

Let Fig. 6.16 represent a cross section of the rotor and let the vector $\overline{OS}$ be the static unbalance of the rotor: *i.e.*, O is the geometric center of the rotor and S is its center of gravity when not rotating. For very slow rotation the rotor will bow out under the influence of the centrifugal force in the direction OS, but at higher speeds the "high spot" will no longer coincide with the "heavy spot" S. The high spot will be given by the dynamic unbalance vector $\overline{OD}$, where D is the location of the geometric center of the rotor while running. Since $\overline{OS}$ is the "force" and $\overline{OD}$ the "displacement," the result of Figs. 2.21 and 2.22b can be applied, from which it is seen that the high spot always trails behind the heavy spot by an angle φ which is less than 90 deg. below resonance and between 90 and 180 deg. above resonance.

The unsymmetry in the direction $\overline{OD}$ caused by the bowing out of the rotor may be the cause of local heating at D. This may be in the form of actual rubbing on the periphery at D or, in the case of a hollow rotor, may be the result of condensation. The water droplets from the condensing steam will be moved by centrifugal force to D, thus causing further condensation and heating at that point. The heating at D in turn causes the rotor fibers to expand, thus producing an elastic bowing out of the rotor with a consequent shift in the location of the center of gravity. The point S therefore shifts to S', the vector $\overline{SS'}$ having the direction $\overline{OD}$. The new static unbalance $\overline{OS'}$ is angularly displaced with respect to $\overline{OS}$; the angle φ remains the same, so that $\overline{OD}$ also shifts clockwise. In this manner we see a slow rotation of the unbalance in a direction opposite to that of rotation. Also $\overline{OS'}$ is slightly greater than $\overline{OS}$, so that the result will be that the point S describes a spiral within the rotor. We have thus seen that for local heating at the high spot below resonant speed we get a retrograde and increasing spiral. In a similar manner it is shown that above resonant speed ($\varphi > 90$ deg.) the spiral is still retrograde but now decreasing. If there happens to be cooling at the high spot, instead of heating, the spiral is forward and decreasing below resonance, forward and increasing above that speed.

For very flexible rotors, running well above their first critical speed and close to the second critical, the phase angle φ becomes greater than 180 deg. and the analysis in terms of a single degree of freedom can no longer be applied. However, the general reasoning is the same; only the value of φ is different.

In this connection it may be of interest to mention another temperature effect observed in steam turbines. After a turbine has stood still for some time, the temperature of the top fibers of the rotor is usually somewhat higher than that of the bottom fibers, so that the rotor is "humped up." When rotating the unit, this evidently corresponds to a huge unbalance, since a bend of 0.001 in. in the center line of a 20-ton rotor means an unbalance of 40 in.-lb. Thus an attempt to bring the machine to full speed at once would end in disaster. It is necessary to rotate the spindle at a low speed for some time before the temperature differences are sufficiently neutralized and the machine can be put in operation.

6.6. Secondary Critical Speeds. Besides the main or ordinary critical speed caused by the centrifugal forces of the unbalanced masses, some disturbance has been observed at half this critical speed, *i.e.*, for the single disk of Fig. 6.1 at $\omega = \frac{1}{2} \sqrt{k/m}$.

This effect has been observed on horizontal shafts only. On vertical shafts it is absent, indicating that gravity must be one of the causes of it. There exist two types of this disturbance, caused by gravity in combination with unbalance and by gravity in combination with a non-uniform bending stiffness of the shaft.

These phenomena are known as "secondary critical speeds," and, as the name indicates, their importance and severity are usually less than for the ordinary or "primary" critical speeds. The theory of the actual motion is very complicated, and its detailed discussion must be postponed to the last chapter, pages 335 to 350. Here, we propose to give merely a physical explanation of the phenomena and a calculation of the amplitude of the disturbing forces involved.

To this end we imagine the simple shaft of Fig. 6.1 to be rotating without any vibration or whirl, and then we calculate which alternating forces are acting on the disk. For the ordinary critical speed of page 226 we have a rotating centrifugal force $m\omega^2 e$ (m = mass of the entire disk, e = eccentricity of its center of gravity), and this force can be resolved into its horizontal and vertical components. Each of these is an alternating force of frequency ω and amplitude $m\omega^2 e$.

Consider next the case of a perfectly balanced disk ($e = 0$) running on a shaft which is not equally stiff in all directions. Since a shaft cross section has two principal axes about which the moment of inertia is maximum and minimum, it is seen that for each quarter revolution the stiffness of the shaft in the vertical direction passes from a maximum to a minimum (Fig.

FIG. 6.17. Shaft cross sections of non-uniform flexibility.

6.17). For a full revolution of the shaft the stiffness is twice a maximum and twice a minimum, or for each revolution the stiffness variation passes through two full cycles.

If the spring constant of the shaft varies between the minimum value $k - \Delta k$, and the maximum value $k + \Delta k$, with an average value of k, then for uniform rotation ω the stiffness can be expressed by

$$k + \Delta k \cdot \sin 2\omega t$$

If the disk is not vibrating and its downward deflection during rotation is δ, there are two vertical forces acting on it, *viz.*:

The weight mg downward.

The spring force $(k + \Delta k \cdot \sin 2\omega t)\delta$ upward.

Naturally the weight and the constant part of the spring force are in equilibrium, so that we have a vertical disturbing force of frequency 2ω and of amplitude

$$\Delta k \cdot \delta = \Delta k \cdot \frac{mg}{k} = \mathbf{W} \cdot \frac{\Delta k}{k}$$

If the shaft is running at half its critical speed, the impulses of this force occur at the natural frequency so that we expect vibration.

The next case, that of an unbalanced disk on a uniform shaft, is somewhat more difficult to understand. Assuming no vibration, *i.e.*, the center S of the shaft being at rest and coinciding with B, and assuming an eccentricity e, the center of gravity G describes a circular path of radius e (Fig. 6.18). The weight $\mathbf{W}$ of the disk exerts a torque on the shaft which retards the rotation when G is in the left half of Fig. 6.18 and accelerates it when G is in the right half. The magnitude of this torque is $\mathbf{W}e \sin \omega t$. If the moment of inertia of the disk about the shaft axis is $m\rho^2$ (ρ = radius of gyration), the angular acceleration of the shaft caused by this torque is $(\mathbf{W}e/m\rho^2) \sin \omega t$. The point G in its circular path has an acceleration of which the radial or centripetal component is of no interest to us in this case, since it will lead to the ordinary (primary) critical speed. However, on account of the angular acceleration, G has a tangential component of acceleration of magnitude

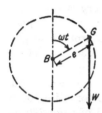

FIG. 6.18. Explains the secondary critical speed caused by unbalance and gravity.

$$\frac{\mathbf{W}e^2}{m\rho^2} \sin \omega t$$

This means that there must be a tangential force acting on G of value $(\mathbf{W}e^2/\rho^2) \sin \omega t$. The vertical component of this force is $\sin \omega t$ times as large, or

$$\left(\frac{\mathbf{W}e^2}{\rho^2}\right) \sin^2 \omega t = \text{const.} - \frac{\mathbf{W}e^2}{2\rho^2} \cos 2\omega t$$

The constant part of this force is taken up as a small additional constant deflection of the shaft and is of no interest. However, the variable part has the frequency 2ω and the amplitude $\mathbf{W}e^2/2\rho^2$.

Summarizing, we have for amplitudes of the disturbing forces the following expressions:

$$\left.
\begin{array}{ll}
\text{At the ordinary critical speed,} & m\omega^2 e \\[2mm]
\text{At the ``unbalance'' secondary speed,} & \dfrac{\mathbf{W}}{2} \cdot \left(\dfrac{e}{\rho}\right)^2 \\[2mm]
\text{At the ``flat-shaft'' secondary speed,} & \mathbf{W} \cdot \dfrac{\Delta k}{k}
\end{array}
\right\} \quad (6.6)$$

In practice, the order of magnitude of e/ρ to be expected in a machine is about the same as that of $\Delta k/k$, both being very small, say 0.001. It is seen that the disturbing force of the "unbalance" secondary critical speed is of a much smaller order than that of the "flat-shaft" critical speed, since e/ρ appears as a square. Therefore, in most cases where the secondary critical is observed, it is due to non-uniformity of the shaft rather than to unbalance. The nature of the trouble can be established by balancing the machine at its primary critical speed. If the amplitude of the secondary critical speed is not affected by this procedure, that speed is clearly due to shaft flatness.

A more detailed analysis of this problem is given on pages 335 to 350.

6.7. Critical Speeds of Helicopter Rotors. About the year 1940, helicopter rotors with the usual hinged-blade construction were observed to come to a violent critical condition at a speed very much lower than that calculated from the $\omega^2 = k/M$ formula. This happens while the aircraft is standing still on the ground prior to take-off and consequently is called the "ground critical." The phenomenon was explained by R. P. Coleman of Langley Field in N.A.C.A. reports of 1942 and 1943, and the simpler portion of his results are here reproduced for the great interest attached to them.

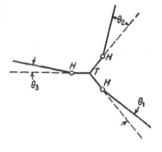

The system is as shown in Fig. 6.19. The blades of a helicopter rotor are hinged at H, so that they can swing freely about H in the plane of rotation. The hub of the rotor coincides with the top T of a "pylon" OT, which at O is supposed to be built-in into the helicopter structure. If k be the

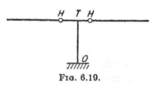

Fig. 6.19.

stiffness of this pylon against a force at T in the plane of the rotor and if M be the total mass of the hub and all attached blades, then the observed critical speed ω^2 was very much smaller than k/M.

Consider the three-bladed rotor of Fig. 6.20, where O is the bottom of the pylon seen from above and T is the top of the pylon, displaced to the right through the distance $OT = e$, the eccentricity. The pylon is supposed to be bent elastically through distance e, and the entire figure as a solid body rotates or whirls at speed ω about the vertical axis O. The blades will turn about their hinge axes H through small angles ϵ, so that the blade lines up with the centrifugal field through the center of rotation O. During the whirling motion these angles ϵ are constant and no relative motion takes place across any of the hinges H. We now calculate the centrifugal forces of all three blades and of the hub and set their sum equal to ke, the elastic homing force of the pylon. This will give the critical speed.

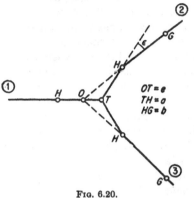

FIG. 6.20.

In the triangle OTH the angle OTH is 120 deg., the angle $THO = \epsilon$ is considered "small," the hinge radius $TH = a$, and the eccentricity $OT = e$ is again "small" with respect to a. From the geometry of this triangle the reader should derive as follows:

$$\sin \epsilon = \epsilon = \frac{\sqrt{3}}{2} \cdot \frac{e}{a} \qquad OH = a + \frac{e}{2}$$

Thus the centrifugal force of blade 2, Fig. 6.20 is $m_b \omega^2 \left(a + b + e/2 \right)$, directed along GH. This force is now resolved into components parallel and perpendicular to OT. The component parallel to OT (to the right) is

$$m_b \omega^2 \left(a + b + \frac{e}{2} \right) \cos (60° - \epsilon)$$

$$= m_b \omega^2 \left(a + b + \frac{e}{2} \right) (\cos 60° + \epsilon \sin 60°)$$

$$= m_b \omega^2 \left(a + b + \frac{e}{2} \right) \left(\frac{1}{2} + \frac{3}{4} \frac{e}{a} \right)$$

$$= m_b \omega^2 \left[\frac{1}{2} (a + b) + e \left(1 + \frac{3}{4} \frac{b}{a} \right) \right] \quad (6.7)$$

For blade 3 the result is the same for reasons of symmetry, while the components of centrifugal force perpendicular to OT for blades 2 and 3 cancel each other. The centrifugal force of blade 1 in the direction OT (to the left) is

$$m_b \omega^2 (a + b - e) \tag{6.8}$$

The centrifugal force of the hub itself (to the right) is

$$m_{hub} \omega^2 e \tag{6.9}$$

Thus the total centrifugal force to the right is twice Eq. (6.7) less Eq. (6.8) plus Eq. (6.9):

$$\omega^2 e \left[m_{hub} + m_b \left(3 + \frac{3}{2} \cdot \frac{b}{a} \right) \right]$$

Let $m_{hub} + 3m_b = M$, the total mass

and $\mu = \dfrac{3m_b}{M}$, ratio of hinged mass to total mass

Then the total centrifugal force can be written:

$$M\omega^2 e \left(1 + \frac{\mu b}{2a}\right)$$

Equate this to the elastic force ke, and the critical frequency comes out:

$$\omega^2 = \frac{k}{M} \cdot \frac{1}{1 + \mu b/2a} \tag{6.10}$$

It is seen that for the case of no hinged mass, $\mu = 0$, the natural frequency is k/M: the presence of the hinged mass diminishes this frequency.

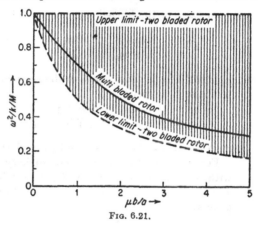

FIG. 6.21.

The relation is shown graphically by the fully drawn curve of Fig. 6.21. Although the above analysis was carried out for a three-blade rotor, the result is good also for a rotor with more than three blades, which is shown in small type on page 252.

In the case of a *two*-bladed rotor, however, the result comes out differently. Figure 6.22 shows the equivalent of Fig. 6.20, this time for two blades. Before repeating the analysis for this case, we notice that in Fig. 6.22 the eccentricity OT has been drawn perpendicular to the line HH connecting the two hinges. If we had assumed the other extreme case, that of an eccentricity OT in the direction of the hinge line HH, the angle ϵ would have been zero, the hinges would not have deflected at all, and consequently the frequency would have come out just $\omega^2 = k/M$, without any hinge effect. With the position shown in Fig. 6.22 the hinge effect is as great as it can be. The principal steps in the calculation of Fig. 6.22 are

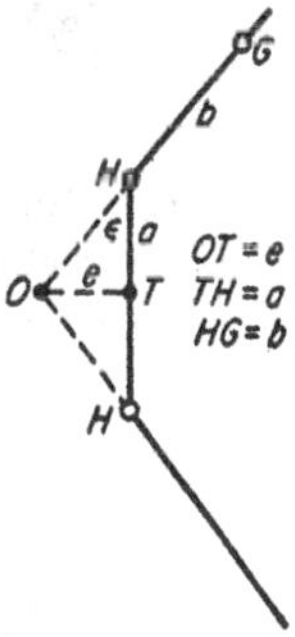

$$\sin \epsilon = \epsilon = \frac{e}{a} \qquad OH = a$$

FIG. 6.22.

$OT = e$
$TH = o$
$HG = b$

Centrifugal force of one blade $= m_b \omega^2 (a + b)$

Component parallel to OT $= m_b \omega^2 (a + b) \dfrac{e}{a}$

Total centrifugal force to right $= 2 m_b \omega^2 (a + b) \dfrac{e}{a} - m_{buh} \omega^2 e$

$$= M \omega^2 e \left[1 + \frac{\mu b}{a} \right]$$

in which μ is again the ratio of the hinged mass to the total mass:

$$\mu = \frac{2 m_b}{(2 m_b + m_{hub})} = \frac{2 m_b}{M}$$

Setting the total centrifugal force again equal to the spring force ke leads to the critical speed:

$$\omega^2 = \frac{k}{M} \frac{1}{1 + \mu b / a} \qquad (6.11)$$

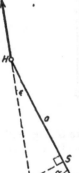

FIG. 6.23.

represented by the dotted line in Fig. 6.21.

For the case that the whirl eccentricity is at angle α_0 with respect to the hinge line H, H, it will be shown in small type below that Eq. (6.11) modifies to the more general

$$\omega^2 = \frac{k}{M} \frac{1}{1 + \mu b \, (\sin^2 \alpha_0 / a)} \qquad (6.12)$$

which reduces to Eq. (6.11) for $\alpha_0 = 90$ deg., and which gives plainly $\omega^2 = k / M$ for $\alpha_0 = 0$ deg. and a value for the frequency in between these two extremes for α_0 between zero and 90 deg. It must be concluded then that for a two-bladed rotor a large amplitude whirl at some value of α_0 is possible for any speed of rotation in the shaded region of Fig. 6.21. Thus the two-bladed rotor has a *region* of instability, shown shaded in Fig. 6.21, whereas a multibladed rotor just has a simple critical speed above which it becomes stable again. All of this is in good agreement with experiment.

In order to write the general theory for a multibladed rotor we start with a single blade located at an arbitrary angle α with respect to the direction of eccentricity OT as shown in Fig. 6.23. With the same assumption as before, that the eccentricity e is small with respect to a, we have in the triangle OTH:

$$OS = e \sin \alpha; \qquad \epsilon = \frac{e}{a} \sin \alpha; \qquad OH = HS = HT - ST = a - e \cos \alpha$$

The centrifugal force of this blade thus is

$$m_b\omega^2(a + b - e \cos \alpha)$$

The component of this force in the direction OT of the eccentricity is

$$-m_b\omega^2(a + b - e \cos \alpha) \cos (\alpha + \epsilon)$$
$$= -m_b\omega^2 \left[(a + b) \cos \alpha - e \left(1 + \frac{b}{a} \sin^2 \alpha \right) + \cdots \right]$$
$$= -m_b\omega^2 \left[(a + b) \cos \alpha - e \left(1 + \frac{b}{2a} \right) + e \frac{b}{2a} \cdot \cos 2\alpha \right]$$

Now let the rotor have N equally spaced blades. The angle between blades is $2\pi/N = \Delta$, and if the angle α of the first blade be α_0, then the angle α of the $(p + 1)$st blade is $\alpha_0 + p\Delta$. Substituting this value for the angle α and adding for all blades we find for the component along OT of the centrifugal forces of all blades:

$$-m_b\omega^2 \left[(a + b) \sum_{p=0}^{p=N-1} \cos (p\Delta + \alpha_0) - Ne \left(1 + \frac{b}{2a} \right) \right.$$
$$\left. + \frac{be}{2a} \sum_{p=0}^{p=N-1} \cos (2p\Delta + 2\alpha_0) \right] \quad (6.13)$$

The first of the sums appearing in this expression can be interpreted as the sum of the horizontal projections of the individual vectors of the star of the blades. Since the total resultant of the vectors themselves is zero, so is the horizontal projection. The second sum is a star of vectors with double angles 2Δ between them, which for a multi-bladed rotor again has a zero resultant. Thus both sums disappear and the OT component of centrifugal force is simply

$$m_b\omega^2 Ne \left(1 + \frac{b}{2a} \right)$$

Adding to this the hub force $m_{h\omega^2}e$ and setting the sum equal to the elastic force ke leads to the result equation (6.10), independent of the number of blades N or of the direction of the whirl α_0.

For a two-bladed rotor the summations in Eq. (6.13) come out differently. The first sum is $\cos \alpha_0 + \cos (180 + \alpha_0)$, which is zero as before. The second sum however becomes $\cos 2\alpha_0 + \cos (360 + 2\alpha_0) = 2 \cos 2\alpha_0$. This makes the centrifugal force component for a two-bladed rotor equal to

$$m_b\omega^2 \left[2e \left(1 + \frac{b}{2a} \right) - \frac{be}{a} \cdot \cos 2\alpha_0 \right] = 2m_b\omega^2 e \left[1 + \frac{b}{a} \sin^2 \alpha_0 \right]$$

Adding to this the centrifugal force of the hub and equating the sum to ke leads to the result equation 6.12.

6.8. Gyroscopic Effects. The disk of Fig. 6.1, being in the middle of the span, will vibrate or whirl in its own plane. When the disk is placed near one of the bearings, and especially when it is located on an overhung shaft, it will *not* whirl in its own plane. Then the system of Fig. 6.24b will have a (primary) critical speed different from the one of Fig. 6.24a;

the mass and shaft stiffness being the same in both cases. This is due to the fact that the centrifugal forces of the various particles of the disk do not lie in one plane (Fig. 6.25) and thus form a couple tending to straighten the shaft. Before calculating this moment, it is necessary to have a clear picture of the mode of motion.

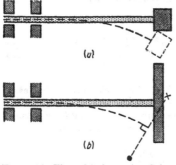

(a)

(b)

Fig. 6.24. The critical speeds of (a) and (b) are *not* equal if the shafts are identical and the masses at the end are equal.

We assume the machine to be completely balanced and whirling at its critical speed in some slightly deflected position. The angular velocity of the whirl of the center of the shaft is assumed to be the same as the angular velocity of rotation of the shaft. This implies that a particular point of the disk which is outside (point · in Fig. 6.24b) will always be outside; the inside point × always remains inside; the shaft fibers in tension always remain in tension while whirling, and similarly the compression fibers always remain in compression. Thus any individual point of the disk moves in a circle in a plane perpendicular to the undistorted center line of the shaft.

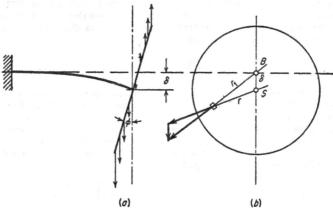

(a)   (b)

Fig. 6.25. The centrifugal forces on the disk tend to bend the disk to a plane perpendicular to the equilibrium position of the shaft. Thus these forces act as an additional spring.

Figure 6.25 shows the centrifugal forces set up by this motion. In Fig. 6.25b we see that the centrifugal force of a mass element dm is $\omega^2 r_1 \, dm$ directed away from the point B. This force can be resolved into two components: $\omega^2 \delta \cdot dm$ vertically down and $\omega^2 r \, dm$ directed away from the disk or shaft center S. The forces $\omega^2 \delta \cdot dm$ for the various mass elements add together to a single force $m\omega^2 \delta$ (where m is the total mass of the disk) acting vertically downward in the point S of Fig. 6.25b. The forces $\omega^2r\,dm$ all radiate from the center of the disk S, and their influence becomes clear from Fig. 6.26, as follows. The y-component of the force $\omega^2r\,dm$ is $\omega^2y\,dm$. The moment arm of this elemental force is $y\varphi$, where φ is the (small) angle of the disk with respect to the vertical. Thus the moment of a small particle dm being $\omega^2y^2\varphi\,dm$, the total moment $\mathbf{M}$ of the centrifugal forces is

$$\mathbf{M} = \omega^2\varphi\int y^2\,dm = \omega^2\varphi I_d$$

where I_d is the moment of inertia of the disk about one of its diameters.

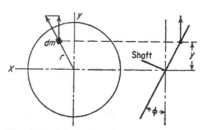

FIG. 6.26. Calculation of the moment of the centrifugal forces.

Thus the end of the shaft is subjected to a force $m\omega^2\delta$ and to a moment $\omega^2I_d\varphi$, under the influence of which it assumes a deflection δ and an angle φ. This can happen only at a certain speed ω, and the calculation of the critical speed is thus reduced to a static problem, namely that of finding at which value of ω a shaft will deflect δ and φ under the influence of $P = m\omega^2\delta$ and $\mathbf{M} = I_d\omega^2\varphi$.

For a rotating overhung cantilever shaft of stiffness EI and length l, this calculation will now be carried out in detail. From the strength of materials the formulas for the deflection and angle of the end of a cantilever due to a force P or a moment $\mathbf{M}$ are

$$\delta_P = \frac{Pl^3}{3EI}; \qquad \varphi_P = \frac{Pl^2}{2EI}; \qquad \delta_M = \frac{Ml^2}{2EI}; \qquad \varphi_M = \frac{Ml}{EI}$$

With these formulas we write

$$\delta = (m\omega^2\delta)\frac{l^3}{3EI} - (I_d\omega^2\varphi)\frac{l^2}{2EI}$$

$$\varphi = (m\omega^2\delta)\frac{l^2}{2EI} - (I_d\omega^2\varphi)\frac{l}{EI}$$

which after rearranging become

$$\left(m\omega^2\frac{l^3}{3EI} - 1\right)\delta + \left(-I_d\omega^2\frac{l^2}{2EI}\right)\varphi = 0$$

$$\left(-m\omega^2\frac{l^2}{2EI}\right)\delta + \left(\omega_dI_d\frac{l}{EI} + 1\right)\varphi = 0$$

This homogeneous set of equations can have a solution for δ and φ only when the determinant vanishes (see page 124 or 132), which gives the following equation for ω^2:

$$\omega^4 + \omega^2\frac{12EI}{mI_dl^3}\left(\frac{ml^2}{3} - I_d\right) - \frac{12E^2I^2}{mI_dl^4} = 0$$

This can be solved for ω^2. Before doing so we prefer to bring the equation to a dimensionless form with the variables:

$$K = \omega \sqrt{\frac{ml^3}{EI}} \qquad \text{(the critical speed function)}$$

and $$D = \frac{I_d}{ml^2} \qquad \text{(the disk effect)}$$

The equation then becomes

$$K^4 \mid K^2 \left(\frac{4}{D} - 12\right) - \frac{12}{D} = 0$$

with the solution

$$K^2 = \left(6 - \frac{2}{D}\right) \pm \sqrt{\left(6 - \frac{2}{D}\right)^2 + \frac{12}{D}} \qquad (6.14)$$

of which only the plus sign will give a positive result for K^2 or a real result for K.

The formula (6.14) is plotted in Fig. 6.27 of which the ordinate K^2 is the square of the "dimensionless natural frequency," *i.e.*, the square of

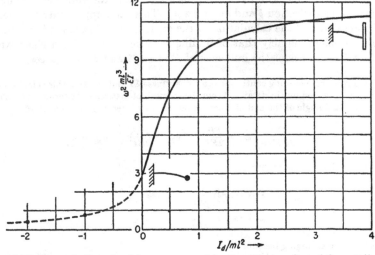

FIG. 6.27. Change in the natural frequency caused by the stiffening effect of the centrifugal forces in the system Fig. 6.24.

the factor by which $\sqrt{EI/ml^3}$ must be multiplied to obtain that frequency. The abscissa is the "disk effect" D, which is zero for a concentrated mass. In that case the frequency of Fig. 6.24a is $\omega^2 = 3EI/ml^3$. On the other hand for $I_d = \infty$ (a disk for which all mass is concentrated at a large radius), no finite angle φ is possible, since it would require an infinite torque, which the shaft cannot furnish. The disk remains parallel to itself and the shaft is much stiffer than without the disk effect. The frequency is $\omega^2 = 12EI/ml^3$.

So far we have discussed a disk which is very thin. When it increases its thickness gradually to, say, one diameter, it looks more like a point mass than like a disk, so that the gyroscopic effect must be small. If the disk becomes even longer, with a length of many diameters, it is a stick, and sketching in the centrifugal forces of Fig. 6.25 for this case it is seen that the couple of these forces is reversed: a disk is urged back home by the centrifugal couple, but a stick is pushed away from home angularly. For the general case we consider a particle in Fig. 6.28. Here the center of gravity of the body is placed already in the axis

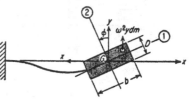

FIG. 6.28. Case of a thick disk.

of rotation x, so that there is no total centrifugal force, only a moment. The force on a particle is $\omega^2 y \, dm$ and its moment arm about G is x, so that the moment is $\omega^2 xy \, dm$. Integrated over the body

$$\omega^2 \int xy \, dm$$

For the thin disk $x = y\varphi$, which checks our previous result. Now consider the two principal axes 1 and 2 of the body in Fig. 6.28, and let the moments of inertia about those axes be I_1 and I_2. This set of axes is at angle φ with respect to the x-y axes. By Mohr's circle we have for the product of inertia about the x-y axes

$$\int xy \, dm = \frac{I_1 - I_2}{2} \sin 2\varphi \approx (I_1 - I_2)\varphi$$

for small angles φ. For a thin disk $I_1 = 2I_d$, and $I_2 = I_d$, so that we check the previous result. Now for a disk of diameter D and thickness b we have

$$I_1 = \frac{mD^2}{8} \qquad I_2 = \frac{mD^2}{16} + \frac{mb^2}{12}$$

so that the gyroscopic moment is

$$\omega^2(I_1 - I_2)\varphi = \left(\frac{mD^2}{16} - \frac{mb^2}{12}\right)\omega^2\varphi$$

Hence the moment of the centrifugal forces is zero for $b = 0.87D$, and it becomes negative for greater lengths than this. Figure 6.27 remains applicable to such a case if the letter I_d is interpreted as the parentheses above, and the dotted part on the left acquires meaning for a thin stick. Also, it is necessary that the shaft extend to the center of the cylinder without interference. If the shaft attaches to the end of the cylinder, as usual, the elastic-influence coefficients are modified.

The phenomenon just described is generally referred to in the literature as a *gyroscopic effect*. The name is unfortunate since in the usual sense of the word a gyroscope is a body which rotates very fast and of which the axis of rotation moves slowly. In the disk just considered the whirl of the axis of rotation is just as fast as the rotation itself, so that it could hardly be called a gyroscope.

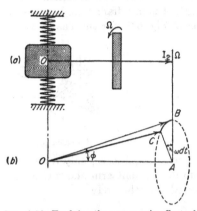

A true gyroscope effect occurs in the experimental set-up of Fig. 7.10, page 294, where a small motor is suspended practically at its center of gravity by three very flexible springs. We want to calculate the natural frequencies of the modes of motion for which the center of gravity O remains at rest and the shaft whirls about O in a cone of angle 2φ (Fig. 6.29b). The disk on

FIG. 6.29. Explains the gyroscopic effect of the apparatus shown in Fig. 7.10.

the motor shaft rotates very fast, and, as the springs on which the motor is mounted are flexible, the whirl takes place at a very much slower rate than the shaft rotation.

Let

Ω = (fast) angular velocity of disk rotation,
ω = (slow) angular velocity of whirl of the shaft center line,
I_1 = moment of inertia of the stationary and rotating parts about an axis through O perpendicular to the paper,
I_2 = moment of inertia of rotating parts about shaft axis,
k = torsional stiffness of the spring system, *i.e.*, the torque about O for $\varphi = 1$ radian.

Further let the direction of rotation of the disk be counterclockwise when viewed from the right so that the angular momentum vector $I_2\Omega$ is as shown in Fig. 6.29a. In case the whirl is in the same direction as the rotation, the time rate of change of the angular momentum of the disk is directed from B toward C in Fig. 6.29b, *i.e.*, out of the paper toward the reader. This is equal to the moment exerted *by* the motor frame on the disk. The reaction, *i.e.*, the moment acting *on* the motor, is pointing into the paper and therefore tends to make φ smaller. This acts as an addition to the existing spring stiffness k, so that it is seen that a whirl in the direction of rotation makes the natural frequency higher. In the same manner it can be reasoned that for a whirl opposite to the direction of rotation the frequency is made lower by gyroscopic effect.

To calculate the magnitude of the effect we see in Fig. 6.29b that

$$\frac{d(I_2\Omega)}{I_2\Omega} = \frac{BC}{OB} = \frac{BC}{AB} \cdot \frac{AB}{OB} = \omega \, dt \cdot \varphi$$

Consequently

$$\frac{d}{dt}(I_2\Omega) = \omega\varphi I_2\Omega$$

is the gyroscopic moment. The elastic moment due to the springs k is $k\varphi$, and the total moment is

$$(k \pm \omega\Omega I_2)\varphi$$

where the plus sign holds for a whirl in the same sense as the rotation and the minus sign for a whirl in the opposite sense. Since the parenthesis in this last expression is the equivalent spring constant, we find for the natural frequencies

$$\omega_n^2 = \frac{k \pm \omega\Omega I_2}{I_1}$$

or

$$\omega_n^2 \mp \frac{I_2}{I_1}\Omega\omega_n - \frac{k}{I_1} = 0$$

or

$$\omega_n = \pm\frac{I_2}{2I_1}\Omega \pm \sqrt{\left(\frac{I_2\Omega}{2I_1}\right)^2 + \frac{k}{I_1}}$$

(6.15)

Of the $\pm$ ambiguity before the square root, only the plus sign need be retained since the minus sign gives two negative values for ω_n which are equal and opposite to the two positive values obtained with the plus sign before the square root.

The result (6.15) is shown in Fig. 6.30, where the ordinate is the ratio

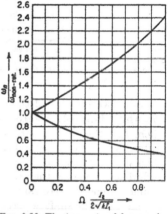

FIG. 6.30. The two natural frequencies of Fig. 6.29; with the faster one of the two, the precession ω has the same direction as the rotation Ω; with the slower frequency these directions are opposite.

between the actual natural frequency and the one without gyroscopic effect, i.e., with a non-rotating shaft. The abscissa is the rotor speed multiplied by some constants so as to make the quantity dimensionless. It is seen that the natural frequency is split into two frequencies on account of the gyroscopic effect: a slow one whereby the whirl is opposed to the rotation, and a fast one where the directions are the same.

In the analysis of pages 253 to 258 the machine was whirling (the circular motion of the deflected shaft center line about its undeflected position with a small amplitude) and rotating at the same angular speed and in the same direction. Cases have been observed where the whirling and the rotation occur at different frequencies, and sometimes in opposite

directions. In order to understand this, we now start to calculate the
natural whirling frequencies of a shaft with a single disk on it at any speed
of rotation Ω, in the most general manner.

In Fig. 6.31 the disk is shown on the end of a rotating and whirling
cantilever shaft, but the analysis we now start will be sufficiently general

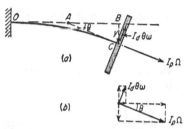

to be applicable to other kinds of
shafts as well. The shaft center line
in its deflected position shown is supposed to rotate about the undeflected
position OA at angular speed ω.
Simultaneously and independently
the disk and shaft are supposed to rotate bodily about the deflected center
line OC with angular speed Ω. The

Fig. 6.31. A shaft with a disk on it at
location C. The shaft is shown in the
whirling position with amplitudes y and
θ at the disk. The arrows are instantaneous components of angular momentum of the disk.

reader should visualize the cases of
$\omega = \omega_0$ with $\Omega = 0$ and, vice versa,
$\omega = 0$ with $\Omega = \Omega_0$. Also, the case
of $\omega = \Omega$ is the motion discussed on
page 254. With this combined ω,
Ω motion we try to find the angular momentum of the disk. If it does
not whirl, but only rotates, the angular momentum is a vector $I_p\Omega$,
where I_p is the disk's polar moment of inertia about the (deflected) shaft
center line. The arrow indicates that the disk rotates counterclockwise
when viewed from the right. Now assume no rotation $\Omega = 0$, but only
a whirl ω. The disk then wobbles in space, and it is difficult to visualize
its angular speed. However, this visualization is made easier by remarking that at C the shaft is perpendicular to the disk always, so that we are
permitted to study the angular motion of the shaft instead of the disk.
The line CA is tangent to the shaft at C, and the piece ds of shafting at
the disk moves with the line AC, describing a cone with A as apex.
The speed of point C (for a whirl counterclockwise seen from the right, in
the same direction as the rotation Ω) is perpendicular to the paper into
the paper, and its value is ωy. The line AC lies in the paper now, but at
time dt later, point C is behind the paper by $\omega y\, dt$. The angle between
the two positions of line AC then is $\omega y\, dt/AC$, and since $y/AC = \theta$ if θ
is small, the angle of rotation of AC in time dt is $\omega\theta\, dt$, and hence the
angular speed of AC (and of the disk) is $\omega\theta$. The disk rotates about a
diameter in the plane of the paper perpendicular to AC at C, so that the
appropriate moment of inertia is $I_d = \frac{1}{2}I_p$ for a thin disk. The angular-
momentum vector of the disk due to whirl then is $I_d\theta\omega$ in the direction
shown in Fig. 6.31. The total angular momentum is the vector sum of
$I_p\Omega$ and $I_d\theta\omega$. We now want to calculate the rate of change of this
angular-momentum vector, and with that in view we resolve, in Fig.

6.31b, the vector into components parallel and perpendicular to the center line OA. The component parallel to OA rotates parallel to itself around OA in a circle with radius y, and it keeps its length during the process so that its rate of change is zero. However, in Fig. 6.31b the component perpendicular to OA is a vector along the direction CB, and it is the rotating radius of a circle with center B. From Fig. 6.31b we see that this component (for small θ) is

$$I_p\Omega\theta \,-\, I_d\theta\omega \,=\, I_d\theta(2\Omega \,-\, \omega)$$

in the direction from B to C. At time $t = 0$ this vector lies in the plane of the paper; at time dt it is behind the paper at angle $\omega\,dt$. The increment in the vector (directed perpendicular to the paper and into it) is the length of the vector itself multiplied by $\omega\,dt$, or

$$I_d\theta(2\Omega \,-\, \omega)\omega\,dt$$

and the time rate of change of angular momentum is

$$I_d\theta(2\Omega \,-\, \omega)\omega$$

By the main theorem of mechanics this is the moment exerted on the disk by the shaft, and by action and reaction the moment exerted by the disk on the shaft is the equal and opposite, $i.e.$, a vector directed out of the paper and perpendicular to it at C, as shown in Fig. 6.32. Besides this couple there is a centrifugal force $m\omega^2 y$ acting on the disk (Fig. 6.32). Now we are ready

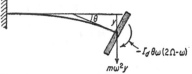

FIG. 6.32. The inertia force and inertia couple acting from the disk on the shaft caused by a shaft rotation Ω and a shaft whirling ω.

to discuss the elastic properties of the shaft at the disk location. These are described by three influence numbers:

α_{11} is the deflection y at the disk from 1 lb. force.

α_{12} is the angle θ at the disk from 1 lb. force, and, by Maxwell,

α_{12} also is the y at the disk from 1 in.-lb. moment.

α_{22} is the angle θ at the disk from 1 in.-lb. moment.

This general terminology takes care of all cases; as an example, for the cantilever shown we have

$$\alpha_{11} = \frac{l^3}{3EI} \qquad \alpha_{12} = \frac{l^2}{2EI} \qquad \alpha_{22} = \frac{l}{EI}$$

Now the shaft equations can be written, stating that the shaft deflection y is caused by a force $m\omega^2 y$ and by the moment of Fig. 6.32:

$$y = \alpha_{11}m\omega^2 y \,-\, \alpha_{12}I_d\omega(2\Omega \,-\, \omega)\theta$$
$$\theta = \alpha_{12}m\omega^2 y \,-\, \alpha_{22}I_d\omega(2\Omega \,-\, \omega)\theta$$

These, as usual, are homogeneous in y and θ, and the frequency equation is found by calculating y/θ from the first equation, then from the second equation, and equating the two answers. Rearranging the terms by powers of ω, it is

$$\omega^4(-m\alpha_{11}\alpha_{22}I_d + m\alpha_{12}^2I_d) + \omega^3(m\alpha_{11}\alpha_{22}I_d2\Omega - m\alpha_{12}^2I_d2\Omega) + \omega^2(\alpha_{22}I_d + m\alpha_{11}) + \omega(-\alpha_{22}I_d2\Omega) - 1 = 0$$

This equation contains seven system parameters: ω, Ω, m, I_d, α_{11}, α_{12} and α_{22}, which makes a good intuitive understanding of the solution very difficult. It is worthwhile to diminish the number of parameters as much as possible by dimensional analysis. Here we can reduce the number from seven to four by introducing the new variables:

$$\left.\begin{array}{l} F = \omega\sqrt{\alpha_{11}m}, \text{ the dimensionless Frequency} \\[2mm] D = \dfrac{I_d\alpha_{22}}{m\alpha_{11}}, \text{ the Disk effect} \\[2mm] E = \dfrac{\alpha_{12}^2}{\alpha_{11}\alpha_{22}}, \text{ the Elastic coupling} \\[2mm] S = \Omega\sqrt{\alpha_{11}m}, \text{ the dimensionless Speed} \end{array}\right\} \quad (6.16)$$

With these the frequency equation becomes

$$F^4 - 2SF^3 + \frac{D+1}{D(E-1)}F^2 - \frac{2S}{E-1}F - \frac{1}{D(E-1)} = 0 \quad (6.17)$$

This is seen to be a fourth-degree algebraic equation in F, so that for a given shaft E, carrying a given disk D and rotating at a certain speed S, there will be *four natural frequencies of whirl*.

For a better understanding of the important equation (6.17) we first consider some special cases of it. Let the mass be concentrated at a point; therefore $D = 0$. Then the third and fifth terms become very much larger than the others, so that they alone count, and the equation reduces to $F^2 = 1$ or $F = +1$ and $F = -1$, which means $\omega = \pm 1/\sqrt{\alpha_{11}m}$. Since α_{11} equals $1/k$ in the usual notation, this result is very familiar.

Next consider the case of no elastic coupling, $E = 0$, or a force causes a deflection only without angle θ, while a couple causes an angle θ only without deflection y. This is the case of a shaft on two simple supports with a disk in the mid-span point. The equation reduces to

$$F^4 - 2SF^3 - \frac{D+1}{D}F^2 + 2SF + \frac{1}{D} = 0$$

From the physical case we know that $F = 1$ must be a solution, independent of the running speed S, and so it proves to be. The equation can be factored as follows:

$$(F + 1)(F - 1)\left(F^2 - 2SF - \frac{1}{D}\right) = 0$$

The frequency F is plotted against the speed S for the numerical value $D = 1$ [which means the disk is so dimensioned that the parallel up-and-down frequency is the same as the wobbling frequency without up-and-down motion (at no rotation $S = 0$)]. Figure 6.33 shows the plot, and it is seen that there are four natural frequencies at any speed. For no rotation $S = 0$ the four frequencies reduce to two $F = \pm 1$, which really is one frequency only because $F = +1$ means a counterclockwise whirl and $F = -1$ a clockwise one.

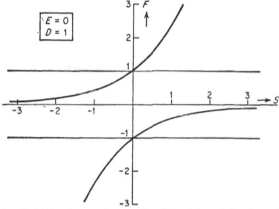

Fig. 6.33. Natural whirling frequencies as a function of the rotational speed S for a symmetrical shaft on two supports with a central disk.

Now we are ready to plot a general case of Eq. (6.17), and for it we take a disk on a cantilever shaft (Fig. 6.31). For such a shaft the influence numbers lead to $E = \frac{3}{4}$, and again we take a disk $D = 1$. Then the equation becomes

$$F^4 - 2SF^3 - 8F^2 + 8SF + 4 = 0 \tag{6.18}$$

This time there are no symmetries in the equation, and it is truly a fourth-degree one with four roots, which require much numerical work for their determination. However, it is our object to construct $F = f(S)$ in Fig. 6.34. We can do this in several ways, one of which is to take a value of S and solve the fourth-degree equation in F, and another of which is to take a value for F and solve the *linear* equation in S. Going about it in the latter way, we have from (6.18)

$$S = \frac{F^4 - 8F^2 + 4}{2F^3 - 8F}$$

and for $F = 1$ we get $S = 0.500$, etc. In this way Fig. 6.34 has been constructed, and a similar figure can be made for any other values of D and E, that is, for other disks and other shafts. It is seen that for zero rotation $S = 0$ there are only two natural frequencies $F = \pm 0.74$ and $F = \pm 2.73$ corresponding to the two degrees of freedom y, θ of a non-rotating disk. When the disk rotates, however, all four natural frequencies are different. The curves are symmetrical about the vertical F-axis, which means that for $+S$ and $-S$ the same F-values occur, in

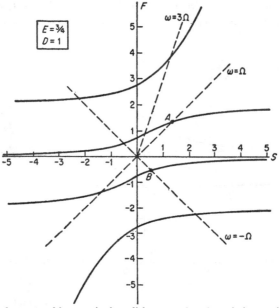

Fig. 6.34. The four natural frequencies for a disk mounted on the end of a rotating cantilever.

other words, the four natural frequencies do not care whether the disk rotates clockwise or counterclockwise.

Figure 6.34 contains a number of thin, dotted straight lines radiating from the origin. Along any of these there is a constant ratio $F/S = \omega/\Omega$. The principal one among these is the 45-deg. radius $\omega = \Omega$, and it intersects the curve at A. At the r.p.m. S determined by this point A there is a "forward whirl" at the same speed as the rotation, and Fig. 6.27 (page 256) is a plot of these intersections A for various values of the disk effect D. This kind of disturbance is obviously excited by unbalance; it is a resonance phenomenon, and the vibration amplitude at this critical is proportional to the amount of unbalance. If we imagine the amplitude to rise in this condition, the centrifugal force of the unbalance does work on the increase in amplitude and, thus, pumps energy into the vibration.

Now we come to a curious and still partly unexplained phenomenon. It was reported in the classical book on steam turbines of Stodola that a rough critical had been observed at the intersection B where $\omega = -\Omega$. This was called a "reverse whirl," or more eruditely a "retrograde precession." No explanation was given of where the energy comes from. A calculation of the work done by unbalance forces or even by a flat shaft on the increase in amplitude leads to a zero answer when integrated over a complete cycle. Whereas the forward whirl or ordinary critical speed can be observed on every machine, the reverse whirl is extremely rare. The author of this book looked around and asked his friends for fifteen years about it without results and was just about ready to conclude that the reverse critical was imaginary, when a case actually occurred. A model was constructed which showed roughness at the calculated B-point, and stroboscopic observation showed the whirl to be actually opposite in direction to the rotation. With this model tests were conducted determining the amplitude of vibration at this critical as a function of (1) unbalance and (2) flatness of shaft (page 247), and it was conclusively established that neither unbalance nor shaft flatness affects the amplitude, which remained constant throughout. During the tests the apparatus was disassembled several times, and new shafts and disks were used. After every such reassembly the reverse-whirl amplitude was different and often entirely absent. Also, roughnesses were observed at other ratios ω/Ω than -1, in particular at $\omega/\Omega = +2$ and $+3$. No rational explanation for this behavior is available, and it is suspected that it is determined by damping or internal friction (page 294 and Fig. 7.11, page 295). In any case, the practical importance of any critical other than $\omega/\Omega = +1$ is not great.

6.9. Frame Vibration in Electrical Machines. Between the stator and rotor of any electric motor or generator magnetic forces exist which have a small rapid variation in intensity with a frequency equal to the number of rotor teeth passing by the stator per second. These alternating forces may cause vibrations in the stator frame if they are in resonance with one of its natural frequencies. For constant-speed machinery such trouble, if it ever appears, can easily be corrected by changing the stiffness of the frame and thus destroying the resonance. If, however, the machine is to run satisfactorily over a wide range of speeds, it is necessary to look for other means of avoiding the trouble. The situation in this respect is quite analogous to that for Diesel engines as discussed in Secs. 5.7 to 5.9.

The number of teeth in the rotor multiplied by its r.p.m. usually leads to a very high frequency, and the amplitudes of vibration observed in practice are invariably so small that no danger for the structural safety of the machine need be feared. The frequency, however, is in the range of greatest sensitivity of the human ear so that noise considerations become of importance. In submarine motors, which have very light frames and thus are apt to be noisy, the problem is of special interest since such noise may be picked up by the enemy sound detectors.

The details of the phenomenon will first be explained with the help of Fig. 6.35, which represents one stator pole and a part of the rotor. The magnetic force R_1 acting between the stator and the rotor can be conveniently resolved into its normal and tangential components N_1 and T_1.

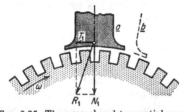

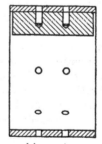

FIG. 6.35. The normal and tangential components of the force R_1 exerted by the rotor on a pole of the stator.

FIG. 6.36. Variation of the magnetic force with the time.

These forces are nearly constant with respect to time; however, they are subjected to small variations of amplitude N and T with the frequency of the rotor teeth passing by the stator (Fig. 6.36). A calculation of the exact phase relation of this variation (i.e., a calculation of the position of the rotor teeth with respect to the pole at the instant that N or T becomes zero) requires electrical theory which is not necessary for our

FIG. 6.37. Stator of submarine propulsion motor.

present purpose. It is sufficient to know that both N and T go through a full cycle of variation each time a tooth goes by, i.e., each time that the relative position between rotor and pole passes from b to a in Fig. 6.35.

Before investigating how the variation of N and T can excite vibration, we shall discuss the possible motions of the frame.

Consider an eight-pole machine for submarine application (Fig. 6.37). In such a construction the poles are comparatively heavy and the "frame" consists of a rolled-up steel plate to which the poles are bolted. Thus the poles practically form the "masses" and the frame shell the "elasticity" of the system. Since each pole as a solid body has six degrees of freedom (page 23), the whole frame must have 48 different natural modes of motion. Some of these are trivial (*e.g.*, the six possible motions of the whole frame as a solid body), and many of the others possess natural frequencies which are far removed from the frequency of the variation of the magnetic forces **N** and **T**. Four of the natural modes that have been causing trouble in an actual installation are shown in Figs. 6.38*a* to *d*.

In the first of these figures the poles move parallel to themselves in a radial direction, while the frame ring alternates between the purely

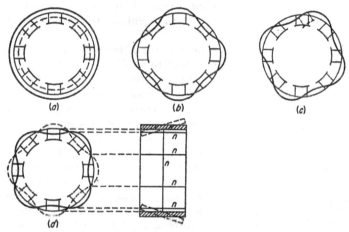

(a)   (b)   (c)

(d)

Fig. 6.38. Four natural modes of motion of the frame shown in Fig. 6.37.

extended and compressed states. In Fig. 6.38*b* the motion of the individual poles is the same as before, but the consecutive poles are 180 deg. out of phase and the frame is in bending. Figure 6.38*c* shows a rotation of the poles about their longitudinal axes with bending in the frame. These three cases have the common property that all cross sections of a pole lying in planes perpendicular to the axis of rotation have the same motion. This is not the case in Fig. 6.38*d*. Here the poles rotate about their transverse axes, and the frame ring is in combined twist and bending. There are eight nodal generators and one central nodal circle on the cylinder, all denoted by n in the figure.

Assume first that the rotor teeth and slots are parallel to the axis of rotation, the forces **T** or **N** reaching their maximum value at the same instant all along a pole. It is clear that the motions of Fig. 6.38*a* and

b may be affected by the normal force N; the tangential force T will act only on Fig. 6.38c, while the motion of Fig. 6.38d will not be excited at all, because, if the normal force helps the motion at one end of a pole, it opposes the motion at the other end of the same pole.

Even if there is a large variation in N of the same frequency as the natural frequency of mode 6.38a or b, these modes are not necessarily excited. If the number of rotor teeth per pole (total number of teeth divided by 8) is an integer, the force N_1 becomes a maximum at all poles at the same time. Then, of course, Fig. 6.38a is excited, but the work input for Fig. 6.38b is zero over a full cycle of the vibration. (While the force N pulls the four downgoing poles downward and does positive work, the same force pulls down on the upgoing poles and there does equal negative work.) On the other hand, if there are $n + \frac{1}{2}$ teeth per pole, Fig. 6.38b is excited and Fig. 6.38a is not. A similar consideration holds for Fig. 6.38c which is excited by the tangential variation T if there are $n + \frac{1}{2}$ teeth per pole.

It is clear that a changing of the number of teeth per pole is not alone sufficient to avoid an excitation of all four modes of motion at once.

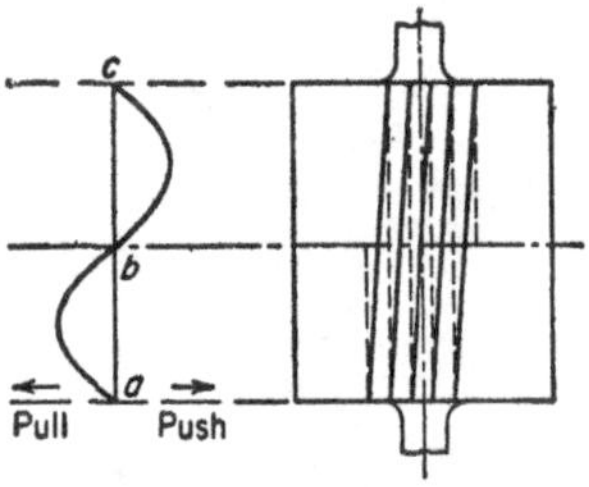

FIG. 6.39. Variation of the time-variable part of the magnetic force along a generating line of a rotor with whole-pitch skewing.

Another possibility of affecting the phenomenon consists in "skewing" the slots or teeth of the armature with respect to the axis of rotation. Figure 6.39 shows how the teeth are skewed by one full tooth pitch over the length of the rotor. In this case the forces N or T at any one instant vary from point to point along the length of the pole, and it can be seen that the diagram of the force as a function of position along the pole must be identical with the diagram of the force at one point of the pole as a function of the time. At the side of Fig. 6.39 the diagram of force as function of position is drawn, the force variation not being necessarily sinusoidal.

Since N and T are the variable parts of N_1 and T_1, their integrated values over a full cycle are zero (Fig. 6.36). In particular, in Fig. 6.39 it is seen that the pull between pole and rotor along one half of the pole length is compensated by a push on the other half of the pole length.

With a machine built as in Fig. 6.39, it is clear that no excitation at all is given to the modes of Figs. 6.38a, b, and c, irrespective of the number of teeth per pole. Now, however, trouble is to be expected from the motion of Fig. 6.38d. It is true that the total integrated value of the forces N and T over the whole pole length is zero, but that is of no consequence in connection with Fig. 6.38d. The total force is zero only because it is pulling down on one side and pushing up on the other side. The motion 6.38d, however, is also up on one side and down on the other side, which creates the possibility of a great input of energy.

In order to circumvent this difficulty, "herringbone" skewing has been proposed in the fashion of Fig. 6.40b, where the slope of the teeth is at the rate of one full tooth pitch over *half* the rotor length. In this arrangement the force diagram (which again may or may not be sinusoidal) is shown in Fig. 6.40a. The radial velocity diagram of the various points along the pole is a straight line (Fig. 6.38d). It can be easily verified that the work input per cycle, which is proportional to the integrated product of the two curves of Fig. 6.40a, is zero. It is also seen that the force by itself, integrated over the pole length, is zero. A herringbone skewed rotor of full tooth pitch over half the rotor length will make the frame free from vibration in any of the four modes of Fig. 6.38, independent of the number of teeth per pole.

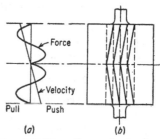

FIG. 6.40. Force diagram for herringbone skewing.

6.10. Vibration of Propellers. Since the introduction of aluminum-alloy propellers in airplanes, a number of fatigue failures have occurred. Some of these were noted in time to avoid failure, being seen in the form of cracks, but in other cases either an entire blade or the tip of a blade has blown away in mid-air. The fact that these failures were unmistakably due to fatigue makes it certain that they were caused by vibration. Before entering into the possible excitations to which a propeller blade may be subjected, it is of interest to consider the determination of its natural frequencies. These are different for various running speeds, because the centrifugal force tends to force the vibrating beam back to its middle position, thus acting like a spring force.

A propeller blade is a complicated system. It can be idealized as a twisted cantilever beam, but the mass per unit length and the bending stiffness EI vary along the length. Exact calculation of the natural frequency, even without the centrifugal effect, is difficult. For each particular case we can calculate the frequency by Rayleigh's method by choosing some probable shape of the deformation and determining the potential and kinetic energies. In this case the potential energy will consist of two parts, one due to bending and one due to the centrifugal effect. As is always true with Rayleigh's method, the answer thus found for the natural frequency is somewhat *too large* (pages 144 and 161).

The actual evaluation of the frequency in this manner requires involved calculations, which can be avoided by applying the following theorem:

Theorem of Southwell: If in an elastic system the spring forces can be divided into two parts so that the total potential energy is the sum of the two partial potential energies, then the natural frequency ω of that system can be calculated approximately from

$$\omega^2 = \omega_1^2 + \omega_2^2 \qquad (6.19)$$

where ω_1 and ω_2 are the *exact* natural frequencies of the (modified) system in which one of the spring effects is absent. The value ω thus found is somewhat *too small.*

A very simple case illustrating this statement is that of a single mass m connected to a wall with two coil springs k_1 and k_2 in parallel (Figs. 2.12a, b, page 35). The natural frequency of this system is $\omega^2 = (k_1 + k_2)/m$ which is exactly equal to $\omega_1^2 + \omega_2^2 = \dfrac{k_1}{m} + \dfrac{k_2}{m}$. The answer is exact in this case because the configuration of the vibration is not changed by omitting one of the springs. Another example is provided by Problem 21.

Applied to the propeller blade, the theorem states that a good approximation for the natural frequency when rotating (ω) can be derived by the relation (6.19) from the exact natural frequency at standstill (ω_1) and the exact natural frequency of a chain without bending stiffness of the same mass distribution as the blade and rotating at full speed (ω_2).

For the proof of Southwell's theorem we apply Rayleigh's procedure to the exact shape of the vibrating blade while rotating. In that shape let

P_{ben} = potential energy due to bending,

P_{cen} = potential energy due to centrifugal forces,

$\omega^2 K$ = kinetic energy, where ω^2 is the (exact) natural frequency of vibration.

Then,
$$\omega_{exact}^2 = \frac{P_{ben} + P_{cen}}{K} = \frac{P_{ben}}{K} + \frac{P_{cen}}{K}$$

We find the *exact* answer for the natural frequency because the exact configuration was assumed (see page 143). But the exact shape of vibration while rotating is different from the exact shape at standstill and also differs from the shape of the rotating chain. Yet the first shape may be considered as an approximation to the latter two. Thus the two terms on the right side of the above equation are Rayleigh approximations of ω_1 and ω_2 (*i.e.*, of the exact standstill frequency and the exact chain natural frequency). Since Rayleigh's approximations are always *too large,*

$$\frac{P_{ben}}{K} + \frac{P_{cen}}{K} \geq \omega_1^2 + \omega_2^2$$

or
$$\omega^2 \geq \omega_1^2 + \omega_2^2$$

the error being of the same order as usually obtained with Rayleigh's method.

The usefulness of the theorem lies in the fact that the standstill frequency ω_1 can be easily determined by experiment on the actual propeller. The chain frequency ω_2, which expresses the effect of rotation, can be calculated without much difficulty. In the case of a central hub of negligible dimension compared with the blade length we find for the chain frequency the remarkably simple result

$$\omega_2 = \Omega \qquad (6.20)$$

i.e., the natural frequency of vibration of a chain rotating about one of its ends O as a center is equal to the angular velocity of its rotation. This is true independent of the mass distribution along the chain, which can be understood as follows.

Since the flat side of a propeller blade lies practically in the plane of its rotation, in the slowest type of vibration the various particles will move nearly perpendicular to the plane of rotation, *i.e.*, parallel to the axis of rotation. Assume that the deflection curve of the chain is a straight line at a small angle φ with respect to the plane of rotation OA (Fig. 6.41). Consider an element dm at a distance r from O. On this element are acting the tensions above and below (which are in line with the chain) and the centrifugal force $\Omega^2 r \, dm$. If φ is small, equilibrium in the vertical direction requires that the tension below exceeds the tension above by this amount. In the horizontal direction there is a resultant force of $\varphi \cdot \Omega^2 r \, dm$ toward the equilibrium position. The deflection of the element dm from equilibrium is φr, since the "curve" was assumed straight. Thus this excess force can be considered as a spring force with a spring constant $k = \Omega^2 \, dm$. The frequency of vibration of *this particle* is $\omega_n = \sqrt{k/m} = \sqrt{\Omega^2 \, dm/dm} = \Omega$. The same answer is found for any particle along the line. Hence we may conclude that the assumed straight line is the exact deformation curve, for, if this were not so, we should have found different frequencies for the individual particles. (In Rayleigh's procedure a nonexact curve is presupposed; in this case the individual particles have different calculated frequencies. In integrating the energies over all the particles, Rayleigh finds some sort of average of all these frequencies.)

FIG. 6.41. Calculating the natural frequency of a rotating heavy chain.

We have proved that the first natural frequency of the small vibrations of a chain, as shown in Fig. 6.41, equals its angular speed, and, since in the proof no mention was made of the mass distribution, the result is true for any distribution of the mass.

Another manner of showing this is by means of Rayleigh's method. Again assume a straight line for the deformation curve. On a particle dm the centrifugal force is $\Omega^2 r \, dm$. When moving from the equilibrium position A to the position C of Fig. 6.42, the particle travels against the centrifugal force over a distance $AB = r\varphi^2/2$. Thus the potential energy in the element is $\Omega^2 r \, dm \cdot r\varphi^2/2$ and the total potential energy of the chain is

$$Pot = \frac{\Omega^2 \varphi^2}{2} \int_0^l r^2 \, dm = \frac{1}{2} \Omega^2 \varphi^2 I_O$$

If the chain is vibrating harmonically with a frequency ω_2, the kinetic energy of the particle dm is $\frac{1}{2}\,dm \cdot v^2 = \frac{1}{2}\,dm \cdot \overline{BC}^2 \cdot \omega_2^2 = \frac{1}{2}\,dm\,\varphi^2 r^2 \omega_2^2$, and for the whole chain

$$Kin = \frac{\omega_2^2 \varphi^2}{2} \int_0^l r^2\,dm = \frac{1}{2}\,\omega_2^2 \varphi^2 I_0$$

Equating the two energies gives the desired result, $\omega_2 = \Omega$, which is independent of mass distribution.

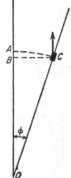

Fig. 6.42. Potential energy of an element of a rotating chain.

We obtain finally as an approximation for the first natural frequency of the rotating propeller blade

$$\omega^2 = \omega_1^2 + \Omega^2$$

For higher modes the result is quite similar; it can be expressed generally by

$$\omega_{rot}^2 = \omega_{non\text{-}rot}^2 + a\Omega^2 \tag{6.21}$$

where Ω is the speed of rotation and a a numerical factor which differs for the different modes and which has been found to be approximately as follows:

Mode 1, $a \approx 1.5$
Mode 2, $a \approx 6$
Mode 3, $a \approx 12$

The principal source of excitation of blade vibration is found in torsional impulses on the crank shaft of the engine. A common manner in which the relation (6.21) is plotted is shown in Fig. 6.43, in which two families of curves are given. The first set are hyperbolas showing the relation between the natural frequencies of a blade in its various modes and the speed of rotation as expressed by Eq. (6.21). The other set is a star of straight lines passing through the origin, expressing the relation between the exciting frequency and the speed. These straight lines have slopes

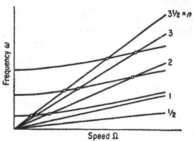

Fig. 6.43. Diagram showing straight lines of exciting frequency and hyperbolas of natural frequency. Intersections are points of resonant speed.

equal to the order of vibration, i.e., to the number of oscillations per revolution. For four-cycle, internal-combustion engines, such as are commonly used on aircraft, the orders occurring are either integer or half-integer numbers, as shown in the figure. Any intersection of one of the straight lines of exciting frequency with one of the curves of natural frequency indicates a possible condition of resonance in torsional vibration.

The determination of the natural frequencies of the non-rotating blade, *i.e.*, the intersections of the hyperbolas with the ordinate axis of Fig. 6.43, is not so simple as appears on first sight.   This is because the bending frequency of a blade cannot be considered separately from the torsional oscillation of an engine crank shaft; the two effects are coupled together. Thus the frequency of vibration of a blade is different for different engines attached.   Still it is desirable to have a criterion by which to determine the characteristics of an arbitrary propeller, independently of the engine to which it is attached.   This is possible by means of the reasoning shown in connection with Fig. 6.44*a*, *b*, and *c*.

Imagine the shaft cut at the hub of the propeller, as in Fig. 6.44*c*. The amplitude of torque transmitted through the cut is $M_0 \sin \omega t$ and the

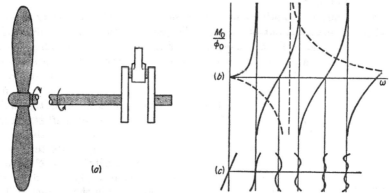

Fig. 6.44. (*a*) The system divided into two parts for purpose of analysis.   (*b*) Impedance diagram: full line for the left half of (*a*), dashed line for the right half of (*a*).   (*c*) Shapes of blade vibration at various frequencies.

amplitude of torsional oscillation at the cut is $\varphi_0 \sin \omega t$.   Looking at the propeller alone, *i.e.*, at the left-hand side of Fig. 6.44*a*, there is a definite ratio between M_0 and φ_0 which is independent of the magnitude of φ_0 but which will be a function of the frequency of oscillation.   This ratio M_0/φ_0, sometimes known as the *mechanical impedance* of the propeller, is plotted in the full line of Fig. 6.44*b*.   The various shapes of natural vibration belonging to various frequencies are shown in their proper positions in Fig. 6.44*c*.   The curve of Fig. 6.44*b* shows a number of points of zero ordinate and another series of points of infinite ordinate.   In the first set of points the torque at the propeller hub is zero, so that these points are the natural frequencies of the propeller with a free hub.   The other series of points of infinite ordinate show a zero angle for a finite torque and therefore are the natural frequencies with clamped hub.   The actual condition of the hub lies between that of completely free and that of completely clamped and depends on the properties of the engine to which it is attached.

Now turning our attention to the right-hand half of Fig. 6.44a and again plotting the ratio of torque to angle (with a negative sign), we obtain the dotted line of Fig. 6.44b. This is the curve for the (negative) "mechanical impedance" of the engine and is the usual resonance curve for the single-degree-of-freedom system of Fig. 2.20. For a natural frequency of the combined system, Fig. 6.44a, it is necessary that the moment-angle ratios to the left and to the right are the same. In other words, the natural frequencies of the combined system are the intersections between the dotted curve and the full curve of Fig. 6.44b. These are the frequencies that must be inserted on the ordinate axis of Fig. 6.43, and then Fig. 6.43 determines the critical speeds of the system caused by a purely torsional excitation.

Vibrations due to other causes have been observed in propeller blades. When the engine is out of balance, the center of the propeller hub may move back and forth laterally, which is a motion entirely independent of torsion, and associated with a displacement of the center of gravity of the engine. The primary cause of such a condition is of course unbalance, but it has also been found as a direct result of torque variation. If, for instance, there is a certain clearance in the main bearings of the engine, or if the crank-shaft structure is flexible, the periodic thrust variations on the crank pin due to the firing cylinders may cause the crank shaft to deform and move within its bearing clearance so that the center of gravity is displaced. Since all of this is due to internal forces in the system, a displacement of the center of gravity of the rotating parts must be associated with a displacement of the center of gravity of the stationary parts, which include the bearing near the hub of the propeller. In this manner, lateral motions of the center of the propeller hub with the firing frequency are possible.

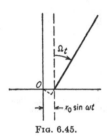

Fig. 6.45.

A lateral motion of the center of the propeller hub with a frequency ω does not, however, cause stress variations in the propeller of that same frequency, but rather of the frequency $\omega + \Omega_{prop}$ or $\omega - \Omega_{prop}$, as will be explained presently. If, as usual, the propeller is geared to the engine so that the propeller speed is related to the engine speed by a fairly complicated fraction such as $\%_{6}$, then an observation of the frequency of stress variation in the propeller makes it possible to distinguish between a vibration caused by pure torsional excitation and one caused by lateral excitation. In order to understand the frequency relation just mentioned, consider Fig. 6.45. A propeller blade is shown to rotate with angular speed Ω_{prop}, while its hub is moving back and forth laterally through the distance $x_0 \sin \omega t$. The displacement $x_0 \sin \omega t$ is now resolved into its components along the blade and perpendicular to the blade. The displacement along the blade does not excite any bending in it, but the displacement across the blade is entirely responsible for just that. Thus the displacement of the blade root in a direction perpendicular to that of the blade is

$$x_0 \sin \omega t \cdot \cos \Omega_{prop} t$$

which, by means of the trigonometric relations of page 15, is equal to

$$\frac{x_0}{2} \sin (\omega - \Omega_{prop})t - \frac{x_0}{2} \sin (\omega + \Omega_{prop})t$$

This lateral motion of the blade root will cause bending vibrations in the blade of the same frequencies as the root displacement, which proves the contention made above. If the lateral displacement of the blade root has been assumed vertically or in a different phase with respect to the rotation, exactly the same result would have been obtained, as can be easily verified.

Still another possible excitation of bending vibrations in propeller blades is that due to aerodynamic forces. In the usual construction of large airplanes a propeller is mounted in front of a wing and consequently each blade comes close to the wing twice in the course of one revolution. The velocity field of the air close to the wing is different from that at some distance from it so that the aerodynamic forces acting on the propeller blade will pass through a periodic change twice per revolution. This has been found to cause bending vibrations in the blade.

Summarizing, it may be stated that bending vibrations in a propeller blade caused by torsional excitation have a frequency equal to an integer or a half-integer multiple of the engine speed; those caused by lateral vibration of the propeller hub have a frequency equal to an integer or a half-integer multiple of the engine speed ± the propeller speed; and finally, the bending vibrations caused by aerodynamic excitation have a frequency which is a multiple of the propeller speed.

The internal friction in propeller blades of steel or aluminum is very small and the only damping that the vibrations experience is aerodynamic and is of the same nature as that discussed with reference to ship propellers on page 209. In Fig. 5.29, a vibratory motion of the blade in its limber direction causes periodic variations in the angle of attack α and consequently periodic variations in the aerodynamic lift force. The reader is urged to follow this phenomenon in detail and to verify the statement that the lift-force variation caused by this motion will be directed against the velocity of the motion and thus constitutes a true damping. This is true only for relatively slow frequencies, for the reasoning leading to this conclusion concerning damping presupposes a "succession of steady states." It will be seen in Chap. 7, page 321, that for very fast frequencies and high air speeds this reasoning is no longer valid and that under such circumstances the blade may experience "negative damping" and get into a state of "flutter." When such a condition exists, the aerodynamic forces become very large, of the same order almost as the spring forces and the inertia forces, so that even the frequency of the fluttering blade is considerably different from the natural bending frequency as calculated without air forces.

Not only aircraft propellers but also ship propellers have been responsible for serious cases of vibration. The excitation of a ship propeller falls into two classes: torsional and linear. When an individual blade passes close by the hull of the ship or by the "bossing" that holds the propeller tube in place, it finds itself in a region of flow which is different from that in the more or less open water. Consequently the hydrodynamic forces are different so that these forces experience variations with the blade frequency, i.e., the frequency of revolution of the propeller multiplied by its number of blades. The torque variation caused by this effect is more serious when the bossings are close to the propeller than when they are cut away. At present there is not a great deal of detailed knowledge available on the subject but a figure which represents a good average condition is a torque variation equal to 7.5 per cent of the total propeller torque. This effect is responsible for the fact that even in steam-turbine drives the main propulsion shafting of a ship has been found to experience definite resonant speeds. It has become standard practice to precalculate these speeds, as discussed on page 195 and in Problem 167.

The other effect caused by the variation in the hydrodynamic forces of the propeller is found in their reactions on the ship's hull and on the bossings. These force variations were determined by F. M. Lewis on an experimental model in a tank and were found to be as large as 12.5 per cent of the total propeller thrust. Naturally, this figure is very much dependent on the bossing clearance and the tip clearance of the propeller but it represents a good average figure for ships of conventional design up to date. These hull forces are responsible for the vibrations usually observed on the afterdecks of steamships. They were not considered to be of any great importance until the great French liner "Normandie" brought the matter into the limelight. In that case it happened that the propeller forces were of the same frequency as one of the natural frequencies of the entire ship as a "free-free" beam (page 154) so that oscillations of considerable magnitude were set up. The trouble was cured primarily by replacing the three-bladed propellers by four-bladed ones which eliminated this resonant condition.

6.11. Vibration of Steam-turbine Wheels and Blades. In the mechanical construction of large reaction steam turbines we can distinguish two principal types, which may be designated as the disk type and the drum type. In the disk type the rotor or spindle consists of a shaft on which a number of disks are shrunk. The diameter of these disks is about four times as large as the shaft diameter, and the turbine blades are attached to the rim of the disks. With the drum type the spindle consists of a hollow forging of a diameter equal to the outside diameter of the disks in the disk type, and the blades are fastened directly to the outside of this spindle.

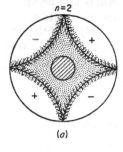
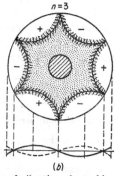

FIG. 6.46. The four- and six-noded modes of vibration of a turbine disk.

In both types fatigue failures of the blades have occurred. Whereas in the drum type the failures have been restricted to the blades themselves, in the disk type the breaks have been found to extend into the solid parts of the disks as well.

As in the case of the airplane propeller we have a resonance phenomenon between the natural frequency of vibration of the disk and some multiple of the running speed. Before proceeding to an explanation of the origin of the disturbing forces, it is necessary to have a clear understanding of the natural modes of vibration. First consider a disk at standstill (*i.e.*, without rotation). The center is clamped on the shaft, and the periphery with the blades is free to vibrate. In such a system there are infinitely many modes of natural motion, of which only a few are of importance for this problem. The four modes which have caused failures in the past are those in which the periphery bends into a sine wave with 4, 6, 8, or 10 nodes designated as the flexural modes of $n = 2, 3, 4, 5$, respectively.

The first two of these, being the most important ones, are illustrated in Fig. 6.46. In producing these figures the disk is held in a horizontal plane and evenly covered with light sand. Vibration is excited (usually by an alternating-current magnet of variable frequency), and at resonance the sand is thrown away from the vibrating parts of the disk and accumulates on the nodal lines. The circumference of the disk thereby divides up into an even number of equal parts which alternate in moving up and down. The signs + and − written in at the locations of the antinodes pertain to a certain instant of time. At that instant, the plus sign indicates that the disk is deflected upward and the minus sign that it is deflected downward. After half a vibration period these signs are reversed. At the nodes, of course, no motion takes place at all. The 8- and 10-noded motions are not illustrated but can be easily visualized. The deflections along the circumference are such that, if the perimeter is developed into a straight line, the deflections appear approximately as sine curves with n full cycles for the $2n$-noded vibration.

In the *rotating* disk the conditions are only slightly different. The whole Fig. 6.46 now revolves with the angular velocity Ω of the wheel. Moreover, the centrifugal forces which are set up by the rotation will

Fig. 6.47. A steady, stationary force F can put work into a rotating and vibrating disk.

raise the frequency of the vibration and also alter the shape of the natural mode. The latter effect is of little importance and will not be considered. The rise in the natural frequency ω follows the same trend as in the propeller blade of the previous section, *i.e.*, it is expressed approximately by the hyperbolic relation

$$\omega^2 = \omega^2_{\text{non-rot}} + B\Omega^2 \tag{6.22}$$

where Ω is the angular velocity of rotation and B is a numerical factor which is greater than unity and has different values for different modes of vibration, as discussed on page 272. The derivation of this formula is very similar to the derivation of Eq. (6.21); only it is much more complicated on account of the substitution of a plate for the beam of the previous problem.

A vibration in the modes of Fig. 6.46 may be excited in the rotating disk by a constant force standing still in space, *e.g.*, by the steam jet of a stationary nozzle playing on the disk. This can be understood from Fig. 6.47, which represents one nth part of the developed perimeter of the disk vibrating in its $2n$-noded mode. The amplitude varies periodically with the time between the full-drawn and the dotted curves. Simultaneously the curve (with its nodal points A, C, and E) moves to the right with the circumferential speed of the wheel. The force F remains fixed in space. Let the period of vibration and the circumferential speed be related in such a manner that when the point C has arrived at A one-half period of oscillation has passed so that the periphery will then have the dotted shape. To be more precise, let the relation be such that, when the force F is

opposite A, the full curve exists;
opposite B, no deflections exist anywhere;
opposite C, the dotted line is the shape;
opposite D, no deflections, etc.

Thus while the piece AC of the curve passes by the force F, that curve goes from its full-drawn to its dotted shape. During this time all points of the curve AC have an upward velocity so that F does positive work. But while CE passes by F, the shape goes from the dotted to the full line, which again is associated with upward velocities in the stretch CE, so that F again does positive work. The stretch AC in that interval of time goes downward to its full line position, but then it is not situated opposite F.

The speed at which this relation holds is called the "critical speed" of the disk; it exists when $1/n$th revolution occurs during one vibration period:

$$\text{r.p.s.} = \frac{f}{n}; \qquad \Omega = \frac{\omega}{n} \left.\right\} \tag{6.23}$$
$$\omega^2 = \omega^2_{\text{non-rot}} + B\Omega^2 = n^2\Omega^2 \left.\right\}$$

As was stated before, B is larger than unity, so that this equation coincides with (6.21). Therefore, it can be represented by Fig. 6.43, with the understanding that resonance in the $2n$-noded mode occurs at the intersection point of the hyperbola with the line of slope n. In particular we see that the two-noded mode ($n = 1$, one nodal diameter) can never be excited by a constant force F. Also, excitations of half-integer order do not occur in the turbine.

It is clear from Fig. 6.43 that a great number of critical speeds are possible. The disks in a turbine vary considerably in size from the high-pressure to the low-pressure ends, and in most cases there will be one or more disks among them in which the cluster of critical speeds ranges around the operating speed. This accounts for the great number of failures which occurred before the cause was understood.

To overcome the trouble, the disks are so designed that their criticals do not coincide with the running speed. Since in the first place the analysis is too crudely developed to permit great accuracy in this calculation, and since the frequency depends quite sensibly on the amount of shrink pressure at the center, the design is carried out in an empirical manner by comparison with previous constructions. After the turbine is built and assembled, the frequencies of those disks in which trouble may be expected are determined by experiment (excitation at variable frequency either by a mechanical vibrator or by means of an alternating-current magnet). In case such a frequency lies too close to the service speed, it is changed by a "tuning" process consisting of machining a thin layer of metal from the disk, usually near its periphery. The minimum difference between the critical and the running speed which is tolerated in practice is given as 15 per cent for the 4-noded mode and as 10 per cent for the 6- and 8-noded vibrations.

With turbines of the *drum* type fatigue failures of the individual blades or of groups of blades have occurred repeatedly. The explanation is exactly the same as for the disks; a drum with a row of blades can be regarded as a disk of which the central portion is infinitely stiff. There exist, however, other possibilities of resonance than the one just described. Imagine a turbine (disk or drum type) in which the blades are attached only at their base and are not connected either by a shrouding ring or by lashing wires, so that each blade can vibrate individually. If there is a single nozzle, the first mode of vibration of the blade (without nodes except at the base) can be excited if the rotational speed is such that an integer number of vibration cycles occurs during one revolution. This is because if the blade passes by the jet, while the blade is receding in its vibratory motion, the jet does positive work on it. If the number of vibrations per revolution is an integer, the phase of the vibratory motion is the same each time the blade comes in contact with the jet. This opens up a large number of possibilities for trouble. In practically all turbines, however, the blades are connected either completely or else in groups of approximately eight blades. Such a group of blades has natural frequencies that may be excited in the manner just described.

This particular phenomenon has been responsible for a series of serious failures. The blades in question were in the first impulse stage of turbines of very high pressure and temperature: 1,200 lb./sq. in. pressure at 900°F. The blades themselves were about 1 in. high and 1 in. deep, and under the influence of the very thick steam at high velocity developed 100 hp. each. They were found to fail in fatigue after an operation of some 5 hours. The natural frequency of the blades was such that approximately 60 full cycles occurred during one revolution. This put the various consecutive critical speeds only 1.5 per cent apart so that it was impossible to avoid resonance by tuning. Ordinarily it would be expected that a blade, after having passed the steam nozzle and having acquired a certain amplitude of vibration by the steam impact of that nozzle, would execute a damped vibration from there on and in the ensuing 60 cycles practically lose all of its amplitude. Then, coming back to the nozzle, it would get a new impact. The blades in question were calculated to be sufficiently strong to stand this variable loading.

It was found, however, that the internal damping in the blades was so small that at the end of 60 cycles, *i.e.*, at the end of a full revolution, the amplitude of vibration had hardly diminished so that the blade would enter the jet with a substantial amplitude. Thus with proper phase conditions the amplitude could be pushed up to a value many times greater than that caused by a single exposure to the jet. The surprising fact was brought to light that the internal hysteresis at temperatures approaching that of the red-hot state was smaller than the hysteresis at room temperature. A partial cure for the trouble consists in rounding off the edges of the steam jet by providing suitable leakage passages in the nozzle.

Researches carried out on the internal hysteresis at high temperatures have resulted in the development of a special molybdenum steel having many times the damping of carbon steel. This type of molybdenum steel is now being used almost exclusively in steam turbines. More recently similar troubles have occurred in gas turbines in which the operating temperatures are much higher, so that only very special high-temperature-resistant steels can be used. Unfortunately, there exists at present no steel having good strength at high temperatures and simultaneously having a high hysteresis loss, but the metallurgists are quite busy trying to concoct something.

In reaction turbines no actual nozzles exist in the blading such as would account for the definite force F of Fig. 6.47. However, any deviation from radial symmetry of the pressure distribution acts in the same manner as a nozzle. While rotating, the blade passes through a periodic pressure field of which the fundamental component has the frequency of revolution, and in which most of the higher harmonics are present as well. Consider as an example the nth harmonic of this field. It is capable of exciting vibration, if the blade rotates at the rate of n natural periods per revolution. The phase of the motion will be such that while passing through the regions of great nth harmonic pressure, the blade recedes in its vibratory motion, whereas in regions of low pressure it is coming forward. We see that, in principle, resonance can occur if any natural frequency of a blade or blade group is an integer multiple of the speed of rotation, provided the pressure is unevenly distributed around the circumference.

Problems 173 *to* 195.

CHAPTER 7

SELF-EXCITED VIBRATIONS

7.1. General. The phenomena thus far discussed were either free vibrations or forced vibrations, accounting for the majority of troublesome cases which occur in practice. However, disturbances have been observed which belong to a fundamentally different class, known as *self-excited vibrations*. The essence of the difference can best be seen from a few examples.

First consider an ordinary single-cylinder steam engine, the piston of which executes a reciprocating motion, which may be considered a "vibration." Evidently the force maintaining this vibration comes from the steam, pushing alternately on the two sides of the piston.

Next consider an unbalanced disk mounted on a flexible shaft running in two bearings (Fig. 7.1). The center of the disk vibrates, the motion being maintained by the centrifugal force of the unbalance pushing the disk alternately up and down.

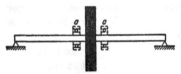

Fig. 7.1. Unbalanced shaft as an example of forced vibration.

The steam engine is a case of *self-excited* vibration, while the disk executes an ordinary *forced* vibration. Imagine that the piston is prevented from moving by clamping the crosshead or the flywheel. Then the valves do not move either, and hence *no alternating* steam *force* acts on the piston.

On the other hand, let us prevent the disk from vibrating. This can be done, for example, by mounting two ball bearings a, a on the shaft adjacent to the disk and attaching their outer races to a solid foundation, thus preventing vibration of the disk but leaving the rotation undisturbed. Since the unbalance is still rotating, *the alternating force remains.*

Thus we have the following distinction:

In a self-excited vibration the alternating force that sustains the motion is created or controlled by the motion itself; when the motion stops the alternating force disappears.

In a forced vibration the sustaining alternating force exists independently of the motion and persists even when the vibratory motion is stopped.

282

Another way of looking at the matter is by defining a self-excited vibration as a free vibration with *negative damping*. It must be made clear that this new point of view does not contradict the one just given. An ordinary positive viscous damping force is a force proportional to the velocity of vibration and directed opposite to it. A negative damping force is also proportional to the velocity but has the same direction as the velocity. Instead of diminishing the amplitudes of the free vibration, the negative damping will increase them. Since the damping force, whether positive or negative, vanishes when the motion stops, this second definition is in harmony with the first one.

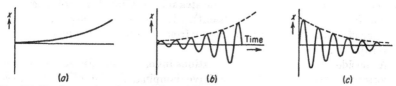

FIG. 7.2. The free motion of a system in various states of stability. (a) Statically unstable; (b) statically stable but dynamically unstable; (c) statically and dynamically stable.

Examine the differential equation of a system having a single degree of freedom with negative damping:

$$m\ddot{x} - c\dot{x} + kx = 0 \qquad (7.1)$$

Since this equation differs from (2.12) on page 37 only in the sign of c, its solution can be written as

$$x = e^{+\frac{c}{2m}t}(A \cos qt + B \sin qt) \qquad (7.2)$$

which is clearly a vibration with exponentially increasing amplitude (Fig. 7.2).

A system with positive damping is sometimes said to be *dynamically stable*, whereas one with negative damping is known as *dynamically unstable*. There is a difference between *static* and *dynamic* stability. A mechanical system is statically stable if a displacement from the equilibrium position sets up a force (or couple) tending to drive the system back to the equilibrium position. It is statically unstable if the force thus set up tends to increase the displacement. Therefore static instability means a *negative spring constant k* or, more generally, a negative value of one of the natural frequencies ω^2.

Figure 7.2 shows the behavior of a system in three different stages of stability. It is to be noted that dynamic stability always presupposes static stability (Fig. 7.2c), but that the converse is not true: a statically stable system may yet be dynamically unstable (Fig. 7.2b).

Regarding the *frequency* of the self-excited vibration, it may be said that in most practical cases the negative damping force is very small in comparison to the elastic and inertia forces of the motion. If the damping force were zero, the frequency would be the *natural* frequency. A damping force, whether positive or negative, lowers the natural frequency somewhat, as expressed by Fig. 2.16, page 40. However, for most practical cases in mechanical engineering this difference is negligible, so that then *the frequency of the self-excited vibration is the natural frequency of the system.* Only when the negative damping force is large in comparison with the spring or inertia forces does the frequency differ appreciably from the natural frequency. Such cases, which are known as "relaxation oscillations," are discussed on page 363. The steam engine is an example, as the negative damping force of the steam is very much greater than the spring force (which is wholly absent). Hence, for the engine, the frequency of vibration differs appreciably from the natural frequency (which is zero).

A consideration of the energy relations involved will also serve to give a better understanding. With positive damping, the damping force does negative work, being always opposed to the velocity; mechanical energy is converted into heat, usually in the dashpot oil. This energy is taken from the vibrating system. Each successive vibration has less amplitude and less kinetic energy, and the loss in kinetic energy is absorbed by the damping force. In the case of negative damping the damping force (which is now a driving force) does positive work on the system. The work done by that force during a cycle is converted into the additional kinetic energy of the increased vibration. It is clear that self-excited vibration cannot exist without an extraneous source of energy, such as the steam boiler in our first example. The source of energy itself should not have the alternating frequency of the motion. In most cases the energy comes from a source without any alternating properties whatever, for example, a reservoir of steam or water under pressure, a steady wind, the steady torque of an engine, etc. However, there are a few cases (discussed on pages 378 and 379) where the source is alternating with a high frequency, much higher than that of the vibration it excites.

With a truly *linear* self-excited system the amplitude will become infinitely large in time, because during each cycle more energy is put into the system (Fig. 7.2b). This infinitely large amplitude is contrary to observation. In most systems the mechanisms of self-excitation and of damping exist simultaneously and separately. In Fig. 7.3 the energy per cycle is plotted against the amplitude of vibration. For a linear system this energy follows a parabolic curve since the dissipation per cycle is $\pi c \omega x_0^2$ (see page 52). If the negative damping force is also linear, another parabola will designate the energy input per cycle. The system is self-excited or is damped according to which parabola lies higher. In all practical cases, however, either the input or the damping, or both, are non-linear and the input and dissipation curves intersect.   If in Fig. 7.3 the amplitude happens to be OA, more energy is put in than is dissipated, so that the amplitude grows.   On the other hand, if the amplitude happens to be OC, there is more damping than self-excitation and the vibration will decrease.   In both cases the amplitude tends toward OB where energy equilibrium exists.   The motion thus executed is an undamped steady-state free vibration.

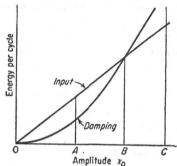

Since the non-linearity of the damping or input forces leads to great mathematical complication (see Chap. 8), we usually assume linear systems of very small amplitude and determine whether the damping or the energy input is the stronger.   If the system

FIG. 7.3. Work per cycle performed by a harmonic force and by a viscous damping force for various amplitudes.

is found to be unstable, it means merely that the amplitude will *begin* to build up; how far this building up will develop depends on the nature of the non-linearity.

In electrical engineering, self-excited vibrations are of even greater importance than in the mechanical field.   The electrical analogue of a forced vibration was seen to be an LC circuit with an alternator all in series (Fig. 2.5, page 27).   An electrical self-excited system is exemplified by an oscillating vacuum-tube circuit.   The B-battery is the non-alternating source of energy; the frequency is determined by the L and C values of the plate circuit, and the negative damping or feed-back is supplied by the grid.

7.2. Mathematical Criterion of Stability.   For single-degree-of-freedom systems, such as are discussed in Secs. 7.3 to 7.6, a simple physical reasoning usually suffices to show the negativity of the damping constant c.   Thus the criterion of dynamic stability can be derived by physical rather than by mathematical means.   With systems of two or three degrees of freedom, a physical conception is always very helpful but usually does not give a complete interpretation of what happens.   A mathematical approach is necessary, and this involves at first the setting up of the differential equations of the problem.   As long as we deal with small vibrations (and thus disregard any non-linearities that may exist), the equations are all linear and of the second order, of the type (3.18) or (4.16).   Their solution, as usual, is found by assuming

$$\left.\begin{aligned} x_1 &= x_{1max}e^{st} \\ x_2 &= x_{2max}e^{st} \\ &\cdots\cdots\cdots \\ x_n &= x_{nmax}e^{st} \end{aligned}\right\} \tag{7.3}$$

where s is a complex number the real part of which determines the damping and the imaginary part of which is the natural frequency. Substituting (7.3) into the differential equations of the free vibration transforms these equations into a set of n homogeneous, linear algebraic equations in the (complex) unknowns $x_{1max} \ldots x_{nmax}$. A process of algebraic elimination is then performed with the result that one equation is obtained which does not contain any of these variables. This equation, known as the "frequency equation," is generally of the degree $2n$ in s. Thus, for a two-degree-of-freedom system we obtain a quartic; for a three-degree-of-freedom system we obtain a sixth-degree equation, etc.

An algebraic equation of degree $2n$ in the variable s has $2n$ roots or $2n$ values of s. Real roots of s would lead to terms e^{st} in the solution, which rarely occur in ordinary vibrating systems (Fig. 2.14, page 38). The roots of s are usually complex and then they always occur in conjugate pairs:

$$s_1 = p_1 + jq_1$$
$$s_2 = p_1 - jq_1$$
$$s_3 = p_2 + jq_2$$
$$s_4 = p_2 - jq_2$$

The solution of the first differential equation is

$$x_1 = C_1 e^{s_1 t} + C_2 e^{s_2 t} + C_3 e^{s_3 t} + \cdots$$

From Eqs. (2.15), (2.18), and (2.19), page 39, we know that these terms can be combined by pairs as follows:

$$C_1 e^{s_1 t} + C_2 e^{s_2 t} = e^{p_1 t}(A \sin q_1 t + B \cos q_1 t)$$

so that the imaginary part of s is the frequency, and the real part of s determines the rate of damping. *If the real parts of all the values of s are negative, the system is dynamically stable; but if the real part of any one of the values of s is positive, the system is dynamically unstable.*

Therefore the stability can be determined by an examination of the signs of the real parts of the solutions of the frequency equation. It is not necessary to solve the equation, because certain rules exist by which from an inspection of the coefficients of the equation a conclusion regarding the stability or instability can be drawn. These rules, which were given by *Routh* in 1877, are rather complicated for frequency equations of higher degree, but for the most practical cases (third and fourth degree) they are sufficiently simple.

Let us consider first the cubic equation

$$s^3 + A_2 s^2 + A_1 s + A_0 = 0 \tag{7.4}$$

which occurs in the case of two degrees of freedom where one mass or spring is zero (in a sense one and one-half degrees of freedom). If its roots are s_1, s_2, and s_3, (7.4) can be written

$$(s - s_1) \cdot (s - s_2) \cdot (s - s_3) = 0$$

or, worked out,

$$s^3 - (s_1 + s_2 + s_3)s^2 + (s_1 s_2 + s_1 s_3 + s_2 s_3)s - s_1 s_2 s_3 = 0 \quad (7.5)$$

A comparison with (7.4) shows that

$$\left.\begin{aligned}
A_2 &= -(s_1 + s_2 + s_3) \\
A_1 &= s_1 s_2 + s_1 s_3 + s_2 s_3 \\
A_0 &= -s_1 s_2 s_3
\end{aligned}\right\} \quad (7.6)$$

One of the three roots of a cubic equation must always be real, and the other two are either real or conjugate complex. Separating the roots s_1, s_2, s_3 into their real and imaginary parts, we may write

$$\begin{aligned}
s_1 &= p_1 \\
s_2 &= p_2 + jq_2 \\
s_3 &= p_2 - jq_2
\end{aligned}$$

Substituted into (7.6) this leads to

$$\left.\begin{aligned}
A_2 &= -(p_1 + 2p_2) \\
A_1 &= 2p_1 p_2 + p_2^2 + q_2^2 \\
A_0 &= -p_1(p_2^2 + q_2^2)
\end{aligned}\right\} \quad (7.7)$$

The criterion of stability is that both p_1 and p_2 be negative. It is seen in the first place that *all coefficients A_2, A_1, and A_0 must be positive*, because, if any one of them were negative, (7.7) requires that either p_1 or p_2, or both p_1 and p_2, must be positive. This requirement can be proved to hold for higher degree equations as well. Hence a frequency equation of any degree with one or more negative coefficients determines an unstable motion.

Granted that the coefficients A_0, A_1, and A_2 are all positive, the third equation (7.7) requires that p_1 be negative. No information about p_2 is available as yet. However, on the boundary between stability and instability, p_2 must pass from a positive to a negative value through zero. Make $p_2 = 0$ in (7.7) and

$$\left.\begin{aligned}
A_2 &= -p_1 \\
A_1 &= q_2^2 \\
A_0 &= -p_1 q_2^2
\end{aligned}\right\} \quad (7.8)$$

These relations must be satisfied on the boundary of stability. By eliminating p_1 and q_2, we find

$$A_0 = A_1 A_2$$

We do not know yet on which side of this relation stability exists. That can be found in the simplest manner by trying out one particular case. For example let $s_1 = -1$ and $s_{2,3} = -1 \pm j$, which obviously is a stable solution. Substitution in (7.7) gives

$$A_2 = 3 \qquad A_1 = 4 \qquad A_0 = 2$$

so that

$$A_0 < A_1 A_2$$

The complete criterion for stability of the cubic (7.4) is that all coefficients A are positive and that

$$A_1 A_2 > A_0 \tag{7.9}$$

Practical examples of the application of this result are given in Secs. 7.7 and 7.8.

Next consider the quartic

$$s^4 + A_3 s^3 + A_2 s^2 + A_1 s + A_0 = 0 \tag{7.10}$$

for which the procedure is similar. Since a quartic can be resolved into two quadratic factors, we may write for the roots

$$\left.\begin{aligned} s_1 &= p_1 + jq_1 \\ s_2 &= p_1 - jq_1 \\ s_3 &= p_2 + jq_2 \\ s_4 &= p_2 - jq_2 \end{aligned}\right\} \tag{7.11}$$

and substitute in (7.10), which leads to

$$\left.\begin{aligned} A_3 &= -2(p_1 + p_2) \\ A_2 &= p_1^2 + p_2^2 + q_1^2 + q_2^2 + 4p_1 p_2 \\ A_1 &= -2p_1 \cdot (p_2^2 + q_2^2) - 2p_2(p_1^2 + q_1^2) \\ A_0 &= (p_1^2 + q_1^2) \cdot (p_2^2 + q_2^2) \end{aligned}\right\} \tag{7.12}$$

The requirement for stability is that both p_1 and p_2 be negative. Substitution of negative values of p_1 and p_2 in (7.12) makes all four A's positive, so that the first requirement for stability is that all coefficients A be positive. Granted that this is so, the first equation of (7.12) requires that at least one of the quantities p_1 or p_2 be negative. Let p_1 be negative. We still need another requirement to make p_2 also negative. On the boundary between stability and instability, $p_2 = 0$, which substituted in (7.12) gives

$$\left.\begin{aligned} A_3 &= -2p_1 \\ A_2 &= p_1^2 + q_1^2 + q_2^2 \\ A_1 &= -2p_1 q_2^2 \\ A_0 &= (p_1^2 + q_1^2)q_2^2 \end{aligned}\right\} \tag{7.13}$$

being four equations in the three variables p_1, q_1, and q_2.

Elimination of these variables leads to a relation between the A's:

$$A_1 A_2 A_3 = A_1^2 + A_3^2 A_0$$

To find out on which side of this equality stability exists, we try out a simple stable case, for example,

$$s_{1,2} = -1 \pm j \qquad s_{3,4} = -2 \pm 2j$$

which, on substitution in (7.12), gives

$$A_3 = 6 \qquad A_2 = 18 \qquad A_1 = 24 \qquad A_0 = 16$$

so that

$$A_1 A_2 A_3 > A_1^2 + A_3^2 A_0$$

The complete criterion for stability of the quartic (7.10) *is that all coefficients A are positive and that*

$$A_1 A_2 A_3 > A_1^2 + A_3^2 A_0 \tag{7.14}$$

Applications of this relation are made in Secs. 7.7, 7.9, and 7.10.

Systems with three degrees of freedom generally have a sextic for their frequency equation and in degenerated cases a quintic. In such cases there are three real parts of the roots s, and besides the requirement of positive signs for all coefficients A there are *two* other requirements, each of which is rather lengthy. For further information in this field the reader is referred to the original work of Routh.

7.3. Instability Caused by Friction. There are a number of cases where friction, instead of being responsible for positive damping, gives rise to negative damping. One of the well-known examples is that of the violin string being excited by a bow. The string is a vibrating system and the steady pull of the bow is the required source of non-alternating energy. The friction between the string and the bow has the characteristic of being greater for small slipping velocities than for large ones. This property of dry friction is completely opposite to that of viscous friction (Fig. 7.4a). Consider the bow moving at a constant speed over the vibrating string. Since the string moves back and forth, the relative or slipping velocity between the bow and the string varies constantly. The absolute velocity of the bow is always greater than the absolute vibrating velocity of the string, so that the direction of slipping is always the same. However, while the string is moving in the direction of the bow, the slipping velocity is small and consequently the friction force great; but during the receding motion of the string, the slipping velocity is large and the friction small.

We note that the *large* friction force acts in the direction of the motion of the string, whereas the *small* friction force acts against the motion of the string. Since the string executes a harmonic motion, the work done by the friction force during one-half stroke in $2F x_0$, where F is an average value of the friction force and x_0 the amplitude of vibration. Since F if greater during the forward stroke (when the friction does positive work on the string) than during the receding stroke (when negative work is done), the total work done by the friction over a full cycle is positive and hence the vibration will build up.

In mechanical engineering certain vibrations, usually referred to as "chatter," can be explained in the same manner. The cutting tool of a lathe may chatter and also the driving wheels of a locomotive. When starting a heavy train these drivers are sometimes seen to slip on the rails. While, as a rule, the slipping takes place in a uniform manner, "chattering slip" has been sometimes observed. Besides the major slipping rotation, the wheels then execute torsional oscillations which may cause very large

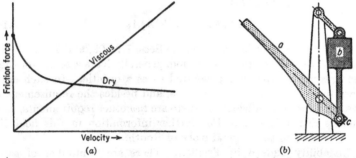

    (a)   (b)

Fig. 7.4. (a) Damping forces with positively and negatively sloping characteristics. (b) Drawbridge which failed structurally because of a negative friction characteristic in the pin bearing c.

alternating stresses in the crank pins and side rods. A negative slope (Fig. 7.4a) of the friction-velocity characteristic between the wheels and the rails is essential for this phenomenon.

The phenomenon may be observed in many homely examples such as the door that binds and screeches when opened, and the piece of chalk that is held perpendicular to the blackboard while writing. Another case is the familiar experiment in the physics laboratory of rubbing the rims of water glasses with a wet finger to cause them to sing.

A torsional vibration of this type has been observed in ships' propeller shafts when rotating at very slow speeds (creeping speeds). The shaft is usually supported by one or two outboard bearings of the lignum-vitae or hard-rubber type, which are water-lubricated. At slow speeds no water film can form and the bearings are "dry," causing a torsional vibration of the shaft at one of its natural frequencies, usually well up in the audible range. The propeller blades have natural frequencies not too far removed and act as loud-speakers, making this "starting squeal" detectable at great distances under water.

A striking technical example of the self-excited vibrations caused by dry friction is shown in Fig. 7.4b which represents a drawbridge of rather large dimensions. The bridge deck *a* is counterbalanced by a large concrete counterweight *b* which, together with its guiding links and the supporting tower, forms a parallelogram as shown.

After about a year's operation one of the towers of this bridge broke and on inspection the failure proved to be unmistakably caused by fatigue. Experiments with the other half of the bridge, still standing, showed that, when the deck was raised and lowered, violent vibrations of the whole structure took place at a very slow frequency, of about six cycles during the entire time of raising the bridge deck. The explanation was found in the bearing *c* which carries the tremendous load of the counterweight *b*. Whatever grease happened to be in this bearing at the beginning of the life of the bridge was soon squeezed out and the bearing

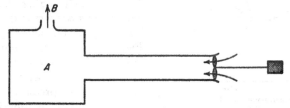

FIG. 7.5. Fan blowing air through a long tube into a chamber *A*.

was found to be entirely dry. The dry-friction chatter thus caused was sufficiently violent to cause the failure. Obviously the remedy for this case consists in proper grease cups, which have to be kept in proper order and must be inspected daily.

Another interesting phenomenon caused by a "negative characteristic" is shown in Figs. 7.5 and 7.6. A fan is blowing air into a closed chamber *A* of fairly large dimensions and the air is leaking out of that chamber through definite orifices *B*. The practical case of which Fig. 7.5 is a schematic representation was a boiler room in a ship which was kept under a slight pressure by the fan, and the orifices *B* were the boilers and stacks through which the air was forced out. It was observed that for a certain state of the opening *B*, *i.e.*, for a certain steam production, violent pressure variations of a frequency of about one cycle per second took place in the boiler room.

The explanation is partly given by Fig. 7.6, which is the characteristic curve of a blower. The volume delivered by the blower is plotted against the pressure developed by it. The point *P* of the characteristic obviously refers to the condition where the orifice *B* is entirely closed so that no volume is delivered, but a maximum of pressure is developed. The point *Q* of the characteristic refers to operation of the fan in free air where no pressure is developed but a large volume is delivered.   By changing the opening at B in Fig. 7.5, operation of the fan can be secured over a range in Fig. 7.6 from the point P almost down to Q.   It is seen that most of this curve has a slope descending from P to Q but that there is a short section between C and D in which the slope is reversed.   This is a characteristic of the construction of the fan and it is very difficult to build a fan in which the characteristic curve drops from P to Q with the slope in one direction only, and at the same time have good efficiency in the region between Q and D, for which the fan is built primarily.

It can be shown that operation near the point A in Fig. 7.6 is stable, whereas operation near the point B is unstable and will lead to the surging

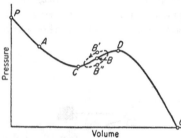

Fig. 7.6. Pressure-volume characteristic of a fan.   At point P the discharge opening is closed, while at Q the fan discharges freely into the open atmosphere.   Between C and D the slope of the characteristic is reversed, causing unstable operation.

condition just described.   Imagine operation near the point A and let the pressure in the chamber of Fig. 7.5 be slightly higher than normal. This means a decreased volume delivered by the fan, as can be seen from Fig. 7.6.   Thus an output of the fan less than normal will cause the pressure in the chamber to drop again, and since the pressure was higher than normal, the equilibrium condition tends to be restored.   Similarly, if by an accident the pressure were temporarily lower than normal at A, the volume delivered would be increased, which tends to boost the pressure and restore the equilibrium.

On the other hand, consider operation near the point B of Fig. 7.4. If now the pressure in the chamber is higher than normal instantaneously, the fan delivers more volume than in the normal condition and thus increases the pressure in the chamber still more.   Therefore, if the pressure in the chamber is increased by accident, the fan operation will immediately increase it still more, which means an unstable condition.

An important case of dry-friction excitation, which repeatedly has led to serious trouble in practice, is the so-called "shaft whipping" caused by a loose guide or by a poorly lubricated bearing with excessive clearance.   In Fig. 7.7 let the circle A designate the inside of a bearing or guide and B the cross section of the vertical shaft rotating in it.   Let the shaft be

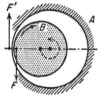

Fig. 7.7. Shaft whirl caused by dry friction.

rotating clockwise and be temporarily deflected from its equilibrium position in the center of A so that it strikes A at the left.   On account of its rotation the shaft sets up friction forces F and F', of which F is the force acting on the shaft and $F' = -F$ acts on the guide or bearing. The force F can be replaced by a parallel force of equal magnitude through the center of the shaft B and a couple Fr. The couple acts merely as a brake on the shaft, which is supposed to be driven at uniform speed, so that the only effect of the couple is to require some increase in the driving torque and is inconsequential. The force F through the center of the shaft, however, drives it downward or rather in a direction tangent to the circle A. The direction of F changes with the position of the shaft B in A, so that the shaft will be driven around as indicated by the dotted circle. It will be noticed that the shaft is driven around the clearance in a direction opposite to that of its own rotation. If the shaft rotates in the center of the guide without touching it, the shaft is stable; but as soon as it strikes the guide for any reason, the shaft is set into a violent whirling vibration.

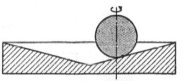

FIG. 7.8. The ball whirls around on account of the friction at the point of contact.

This effect is present in many modifications. A very simple model for demonstrating it is as follows. Take a shallow conical cup (Fig. 7.8) and a steel ball of about 1 in. diameter. Spin the ball between the fingers at the bottom of the cup. This position is an unstable one for the rotating ball because, if it is accidentally displaced a very small distance from the center of the cup, the point of contact with the cup no longer coincides with the (vertical) axis of rotation. There will be slip and a friction force perpendicular to the paper tending to drive the ball around in a circle. The direction of rolling of the ball will be opposite to the direction of spin.

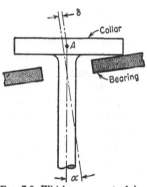

FIG. 7.9. Whirl on account of dry friction in a thrust bearing.

The phenomenon is not restricted to cylindrical guides or bearings but has also been observed on thrust bearings. Figure 7.9 represents schematically a thrust bearing and shaft, of which the equilibrium position is central and vertical. Suppose that the elastic system of which the shaft forms a part is capable of a natural mode of motion whereby the shaft center line whirls around the vertical with an eccentricity δ and an inclination α. The center A of the collar disk describes a circle of radius δ and the shaft a cone of apex angle 2α. This mode of motion will be self-excited by friction, because during the vibration the collar rests on the bearing on one side only. This causes a tangential friction force on that side urging the point A around the center line in a direction opposite to that of the rotation of the collar disk. The obvious way to prevent this sort of disturbance is to make the bearing support so flexible that, in spite of the angular deviation, the pressure on the various parts of the bearing remains uniform.

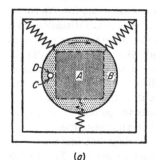

A very instructive model demonstrating this effect may be built as follows (Fig. 7.10). A small motor A carries a disk B on the end of its horizontal shaft and is supported very flexibly on three springs lying in a plane through the center of gravity and perpendicular to the shaft. When running, this motor is capable of a large number of natural modes of vibration, two of which are particularly interesting. They are illustrated by Fig. 7.10c and also by Fig. 6.29 (page 258). The shaft describes a cone characterized by δ and α and whirls either in the direction of rotation or opposite to it. The natural frequencies of these two modes of motion are shown in Fig. 6.30.

Imagine a piece of felt or paper C held against the front side of the disk near its circumference. It will strike (or press hard) when α (and consequently δ) is just in the position shown in Fig. 7.10c. Assume B to be rotating clockwise in Fig. 7.10a. The obstacle C will cause a friction force tending to push the disk *down*. As in the argument given with Fig. 7.7, this friction force is replaced by a retarding couple and a force

Fig. 7.10. Self-excited whirl caused by friction on the disk B.

through the shaft center. The retarding couple merely retards the motion slightly, but the force through the center of the disk pushes that center down, *i.e.*, in a direction of clockwise whirl. Thus friction on the *front* side C of the disk will encourage a precessional motion in the *same* direction as the rotation.

On the other hand, if D is pressed against the back of the rotating disk B, it will strike and cause friction when α and δ have reached the position just opposite to that of Fig. 7.10c. The friction again kicks the disk down, because the direction of rotation is still clockwise. This downward force now excites a counterclockwise whirl, because the deflection δ is opposite to that shown in Fig. 7.10c.

The experiment consists of rubbing the front of the disk and noticing the self-excitation of the mode of vibration of rather high frequency with the precession in the same direction as the rotation. Then, taking the rub from the front and applying it to the back side, this motion is seen to damp out very fast, and the second mode (precession against the rotation) with a much slower frequency is seen to build up. This latter motion can be damped very effectively by again rubbing the front of the disk. The difference in the two frequencies is caused by the gyroscopic action of the disk as explained on page 258.

7.4. Internal Hysteresis of Shafts and Oil-film Lubrication in Bearings as Causes of Instability. Another highly interesting case of self-excited vibration is that caused by internal hysteresis of the shaft metal. *Hys-*

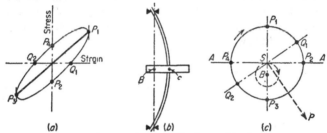

Fig. 7.11. Shaft whirl caused by internal hysteresis.

teresis is a deviation from Hooke's stress-strain law and appears in most materials with alternating stresses. In the diagram 7.11a Hooke's law would be represented by a straight line, and a fiber of a vibrating shaft, which experiences alternately tension and compression, should move up and down that line between P_1 and P_3. Actually the stress-strain relation is represented by a long narrow elliptic figure which is always run through in a clockwise direction. The ellipse as shown in Fig. 7.11 has its width greatly exaggerated; in reality it is so narrow that it can hardly be distinguished from the straight line P_1P_3.

Consider a vertical rotating shaft in two bearings with a central disk as shown in Fig. 7.11b. During the whirling motion the center of the shaft S describes a circle about the point B on the bearing center line. The point B is the normal or equilibrium position of S when no whirl exists. Figure 7.11c shows a cross section of the middle of the shaft, $P_1Q_1P_2P_3P_4$ being the outline of the shaft and the dotted circle being the path of S during the whirl. The deflection BS of Fig. 7.11c is a practical possibility; that of Fig. 7.11b is enormously exaggerated.

It is assumed that the rotation of the shaft and the whirl are both clockwise as shown. The shaft is bent, and the line AA divides it in two parts so that the fibers of the shaft above AA are elongated and those below AA are shortened. The line AA may be described as the *neutral line of strain*, which on account of the deviation from Hooke's law does not coincide with the *neutral line of stress*.

In order to understand the statement just made consider the point P_1 in Fig. 7.11c which may be thought of as a red mark on the shaft. In the course of the shaft rotation that red mark travels to Q_1, P_2, P_3, etc. Meanwhile the shaft whirls, whereby S and the line AA run around the dotted circle. The speed of rotation and the speed of whirl are wholly independent of each other. In case the speed of rotation is equal to the speed of whirl, the red mark P_1 will always be in the elongation of the line BS, or in other words, P_1 will always be the fiber of maximum elongation. In case the rotation is faster than the whirl, P_1 will gain on S and consecutively reach the position P_2 (of no elongation), P_3 (of maximum shortening), etc. On the other hand, if the rotation is slower than the whirl, P_1 will go the other way (losing on S) and go through the sequence P_1, P_4, P_3, P_2, etc.

First, investigate a rotation faster than the whirl. The state of elongation of the shaft fibers of the various points P_1, P_2, P_3, P_4 of Fig. 7.11c is indicated by the same letters in Fig. 7.11a. In Fig. 7.11a the point Q_1 of *no stress* lies between P_1 and P_2. The point Q_1 is now drawn in Fig. 7.11c and the same is done with Q_2 between P_3 and P_4. Thus the line Q_1Q_2 is the line of no stress (neutral line of stress) and all fibers above Q_1Q_2 have tensile stress while those below Q_1Q_2 have compressive stress. The stress system described sets up an elastic force $\bar{P}$, as shown. This elastic force $\bar{P}$ has not only a component toward B (the usual elastic force) but also a small component to the right, tending to drive the shaft around in its path of whirl. Thus there is a self-excited whirl.

The reader will determine for himself the truth of the statement that, if the rotation is slower than the whirl, the inclination of Q_1Q_2 reverses and the elastic force has a damping instead of a driving component.

The whirling motion is determined primarily by the elastic force of the shaft toward the center B combined with the inertia forces of the disk (see page 258) and therefore takes place with the natural frequency. The very small *driving* component of the elastic force merely overcomes damping. Internal hysteresis of the shaft acts as damping on the whirl below the critical speed, whereas above that speed a self-excited whirl at the critical frequency may build up.

Internal hysteresis in the shaft material is usually very small, but a more pronounced hysteresis loop is found in cases where actual slipping occurs, as in loose shrink fits or other joints. Thus a shaft with a loosely shrunk disk will probably develop a whirl at the natural frequency above the critical speed.

A self-excited vibration known as *oil whip* is caused by certain properties of the oil film in generously lubricated sleeve bearings. In order to understand this phenomenon, it is necessary to know that a horizontal shaft rotating in a counterclockwise direction in an oil-film lubricated bearing does not seek a central position but is deflected somewhat to the right (Fig. 7.12). The direction of this deviation can easily be remembered by noting that it is opposite to the direction in which one would expect the journal to climb. Since on such a journal the load or weight **W** is acting downward, as indicated in the figure, the resultant P of the oil pressures on the journal is equal and opposite to **W** and makes a certain angle α with the line OA connecting the center of the bearing and the center of the journal.

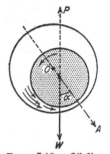

Fig. 7.12. Oil-film lubrication in a bearing causes whirling because the weight **W** and the axis of symmetry OA are not in line.

Consider a *vertical* guide bearing with a shaft in it. If there are no lateral loads acting, the shaft will seek the center of the bearing. If, for some reason, the shaft starts whirling around in the bearing, it will occupy an eccentric position at any instant. Moreover, if during that whirling the oil pressures are the same as in Fig. 7.12 (where **W** now must be replaced by a centrifugal force in the direction OA), there is no equilibrium between P and the centrifugal loading, but there is a small resultant force tending to drive the journal around in the bearing in a counterclockwise direction. Thus the oil-pressure distribution will encourage or self-excite a whirl in the direction of rotation but will damp a counterrotating whirl, if one ever sets in.

There remains to be considered the condition under which the oil-film pressures during the whirling will be the same as in the steady-state case for a horizontal bearing with gravity loading. Consider two extreme cases, namely those in which the ratio of the angular velocity of whirl to the angular velocity of rotation is either very small or very great.

In the first case the shaft makes say 100 revolutions while the whirl moves forward 5 deg. It is clear that such a slow drift can have no effect on the pressure distribution, so that for a *slow* whirl the succession of steady states actually occurs and the oil whip will develop. In the second case the journal center whirls around while the journal itself hardly rotates. Then, of course, no oil film develops at all and the shaft merely vibrates in a bath of oil, which effectively damps the motion.

Therefore we recognize that for whirl frequencies which are slow with respect to the angular velocity of rotation the oil whip develops, while for comparatively fast whirls all vibratory motions are damped. The ratio of $\omega_{\text{rotation}}/\omega_{\text{whirl}}$, at which the damping passes from a positive value to a negative one, can be determined only by experiment.

It has been found in this manner that if the ω of the whirl is equal to or smaller than half the ω of the shaft (*i.e.*, if the shaft runs faster than twice its critical speed), the oil whip develops. This constitutes a serious trouble for high-speed machines with vertical shafts in oil-lubricated guide bearings, which is very difficult to overcome.

An interesting justification for this result is due to Hagg. In Fig. 7.13

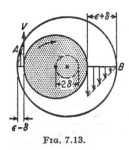

FIG. 7.13.

let the radial clearance be ϵ and the radius of whirl be δ, the diameter of the journal D. For a slow whirl, the velocity distribution across the oil film is linear, and, with the journal peripheral velocity being V, the volume of oil (per unit shaft length) transported up at A is $\frac{V}{2} \cdot (\epsilon - \delta)$, while the volume passing downward through B is

$$\frac{V}{2} \cdot (\epsilon + \delta)$$

Thus assuming no end leakage, the net transport of oil into the lower half of the film is $V\delta$. Now the journal whirls with a frequency f and the whirl velocity v of the journal center is $v = f \cdot 2\pi\delta$. The area of the lower half of the oil film increases at the rate $vD = f2\pi\delta D$. If the whirl frequency is slow enough, the rotation of the shaft will wipe enough oil into the lower half of the film to fill the cavity caused there by the upward whirling motion. For faster whirl the rotation will not transport enough oil and the film breaks. This occurs therefore at $V\delta = f2\pi\delta D$. The peripheral velocity V is related to the shaft speed by $V = \pi D \cdot \text{r.p.s.}$ Substituting this we obtain

$$f = \frac{\text{r.p.s.}}{2}$$

This shows that if the whirl is faster than half the shaft speed, the oil film breaks down and no self-excitation can take place. In the presence of end leakage this breakdown will occur at a whirl frequency below half the shaft rotation.

Comparing Fig. 7.13 with Fig. 7.7 we note that while for dry friction the direction of whirl is opposite to that of shaft rotation, the two directions are the same for oil-film excitation.

For horizontal bearings with a certain loading, the oil whip also appears for speeds above twice the critical. The explanation is along the same lines as for the vertical shaft. During the whirl the oil pressures will not have a purely radial direction but will have a tangential compo-

nent as well. That tangential component may be driving during a part
of the whirling cycle and retarding during another part. For excitation
it is necessary merely that the total work done by the tangential force
component on the motion during a *whole* whirling cycle be positive, *i.e.*,
that the average value of the tangential force component be positive or
driving.

7.5. Galloping of Electric Transmission Lines. High-tension electric
transmission lines have been observed under certain weather conditions
to vibrate with great amplitudes and at a very slow frequency. The
line consists of a wire, of more or less circular cross section, stretched
between towers about 300 ft. apart. A span of the line will vibrate
with one or two half waves (Fig. 4.16a or b) with an amplitude as great
as 10 ft. in the center and at a rate of 1 cycle per second or slower. On
account of its character this phenomenon is hardly ever described as a
vibration but is commonly known as "galloping."

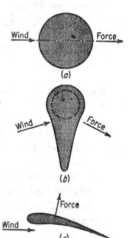

It has never been observed in countries with a
warm climate, but it occurs about once every
winter in the Northern states and in Canada,
when the temperature hovers around 32°F. and
when a rather strong transverse wind is blowing.
In most cases sleet is found on the wire. A rough
calculation shows that the natural frequency of
the span is of the same order as the observed
frequency. The fact that, once started, the disturbance is very persistent and continues sometimes for 24 hr. with great violence makes an
explanation on the basis of "forced" vibration
quite improbable. Such an explanation would
imply gusts in the wind having a frequency equal
to the natural frequency of the line to a mirac-
ulous degree of precision. For example, letting
$T = 1$ sec., if in 10 min. there were not exactly
600 equally spaced gusts in the wind but 601
instead, the vibration would build up during 5
min. and then be destroyed during the next 5 min.

FIG. 7.14. The directions
of the wind and the force
it causes include an angle
for nonsymmetrical cross
sections.

To keep the line vibrating for 2 hr. would require an error in the gustiness of
the wind of less than 1 part in 7,200, so that this explanation may be
safely dismissed.

We have a case of self-excited vibration caused by the wind acting on
the wire which, on account of the accumulated sleet, has taken a non-
circular cross section. The explanation involves some elementary aerodynamic reasoning as follows.

When the wind blows against a circular cylinder (Fig. 7.14a), it exerts

a force on the cylinder having the same direction as the wind.   This is
evident from the symmetry.   For a rod of non-circular cross section
(Fig. 7.14b) this in general does not hold true, but an angle will be included
between the direction of the wind and that of the force.   A well-known
example of this is given by an airplane wing where the force is nearly
perpendicular to the direction of the wind (Fig. 7.14c).

Let us visualize the transmission line in the process of galloping and
fix our attention on it during a downward stroke.   If there is no wind,
the wire will feel air blowing from below because of its own downward
motion.   If there is a horizontal side wind of velocity V, the wire, moving
downward with velocity v, will experience a wind blowing at an angle
$\tan^{-1} v/V$ slightly from below.   If the wire has a circular cross section,

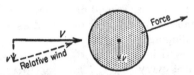

the force exerted by that wind will
have a small *up*ward component
(Fig. 7.15).   Since the wire was moving *down*ward this upward component of the wind exerts a force in opposition to the direction of motion of
the wire and thus damps it.   However, for a *non*-circular cross section,

FIG. 7.15. A horizontal side wind appears
to come from below if the line moves
in a downward direction.

it may well be that the force exerted by the wind has a downward component and thus furnishes negative damping (Fig. 7.14b).

Considering the conditions during the upward stroke of the vibration,
it can be seen in a similar manner that the relative wind felt by the wire
comes obliquely from above, and the force caused by it on a circular wire
has a downward component which causes damping.   For a non-circular
section, it may be that the force has some upward component, and this
component being in the direction of the motion acts as a negative damping.

If the sleet on the wire gives a cross section exhibiting the relation
between the wind and force directions shown in Fig. 7.14b, we have a
case of dynamic instability.   If by some chance the wire acquires a small
upward velocity, the wind action pushes it even more upward, till the
elastic or spring action of the wire stops the motion.   Then this elastic
force moves the wire downward, in which process the wind again helps,
so that small vibrations soon build up into very large ones.

There remains to be determined which cross sections are dynamically
stable (like the circular one) and which are unstable.   This brings us
into the domains of aerodynamics and of irregular cross sections, where
little general knowledge exists.   Usually we can only make a direct
test, but in some very pronounced cases qualitative reasoning gives
information.   The most "unstable" section so far known is the semicircle turned with its flat side toward the wind.   Figure 7.16 shows such
a section in a wind coming slightly from above, corresponding to the

upward stroke of a galloping transmission line. The air stream leaves
the wire at the sharp edge at the bottom but can follows around the upper
sharp edge for some distance on account of the wind coming from above
in a slightly inclined direction.

The region indicated by dots is filled with very irregular turbulent
eddies, the only known property of which is that in such a region the
average pressure is approximately equal to atmos-
pheric. On the lower half of the circular surface
of the cylinder we have atmospheric pressure, *i.e.*,
the pressure of the air at some distance away from
the disturbance created by the line. Above the
section the streamlines curve downward. This
means that the pressure decreases when moving
from *a* to *b*, which may be seen as follows. Consider an air particle in a streamline. If no force
were acting on it, the particle would move in a
straight line. Since its path is curved downward,
a force must be pushing it from above. This force
can be caused only by a greater pressure above
the particle than below it, so that the pressure at
a must be larger than at *b*. Since at *a* there is atmospheric pressure

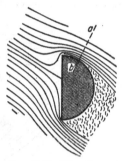

Fig. 7.16. The flow of
air around a semicircular
cylinder.

(being far away from the disturbance), the pressure at *b* must be below
atmospheric. Thus the semicircular section experiences an upward force
on account of the pressure difference below and above, and, since the up-

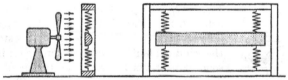

Fig. 7.17 Apparatus for demonstrating transmission line galloping.

ward force is caused by a wind coming from above, we recognize the case
of Fig. 7.16 as definitely unstable.

This may be shown by a simple experiment. A semicircular bar of
very light wood (2 in. diameter and 15 in. long) is suspended by four
springs so as to have a vertical natural frequency of about 6 cycles per
second (Fig. 7.17). If sufficient care is taken to reduce damping to a
minimum in the connections between the springs and the frame or bar,
the apparatus will build up vibrations with more than one radius amplitude when placed in front of an ordinary desk fan. The bar in this device
is made as light as possible which, for a given frequency and amplitude,
makes both the spring force and inertia force small. The input force of
the wind is determined only by the shape and size of the bar and is inde-

pendent of the weight.   Thus, by making the bar light, the ratio between the wind force and the spring force is made as great as possible.

Another cross section which is known to be unstable is an elongated rectangle exposed with its broad side to the wind.   The explanation is the same as for the semicircular bar (Fig. 7.18), only the effect is less pronounced.   It can be observed easily by means of any flat stick held in the hand at one end and dipped vertically into a tub of water.   When

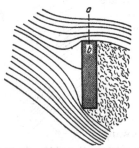

FIG. 7.18. The effect for a rectangle is less pronounced than that for a semicircle.

FIG. 7.19. The Lanchester tourbillion.

the stick is pulled through the water with the broad side of the rectangle perpendicular to the motion, it moves in zigzag fashion.   On the other hand, when pulled with the narrow side perpendicular to the motion, it moves forward quite steadily.

If, instead of mounting the unstable section in springs as shown in

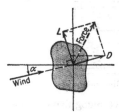

FIG. 7.20. The total wind force resolved into a lift L and a drag D.

Fig. 7.17, it is pivoted in the middle and placed before a fan (Fig. 7.19), we have a case of self-excited *rotation*.   While the apparatus stands still, the wind evidently exerts no torque on it, but, as soon as it starts rotating, the torque of the wind urges it on in the same direction.   The direction of rotation naturally is determined only by the direction at the start, *i.e.*, by accident.   This very interesting toy is known as Lanchester's "aerial tourbillion."

In aerodynamic work it is customary to resolve the total air force on an object into two components:

a. In the direction of the wind (the *drag* or resistance D).

b. Perpendicular to the wind (the *lift* L).

These two forces can be measured easily with the standard windtunnel apparatus.

Let Fig. 7.20 represent a section moving downward in its vibrational motion so that the wind appears to come from below at an angle $\alpha = \tan^{-1} v/V$.   The lift and drag forces L and D have vertical upward components (*i.e.*, components opposite to the direction of the motion) of **L** cos α and **D** sin α. The total upward damping force F of the wind is

$$F = \mathbf{L} \cos \alpha + \mathbf{D} \sin \alpha \qquad (7.15)$$

We are not interested in the force F itself but rather in $dF/d\alpha$, *i.e.*, in the variation of the upward force with a variation in α or in v/V. Assume that F has a large value and that $dF/d\alpha$ is zero. The result would be that part of the weight of the line would not be carried by the towers but by the wind directly. Any vibration or galloping of the line would not change the wind-carried weight ($dF/d\alpha = 0$) so that the vibration would not be affected. On the other hand, assume that $dF/d\alpha$ is negative, which means that the upward wind force increases for negative α and decreases for positive α. Then clearly we have the case of an encouraging alternating force as already explained. The criterion for dynamic stability is

$$\frac{dF}{d\alpha} < 0 \qquad \text{(unstable)}$$

and $$\frac{dF}{d\alpha} > 0 \qquad \text{(stable)}$$

In performing the differentiation on (7.15), it is to be noted that, for small vibrations, v is small with respect to V, so that α is a small angle of which the cosine equals unity and the sine is negligible with respect to unity:

$$\frac{dF}{d\alpha} = \frac{d\mathbf{L}}{d\alpha} \cdot \cos \alpha - \mathbf{L} \sin \alpha + \frac{d\mathbf{D}}{d\alpha} \sin \alpha + \mathbf{D} \cos \alpha$$

$$= \sin \alpha \left(-\mathbf{L} + \frac{d\mathbf{D}}{d\alpha} \right) + \cos \alpha \left(\frac{d\mathbf{L}}{d\alpha} + \mathbf{D} \right)$$

$$\approx \frac{d\mathbf{L}}{d\alpha} + \mathbf{D}$$

Thus the system is *unstable* when

$$\frac{d\mathbf{L}}{d\alpha} + \mathbf{D} < 0 \qquad (7.16)$$

The values of the lift and drag of an arbitrary cross section cannot be calculated from theory but can be found from a wind-tunnel test. The results of such tests are usually plotted in the form of a diagram such as Fig. 7.21. In words, (7.16) states that

A section is dynamically unstable if the negative slope of the lift curve is greater than the ordinate of the drag curve.

In Fig. 7.21 it is seen that an elongated section is always stable when held "along" the wind ($\alpha = 0$), whereas it is usually unstable when held "across" the wind ($\alpha = 90$ deg.). A transmission line which is being coated with sleet at approximately freezing temperature has the tendency to form icicles that are more or less elongated in a vertical direction, corresponing to $\alpha = 90$ deg. in the diagram.

At this angle, for small amplitudes of vibration (say varying between 89 and 91 deg.), there is energy input during a cycle. This will increase the amplitude, and the increase will continue so long as there is an excess of energy furnished by the wind. At some large amplitude this excess of energy will become zero so that we have energy balance and reach the final amplitude. In Fig. 7.21 this will take place presumably at α varying between 30 and 150 deg., say. Near the ends of each stroke, energy

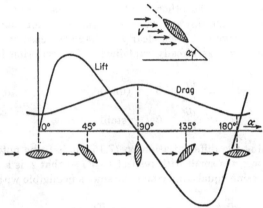

FIG. 7.21. Lift and drag as a function of the angle of attack for an elongated, symmetrical cross section.

is put in; but in the middle of the stroke, energy is destroyed by damping, since $\dfrac{d\mathbf{L}}{d\alpha} + \mathbf{D}$ is larger than zero at these places (see also Fig. 8.32, page 363). The final amplitude can be found by a process of graphical or numerical integration over the known curve of the diagram, in the manner already indicated.

Thus far in the discussion the system has been assumed to be one of a single degree of freedom, which certainly is not the case with a span of transmission line, of which each point vibrates with a different amplitude (large in the center of the span and small near the towers). Since the wind force is small in comparison to the elastic and inertia forces of the vibration, the form of the motion is the same as if the wind force were absent; in other words, the line vibrates in its first natural mode. The final amplitude can be determined by finding the energy input for the whole span. If for a certain assumed amplitude this energy comes out positive, the amplitude assumed was too small; whereas, if the energy comes out negative (damping), the assumed amplitude was too great.

The determination of the energy involves a *double* graphical integration, first with respect to α for each point of the line and then with respect to the position x along the line. This process is straightforward and involves no difficulties, though it may require much time.

The phenomenon discussed so far is one of very *slow* frequency and *large* amplitude in the transmission line. It has been observed but rarely, where the weather conditions brought together sleet deposits as well as a lateral wind of considerable strength. There is another case of vibration of transmission lines characterized by *high* frequency and *small* amplitude which is much more common and for the occurrence of which only a lateral wind is necessary. The explanation of this phenomenon is found in the so-called "Kármán vortex trail."

7.6. Kármán Vortices. When a fluid flows by a cylindrical obstacle, the wake behind the obstacle is no longer regular but in it will be found distinct vortices of the pattern shown in Fig. 7.22. The vortices are

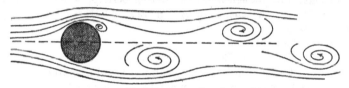

FIG. 7.22. Kármán vortices in a wake.

alternately clockwise and counterclockwise, are shed from the cylinder in a perfectly regular manner, and are associated with an alternating sidewise force. This phenomenon has been studied experimentally and it has been found that there is a definite relation among the frequency f, the diameter of the cylinder D, and the velocity V of the stream, expressed by the formula

$$\frac{fD}{V} = 0.22 \qquad (7.17)$$

or the cylinder moves forward by about $4\frac{1}{2}$ diameters during one period of the vibration. It is seen that this fraction is dimensionless and that therefore the value 0.22 is independent of the choice of units. It is known as the *Strouhal number*. The eddy shedding on alternate sides of the cylinder causes a harmonically varying force on the cylinder in a direction perpendicular to that of the stream. The maximum intensity of this force can be written in the form usual for most aerodynamic forces (such as lift and drag) as follows:

$$F_K = (C_K \cdot \frac{1}{2}\rho V^2 \cdot A) \sin \omega t \qquad (7.18)$$

The subscript K is for Kármán, F_K being the Kármán force and C_K the (dimensionless) Kármán-force coefficient. The value of C_K is not known with great accuracy; however, a satisfactory figure for it is $C_K = 1$ (good for a large range of Reynolds numbers from 10^2 to 10^7), or in words: The intensity of the alternating force per unit of sidewise projected area is about equal to the stagnation pressure of the flow.

The mechanism of eddy shedding from a cylinder at rest is truly a self-excited one, because there is no alternating property in the approaching stream, and the eddies come off at the natural Strouhal frequency. Whereas for all previous cases discussed in this chapter the self-excited vibration in itself was dangerous, this is not the case here: we do not much care what happens in the stream; all that we care about is to know that a force of magnitude (7.18) is exerted on the cylinder. As a rule this is without much consequence, and trouble occurs only if the self-excited frequency of eddy shedding (7.17) coincides with the natural frequency of the structure on which it acts. Then a resonance occurs which may be destructive. Objects on which this has been observed are many and various: transmission lines, submarine periscopes, industrial smokestacks, a famous suspension bridge, a large oil storage tank, and tiny rain droplets.

First consider a transmission line of 1 in. diameter in a wind of 30 m.p.h. The Strouhal frequency [Eq. (7.17)] gives 116 cycles per second. Vibration of lines at such high frequencies and at small amplitudes have often been observed and have repeatedly ended in fatigue failures. Resonance obviously takes place at a high harmonic of the line in which the span is subdivided in many half sine waves, of the order of 20 to 30. Because of the high frequency and small amplitude of this motion, it has been found possible and practical to bring it under control by damped vibration absorbers. The simplest and most common construction, known as the "Stockbridge damper," is shown in Fig. 7.23. A piece of steel cable with weights at its ends is clamped to the line. The cable acts as a spring and is roughly tuned to the frequency of the expected vibration. Any motion of the line at point A will cause relative motion in the pieces of cable, and the friction between its strands dissipates energy. The point A of attachment is so chosen along the line that it cannot coincide with a node of the motion, where the damper would be useless. Since the length of the half sine wave varies between 8 ft. and 20 ft., a location at about 6 ft. from the point of support on a tower will fit most frequencies and wave lengths. This device, crude as it is, has proved to be entirely effective in protecting the lines against damage

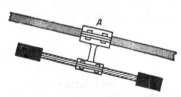

FIG. 7.23. Stockbridge damper for transmission lines, consisting of a piece of stranded steel cable about 12 in. long, carrying cast iron weights of about 1 lb. at each end.

from the Kármán-vortex vibration. An attempt to design one of these dampers for the galloping line of page 299, where the frequency is 100 times as slow and the amplitude 100 times as large, would lead to weights of several tons, which is entirely impractical.

Submarine periscopes, being cantilevers of some 20 ft. or more in length (when extended) and of diameters of the order of 8 in., have shown resonance with Kármán vortices at speeds of about 5 m.p.h. Since the fluid here is water with a high density ρ, the exciting force [Eq. (7.18)] is very large and the vibration is very severe, leading to a blurring of the view seen through the device. A cure would consist in changing the cylindrical cross section into a streamlined one, but the periscope tube must be free to rotate in its guide if views in all directions are wanted, and the constructional complications are so great that this has never yet been attempted.

Steel industrial smokestacks regularly show Kármán-induced resonant vibrations at wind speeds of the order of 30 m.p.h. Brick or concrete stacks have not exhibited this phenomenon, and in steel stacks it has become worse when riveting was replaced by welding, thus decreasing the internal damping of the stack. Recently (1953) a welded stack of 16 ft. diameter and 300 ft. height came to resonance at its natural frequency of about 1 cycle per second (in a wind of some 50 m.p.h.) with such violence that it buckled and developed a crack in the steel extending over 180 deg. of circumference. The damaged upper half had to be torn down and rebuilt, and the new stack was provided with vibration dampers. These were mounted in guy cables between the top of the stack and the ground, as shown in Fig. 7.24. There are two large springs A of several feet length, and the entire assembly of springs and guy wire is put in initial and permanent tension, so that the springs permanently

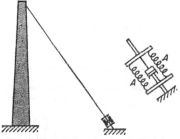

FIG. 7.24. Friction damper built into a guy wire of a smokestack.

are some 8 in. longer than when unstressed. Between the springs is a dashpot in the form of a large automotive-truck shock absorber. When the stack starts vibrating, a relative motion occurs across the dashpot, thus dissipating energy. The design is such that the amount of energy dissipated in the damper is about the same as that dissipated in the stack itself, thus doubling the natural damping, which is sufficient to prevent damage to the stack.

In the autumn of 1940 the great suspension bridge across the Tacoma Narrows in Washington, after being in operation for about a year, was completely wrecked by a moderate wind. The precise details of the cause were not known at the time; the explanations offered were based either on Kármán vortices or on flutter (page 321). In the years after the disaster elaborate experiments were made at the University of Washington in a specially constructed wind tunnel, and it is now clear that the cause of failure was a Kármán-vortex trail. The object this time is not a cylinder, but an I-shaped section. The phenomenon under discussion occurs for all manner of shapes; the frequency and intensity formulas (7.17) and (7.18) hold for all sections with different numerical values of the Strouhal and Kármán dimensionless coefficients. In this case the wind blew steadily for more than an hour at 42 m.p.h., and the eddy frequency coincided with the torsional frequency of the bridge deck on its suspension cables, resulting in resonance whereby the angles of motion of the deck went to 45 deg. on either side of horizontal, the distribution along the span being in two half sine waves. This wrecked the bridge completely. After elaborate studies it was rebuilt with three major changes in the construction. First, the solid plate side girders were replaced by open trusswork, so that the wind could blow through without forming large eddies. Second, longitudinal slits (with grillwork between traffic lanes) were provided in the deck, so that, even if large eddies should form, no large pressure difference between the top and bottom side of the horizontal deck could be maintained. Third, the "open" section of Fig. 7.25a was replaced by the "closed" box section of Fig. 7.25b, by lowering the side trusses and tying them together at the bottom end by another horizontal truss. Such a closed box section is of the order of 100 times as stiff against torsion as an open section, and thus the

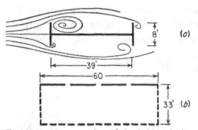

FIG. 7.25. Cross section of the bridge deck at Tacoma Narrows. (a) The old bridge that blew down had a solid deck and solid vertical side plates; (b) the new bridge has open side trusses, slits in the deck, and in addition a bottom truss, making a box section of it. (*University of Washington, Engineering Experiment Station, Bulletin No. 116 Parts III to V.*)

torsional natural frequency is raised to such a point that the resonant wind speed would be out of range on the high side.

The Tacoma bridge deck was 39 ft. wide, and still this is not the biggest Kármán vortex observed. That occurred in an oil storage tank in Venezuela. The tank was roofless and of the usual cylindrical construction, about 150 ft. in diameter and 48 ft. high. When new it was filled with sea water to a level 4 ft. under the top. Under the influence of the steadily blowing local trade wind of about 30 m.p.h., a standing wave developed of 8 ft. height between crest and bottom. The frequency of these waves was not observed, but there exists a formula

$$V = \sqrt{\frac{gl}{2\pi}}$$

for the propagation of ocean gravity waves in which V is the velocity of propagation and l the wave length. Photographs of the wave pattern showed the wave length to be the diameter $l = 150$ ft., so that $V = 27.5$ ft./sec. A full period occurs when the wave crest travels from one end of the tank to the other through 150 ft. This makes the period $150/27.5 = 5\frac{1}{2}$ sec. Applying Eq. (6.17) and taking the tank height of 48 ft. for the "diameter," we find a wind speed of 37 ft./sec. or 25 m.p.h., which checked roughly with local observations at the time. It is suggested that the reader check this wind speed with Eq. (7.18) and deduce that the excitation on the water surface is $\frac{1}{3}$ in. of water pressure. The observed amplitude was 48 in., so the resonant magnification factor was about 150. The remedy consists in mounting perforated vertical plates in the tank parallel to the outside wall about a foot away from it.

The smallest manifestation of Kármán vortices is in connection with waterdrops of about 1 mm. diameter. It was observed by researchers in meteorology that waterdrops fall vertically in still air, with the exception of drops of just 1 mm. diameter, which drift sidewise in odd directions. When the velocity of fall of such a droplet is calculated and Eq. (6.17) is applied, the eddy frequency is determined. It comes out that for this size of droplet the natural frequency of vibration of the drop from a sphere to an ellipsoid under the spring effect of capillary skin tension is just the same as the eddy frequency. Thus the drops are pulsating while falling, and their path of descent is no longer simply vertical, but becomes irregular.

7.7. Hunting of Steam-engine Governors. Quite interesting self-excited vibration phenomena have been observed in steam engines or turbines operating in conjunction with an inertia governor of the direct-acting type. By this is meant that the speed-sensitive part of the governor, *i.e.*, the flyballs, is in direct mechanical connection with the steam-supply throttle valve. In very large engines or turbines too much power is required to open and close the throttle directly, so that the governor merely operates electric contacts or oil valves (a relay) which in turn set the throttle valve in motion. Such *indirect* governor systems will not be discussed here.

In Fig. 7.26 the system is shown schematically. When the speed of the engine a increases for any reason, the flyballs lift the sleeve b of the governor somewhat higher, thereby reducing the opening of the main steam valve c. In this fashion less steam is admitted to the engine, and its speed falls. Since there is inertia in the system, the speed will decrease below normal, which will result in the governor opening the valve more than normal. In this manner, oscillations in the speed of the engine occur, which may be damped or self-excited, depending on the circumstances. The unstable case has occurred repeatedly. Technically it is known as *hunting*, and if such a hunting engine drives an electric generator, its voltage will fluctuate so that a marked flicker in the lights is observed.

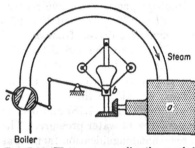

In order to understand this phenomenon in greater detail, it is convenient to start from the differential equations. In the first place, the governor is an ordinary vibrating system consisting of a mass, a spring, and a dashpot. This will give three terms in the differential equation. Moreover, the governor is coupled to the engine in such a manner that, when the engine speed $\dot{\varphi}$ increases, an additional upward force on the governor mass ensues, caused by the centrifugal action of the flyballs. This gives the equation

FIG. 7.26. Watt governor directly coupled to a throttle valve regulating the steam supply to a turbine. Without damping this system is unstable and will hunt.

$$m\ddot{x} + c\dot{x} + kx = C_1\dot{\varphi} \qquad (7.19)$$

where x = upward displacement of the governor sleeve, measured from the normal position at a certain load.

m = equivalent mass of the governor sleeve.

c = damping coefficient at the governor sleeve.

k = stiffness of governor spring.

$\dot{\varphi}$ = difference between the instantaneous engine speed and the normal or average speed at a certain load.

C_1 = increase in the upward force on the governor sleeve (from centrifugal action) caused by an increase in engine speed of 1 radian per second.

It is seen that the two coordinates x and $\dot{\varphi}$ are based on a certain "normal" condition which exists when the engine is running at a constant speed with a constant load and constant throttle opening, while the governor sleeve stands vertically still. In this condition $x = 0$, so that x is positive when the governor sleeve is higher than normal and negative when lower than normal; similarly, $\dot{\varphi}$ is negative while the engine is temporarily running at a speed slower than normal.

Properly speaking, the engine in itself is not a vibrating system since it has no spring pulling it back into an equilibrium position. There is, however, a mass or rather a moment of inertia I. The damping torque of the engine will be neglected in this investigation. The engine is coupled to the governor in the sense that, when the governor sleeve is lower than normal (negative x), the throttle is opened wider than usual so that an extra positive or driving torque is exerted on the engine. Its equation of motion becomes

$$I\ddot{\varphi} = -C_2 x \tag{7.20}$$

where I = equivalent moment of inertia of the engine.

C_2 = increase in steam torque of engine caused by a lowering of the governor sleeve by 1 in.

Equations (7.19) and (7.20) represent the *free* vibrations of the engine-governor system, since no periodic force is present. The solution therefore must be a function of the shape

$$e^{pt} \cos qt \tag{7.21}$$

where q is the (damped) natural frequency and p is a measure for the damping, which may be positive or negative. Instead of writing the solution in the form (7.21), we may write

$$e^{(p+iq)t}$$

the real part of which is the same as (7.21); or, still shorter, we may assume that

$$\left. \begin{array}{l} x = x_{\max} e^{st} \\ \varphi = \varphi_{\max} e^{st} \end{array} \right\} \tag{7.3}$$

where s is a complex number (the "complex frequency").

Substitute (7.3) in the differential Eqs. (7.19) and (7.20), which then can be divided by e^{st}, giving

$$(ms^2 + cs + k)x_{\max} - C_1 s \varphi_{\max} = 0$$
$$C_2 x_{\max} + I s^2 \varphi_{\max} = 0$$

These are a set of homogeneous algebraic equations which have a solution for $x_{\max}$ and $\varphi_{\max}$ only if

$$\frac{ms^2 + cs + k}{C_2} = \frac{-C_1 s}{I s^2}$$

or if

$$s^3 + \frac{c}{m} s^2 + \frac{k}{m} s + \frac{C_1 C_2}{mI} = 0 \tag{7.22}$$

Equation (7.22) is the frequency equation of the set (7.19) and (7.20). On account of the absence of a "spring" in the engine system, the equation is a cubic and not a quartic as would be expected for an ordinary two-degree-of-freedom system.

Of the two criteria for stability on page 288, the first one, requiring all coefficients to be positive, is satisfied. The other criterion, expressed by (7.9), becomes

$$\frac{c}{m} \cdot \frac{k}{m} > \frac{C_1C_2}{mI}$$

or
$$c > \frac{mC_1C_2}{kI} \tag{7.23}$$

If the damping in the governor dashpot is greater than the value indicated by this formula, the system will come to rest after a sudden change in load, but for any damping less than this the system is inoperative (Fig. 7.2b).

In case the engine is rigidly coupled to an electric generator feeding a large network, the problem becomes more complicated. Then, there is an "engine spring," since the network tends to keep the generator rotor in a definite angular position. Any deviation from that synchronous position is opposed by a torque caused by the magnetic spring in the air gap of the generator. In such cases Eq. (7.20) contains one more term $k_e\varphi$, and, if there is generator damping in an electric damper winding, Eq. (7.20) contains two more terms. The two simultaneous differential equations of the problem are

$$\left.\begin{array}{l} m\ddot{x} + c_g\dot{x} + k_gx = C_1\dot{\varphi} \\ I\ddot{\varphi} + c_e\dot{\varphi} + k_e\varphi = -C_2x \end{array}\right\} \tag{7.24}$$

where the subscript g stands for governor and e for engine. Clearly I means the inertia of all the rotating parts, i.e., of the engine and generator combined.

The damper winding in the generator just mentioned is a device invented by Leblanc in 1901 with the object of alleviating the hunting trouble. It consists of a short-circuited copper winding in the pole faces of the rotating part of the generator. As long as the generator runs at constant (synchronous) speed, no current flows in this winding and consequently it does not impede the motion. With changes in speed, however, currents are induced in the winding, which together with the magnetic field in the air gap produce a torque proportional to the deviation of the angular velocity from synchronous ($\dot{\varphi}$) and directed opposite to $\dot{\varphi}$, i.e., braking while the engine temporarily runs too fast and driving while it runs too slow.

Assuming the solution of (7.24) in the form (7.3) and substituting in (7.24), the frequency equation becomes

$$s^4 + s^3\left(\frac{c_e}{I} + \frac{c_g}{m}\right) + s^2\left(\frac{k_e}{I} + \frac{k_g}{m} + \frac{c_e \cdot c_g}{Im}\right)$$
$$+ s\left(\frac{c_g}{m} \cdot \frac{k_e}{I} + \frac{c_e}{I}\frac{k_g}{m} + \frac{C_1C_2}{Im}\right) + \frac{k_ek_g}{Im} = 0$$

in which all coefficients are seen to be positive. The criterion of stability (7.14) becomes

$$
\left(\frac{c_e}{I} + \frac{\dot{c}_g}{m}\right) \cdot \left(\frac{k_e}{I} + \frac{k_g}{m} + \frac{c_e c_g}{Im}\right)\left(\frac{c_g}{m} \cdot \frac{k_e}{I} + \frac{c_e}{I} \cdot \frac{k_g}{m} + \frac{C_1 C_2}{Im}\right)
$$
$$
> \left(\frac{c_g}{m} \cdot \frac{k_e}{I} + \frac{c_e}{I} \cdot \frac{k_g}{m} + \frac{C_1 C_2}{Im}\right)^2 + \frac{k_e k_g}{Im}\left(\frac{c_e}{I} + \frac{c_g}{m}\right)^2 \quad (7.25)
$$

which depends on the governor damping c_g/m, on the engine damping c_e/I, on the natural frequencies $\omega_e^2 = k_e/I$ and $\omega_g^2 = k_g/m$, and on the "coupling" $C_1 C_2/Im$. The only simple conclusion that can be drawn from (7.25) is that, when no damping exists at all ($c_g = c_e = 0$), the left-hand side is zero, while the right-hand side is $(C_1 C_2/Im)^2$, so that the inequality is violated. *Without any damping the system hunts.*

In order to see the physical meaning of (7.25), consider first the special case where engine damping is absent, $c_e = 0$. Equation (7.25) reduces to

$$
c_g \frac{I}{C_1 C_2} (\omega_g^2 - \omega_e^2) > 1 \quad (7.26)
$$

In case the governor frequency ω_g is less than the engine frequency ω_e, the left-hand side is negative and the inequality is violated, which means unstable operation even if c_g is very large. Conversely, when ω_g is greater than ω_e, the left-hand side is positive and stable operation prevails if the governor damping is larger than

$$
c_g > \frac{C_1 C_2}{I(\omega_g^2 - \omega_e^2)}
$$

It is seen that (7.23) is a special case of this more general result.

The second simple case to be considered is when the only damping is in the engine and none is in the governor, $c_g = 0$. The large form (7.25) can then be reduced to

$$
c_e \frac{m}{C_1 C_2} (\omega_e^2 - \omega_g^2) > 1 \quad (7.27)
$$

which shows that instability exists if the governor frequency is greater than the engine frequency. When the opposite is the case, the system may be stable if the engine damping is sufficiently large.

Summarizing, if a system determined by Eqs. (7.24) is found to be unstable, it should be cured by increasing the damping in the governor dashpot in case the governor frequency is greater than the engine frequency; on the other hand, if the governor frequency is the smaller of the two, damping should be introduced in the engine or generator.

7.8. Diesel-engine Fuel-injection Valves. A common construction of a liquid fuel-injection valve and nozzle for Diesel engines is sketched in

Fig. 7.27. The chamber V is permanently filled with liquid fuel oil and is connected to the fuel pump through a short passage B. The normal position of the valve A is on its seat N. At the instant that the engine piston is ready to start on its firing stroke the fuel pump pushes a certain amount of fuel into V, where the pressure rises greatly. Since the valve stem has a greater diameter above than below, this pressure tends to push the valve up. As soon as the pressure is sufficiently large to overcome the force of the set-up in the spring S, the stem will go up and the liquid is forced through the nozzle N into the cylinder head. At the end of the pump stroke the pressure in V falls and the spring S closes the valve again.

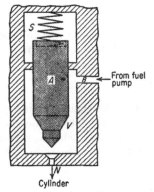

FIG. 7.27. Diesel-engine fuel-injection valve. Without damping, the valve is dynamically unstable.

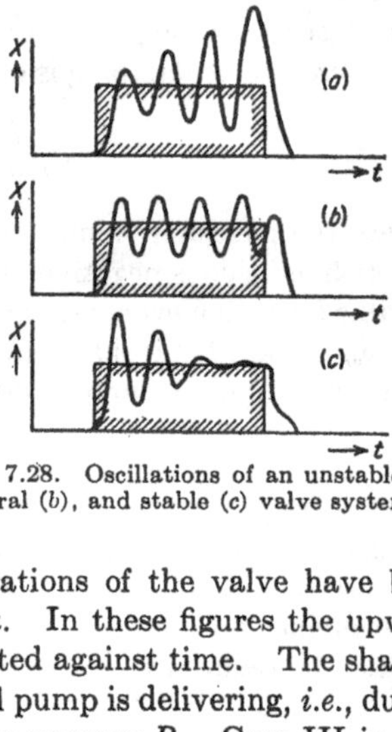

FIG. 7.28. Oscillations of an unstable (a), neutral (b), and stable (c) valve system.

With this mechanism self-excited vibrations of the valve have been observed of the type shown in Fig. 7.28a. In these figures the upward displacement x of the valve has been plotted against time. The shading refers to the interval during which the fuel pump is delivering, $i.e.$, during which fuel is actually flowing through the passage B. Case III is that of positive damping, case II is neutral, and case I shows negative damping.

The physical action may be understood as follows. During the vibration, part of the valve stem retreats from the chamber V, oil flows in at B and out at N, all of which affects the pressure in V. In case the average pressure is greater during the upward stroke than during the downward stroke of the valve stem, there is a feeding of energy into the vibration. If this energy is greater than the friction loss in the gland, the vibration is self-excited.

Indeed, in the absence of gland damping, the system is unstable, which can be seen physically as follows. Consider only the period during which the fuel pump is operating, and assume that the fuel oil is flowing in at a constant rate through the passage B. The outflow of oil through the nozzle is varying, depending on the position of the valve stem. Let the valve stem vibrate about some average position. In this average position the outflow through the nozzle equals the inflow through B; while the stem is $\begin{Bmatrix} \text{higher} \\ \text{lower} \end{Bmatrix}$ than the average position, the outflow is $\begin{Bmatrix} \text{greater} \\ \text{smaller} \end{Bmatrix}$ than the inflow. The pressure in the chamber V depends on the amount of oil in it; the more oil, the greater the pressure. Consider the valve stem in the neutral or average position in the act of going up. During the next two quarter cycles of vibration, the outflow exceeds the inflow and the pressure diminishes. Thus, when the valve stem finds itself in the neutral position going down, the pressure is at a minimum. In the same manner it can be shown that, when the stem is in the middle of its upward stroke, the pressure is a maximum. Thus the pressure does work on the vibration.

In the above argument one fact has not been mentioned, namely, that, on account of its motion, the valve stem changes the volume of oil in the chamber V, thus causing pressure variations. The total pressure caused by the fuel pump is so great that these variations are supposed not to affect the outflow, which is determined by the nozzle opening only. Moreover, these pressure variations are in phase with the valve displacement and thus act as an oil *spring* and not as a damping.

Mathematically we come to the same conclusions. Our two dependent variables are the upward displacement x of the valve stem and the pressure $\mathbf{p}$ in the chamber, both measured as deviations from their average values during a vibration cycle; the independent variable is the time. There are three upward forces acting on the valve stem:

1. The spring force $-F_0 - kx$.
2. The damping force $-c\dot{x}$.
3. The pressure force $+\mathbf{p}A + \mathbf{p}_0A$.

In the first expression, F_0 is the set-up force of the spring S and k is its stiffness; in the third expression, A is the cross section of the stem at the gland and $\mathbf{p}_0$ is the average value of the pressure. The constant forces $-F_0$ and $+\mathbf{p}_0A$ are equal and opposite; they keep each other permanently in equilibrium. Thus the equation of motion of the valve stem is

$$m\ddot{x} + kx + c\dot{x} - \mathbf{p}A = 0 \qquad (7.28)$$

in which both variables x and $\mathbf{p}$ occur.

The second equation is found by considering the change in volume of the oil in the chamber V and correlating it with a change in the pressure. It is assumed that the flow of oil in the passage B occurs at constant speed during the stroke of the pump. It has also been found with a good degree of approximation that the velocity of oil flow through the nozzle is proportional to the distance of the valve from the nozzle. This distance consists of the average distance x_0 with the variation x superposed on it. The amount of fuel flowing out of the nozzle for a valve set at x_0 equals the amount coming in through B. Thus the *excess* volume of fuel oil flowing in per second is $-Cx$, where C is the total volume flowing through the nozzle per second when x_0 equals one unit of length. However, the volume of the chamber V does not remain constant, since the stem moves in and out of it. The change in volume per second on account of this motion of the stem is $A\dot{x}$. The difference

$$-Cx - A\dot{x}$$

is the excess rate of fluid flow inward for a constant volume V. It can be written as $dV/dt = \dot{V}$. The definition of the modulus of elasticity E of a fluid in compression is

$$\frac{dV}{V} = \frac{dp}{E}$$

from which follows that

$$\frac{\dot{V}}{V} = \frac{\dot{p}}{E}$$

so that the second differential equation is

$$\dot{p} = -\frac{E}{V}(Cx + A\dot{x}) \tag{7.29}$$

The variable p can be eliminated between (7.28) and (7.29) by differentiating (7.28) and then substituting (7.29), giving

$$m\ddot{x} + c\ddot{x} + \left(k + \frac{A^2E}{V}\right)\dot{x} + \frac{AEC}{V}x = 0 \tag{7.30}$$

A substitution of (7.3) leads to the frequency equation:

$$s^3 + \frac{c}{m}s^2 + \left(\frac{k}{m} + \frac{A^2E}{mV}\right)s + \frac{AEC}{mV} = 0 \tag{7.31}$$

in which all the coefficients are positive, so that the stability criterion (7.9) becomes

$$\frac{c}{m}\left(\frac{k}{m} + \frac{A^2E}{mV}\right) > \frac{AEC}{mV}$$

or

$$c > \frac{CE}{V} \cdot \frac{mA}{\left(k + \frac{A^2E}{V}\right)} \tag{7.32}$$

Only when the damping in the gland or elsewhere is as large as is shown in this expression is the motion stable.

It is of interest to note that the bracket in the denominator is the combined spring constant due to S and to the oil chamber, and also that the combination CE/V means the rate of increase in the oil pressure caused by a deflection of 1 in. from the average position of the valve stem. In this light it can be seen that the frequency equations (7.22) and (7.31) for the apparently widely different problems of Watt's governor and the Diesel nozzle have exactly the same structure. The coefficient A_2 is a measure of the damping, A_1 is the square of the natural frequency, and A_0 determines the intensity of back-feeding of energy.

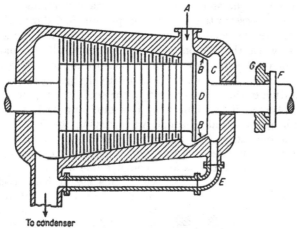

FIG. 7.29. Steam turbine showing dummy piston D, labyrinth B, thrust bearing F, G, and equilibrium pipe E.

7.9. Vibrations of Turbines Caused by Leakage of Steam or Water.

Several cases quite similar to the one just discussed have been observed on steam turbines and hydraulic turbines. The spindle of a large steam turbine and the rigidly coupled electric generator rotor were found to be oscillating in an axial direction in the bearings at a frequency of the order of 20 cycles per second. The explanation of this trouble was found in a pressure variation in the space behind the "dummy piston" and was caused by leakage of steam into this space. As with the Diesel valve, the rate at which this leakage takes place depends on the longitudinal position of the turbine spindle.

The construction is roughly indicated in Fig. 7.29. The high-pressure steam enters through A and passes to the left through the blading to the condenser. On account of the pressure difference between the boiler and the condenser, an appreciable force to the left is exerted on the spindle and this force has to be balanced. This is done partly by the dummy piston D and partly by the thrust bearing F.

A very small quantity of high-pressure steam leaks by the labyrinth B into the chamber C, which is connected by the "equilibrium pipe" E (of some 16 ft. length) to the condenser. Thus the pressure in C is about equal to (slightly above) the condenser vacuum, and this results in a force tending to pull the dummy piston D to the right and thus partly balances the steam thrust. The details of the labyrinth B vary widely in construction, but usually they are such that an axial displacement of the rotor changes the rate of leak. Since the pipe E is rather long, longitudinal oscillations of its steam column are associated with pressure variations in C, which react on the spindle motion. To have damping of the axial oscillation, it is necessary that the average pressure in C during the spindle motion to the left be smaller than during the stroke to the right. The frequency at which the motion takes place is practically the natural one of the spindle on the springs G of the thrust-bearing structure, since the steam forces are usually small compared with the spring forces.

The vibration of the steam column in the equilibrium pipe E becomes rather complicated if the length of this pipe approaches one-quarter wave length of the standing sound wave having the frequency of the axial turbine oscillation (Figs. 4.18b, c). In most cases, however, the length is appreciably less than this, which means that the steam in the pipe surges back and forth as an incompressible body. The spring on which this steam mass oscillates is found in the volume C, where the pressure changes as a result of an alternating motion of the steam column in E. Thus the system is as shown schematically in Fig. 7.30, where the mass m may be regarded as a piston (made

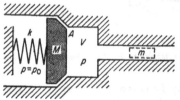

FIG. 7.30. Idealized system of the axially vibrating turbine.

of steam) sliding back and forth in the equilibrium pipe. There is a general drift of steam in the pipe m to the right. For our analysis we subtract from the total steam velocity its average value so that only the variable part of the velocity of m is considered.

In the actual construction, the volume V is very small, and thus some physical reasoning on the behavior of Fig. 7.30 for zero volume V is of interest. Assume the mass M (being the turbine spindle and the generator rotor) to be vibrating back and forth according to Fig. 7.31. Since the volume V is assumed to be zero, the motion of m is directly determined by the amount of steam leaking past M. Thus the velocity of m to the right is maximum when the leakage is maximum or when M is in its extreme left position (point A of Fig. 7.31). While M is in its extreme right position, the leakage is minimum, less than average, so that m has its maximum velocity to the left (point B of Fig. 7.31). In this manner the curve determining the position of m is found. There is no spring acting on m, so that its motion is wholly caused by the steam pressure in (the small volume) **V**. Between A and B the steam column is being accelerated to the left, which means that the pressure in **V** is less than the average. This is turn means that between A and B the steam force in **V** pulls M to the right. But in this interval AB, M is moving toward the right also, so that we conclude that the motion is self-excited.

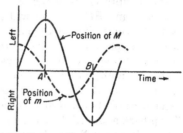

Fig. 7.31. Motion diagrams of Fig. 7.30 for the case of zero volume **V**.

On the other hand, suppose the volume **V** is very large. Then any variation in leakage can hardly affect the pressure in **V**, so that the variation in the steam force on M, be it positive or negative, is very small. A little friction in the system is then certain to neutralize any small negative damping that may exist.

The trouble in the actual machines was cured by the insertion of a chamber of about 2 cu. ft. volume between the space C and the equilibrium pipe E (Fig. 7.29).

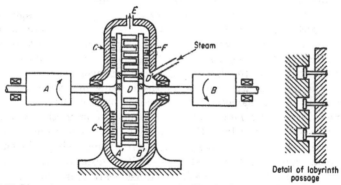

Fig. 7.32. Ljungstrom counterrotating steam turbine having a whirling vibration at the natural frequency (different from the r.p.m.), self-excited by pressure variations in the labyrinth leakage passages F.

Another case is that of Fig. 7.32, illustrating the Swedish "Ljungstrom" counterrotating steam turbine. A and B are two identical electric generators, driven by the turbine rotors A' and B' rotating at equal speeds but in opposite directions. The high-pressure boiler steam is led into the center of the turbine at D and flows radially outward to the space E, which is in direct connection with the condenser. The turbine blades are arranged on the rotors in concentric circular rows. To prevent direct leakage of the steam from the high-pressure region D' to the low-pressure region E, a labyrinth seal F between the rotors and the enveloping housing is built in. This labyrinth consists of a multitude of very narrow passages through which the leakage steam must pass, and it is rotationally symmetrical about the shaft center line of the entire machine. The observed vibration is of the whirling type; i.e., the shaft-disk combinations AA' and BB' are bent and tilted and these deformed shapes rotate about the center line at the natural frequency and not at the rotational speed (see page 284). The gyroscopic effect of the disks influences the natural frequency, but the system is symmetrical between A and B, so that, if both halves whirl in the same direction as their own rotors (and hence the A-whirl is opposite to the B-whirl), then the two natural frequencies of A and B are the same. In this motion the disks A' and B' flap back and forth through a small angle and so change the clearances in the labyrinths F and thereby change the steam pressure in F periodically. In this case the periodic steam-pressure variation on the disks was of such a phase that its moment was in the direction of the flapping velocity of the disks, and hence made the system unstable. The remedy is to be sought in changing the labyrinth passage dimensions, which will affect the phase relation between the pressure variation on the disk and its own motion, or, more effective still, in changing the labyrinth dimensions in such a way that the pressure variations diminish.

An almost identical situation was observed many years ago on a hydraulic turbogenerator of the usual vertical-shaft type where the generator rotor is above the water wheel on the same shaft and that shaft is suspended from its top end by a Kingsbury thrust bearing. This installation was found to whirl violently in cantilever fashion, so that a point of the housing near the top of the housing described a circular path of about $\frac{1}{16}$ in. diameter, while the lower parts of the machine near the foundation hardly moved at all. This whirling motion was at a frequency quite different from the r.p.m. and would occur only at loads greater than half the rated horse power, the frequency being independent of the load. On being diagnosed a self-excited vibration, various modifications were made in the Kingsbury thrust bearing, in the (unloaded) guide bearings of the vertical shaft, in the magnetic field strength of the generator, and in the water wheel in attempts to find the exciting cause. It was finally located in the water wheel and in the part which corresponds to the labyrinth of Fig. 7.32. In Fig. 7.33 A is the rotating Francis wheel, and B is the non-rotating part. The water flows from B into the blades A and down into the tailrace. The water pressure in the space B and in the adjacent portion of A is high, while in the tailrace C and communicating parts C' it is low. Practically all the water flows from B to C through the blades A, but a small fraction leaks through the annular passages D, which constitute the "seal" between the rotating and stationary parts.   The dimensions of the passage D are about $\frac{1}{16}$ in. wide, 3 in. high, and 8π ft. long.   During the whirling motion the rotor A moved laterally relative to the housing B, thus causing periodic varia- tions of the labyrinth width.   Because this width is so small, $\frac{1}{16}$ in., even very small absolute motions of the rotor cause measurable percentage variations in width and, hence, variations in pressure on the seal rings D. Unfortunately, these pressure variations were in phase with the rotor whirl velocity (as actually measured) and caused the instability.   The cure consisted in changing the D-detail from Fig. 7.34a to Fig. 7.34b, which two constructions are about equally effective as a seal, because the pressure drop of Fig. 7.34a, consisting of one large viscous- plus one kinetic-energy part, is replaced in Fig. 7.34b by a smaller viscous- plus

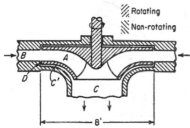

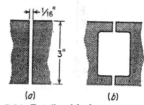

FIG. 7.33. Francis water turbine wheel whirl- ing at its natural frequency, self-excited by pressure variations in the leakage passage D.

FIG. 7.34. Details of leakage passage D of Fig. 7.33. (a) Leading to self-excited vibration; (b) suppressing the whirl.

two kinetic-energy pressure drops.   But while in Fig. 7.34a the pressure variations act on the entire 3-in.-high face of the seal ring, in Fig. 7.34b there is practically no pressure variation in the inner angular chamber, because the percentage width variation there is nearly zero.

Cases of self-excited oscillation caused by leakage variations are extremely rare in turbine practice, but when they do occur, the above case descriptions should make them easily recognizable.

7.10. Airplane-wing Flutter.   In certain airplanes flying at high speed, particularly when diving, the wings have been repeatedly observed to develop a very violent vibration.   On a number of occasions this flutter has been so excessive that it has caused the wing to break off in mid-air.

An explanation on the basis of the phenomenon of Sec. 7.5 might be attempted.   For wings in a "stalled" position the slope of the lift curve is negative (Fig. 7.21), and the up-and-down motion of a cantilever wing is unstable.   This has been observed; however, it is not a condition of actual flight and in the typical "flutter" cases on record the angle of attack of the wing is small and the slope of the lift curve decidedly positive. This, by the argument of Sec. 7.5, leads to a definite positive damping.

Thus we distinguish between *stall flutter* with large angles of attack ($\alpha > 15$ deg.) and *classical flutter* with small angles of attack. Stall flutter has been found of great importance in turbine blades, particularly in the compressor blades of jet engines, where it constitutes a serious problem which is not as yet solved (in 1955). In aircraft wings and rudders, however, the classical type of flutter is of practical importance, especially recently with airplanes reaching or surpassing the velocity of sound.

Any attempt at an explanation in terms of a single degree of freedom (for example, where the wing vibrates up and down only, like a cantilever beam) does not succeed. We have another case of a coupled two-degree-of-freedom system, since the wing not only vibrates up and down, but simultaneously executes a twisting motion. The interplay of the vertical

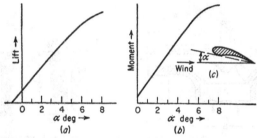

Fig. 7.35. The lift and moment diagrams of an airfoil are practically straight lines for small angles of attack.

and twisting vibrations with the air stream as the source of energy may lead to instability. The possibility of such an occurrence can be explained physically in a rather simple manner.

For a certain value of the *angle of attack* α (defined in Fig. 7.35c) the wing experiences an aerodynamic lift and also a clockwise twisting moment. While the wing executes a twisting vibration, the angle α varies; and therefore we are interested in knowing how the lift and the moment vary with this angle.

Figures 7.35a and b show these relations as obtained by a wind-tunnel test. For the angles α at which flight takes place (0 to 10 deg.), these characteristics are practically straight lines.

Assume that the vertical and the twisting motions of the wing are coupled in such a manner that during the upward stroke the angle α is larger than during the downward stroke. According to Fig. 7.35a the lift during the upward motion is larger than during the downward stroke, which means that the wind feeds energy into the vibration. An energy input is also possible by virtue of Fig. 7.35b. This follows from the fact that even without any twisting motion the angle of attack varies on account of the vertical vibration as explained in Fig. 7.15. Due to this effect the angle of attack and consequently the twisting moment are made larger during the downward stroke and smaller during the upward stroke. Thus, if during the downward motion the wing twists clockwise, energy will be put into the system and the vibration will grow.

A form of flutter which occurred commonly a few years ago was that of bending of the wing associated with flapping of the aileron. Suppose the aileron is hinged about an axis *not* passing through its center of gravity and suppose the wing to be vibrating up and down. Independent of any aerodynamic forces the alternating vertical motion of the hinge axis will force the aileron to execute an angular motion, since the hinge axis does not pass through the center of gravity. The aileron is restrained from doing this by the control wires attached to it, which act as springs, since they are necessarily flexible. Thus the aileron-pendulum has a natural frequency of its own which may be above or below the natural frequency of the flutter motion of the wing, so that the aileron motion may be in phase with or in opposition to the wing motion (when the difference between the two frequencies is great) or the aileron motion may have a phase angle near 90 deg. with respect to the wing motion (when the two frequencies are close together, page 51). In the latter case the aileron motion lags behind the force, so that in the middle of the downward stroke of the wing the aileron is up, causing a downward air force on the wing: hence instability. Trouble of this sort was recognized early and the obvious remedy is to locate the aileron hinge axis through the center of gravity of the aileron by the addition of counterbalance weights if necessary. Even this in itself is not always sufficient to prevent "inertia coupling." To understand this, assume a uniform rectangular aileron hinged about its center line of symmetry. Add to this aileron two equal weights in two opposite corners of the rectangle, leaving the center of gravity where it was. For a purely up-and-down disturbance of the hinge axis the aileron is still balanced and has no tendency to rotate; but if this aileron is placed in an actual wing performing a cantilever vibration with a large amplitude at the tip and a smaller one in the middle, the inertia forces on the two added weights will differ from each other and the aileron will have a turning moment about its hinge axis. Complete balance against all possible motions can be obtained by insisting not only that the center of gravity lies in the hinge axis, but also that the hinge axis is a principal axis of inertia (so that there is zero product of inertia about the hinge axis). This is an ideal that the designer will satisfy as well as design conditions make feasible and as well as flutter difficulties demand. It applies not only to ailerons, but to other movable surfaces (rudder and elevators) as well, which incidentally also have given rise to flutter phenomena in conjunction with the entire fuselage of the airplane.

We now proceed to a more quantitative analysis of the torsion-bending flutter of a solid wing without aileron, and start by setting up the differential equations of motion. In reality the wing acts more or less as a cantilever beam built in at the fuselage, but for simplicity we assume the wing to be a solid body supported on springs so that it can move up and

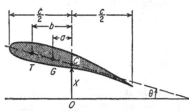

down as well as rotate about its longitudinal axis. In Fig. 7.36 the origin of coordinates O is taken to be at the center point of the span in the position of equilibrium of the wing. The wing departs from this position by the amounts x and θ as shown. Besides the center of the span C, two other points in the wing section are of importance, viz., G and T.

Fig. 7.36. The airplane wing with its two characteristic points: the center of gravity G and the center of twist T.

The point G is the center of gravity, by which the inertia properties are determined. The location of the point T determines the elastic properties of the spring suspension. The point T is known as the "center of twist" and is defined in one of the following manners: T is that point on the wing where a vertical force causes only a vertical displacement and *no rotation*. T is also that point of the wing which does not displace itself if the wing is subjected to a pure torque causing a rotation of the section. These two properties of T always go together as can be shown by Maxwell's theorem of reciprocity.

Let k_x be the up-and-down spring constant and k_θ the torsional spring constant per unit length of wing, let $\mathbf{L}$ be the aerodynamic lift force (a function of x, θ, and the time), and let $\mathbf{M}$ (also a function of x, θ, and t) be the moment of all aerodynamic forces about O, positive when clockwise, again per unit span. Then the equations of motion are

$$\left.\begin{array}{l} m(\ddot{x} + a\ddot{\theta}) + k_x(x + b\theta) = \mathbf{L} \\ I_G\ddot{\theta} + k_\theta\theta + k_x(x + b\theta)(b - a) = \mathbf{M} - \mathbf{L}a \end{array}\right\} \quad (7.33)$$

The combinations $(x + a\theta)$ and $(x + b\theta)$ occurring in these equations are the vertical displacements of G and T, respectively. The symbols m and I_G not only refer to the inertia of the wing itself but include that of the surrounding air as well. Usually we take for this a cylinder of air of radius $c/2$. Although this effect is rather insignificant for propeller blades, it is important for airplane wings, which may weigh not more than three times as much as the cylinder of air around them.

The alternating air force $\mathbf{L}$ in actual wings is quite considerable, of the same order of magnitude as the spring and inertia forces. In practically all previous cases treated in this book, the exciting forces (and damping forces) were small in comparison to the inertia and spring forces, so that the resonant frequency was determined by k/m only and was independent of the exciting force. Here the exciting force L, being of the same order of magnitude as the spring force, *does* affect the frequency and the system will flutter at a frequency distinctly different from any of the natural frequencies of the structure in still air.

The lift force L (per unit length of wing) for a steady-state case is $(\rho V^2/2)cC_L$, and the lift coefficient C_L is proportional to the angle of attack of the air stream. The angle of attack is $\theta - \dot{x}/V$, in which the first term is obvious and the second term is caused by the fact that $\dot{x}$ is the vertical component of the wind speed relative to the wing. Wing theory shows that the proportionality constant is 2π so that

$$\mathbf{L} = \frac{1}{2}\,\rho V^2 c 2\pi \left(\theta - \frac{\dot{x}}{V}\right) = \pi\rho V^2 c \left(\theta - \frac{\dot{x}}{V}\right)$$

Simple steady-state subsonic-wing theory states that this lift force applies at the forward quarter span point, so that the moment of the wind force about the center C of the wing is

$$\mathbf{M} = \mathbf{L}\,\frac{c}{4} = \pi\rho V^2 \frac{c^2}{4}\left(\theta - \frac{\dot{x}}{V}\right)$$

Entering with these two expressions into Eq. (7.33), we are ready to test the stability by Routh's criterion. With the usual assumptions (7.3) (page 285), the two equations (7.33) become

$$x[ms^2 + k_x + \pi\rho V cs] + \theta[mas^2 + k_x b - \pi\rho V^2 c] = 0$$

$$x\left[k_x(b - a) + \pi\rho V\left(\frac{c^2}{4} - ca\right)s\right]$$
$$+ \theta\left[I_G s^2 + k_\theta + k_x b(b - a) + \pi\rho V^2\left(ca - \frac{c^2}{4}\right)\right] = 0$$

By diagonally multiplying the bracketed coefficients with each other the letters x and θ are eliminated and the frequency equation is obtained:

$$s^4 + s^3\left[\frac{\pi\rho V c}{mI_G}\left(I_G + ma^2 - m\frac{ac}{4}\right)\right] + s^2\left[\frac{k_\theta}{I_G} + \frac{k_x}{m} + \frac{k_x}{I_G}(b - a)^2\right.$$
$$\left. + \frac{\pi\rho V^2 c}{I_G}\left(a - \frac{c}{4}\right)\right] + s\,\frac{\pi\rho V c}{mI_G}\left[k_\theta - k_x b\left(b - \frac{c}{4}\right)\right]$$
$$+ \left[\frac{k_x k_\theta}{mI_G} + \frac{\pi\rho V^2 k_x}{mI_G}\,c\left(b - \frac{c}{4}\right)\right] = 0$$

This is a very complicated expression, and several of the coefficients can become negative for certain values of the dimensions a and b in Fig. 7.36.

First we examine the last bracket, designated as A_0, and we see that the second term of A_0 becomes negative when b is less than $c/4$, that is, when the center of twist is close to the center of the wing, as in a thin, symmetrical wing profile. The system is *unstable* when A_0 becomes negative, or when

$$\pi \rho V^2 c \left(\frac{c}{4} - b \right) \geq k_\theta$$

or
$$V^2 \geq \frac{4 k_\theta}{\pi \rho c (c - 4b)} \tag{7.34}$$

This expression has a very simple physical meaning: it is a *static* instability of the nature of Fig. 7.2a (page 283), rather than a true flutter. Imagine a non-vibrating wing in the wind at zero angle of attack so that it has no lift. Imagine that wing now twisted to the angle of attack θ, which brings into play the lift force $\pi \rho V^2 c \theta$ at the quarter cord point. Its moment arm about the center of twist is $(c/4) - b$, so that the aerodynamic twisting couple is

$$\pi \rho V^2 c \theta \left(\frac{c}{4} - b \right)$$

which is resisted by the elastic couple $k_\theta \theta$. For slow wind speeds the aerodynamic twisting couple is less than the elastic restoring couple, so that the wing will snap back to $\theta = 0$. For a certain wind speed, however, the two torques become equal for any value of θ, and then we have indifferent equilibrium in any position θ. The wind speed so found is that of Eq. (7.34), and it is known as the *divergent wind speed*, the phenomenon of static indifferent equilibrium being known as *divergence* among aerodynamicists.

Another instability criterion follows when the coefficient A_2 becomes negative on account of its last term. This occurs when a is small or when the center of gravity comes close to the center of the wing. The critical wind speed from this becomes

$$V^2 \geq \frac{4 k_\theta + 4 k_x [(b - a)^2 + I_G/m]}{\pi \rho c (c - 4a)} \tag{7.35}$$

This speed may be larger or smaller than that of (7.34). When it is the smaller of the two, a true classical flutter occurs at a wind speed lower than that required for divergence.

The analysis so far is applicable only to flutter at slow frequencies (for torsionally very flexible wings) because our expressions for lift and aerodynamic moment were for the steady state. In a slow frequency vibration the lift force still follows the variation of angle of attack, so that we have a "succession of steady states," to which the above analysis applies. For the practical case of a fast flutter (of the order of 20 cycles per second) this is no longer true, and the analysis now becomes extremely complicated, in fact so complicated that flutter analysis has become a specialty in itself to which some engineers devote their full time.

The expressions for the air force and moment per unit length of the harmonically flapping wing for subsonic speeds have been derived by a complicated analysis in the now classical paper by Theodorsen (1935) with the result

$$\left. \begin{array}{l} L = \pi\rho V^2 c \left[Y \left(\theta - \dfrac{\dot{x}}{V} \right) + (1 + Y) \dfrac{\theta c}{4V} \right] \\ M = \pi\rho V^2 \dfrac{c^2}{4} \left[Y \left(\theta - \dfrac{\dot{x}}{V} \right) - (1 - Y) \dfrac{\theta c}{4V} \right] \end{array} \right\}$$ (7.36)

in which the worst complication is that the quantity Y not only is complex but depends on the frequency of flutter as well:

$$Y = F - iG$$ (7.37)

where both F and G are functions of $\omega c/2V$, in which ω is the circular frequency of flutter. The values of F and G are given in Fig. 7.37, taken from Theodorsen's paper. In these expressions for F and G it has already been assumed that the wing is fluttering, $i.e.$, that it is on the borderline between positive and negative damping and hence

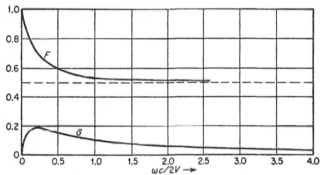

FIG. 7.37. The functions F and G of Eq. (7.37).

executing a purely harmonic undamped motion. The s of page 286 therefore is already assumed to be without real part p, and the imaginary part q is identified with the flutter frequency ω. This, of course, makes stability conditions such as Eq. (7.9) or (7.14) inapplicable.

If still we would proceed to set up the frequency equation, in the manner outlined on page 286, that equation would contain F and G and hence, also, s or ω in a much more complicated manner than a fourth degree algebraic. Theodorsen proceeds to set $x = x_0 e^{i\omega t}$ and $\theta = \theta_0 e^{i\omega t}$ into Eqs. (7.33) and then eliminates x_0 and θ_0 by setting the determinant zero as before. The frequency equation now contains real and imaginary parts, each of which must be zero individually. In this manner two equations are found, which must be true on the border between positive and negative damping. These two equations are in terms of two unknowns: the flutter speed V and the flutter frequency ω, but they are not linear in either V or ω, since they contain the curve Fig. 7.37. The details of Theodorsen's solution of V and ω from this pair of equations are too complicated to be reproduced here and the reader is referred to the original publication. In a subsequent paper by Kassner and Fingado a nomogram is given, based on an analysis similar to Theodorsen's, by which the flutter speed of any individual wing can be determined in a few minutes after the constants have been found.

Another method, originally suggested by Bleakney and Hamm, and now in extensive use, consists of assuming numerical values for the flutter speed V, and for the flutter frequency ω. Then the forces (7.36) can be calculated and substituted into Eqs. (7.33), which of course are not satisfied by the substitution because V and ω are not correct. But they can be made to satisfy Eq. (7.33) by assigning proper values to the stiffnesses k_x and k_θ, which appear linearly in (7.33) and thus can be calculated very easily. This means physically that the arbitrarily chosen V and ω are the true flutter speed and frequency for a wing with stiffnesses different from those of the wing we are considering. The result of this calculation is plotted in Fig. 7.38 in the form of two points marked 1, one each of the V-, ω-diagram and in the k_x-, k_θ-diagram. The entire calculation is then repeated for a different value of the flutter frequency ω with the same flutter speed V (point 2 in the V-, ω-diagram) and the result plotted as point 2 in the k_x-, k_θ-diagram. A third calculation for the points 3 follows.

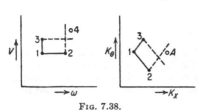

FIG. 7.38.

The actual wing has stiffness values designated by point A in the diagram. Looking at the relative positions of the points, we then pick point 4 in the ω-, V-diagram as a likely next approximation. These trials are continued until we have found a point in the V-, ω-plane whose image in the k_x-, k_θ-plane is sufficiently close to the desired point A.

So far the problem has been one of two degrees of freedom only, i.e., of a wing in which the amplitudes are constant along the span. An approximate value for the flutter speed is obtained by replacing the actual wing by one of the same stiffness but with all its inertia concentrated at a point 70 per cent of the span length distant from the root of the wing, and thus reducing the structure to that of Fig. 7.36. A better result can be found by numerical integration over the length of the wing. Assume, as with the Rayleigh method, a likely shape of deformation. In practice we take for this the shapes of the bending and torsion modes of the wing without air forces, Eq. (7.36), and assume that the bending and torsion motions occur in phase with each other at the same frequency of flutter ω. Assume, next, with Bleakney and Hamm, numerical values for V and ω. Then the air force, Eq. (7.36), the inertia force $\omega^2 y\, dm$, and the elastic force $EIy^{(4)}$ [Eq. (4.36), page 148] are all known numerically. Also the corresponding moments are known numerically. If the assumed shape were the correct one, the sum of all these three forces would be zero for each individual element of the beam, as expressed by the first equation of Eq. (7.33), and likewise the sum of the inertia, elastic, and air moments at each element would be zero as expressed by the second equation (7.33). Since the assumed shape is not the correct one, the equilibrium is violated for each individual element dx, but, with Rayleigh, we can integrate along the entire length of the beam and satisfy the over-all equilibrium. Thus the individual terms of Eq. (7.33) become integrals extending over the length of the wing:

$$\left. \begin{aligned}
\omega^2 \int_0^l y(x)\mu(x)\,dx + EI_0 \int_0^l \frac{EI(x)}{EI_0} \cdot \frac{d^4y}{dx^4} \cdot dx &= \int_0^l L(x)\,dx \\
\omega^2 \int_0^l \theta(x)I_G(x)\,dx + C_0 \int_0^l \frac{C(x)}{C_0} \cdot \frac{d^2\theta}{dx^2}\,dx &= \int_0^l [M(x) - aL(x)]\,dx
\end{aligned} \right\} \quad (7.38)$$

These equations are written for a wing of non-uniform cross section in which the mass μ, the bending stiffness EI, the torsional stiffness C, the deflection of the center of gravity $y = x + a\theta$, the angle θ, and the distance a are all functions of x, variable along the length from 0 to l. The bending stiffness at the root EI_0 and the torsional stiffness at the root C_0 have been brought outside the integrals, Eq. (7.38). Instead of the constants k_x and k_θ we plot in Fig. 7.38 against the stiffness factors EI_0 and C_0 at the root of the wing. Calculations carried out with this procedure come out with errors in the flutter speed of the order of 10 per cent.

Another approach is by model testing. Consider a model of the wing to a reduced scale, made of the same material as the original, but scaled down equally in all dimensions and details. Put this model in a wind tunnel with the *same* air-speed as the original wing. In order to leave the values in the square brackets of Eq. (7.36) unchanged, it is necessary to assume that the time unit is reduced by the same scale as the length unit. But then, as the reader should reason out carefully for himself, the spring force, air force, and inertia force on the model wing all are reduced by a factor l^2, i.e., by the square of the scale ratio for length and time. Thus the flutter speed V (having the dimenison l/t) will remain unchanged, and the flutter frequency ω will go up by the scale ratio. Such a test has the additional advantage of still being valid for air speeds near to or exceeding the velocity of sound, where the expressions (7.36) break down completely. However, a model test as described involves very careful building of the model and elaborate testing apparatus in a wind tunnel. For subsonic air speeds a calculation is simpler and less time-consuming than a test, but for sonic and supersonic air speeds the model test is the only possibility at the present time, because we do not as yet have reliable formulas for the aerodynamic forces in the transonic and supersonic regions.

7.11. Wheel Shimmy. In the early 1930s the air pressure in automobile tires was greatly diminished from previous practice with the object of obtainining smoother riding characteristics. This change introduced as a side effect the front wheel "shimmy," which remained a nuisance for several years until it was cured by proper dimensioning, and by independent spring suspension of the front wheels. This shimmy consisted in a rotational oscillation of the front wheels about vertical axes, and in most cases it was a self-excited vibration. For a proper explanation it is necessary to consider *three* degrees of freedom, so that the problem is more complicated than any of those previously discussed.

Let Fig. 7.39a represent an elevation looking from the front of the car, A being the axle and B the "king-pins." The axle is capable of tilting in a vertical plane (through the angle φ with respect to the road) by virtue of the vertical elasticity in the tires. It is also capable of shifting sidewise with respect to the body or the road (deviation x) on account of the lateral flexibility of the main springs or of the tires. Looking in Fig. 7.39b from the top, the wheels can flutter through an angle ψ, which constitutes the motion usually referred to as shimmy. Since the two wheels are connected to each other by the rigid steering connecting rod C, the angle of flutter ψ has to be the same for both wheels. Several other motions are possible, but these may be neglected for the purpose in hand.

There are thus three degrees of freedom, φ, ψ, and x. In order to show the possibility of self-excited vibration, it has to be demonstrated that these three are coupled to one another and also that some source of energy is available.

The mass of the front wheel and axle is considerably smaller than that of the spring-supported body of the car. Since the shimmy vibration takes place at a rather rapid rate, the body is practically unable to take part in the motion. In the following discussion it will be assumed that the car body moves forward in a straight line along the road while the front wheels and axle vibrate.

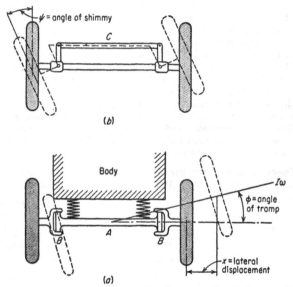

Fig. 7.39. Illustrating the coupling between the shimmy motion ψ, the tramping motion φ, and the lateral vibration x.

Consider the sidewise vibrating motion $x = x_0 \sin \omega t$ of the wheels with respect to the body or to the road. This takes place by distorting the *main springs* laterally and sets up an alternating external force on the axle. A part of this main-spring force is used to accelerate the axle in the x-direction and the rest of it finds a reaction which can occur only in the form of lateral road friction at the tires. This reaction force causes a couple in the plane of Fig. 7.39a, tending to set up an alternating angular motion φ. The motion φ, on the other hand, is coupled to the shimmy ψ by the gyroscopic action of the rapidly rotating wheels. If, for example, the wheel has a "tramping" velocity φ, a gyroscopic couple will be created tending to make ψ smaller. (Incidentally the gyroscopic coupling is responsible for the fact that when *one* wheel rides over a bump in the road the steering wheel gets a rotational jolt.) Finally, the alternating shimmy angle ψ causes the front wheels to follow a wavy path and thus sets up a lateral displacement x. Thus each one of the three degrees of freedom is definitely coupled to the two others.

A source of energy can also be readily found. It was seen that the sidewise motion x is associated with lateral friction forces at the tread of the tires. These forces in turn cause slipping, if not over the *whole* area of the tire in contact with the road, at least over a part of it. Therefore the sidewise displacement x and the lateral force F on the tire do not bear to each other the simple spring relation $F = kx$ but this relation is much more complicated. Without entering into technical details, it is clear that with certain phase relations between the motion x and the road force F this force may do work on the vibration. The ultimate source of energy naturally is the forward kinetic energy of the car.

In case the proper phase relations for instability do exist, the vibration will be all the more violent the smaller the flexibilities and the stronger the coupling. The most important change in the front-wheel construction of the last few years has consisted in the introduction of balloon tires, the great flexibility of which make large φ-motions possible. The general application of superballoon tires, which are very desirable for the riding quality, was retarded for a number of years on account of this shimmy trouble.

A mathematical analysis of the problem is possible, but even in the most elementary case (where many important simplifications have been made) it leads to a sixth-degree frequency equation, whereas a more complete investigation gives an equation of the eighth degree. The complications of such calculations make them hardly worth while.

Though most cases of shimmy are self-excited vibrations, this is not invariably so. The disturbance may be excited by unbalance of the wheels, which always exists to a certain extent, especially with unevenly worn tires. Suppose the unbalance weight at the left wheel to be on top, while at the right wheel it is at the bottom. Then the centrifugal forces of these unbalances will cause a tramping φ-vibration and this in turn causes a shimmy. At a speed such that the frequency of rotation of the wheels coincides with the natural shimmy frequency, the disturbance will be great, as we have an ordinary resonance phenomenon. Since the diameters of the two wheels are different, say by 1 part in 500, the unbalances in the two wheels will be in the same direction after 250 revolutions and then excite only an up-and-down motion, which is not coupled with shimmy. In this manner typical and very slow beats are observed, as indicated in Fig. 7.40.

The most effective method of eliminating shimmy, whether forced or self-excited, is to do away with the gyroscopic coupling. This has been accomplished in most cars by an independent front-wheel suspension. There is no front axle and the wheel is supported in such a manner that it can move only up and down in its own plane parallel to itself and can execute no φ-deviation. With such a construction extremely flexible tires and front springs can be used without any undesirable results.

It is of interest to mention another self-excited phenomenon very similar to that of shimmy, *viz.*, the "nosing" of electric street cars or locomotives. This disturbance occurs frequently with cabs mounted on trucks with some lateral flexibility and consists of a violent lateral sway of the cab with a period of several seconds per cycle. It is obviously a self-excited vibration with the energy furnished ¦by the rail friction. However, there is no gyroscopic coupling as in the automobile. The details of the mechanism of this phenomenon are not fully understood at the present time.

The phenomenon of shimmy is not restricted to pairs of wheels on a common axle as in Fig. 7.39 but has been observed on single castored

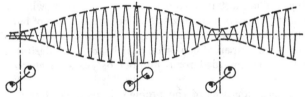

FIG. 7.40. A forced shimmy motion caused by unbalanced wheels.

wheels as well. Hand pushcarts, or "dollies," as they are commonly used in factories, usually have four wheels, of which the two front ones are solidly mounted and the two rear ones can swivel about vertical axes with an offset between the swivel axis and the vertical line through the ground contact. Such castor wheels are often seen to shimmy violently, particularly when the swivel axis is well lubricated and the floor hard and smooth. Nobody cares much about dollies, but when very large and very expensive airplanes become complete wrecks on account of landing-wheel shimmy (which has happened on numerous occasions), it is an altogether different matter. Aircraft landing-gear shimmy has given trouble periodically and has been cured periodically. The last and heaviest airplanes with tail swivel wheels had considerable trouble until means were found to construct these wheels properly. Lately with a nose wheel instead of a tail wheel and with the heaviest modern types, the difficulty has reappeared, and at present it is not yet possible to predict by theory at what speed a given design of wheel will start to shimmy. The problem is being actively studied, theoretically as well as experimentally, but is extremely complicated, almost as bad as that of wing flutter.

A very much simplified analysis will now be given, just to show the principle, though this analysis is too simple to have much practical bearing on an actual case.

We assume the mass and rigidity of the airplane itself to be "large" with respect of that of the swivel landing wheel, so that the point of attachment of the swivel axis to the airplane moves forward at constant speed V. We also assume the tire to be rigid. Then in Fig. 7.41, which is a plan view of the shimmying wheel seen from above, point C is the point where the wheel strut is built into the airplane. Point C moves to the right with constant velocity V. Point B is the bottom point of the strut; normally B lies right under C, but while shimmying the strut is assumed to flex sidewise through distance x. The wheel is behind B with angle φ, the shimmy angle, which is zero for normal ideal operation. A is the center of the wheel, and G is the center of gravity of

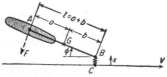

FIG. 7.41. Swivel-wheel shimmy.

the combined landing gear. When shimmying there is a force F from the ground transmitted by friction to the wheel tread. Also, the swiveling mass feels a force kx from the strut at B. The equations of motion of the swiveling mass for small x and φ are

$$\left. \begin{array}{r} m(\ddot{x} + b\ddot{\varphi}) + kx + F = 0 \\ kxb - Fa = I_G\ddot{\varphi} \end{array} \right\} \tag{7.39}$$

two equations with the three unknowns x, φ, F. A third equation is found from the geometry of no sideslip of the wheel, which says that the path of point A must be tangent to the direction AB:

$$-\frac{\dot{x} + l\dot{\varphi}}{V} = \tan \varphi \approx \varphi$$

or

$$\dot{x} + l\dot{\varphi} + V\varphi = 0 \tag{7.40}$$

Elimination of the force F between Eqs. (7.39) gives

$$m(\ddot{x} + b\ddot{\varphi}) + kx + \frac{kb}{a}x - \frac{I_G}{a}\ddot{\varphi} = 0 \tag{7.41}$$

With the usual assumption $x = xe^{st}$ and $\varphi = \varphi e^{st}$ we find

$$\left. \begin{array}{r} x(s) + \varphi(V + ls) = 0 \\ x(ams^2 + kl) + \varphi(-I_G s^2 + mabs^2) = 0 \end{array} \right\}$$

and for the frequency equation, after simplifying,

$$s^3 + s^2 \frac{amV}{I_G + ma^2} + s \frac{kl^2}{I_G + ma^2} + \frac{Vkl}{I_G + ma^2} = 0$$

All coefficients being positive, we apply Routh's test [Eq. (7.9), page 288], and we have stability if

$$mal > I_G + ma^2$$

Writing $I_G = m\rho_G^2$ this can be simplified to

$$\text{Stable for } ab > \rho_G^2 \qquad (7.42)$$

For practically all existing gear constructions this inequality is violated, as will be recognized by reading or working Problems 213 and 214. If in Fig. 7.41 there are concentrated point masses at A and B connected by a weightless bar, we have $ab = \rho_G^2$ and we are just free from shimmy; if A and B are point masses and the connection between them has mass, then $ab > \rho_G^2$. The wheel at A always is the largest mass, and it protrudes to the left of A, so that in all practical cases $ab < \rho_G^2$ and there is instability against shimmy.

In most existing constructions there is a torsional dashpot at point B in Fig. 7.41, and a certain damping in this dashpot then is essential for proper operation (Problem 216). Another way of introducing damping consists in having two wheels parallel to each other on the same shaft, but with the wheels connected together and turning together at the same speed about the shaft. This causes road friction which acts as an effective damping. In general the inequality (7.42) is helped out by making the moment of inertia of the wheel small and by making $a + b = l$ large. In many constructions just the opposite is done, and then a very large dashpot effect is required.

The analysis of page 333 and Fig. 7.41 is an extremely sketchy one. A more truthful analysis must consider two important effects: the dynamics of the rubber tire and the finiteness of the fuselage. The relation between the tire side force F of Fig. 7.41 and the deviation between the tangent to the path of point A and the angle φ is a complicated one, which has been found experimentally for slow frequencies, but which is not fully known at the higher frequencies observed in shimmy (up to 20 cycles per second).

At such high frequencies the point C of the fuselage (Fig. 7.41) certainly will not behave as an infinite mass of infinite stiffness. The fuselage is so complicated that it is advisable to find the "impedance" Z of point C of the fuselage by experiment. A small unbalanced motor can be attached to the bottom of the airplane at C, and the ratio of the sidewise force to the sidewise motion can be determined for a range of frequencies. This impedance then should go into the differential equations instead of the value $Z = \infty$ as before. In fact the $Z = f(\omega)$ curve will usually show several infinite (= large) peaks and several (near) zeros in the practical frequency range.

Problems 196 *to* 216.

CHAPTER 8

SYSTEMS WITH VARIABLE OR NON-LINEAR CHARACTERISTICS

8.1. The Principle of Superposition. All the problems thus far considered could be described by linear differential equations with constant coefficients, or, physically speaking, all masses were constant, all spring forces were proportional to the respective deflections, and all damping forces were proportional to a velocity. In this chapter it is proposed to consider cases where these conditions are no longer true, and, on account of the greater difficulties involved, the discussion will be limited to systems of a single degree of freedom. The deviations from the classical problem Eq. (2.1), page 26, are twofold.

First, in Secs. 8.2 to 8.4, we shall consider differential equations which are linear but in which the coefficients are functions of the time. In the remainder of the chapter non-linear equations will be discussed. The distinction between these two types is an important one. Consider the typical linear equation with a variable coefficient:

$$m\ddot{x} + c\dot{x} + f(t)x = 0 \qquad (8.1)$$

which describes the motion of a system where the spring constant varies with the time. Assume that we know two different solutions of this equation:

$$x = \varphi_1(t) \quad \text{and} \quad x = \varphi_2(t)$$

Then $C_1\varphi_1(t)$ is also a solution and

$$x = C_1\varphi_1(t) + C_2\varphi_2(t) \qquad (8.2)$$

is the *general* solution of Eq. (8.1). Any two known solutions may be added to give a third solution, or

The principle of superposition holds for the solutions of linear differential equations with variable coefficients.

The proof of this statement is simple.

$$m\ddot{\varphi}_1(t) + c\dot{\varphi}_1(t) + f(t)\varphi_1(t) = 0$$
$$m\ddot{\varphi}_2(t) + c\dot{\varphi}_2(t) + f(t)\varphi_2(t) = 0$$

Multiply the first equation by C_1 and the second by C_2 and add:

$$m[C_1\ddot{\varphi}_1(t) + C_2\ddot{\varphi}_2(t)] + c[C_1\dot{\varphi}_1(t) + C_2\dot{\varphi}_2(t)] + f(t)[C_1\varphi_1(t) + C_2\varphi_2(t)] = 0$$

This shows that $[C_1\varphi_1(t) + C_2\varphi_2(t)]$ fits the differential Eq. (8.1) and therefore is a solution.

In mechanical engineering it is usually the *elasticity* that is variable (Eq. 8.1). There is, however, one important case where the *mass* is variable with time (Fig. 5.10, page 184). This case can be discussed on the same mathematical basis as that of variable elasticity, provided damping is absent. We have

$$m(t) \cdot \ddot{x} + kx = 0 \qquad (8.3)$$

where $m(t)$ is the variable mass. Dividing by $m(t)$,

$$\ddot{x} + \frac{k}{m(t)} x = 0 \qquad (8.4)$$

This equation describes a system of unit mass (constant mass) and of variable elasticity.

A *non-linear* equation is one in which the displacement x or its derivatives do not appear any more in the first power, such, for example, as

$$m\ddot{x} + kx^2 = 0 \qquad (8.5)$$

or more generally

$$m\ddot{x} + f(x) = 0 \qquad (8.6)$$

The principle of superposition is not true for the solutions of non-linear equations.

This can be easily verified. Let $x_1 = \varphi_1(t)$ and $x_2 = \varphi_2(t)$ be solutions of (8.5):

$$m\ddot{\varphi}_1(t) + k[\varphi_1(t)]^2 = 0$$
$$m\ddot{\varphi}_2(t) + k[\varphi_2(t)]^2 = 0$$

Hence, by addition,

$$m[\ddot{\varphi}_1(t) + \ddot{\varphi}_2(t)] + k[\{\varphi_1(t)\}^2 + \{\varphi_2(t)\}^2] = 0$$

If $(\varphi_1 + \varphi_2)$ were a solution, the last square bracket should be $(\varphi_1 + \varphi_2)^2$. But the term $2\varphi_1\varphi_2$ is missing, so that $(\varphi_1 + \varphi_2)$ is *not* a solution of (8.5).

The general solution of (8.5) or (8.6) can still be written in a form containing two arbitrary constants C_1 and C_2, since the process of solution is, in principle, a double integration. But although for linear equations a knowledge of two particular solutions immediately leads to the general solution in the form (8.2), this is no longer true for a non-linear equation. Very few non-linear equations exist for which the general solution is known. As a rule all we can do is to find particular solutions and even these in an approximate manner only.

8.2. Examples of Systems with Variable Elasticity. In this section seven cases are discussed physically and a partial explanation of their behavior is given. The more fundamental treatment is necessarily mathematical and will be taken up in the next two sections.

First consider a disk mounted on the middle of a vertical shaft running in two bearings B of which only the upper one is shown in Fig. 8.1. The cross section of the shaft is not completely circular but is of such a nature that two principal directions in it can be distinguished, one of maximum and one of minimum stiffness as, for example, in a rectangular section. Assume the shaft to have two circular spots A, A close to the disk. These round spots A can slide without friction in two straight guides restricting the motion of the shaft to one plane, *e.g.*, to the plane perpendicular to the paper. The disk on the shaft flexibility is a vibrating system of a single degree of freedom. While the shaft is rotating, the spring constant varies with the time, from a maximum $k + \Delta k$ to a minimum $k - \Delta k$, twice during each revolution, so that the equation of motion is

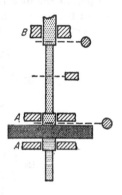

$$m\ddot{x} + (k + \Delta k \sin \omega_k t)x = 0 \qquad (8.7)$$

where ω_k is *twice* the angular speed of rotation of the shaft and the subscript k is used to suggest variation in the elasticity k.

Next, place the same shaft horizontal with the guides A vertical so that the vibration of the disk is restricted to the vertical direction. The weight W of the disk acts as an additional force so that Eq. (8.7) changes to

Fig. 8.1. A disk mounted on a shaft with non-uniform elasticity. The disk is confined to motion in one plane only.

$$m\ddot{x} + (k + \Delta k \sin \omega_k t)x = W \qquad (8.8)$$

If the elasticity were constant, there would be no significant difference between (8.7) and (8.8), because (8.8) could be transformed into (8.7) by merely taking another origin for the coordinate x (the distance between these two origins would be the static deflection of the disk). With variable elasticity, however, this is not so. Let us take a new variable

$$y = x + C$$

where C is a constant to be determined so as to make the result as simple as possible. Substitute in (8.8) which becomes

$$m\ddot{y} + (k + \Delta k \sin \omega_k t)y = W + kC + C\Delta k \sin \omega_k t \qquad (8.9)$$

If the variation in elasticity Δk were zero, we could choose C equal to $-W/k$ and thus transform (8.9) into (8.7). With $\Delta k \neq 0$, this cannot be done. By imagining $W = 0$ in the last result, it is interesting to see that (8.7) by a mere shift of the origin of x can be given a right-hand member which can be classed as an extraneous exciting force of frequency ω_k [see Eq. (6.6c), page 248].

We see that (8.7) and (8.8) cannot be transformed into one another; they are definitely different and have to be so treated.

Assume that the *variations* in k are small with respect to k (Δk is 10 per cent of k or less). Then the elastic force is principally that of k and the motion of the disk is nearly harmonic with the frequency $\omega_n - \sqrt{k/m}$. When this natural frequency of motion ω_n has a proper relation to the frequency of spring variation ω_k and when also a proper phase relation

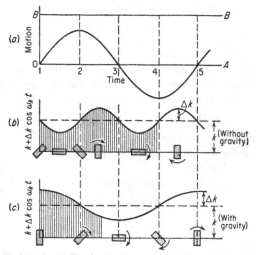

FIG. 8.2. Explains instability in the flat shaft at half and full critical speed.

exists, it is possible to build up large vibrations. Consider the curves Fig. 8.2a and b, illustrating the motion x of the disk with a frequency ω_n and a variation in shaft stiffness taking place at twice that frequency. These diagrams pertain to the vertical shaft (no gravity) so that OA is the equilibrium line where there are no bending stresses in the shaft. The elastic force is therefore the product of the ordinates of Figs. 8.2b and 8.2a, measured from OA. With the phase relation shown in the figure it is seen that, while the disk is moving away from the center position (1-2 and 3-4), the spring force is smaller than its average value, whereas, while the disk is moving toward the center, the spring force is greater than its average value (2-3 and 4-5). Thus the spring force is small while opposing the motion and large while helping the motion. Hence, over a full cycle the spring force does work on the system and the vibration, once started, builds up: we have instability.

With gravity, the spring force is still the product of k and the amplitude of Fig. 8.2a, but this time the ordinate is not measured from base OA, but rather from the base BB, distant from OA by the static deflection δ_{st}. The presence of the extra δ_{st} does not change the previous argument with respect to the k-variation shown in Fig. 8.2b, but it is now possible to obtain work input with another k-variation, shown in Fig. 8.2c, with $2\omega = \omega_n$ (shaft running at *half* critical speed). This is so because the spring force is small (2 to 4) while the disk is going away from its equilibrium position BB and large (1 to 2 and 4 to 5) while it is coming toward BB.

The work input per cycle in general is $\int F\,dx = -\int kx\,dx$, where $x = \delta_{st} - x_0 \sin \omega t$. We write $k = k - \Delta k \sin 2\omega t$, in the case of Fig. 8.2b, and the reader is asked to substitute this into the integral and verify that the work input per cycle is $+\dfrac{\pi}{2}\Delta k x_0^2$, which is independent of δ_{st}. For the case of Fig. 8.2c we write $k = k + \Delta k \cos \omega t$, and the work input becomes $+\pi\,\Delta k x_0 \delta_{st}$, which is seen to exist only in the presence of gravity.

Thus the physical analysis leads to the following conclusions:

1. In the system described by Eq. (8.7), *i.e.*, in the vertical shaft with flats, any small vibrations at the natural frequency $\omega_n = \sqrt{k/m}$ that may exist will be increased to large amplitudes if the shaft runs at its full critical speed ($\omega_k = 2\omega_n$).

2. For the system of Eq. (8.8), *i.e.*, for the horizontal shaft with flats, the same type of instability exists at the full critical speed as well as at half the critical speed ($\omega_k = \omega_n$).

These conclusions are tentative; an analysis of the equation in the following sections will show to what extent they have to be supplemented.

Practical cases in which shafts of non-circular section have given rise to critical speeds of one-half the normal are illustrated in Fig. 8.3. The first of these is a shaft with a keyway cut in it. There the trouble can be corrected by cutting two additional

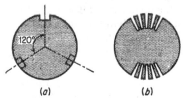

(*a*) (*b*)

Fig. 8.3. Cases of non-uniform flexibility in shafts and rotors.

dummy keyways, symmetrically placed, which makes the stiffness uniform in all directions.

The other example is found in the cross section of a two-pole turbogenerator rotor, in which slots are cut for the electric windings, the solid parts forming the pole faces. In this case the non-uniform elasticity cannot be avoided, so that a two-pole rotor will always be rough at half its critical speed.

A *second* case quite similar to the example of the shaft is that of a string with a mass m in the center. The tension in the string is varied with a frequency ω_k between a maximum $\mathbf{T} + \mathbf{t}$ and a minimum $\mathbf{T} - \mathbf{t}$ by pulling at the end (Fig. 8.4). If we pull hard while m is moving toward the center and slack off while m is moving away from the center, a large

vibration will be built up. While m is describing a full cycle, the end of the string describes two cycles. We havo tho caoo of Fig. 8.2b. If the string is horizontal a gravity effect comes in, making the system subject to Eq. (8.8) and Fig. 8.2c.

The periodic change in tension may also be brought about by a change in temperature. A wire in which an alternating current is flowing has temperature variations and consequently variations in tension having double the frequency of the current. Lateral oscillations will build up if the natural frequency is either equal to or twice as large as the electric frequency.

A *third* case is illustrated in Fig. 8.5. A pendulum bob

FIG. 8.4. String with variable tension as the second example of Eq. (8.7).

is attached to a string of which the other end is moved up and down harmonically. The spring constant k of a "mathematical pendulum" is mg/l, so that a periodic change in the length l means a corresponding change in the spring constant. Thus the sidewise displacements of the bob are governed by Eq. (8.7). In order to build up large oscillations by a length variation of $\omega_k = 2\omega_n = 2\sqrt{g/l}$, the string has to be pulled up in the middle of the swing and let down at the extreme positions, the bob describing a figure eight as indicated in Fig. 8.5. The tension in the string is larger for small angles φ than for great angles on account of two factors. In the extreme position the tension in the string is the weight of the bob multiplied by $\cos \varphi$, which is less than unity. In the center, the tension is the weight plus the centrifugal force of the bob moving in its curved course. Thus

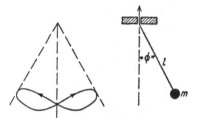

FIG. 8.5. Pendulum of variable length.

the string is pulled up in the center against a large tension and let down at the extreme positions against a small tension. In this way work is put into the system, and this work is converted into the additional kinetic and potential energy of the larger vibrations.

The *fourth* example is more difficult to understand physically. It is nearly the same as the previous one, except that the pendulum is a stiff rod of constant length and the point of support (about which it can turn freely) is given a rapid up-and-down harmonic motion by means of a small electric motor. It will be seen later (on page 348) that such a pendulum has the astounding property of being able to stand up vertically on its support. The spring constant of a pendulum *rod* is again mg/λ, but λ in this case is the "equivalent length." In this experiment the length λ is constant, but the gravity constant g varies periodically. This can be understood by considering the pressure of a man on the floor of an elevator car. While the elevator is standing still or moving at constant speed, this pressure is equal to the weight of the man; in an upward accelerated elevator it is more and in a downward accelerated car it is less. An impartial experimenter in the accelerated elevator may conclude that the value of g differs from its value on the earth. The same is the case

with the pendulum. While it is being accelerated upward, g is apparently larger. Thus a periodically varying spring constant and the validity of Eq. (8.7) are shown. A more satisfactory derivation is given on page 350.

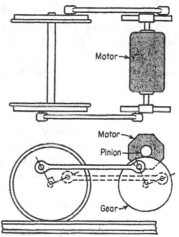

FIG. 8.6. Torsional vibration in electric side-rod locomotive.

The *fifth* case to be discussed is technically the most important one. In electric locomotives of the side-rod type violent torsional vibrations in the drive system have been observed in several speed ranges. They are caused by a periodic pulsation in the spring constant, which can be understood from Fig. 8.6 representing one of the simplest constructions of this type. An electric motor is mounted on the frame and coupled to a driving axle by one connecting rod on each side of the locomotive. The two rods are 90 deg. offset so that the system as a whole does not have any dead center. With the usual operating conditions the wheels are locked to the rails by friction, but the motor can rotate slightly against the flexibility of the two side rods. When a side rod is at one of its dead centers, it does not prevent the motor from turning through a small angle, *i.e.*, its share in k is zero. When it is 90 deg. from its dead center, it constitutes a very stiff spring since it has to be elongated or the crank pins have to be bent to allow a small rotation of the motor. The spring constant of *one* side rod, therefore, varies between a maximum and practically zero and performs two full variation cycles for each revolution of the wheel. The variation in the flexibility of the combination of two side rods is less pronounced and shows four cycles per revolution. The curves 1 and 2, Fig. 8.7, show the torque at the motor necessary for one unit of twisting angle, if only side rod 1 or 2 is attached. Curve 3, being the sum of curves 1 and 2, gives the resultant k for the whole system.

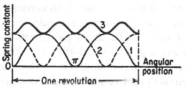

The torsional oscillations of the motor on its side-rod springs will take place superposed on the general rotation of the motor. The phenomenon is represented by Eq. (8.7) where ω_k is four times as large as the angular velocity of the wheels. It is

FIG. 8.7. The torsional spring constant of Fig. 8.6 as a function of the angular position.

to be expected, therefore, that serious vibrations will occur when

$$\omega_n = \tfrac{1}{2} \cdot [4(2\pi \text{ r.p.s.})]$$

A *sixth* example has been found in the small synchronous motors of electric clocks (Fig. 8.8). The rotating part of these motors usually consists of a very light piece of sheet metal A running around the poles B which carry the alterating current. The rotor can slide axially in its bearing but is held in a certain position by the magnetic field of the poles B. These poles act as magnetic springs of which the intensity becomes zero 120 times per second in a 60-cycle circuit, so that the variation in the spring constant is large (100 per cent). The trouble experienced consists in a noisy axial vibration of the rotor.

FIG. 8.8. An electric clock motor is an axially vibratory system on the magnetic springs B.

The *seventh* and last illustration of (8.7) is the electrical analogue. A glance at the table on page 28 shows that we are dealing with a simple inductance-condenser circuit of which the condenser capacity is periodically varying, for instance by means of the crank mechanism of Fig. 8.9. The x of Eq. (8.7) stands for the charge Q on the condenser plates. A constant right-hand member in Eq. (8.8) can be provided by a direct-current battery in the circuit. First consider the system without battery. Two charged condenser plates attract each other mechanically. The current in the L, C circuit will surge back and forth with the frequency $\omega_n = \sqrt{1/LC}$. Let the crank mechanism be so timed that the

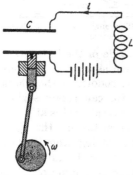

FIG. 8.9. Electric circuit with variable condenser (spring).

plates are pulled apart while the charge Q is large and pushed together again while Q and hence the attractive force are close to zero. Thus the crank mechanism (moving at double the current frequency: $\omega_k = 2\omega_n$) does mechanical work on the system and this work is converted into electrical energy. With a strong battery and small oscillations the charge on the condenser never changes its sign and the crank mechanism has to operate at the frequency $\omega_k = \omega_n$ in the fashion indicated in Fig. 8.2c.

8.3. Solution of the Equation. Most of the problems discussed in the previous section depend for their solution on the differential equation

$$m\ddot{x} + [k + \Delta k \cdot f(t)]x = 0 \qquad (8.10)$$

where $f(t)$ is a periodic function of the time, usually of the form $f(t) = \sin \omega_k t$. It is known as *Mathieu's* equation, and its general solution, containing two arbitrary integration constants, has not yet been found. In fact, there are very few equations with variable coefficients of which solutions are known. However, we are not so much interested in the solution itself, *i.e.*, in the exact shape of the motion, as in the question whether the solution is "stable" or "unstable." The simplest solution of (8.10) is $x = 0$; in other words, the system remains at rest indefinitely. If it is given some initial disturbance ($x = x_0$ or $\dot{x} = v_0$), it cannot remain at rest and the distinction between stable and unstable motion refers to this case. By a stable solution we mean one in which the disturbance dies down with time as in a damped vibration, whereas an unstable motion is one where the amplitudes become larger and larger with time (Fig. 7.2).

If the "ripple" $f(t)$ on the spring constant has the frequency ω_k, the motion, though it may not be periodic, will show certain regularities after each interval $T = 2\pi/\omega_k$.

Suppose that the system starts off at the time $t = 0$ with the amplitude $x = x_0$ and with the velocity $\dot{x} = v_0$. Let the (unknown) solution be $x = F(t)$ and assume that after the end of one period $T = 2\pi/\omega_k$ the amplitude and velocity of the system are given by their values at the beginning, multiplied by a factor **s** (positive or negative):

$$(x)_{t=\frac{2\pi}{\omega_k}} = \mathbf{s}x_0 \qquad (\dot{x})_{t=\frac{2\pi}{\omega_k}} = \mathbf{s}v_0 \qquad (8.11)$$

Whether this assumption is justified remains to be seen. If it is justified, we find ourselves at the beginning of the second cycle with an amplitude and velocity **s** times as great as at the beginning of the first cycle. Then it can easily be proved that the motion throughout the *whole* of the second cycle is **s** times as large as the motion during the corresponding instants of the first cycle and in particular that the third cycle starts with an amplitude $\mathbf{s}^2 x_0$.

The proof is as follows: Let $x = F(t)$ be the solution of (8.10) with the conditions

$$(x)_{t=0} = x_0 \qquad \text{and} \qquad (\dot{x})_{t=0} = v_0 \qquad (8.12)$$

Take as a new variable during the second cycle $y = sx$. The differential equation becomes (after multiplication by s)

$$m\ddot{y} + [k + \Delta kf(t)]y = 0 \qquad (8.13)$$

If the time is now reckoned from the beginning of the second cycle, the initial conditions are

$$(y)_{t=0} = sx_0 = y_0 \quad \text{and} \quad (\dot{y})_{t=0} = sv_0 = \dot{y}_0 \qquad (8.14)$$

It is seen that (8.13) and (8.14) are exactly the same as (8.10) and (8.12) so that the solution is $y = F(t)$. Therefore, $y = sx$ behaves during the second cycle in exactly the same manner as x behaves during the first cycle.

Thus, if the supposition (8.11) is correct, we have solutions that repeat in ω_k-cycles but multiplied by a constant factor in much the same manner as Eq. (2.19) or Fig. 7.2. If s is smaller than unity, the motion is damped or stable; if s is larger than unity, the motion is unstable. For any general periodic $f(t)$, Eq. (8.10) cannot be solved. The particular case of a "rectangular ripple" Δk on the spring constant k, however, is comparatively simple of solution (Fig. 8.10). In most practical cases the ripple is more sinusoidal than rectangular, but the general behavior of a system such as is shown in Fig. 8.10 is much the same as that of a system with a harmonic ripple on the spring force.

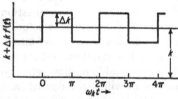

FIG. 8.10. Variation in elasticity for which Eq. (8.10) can be solved.

With the notation $k/m = \omega_n^2$ and $f(t) = \pm 1$, the differential equation (8.10) becomes for $0 < \omega_k t < \pi$,

$$\ddot{x} + \left(\omega_n^2 + \frac{\Delta k}{m}\right)x = 0 \qquad (8.15)$$

and for $\pi < \omega_k t < 2\pi$, or for $-\pi < \omega_k t < 0$.

$$\ddot{x} + \left(\omega_n^2 - \frac{\Delta k}{m}\right)x = 0 \qquad (8.16)$$

Both of these equations are easily solved, since the coefficient of x is now constant. The solution for one half cycle is [see Eqs. (2.7) and (2.8), page 31]

$$x_1 = C_1 \sin p_1 t + C_2 \cos p_1 t \qquad \left(p_1 = \sqrt{\omega_n^2 + \frac{\Delta k}{m}}\right) \qquad (8.17)$$

and for the other half cycle

$$x_2 = C_3 \sin p_2 t + C_4 \cos p_2 t \qquad \left(p_2 = \sqrt{\omega_n^2 - \frac{\Delta k}{m}}\right) \qquad (8.18)$$

These two solutions should be fitted together at $\omega_k t = 0$ with the same amplitude and velocity; moreover, they should describe a motion which at the end of a full cycle is s times as large as at the beginning. Thus

$$\left.\begin{array}{l}
(x_1)_{\omega_k t=0} = (x_2)_{\omega_k t=0} \\
(\dot{x}_1)_{\omega_k t=0} = (\dot{x}_2)_{\omega_k t=0} \\
(x_2)_{\omega_k t=\pi} = s(x_1)_{\omega_k t=-\pi} \\
(\dot{x}_2)_{\omega_k t=\pi} = s(\dot{x}_1)_{\omega_k t=-\pi}
\end{array}\right\} \qquad (8.19)$$

are four equations from which the four arbitrary constants in (8.17) and (8.18) can be determined.

The first two equations of (8.19) are simply

$$C_4 = C_2 \qquad \text{and} \qquad C_3 = \frac{p_1 C_1}{p_2}$$

and the remaining two are

$$C_3 \sin \frac{p_2 \pi}{\omega_k} + C_4 \cos \frac{p_2 \pi}{\omega_k} = -sC_1 \sin \frac{p_1 \pi}{\omega_k} + sC_2 \cos \frac{p_1 \pi}{\omega_k}$$

$$p_2 C_3 \cos \frac{p_2 \pi}{\omega_k} - p_2 C_4 \sin \frac{p_2 \pi}{\omega_k} = sp_1 C_1 \cos \frac{p_1 \pi}{\omega_k} + sp_1 C_2 \sin \frac{p_1 \pi}{\omega_k}$$

or, after substitution of $C_4 = C_2$, etc.,

$$C_1 \left(\frac{p_1}{p_2} \sin \frac{p_2 \pi}{\omega_k} + s \sin \frac{p_1 \pi}{\omega_k} \right) + C_2 \left(\cos \frac{p_2 \pi}{\omega_k} - s \cos \frac{p_1 \pi}{\omega_k} \right) = 0$$

$$C_1 \left(p_1 \cos \frac{p_2 \pi}{\omega_k} - sp_1 \cos \frac{p_1 \pi}{\omega_k} \right) + C_2 \left(-p_2 \sin \frac{p_2 \pi}{\omega_k} - p_1 s \sin \frac{p_1 \pi}{\omega_k} \right) = 0$$

These are homogeneous in C_1 and C_2, so that we can calculate C_1/C_2 from the first equation, then again from the second equation, and equate the two C_1/C_2-values. Rearranging the answer in terms of s, we have

$$s^2 - 2s \left\{ \cos \frac{\pi p_1}{\omega_k} \cos \frac{\pi p_2}{\omega_k} - \frac{p_1^2 + p_2^2}{2 p_1 p_2} \sin \frac{\pi p_1}{\omega_k} \sin \frac{\pi p_2}{\omega_k} \right\} + 1 = 0 \qquad (8.20)$$

If, for brevity, the expression with the braces be denoted by A, the solution of (8.20) is

$$s = A \pm \sqrt{A^2 - 1} \qquad (8.21)$$

In case $A > 1$, one of the two possible values of s is greater than unity and the solution is unstable. After each ω_k-cycle the magnified deflection is in the same direction, so that in each ω_k-cycle there have taken place 1 or 2 or 3 . . . cycles of the free vibration ω_n.

If A lies between -1 and $+1$, the two values for s become complex, which means that the original assumption (8.11) is untenable. However the real part of s is less than unity, so that we expect a motion which does not increase regularly with the time: the system is stable.

Finally, when A is smaller than -1, one of the values of s will also be smaller than -1. This means physically that after one ω_k-cycle, the amplitude and velocity of the system are reversed and are somewhat larger. After two ω_k-cycles they have the same sign and are also larger (multiplied by s^2 which is positive and larger than one). Again we have instability, but during each ω_k-cycle we see $\frac{1}{2}$, $1\frac{1}{2}$, $2\frac{1}{2}$ cycles of the free vibration ω_n.

Thus briefly the system is *unstable* if $|A| > 1$, or if

$$\left| \cos \frac{\pi p_1}{\omega_k} \cos \frac{\pi p_2}{\omega_k} - \frac{p_1^2 + p_2^2}{2p_1 p_2} \sin \frac{\pi p_1}{\omega_k} \sin \frac{\pi p_2}{\omega_k} \right| > 1 \qquad (8.22)$$

where the symbol $|\ |$ means "numerical value of" and the significance of p_1 and p_2 is given by (8.17) and (8.18). In this relation there are two variables, p_1/ω_k and p_2/ω_k, or more significantly, ω_n/ω_k (the ratio of the "free" and "elasticity" frequencies) and $\Delta k/k$ (the "percentage of variation," Fig. 8.10).

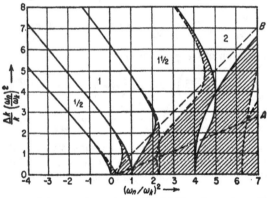

FIG. 8.11. Fundamental diagram determining the stability of a system with variable elasticity. The shaded regions are stable and the blank regions are unstable. (Van der Pol and Strutt.)

The result (8.22) is shown graphically in Fig. (8.11), where for convenience the abscissas are taken as $(\omega_n/\omega_k)^2$ and the ordinates as $(\Delta k/k) \cdot (\omega_n/\omega_k)^2$. The reason for this choice of abscissa is that with the second power, a negative spring constant (such as appears in the vertical pendulum) can be plotted as a negative $(\omega_n/\omega_k)^2 = -k/m\omega_k^2$, whereas with the first power of ω_n/ω_k the abscissa for a negative spring would become imaginary. For the ordinate the case of no steady spring constant, $k = 0$, would lead to an infinite ratio $\Delta k/k$; this defect is avoided by taking $\dfrac{\Delta k}{k}\left(\dfrac{\omega_n}{\omega_k}\right)^2 = \dfrac{\Delta k}{m\omega_k^2}$. In the figure the lines where (8.22) equals $+1$ are drawn in full, while those along which (8.22) is -1 are dashed. In the shaded regions the expression (8.22) is less than unity, which indicates stability, while in the non-shaded regions its value is greater than unity, denoting instability. The numbers $\frac{1}{2}$, 1, $1\frac{1}{2}$, etc., inscribed in the regions of instability indicate the number of vibrations of the system during one ω_k-period of the variation in stiffness.

8.4. Interpretation of the Result. From Fig. 8.11 the behavior of the various systems of Sec. 8.2 can be deduced more accurately than was possible from the simple physical considerations given in Sec. 8.2. The examples come in three groups:

a. The shaft, the string, the locomotive, and the variable condenser all have a frequency of variation ω_k which can vary over a considerable range and have also a *small* variation percentage $\Delta k/k \ll 1$ with a *positive k.*

b. The electric-clock motor has a constant ω_k-frequency, large variations ($\Delta k/k = 1$), and a positive k.

c. The pendulum standing on end has a variable ω_k-frequency and a *negative k, i.e.,* it is statically unstable.

Before discussing any one case in detail, it should be remembered that the diagram, Fig. 8.11, has been derived for a "rectangular ripple," so that only approximate results are to be expected from its use for most actual cases where the variation is nearly harmonic. However, the approximation is a very good one. Moreover, no *damping* has been considered.

First we shall discuss the examples of group *a.* In each case the percentage of variation $\Delta k/k$ and the average natural frequency $\omega_n = \sqrt{k/m}$ are constant. The only variable in the system is the frequency of variation in elasticity ω_k. In the diagram the ordinate is always $\Delta k/k$ times as large as the corresponding abscissa. Each system, therefore, can move only along a straight line through the origin of Fig. 8.11 at an inclination $\tan^{-1} \Delta k/k$ with the horizontal. The line for $\Delta k/k = 0.4$ (40 per cent variation) is drawn and marked *OA.* A slow variation ω_k corresponds to a point on that line *far* from the origin *O*, while the points close to the origin have a small value of $(\omega_n/\omega_k)^2$ and therefore a large ω_k. It is seen that most of the points on *OA* are in stable regions where no vibration is to be feared, but we also note that there are a great (theoretically an infinitely great) number of rather narrow regions of instability. These occur approximately at $\omega_n/\omega_k = \frac{1}{2}$, 1, $1\frac{1}{2}$, 2, $2\frac{1}{2}$, etc.

Now imagine the electric locomotive to start very slowly and to increase its speed gradually, until finally the variation in side-rod elasticity (being four times as fast as the rotation of the wheels) equals twice the natural frequency of torsional vibration. Along *OA* in Fig. 8.11 this means a motion from infinity to the point where $(\omega_n/\omega_k)^2 = \frac{1}{4}$, *and it is seen that an infinite number of critical speeds has been passed.*

From Sec. 8.2 it seems that the two speeds, for $\omega_n/\omega_k = 1$ and $\omega_n/\omega_k = \frac{1}{2}$ are the most significant and that the other critical speeds are much less important.   Nevertheless it is impossible to avoid these low-speed instabilities by changes in the design, unless of course the variation Δk can be made zero.   Vibrations of this sort have caused great trouble in the past.   They were overcome chiefly through introducing torsionally flexible couplings with springs between the motor gear and its crank or between the driving wheel and its crank.   These couplings act in two ways.   First, they shift the natural frequency ω_n to a low value so that all critical regions lie below a rather low speed, say 20 m.p.h.   At these low speeds the intensity of the input cannot be expected to be very great.   Furthermore, the springs, especially if they are of the leaf type, have some internal friction in them so that they introduce damping.

Similar results hold for any of the other examples in group a.   In particular a shaft with two flat sides will pass through a great number of critical-speed regions.   In the actual experiment, however, only the highest two of these critical speeds prove to be of importance, one occurring at half the usual "primary" critical speed and the other at that speed itself.

In group b we have the axial vibrations of the electric-clock motor caused by a periodically vanishing elasticity.   Here $\Delta k/k = 1$, which for variable speed ω_k is represented by a straight line at 45 deg. (shown as OB in the diagram).   In this case, it is seen that the regions of instability are wider than the regions of stability, so that the chance for trouble is far greater than before.

The last case, that of the inverted pendulum, is technically the least important but philosophically the most interesting.

In the first place, the spring constant k for such a pendulum is negative. This will be clear if we remember the definition of k, which is the force tending to bring the system *back* to its equilibrium position from a unit deflection.   The gravity component attempts to *remove* the pendulum from the vertical so that k is negative.   Hence $\omega_n^2 = k/m$ is also negative. For the *hanging* pendulum $\omega_n^2 = g/\lambda$ where λ is the equivalent length ($\lambda = \frac{2}{3}$ of the over-all length in the case of a uniform bar).   For the inverted pendulum,

$$\omega_n^2 = -\frac{g}{\lambda}$$

Let the motion of the supporting point be $e \sin \omega_k t$ which gives an acceleration $-e\omega_k^2 \sin \omega_k t$.   The variation in elasticity amounts to

$$\frac{\Delta k}{k} = \frac{\Delta g}{g} = \frac{-e\omega_k^2}{g}$$

and the ordinate in **Fig. 8.11** becomes

$$\frac{\Delta k}{k}\left(\frac{\omega_n}{\omega_k}\right)^2 = e$$

being the ratio of the amplitude of the base motion to the equivalent length of the pendulum. The abscissa is

$$\left(\frac{\omega_n}{\omega_k}\right)^2 = -\frac{g}{\lambda\omega_k^2}$$

a negative quantity and small for rapid motions of the base.

Figure 8.12 shows a detail of the main diagram of Fig. 8.11 which is important for the inverted pendulum. To be precise, Fig. 8.12 has been taken from the exact solution for a *sinusoidal* ripple (not given in this book), while Fig. 8.11 refers to a rectangular ripple. Incidentally it is seen that very little difference exists between the two.

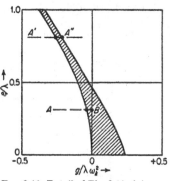

If the pendulum is started with a given base amplitude e and with an increasing frequency ω_k, we move along the horizontal line from A toward B. For slow ω_k the system is clearly unstable, but at a certain speed it enters the stable region and remains there until at B the base speed ω_k becomes

Fig. 8.12. Detail of Fig. 8.11 giving an explanation for the stability of the inverted pendulum. (Van der Pol and Strutt.)

infinitely large. However, when the ratio e/λ is taken greater than about 0.5, there is a large speed at which the pendulum becomes unstable for the second time, as indicated by the point A'' of the line $A'A''$. This has been proved experimentally.

The proof for the statement that the effect of the motion of the point of support of the pendulum is equivalent to a variation in the gravity constant can be given by writing down Newton's laws of motion. In Fig. 8.13 let

a = distance between point of support and center of gravity G,
$s = e \sin \omega_k t$ = displacement of support,
I = moment of inertia about G,
θ = angle with the vertical,
x, y = vertical (up) and horizontal (to right) displacements of G,
X, Y = vertical (up) and horizontal (to right) reaction forces from support on pendulum.

Then the displacements of G are

$$x = s + a \cos \theta \approx s + a \qquad \text{(for small } \theta\text{)}$$
$$y = a \sin \theta \approx a\theta \qquad \text{(for small } \theta\text{)}$$

The three equations of Newton for the vertical and horizontal motion of G and for the rotation about G are

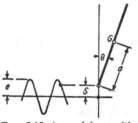

$$X - mg = m\ddot{x} = m\ddot{s}$$
$$Y = m\ddot{y} = ma\ddot{\theta}$$
$$Xa \sin \theta - Ya \cos \theta \approx Xa\theta - Ya = I\ddot{\theta}$$

The reactions X and Y can be eliminated by substituting the first two equations in the third one:

$$I\ddot{\theta} = m\ddot{s}a\theta + mga\theta - ma^2\ddot{\theta}$$

or $$(I + ma^2)\ddot{\theta} - ma(g + \ddot{s})\theta = 0$$

Fig. 8.13. A pendulum with a harmonically moving point of support is equivalent to a pendulum with a stationary support in a space with a periodically varying constant of gravity g.

The expression $I + ma^2$ is the moment of inertia about the point of support and the spring constant is

$$-ma(g + \ddot{s})$$

It is negative and its variation can be interpreted as a variation in g by the amount $\ddot{s}$, the accleration of the support.

Finally we shall discuss the case of variable *mass*, illustrated in Fig. 5.10, page 184. Consider a simple piston and crank mechanism coupled through a flexible shaft k to a flywheel of infinite inertia (Fig. 8.14).

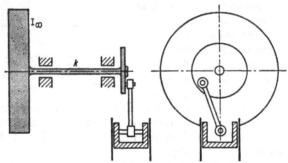

Fig. 8.14. A system with periodically varying inertia.

Let the flywheel be rotating at uniform speed. This system is a torsional one of a single degree of freedom with the constant elasticity k and a variable moment of inertia (mass).

It was seen on page 336 that such a system is mathematically equivalent to one with variable elasticity and constant mass so that Fig. 8.11 applies. According to Fig. 8.11 we ought to experience critical speeds when the average natural frequency $\omega_n = \sqrt{k/I}$ is $\frac{1}{2}$, 1, $1\frac{1}{2}$, 2 . . . times the frequency ω_k of mass variation. It can be easily seen that the main frequency of mass variation is twice the r.p.m., so that the critical speeds should appear for $\omega_n = 1, 2, 3$ . . . times ω_{rpm}. The simple approximate theory culminating in Eq. (5.15), page 178, gives only one critical speed occurring at $\omega_n = 2\omega_{rpm}$ for a connecting rod of infinite length.

8.5. Examples of Non-linear Systems. Non-linearity consists of the fact that one or more of the coefficients m, c, or k depend on the *displacement* x. In mechanical cases the most important non-linearities occur in the damping or in the springs, whereas in electrical engineering the most common case is that of a non-linear inductance (mass).

Let us first consider some examples of non-linear springs. Figure 8.15 shows three cases where the spring force is not proportional to the displacement, but where the individual springs employed are yet ordinary linear coil springs. The first case is the very common one of clearances in the system. The mass can travel freely through the clearance without experiencing any spring force at all, but from there on the force increases

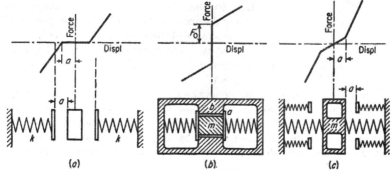

(a) (b) (c)

Fig. 8.15. Combination of linear coil springs which form a non-linear elasticity. (a) Clearances; (b) set-up springs with stops; (c) two sets of springs, one with clearance (sometimes called "stops" also).

linearly. The second case is that in which the springs have an initial compression and are prevented from expanding by the thin washers a resting against the lugs b. The mass m, being loose from the washers, cannot move until a force is applied to it equal to the initial compressive force F of the springs. The third example is that of a spring with so-called "stops." For a small displacement the system is affected only by one set of springs, but after that another set comes into action and makes the combined spring much stiffer. The second set of springs sometimes consists of a practically solid stop, in which case the characteristic becomes nearly vertical after the stop is hit.

All three cases shown in Fig. 8.15 naturally have their torsional equivalents. In particular, set-up springs (Fig. 8.15b) are used often in the construction of torsionally flexible couplings.

Figure 8.16 represents a cantilever spring which, when deflected, lies against a solid guide, thus shortening its free length and becoming stiffer.

Hence its force-deflection characteristic becomes steeper for increasing deflections. More or less curved spring characteristics occur quite often in practice. In fact, most actual springs have a straight characteristic for small deflections only and then become stiffer for larger deflections.

Next we consider some forms of non-linear *damping*. The linear

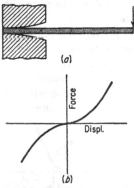

(a)

(b)

FIG. 8.16. Spring with gradually increasing stiffness.

damping force is $c\dot{x}$, proportional to the velocity. It is known also as a *viscous* damping force, because it occurs in a dashpot with a viscous fluid.

Other types of damping which occur frequently are dry-friction or Coulomb damping and air or turbulent-water damping. The first of these is independent of the magnitude of the velocity, but is always opposite in direction to the velocity. The air or turbulent-water damping is approximately proportional to the square of the velocity and also is directed against it. The various forces plotted against the time for a sinusoidal motion are shown in Fig. 8.17.

In practical mechanical problems the *mass* is usually a constant quantity. It is possible, however, to imagine a system when even this coefficient varies with the displacement. In Fig. 8.18, let the piston be very light and the amount of water in the cylinder small in comparison with that in the tank. Evidently the piston with the water column above it

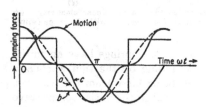

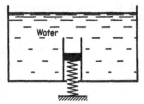

FIG. 8.17. Various damping forces for the case of harmonic motion. (a) Viscous friction $c\dot{x}$; (b) Coulomb friction $\pm F$; (c) turbulent air damping $\pm c\dot{x}^2$.

FIG. 8.18. A non-linear system in which the mass depends on the displacement.

in the cylinder is a vibratory system since the rest of the water in the tank moves very little during the oscillation. But the length of the water column and therefore its mass depend on the displacement x. While for small oscillations of the piston the mass is practically constant, this ceases to be the case for larger motions, so that we have a system with a non-linear coefficient (mass).

This example is of little practical value, and we turn to the electrical field to find important cases where the mass varies with x. Consider the simple L-C-circuit of Fig. 8.19 with or without alternating-current generator. The coil contains a soft iron core, which becomes magnetically saturated for a certain value of the current. This is illustrated in Fig. 8.20 where, for a given frequency, the voltage across the coil is plotted against the current, giving a distinctly non-linear relation for larger values of the current. Since the voltage across the inductance coil is the electrical equivalent of the mechanical inertia force, it is seen that indeed we have before us a case of a mass depending on the displacement.

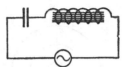

FIG. 8.19. Non-linear electric circuit with a saturated core in the inductance.

8.6. Free Vibrations with Non-linear Characteristics. The most important new fact arising in a discussion of the *free* vibrations of these systems is that with the non-linearity in the *springs* the natural frequency is no longer independent of the amplitude of vibration. With non-linear

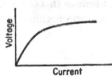

FIG. 8.20. Saturation curve of the inductance of Fig. 8.19.

damping, however (if it is not too great), the frequency depends very little on the amplitude. The reason for this can be readily understood. In a sense the natural frequency is the ratio of the intensity of the spring force to the inertia force for unit frequency. In the linear case these are both proportional to the deflection, and their ratio therefore must be independent of the deflection. If, however, the spring force is not proportional to the amplitude, as with a non-linear system, the natural frequency cannot remain constant.

On account of its 90-deg. phase angle a damping force disturbs the frequency as a second-order effect only (Fig. 2.16, page 40). This is true whether the damping is linear or not. Therefore no appreciable influence of the amplitude on the frequency should be found in the case of non-linear damping.

The general method of investigation consists in plotting a diagram of displacement against velocity, and it is useful to see first how this method works on familiar linear systems. The acceleration $\ddot{x}$ can be written as

$$\ddot{x} = \frac{d\dot{x}}{dt} = \frac{d\dot{x}}{dt} = \frac{d\dot{x}}{dx}\frac{dx}{dt} = \dot{x}\frac{d\dot{x}}{dx} \tag{8.23}$$

This transformation allows us to perform *one* integration of any second-order differential equation of a free system, linear or non-linear. Applying it to the linear undamped free system, we have

$$m\ddot{x} + kx = 0 \qquad\qquad m\dot{x}\frac{d\dot{x}}{dx} + kx = 0$$

$$m\dot{x}\,d\dot{x} + kx\,dx = 0 \qquad m\frac{\dot{x}^2}{2} + k\frac{x^2}{2} = \text{constant}$$

If we plot this result in a diagram with $\dot{x}$ as ordinate and x as abscissa, we find a nest of ellipses with the origin as a common center. It is useful to simplify these to circles by manipulating the coefficients:

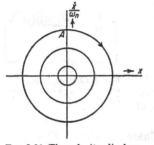

$$\frac{\dot{x}^2}{k/m} + x^2 = \frac{2}{k} \text{ constant} = \text{constant}$$

$$\left(\frac{\dot{x}}{\omega_n}\right)^2 + x^2 = \text{constant}$$

Thus if we plot $\dot{x}/\omega_n$ against x, as in Fig. 8.21, we get a set of concentric circles, representing the result of the first integration. Before working on the second integration, which will lead to $x = f(t)$, we first look at a number of other cases, because this second integration proceeds in the same manner for all systems.

Thus, next we consider the damped free vibration:

$$m\ddot{x} + c\dot{x} + kx = 0$$

Applying the transformation (8.23) to this,

$$m\dot{x}\frac{d\dot{x}}{dx} + c\dot{x} + kx = 0$$

$$\frac{d\dot{x}}{dx} = -\frac{c}{m} - \frac{k}{m}\frac{x}{\dot{x}}$$

or

$$\frac{d(\dot{x}/\omega_n)}{dx} = -\frac{c}{m\omega_n} - \frac{x}{\dot{x}/\omega_n} \tag{8.24}$$

The left-hand part of this is the slope in the $\dot{x}/\omega_n$- versus x-diagram, while the right-hand part can be calculated numerically for each point x, $\dot{x}/\omega_n$ of the diagram. Thus this equation enables us to fill the entire diagram with short directional line stretches at every point (Fig. 8.22), through which the curves must pass. The sketching in of these directional line stretches is facilitated by first drawing in the "isoclinics," i.e., the lines or curves joining all those points in the $\dot{x}/\omega_n$-, x-diagram having the same slope $d(\dot{x}/\omega_n)/dx$. From Eq. (8.24) the equation of such an isoclinic is found by setting the right side of the equation (which is the slope) equal to a constant, which leads us to recognize that the isoclinics in this case are straight lines through the origin. For the horizontal radial line, $\dot{x} = 0$, so that the right-hand side of Eq. (8.24) becomes infinite and the slopes vertical. The slopes are drawn in Fig. 8.22 for the horizontal, vertical, and 45-deg lines. The reader should sketch in for himself the 22.5- and 67.5-deg lines. Then, starting at an arbitrary point, the integral curve is drawn in, following the slopes from point to point.

What happens to Fig. 8.22 when the damping becomes greater and surpasses the critical damping? To investigate this, we ask whether in Fig. 8.22 there are any isoclinics in which the direction of the slope is

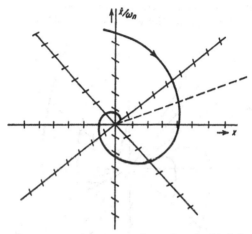

FIG. 8.22. Damped free vibration for $c/m\omega_n = 0.1$ or 5% of critical damping. The isoclinics are straight radial lines, Eq. (8.24), and the integral curve is a converging spiral. For the case of negative damping the same diagram applies with the directional arrow on the spiral reversed, so that the spirals diverge.

along the isoclinic itself. Since the slope of the radial isoclinic itself at point $\dot{x}/\omega_n$, x is $\dot{x}/\omega_n x$, the question is whether there are isoclinics where

$$\frac{\dot{x}/\omega_n}{x} = \frac{d(\dot{x}/\omega_n)}{dx}$$

or with Eq. (8.24)

$$\frac{\dot{x}/\omega_n}{x} = -\frac{c}{m\omega_n} - \frac{x}{\dot{x}/\omega_n}$$

$$\left(\frac{\dot{x}/\omega_n}{x}\right)^2 + \frac{c}{m\omega_n}\left(\frac{\dot{x}/\omega_n}{x}\right) + 1 = 0$$

$$\frac{\dot{x}/\omega_n}{x} = -\frac{c}{2m\omega_n} \pm \sqrt{\left(\frac{c}{2m\omega_n}\right)^2 - 1}$$

It is seen that for subcritical damping the right-hand radical is imaginary, so that there are no real isoclinics with the property that they are directed along the slopes of the diagram. However, for $c > 2m\omega_n$, that is, for supercritical damping, the radical is real, and we find two such isoclinics. This is shown in Fig. 8.23 for the case of $c/2m\omega_n = 1.25$. The integral curves are then sketched in.   The two special isoclinics themselves are integral curves.   No other integral curve can ever cross them, because, upon touching them, an integral curve would be drawn into its own track.

The diagrams of Figs. 8.22 and 8.23 can also be considered to apply to the case of negative damping, because the term $c\dot{x}$ does not change if both c and $\dot{x}$ reverse their signs.   Hence, for negative damping $-c$ we only have to reverse the signs on the vertical axis $\dot{x}$, which amounts to revers-

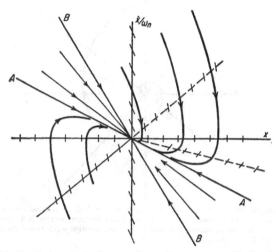

Fig. 8.23. Velocity-displacement diagram for a linear, supercritically damped system $(c/c_c = 1.25)$ There are two isoclinics AA and BB which coincide in direction with the slopes in each of their points.   Again for negative damping the directional arrows along all curves are reversed.

ing the sign of progress on the integral curve, so that we have diverging spirals instead of converging ones.

For a (linear) *negative spring constant,* occurring with the small motions of an upside-down pendulum near its upright unstable equilibrium position, the diagram (Fig. 8.21) does not apply.   By the standard method we have

$$m\ddot{x} - kx = 0$$

$$m\dot{x}\frac{d\dot{x}}{dx} - kx = 0$$

$$\frac{d\dot{x}}{dx} = +\omega_n^2\frac{x}{\dot{x}}$$

$$\frac{d(\dot{x}/\omega_n)}{dx} = +\frac{x}{\dot{x}/\omega_n} \qquad (8.25)$$

The isoclinics, found by setting the right-hand side equal to a constant, are again seen to be radial straight lines. Now we ask again for isoclinics with the slopes along their own direction, or for

$$\frac{d(\dot{x}/\omega_n)}{dx} = \frac{\dot{x}/\omega_n}{x}$$

Combining this with the last equation, we have

$$\frac{x}{\dot{x}/\omega_n} = \frac{\dot{x}/\omega_n}{x} \quad \text{or} \quad \left(\frac{\dot{x}/\omega_n}{x}\right)^2 = 1$$

$$\frac{\dot{x}/\omega_n}{x} = \pm 1, \text{ the two } \pm 45\text{-deg. radii}$$

The diagram is shown in Fig.8.24. Again the two special isoclinics are integral curves themselves; all other integral curves are hyperbolas, which we see from integrating Eq. (8.25):

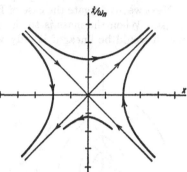

FIG. 8.24. Velocity-displacement diagram for a linear, undamped system with a negative spring constant.

$$\frac{\dot{x}\, d\dot{x}}{\omega_n^2} = x\, dx$$

$$\frac{1}{2}\left(\frac{\dot{x}}{\omega_n}\right)^2 = \frac{1}{2}x^2 + \text{constant}$$

$$\left(\frac{\dot{x}}{\omega_n}\right)^2 - x^2 = \text{constant}; \text{hyperbolas}$$

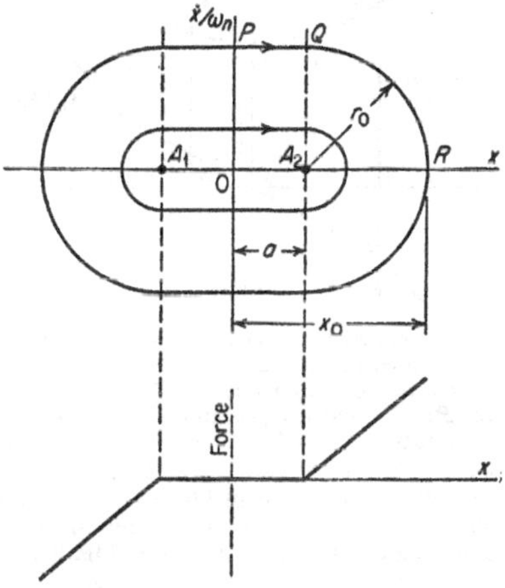

Figures 8.21 to 8.24 illustrate the general method on familiar, linear systems. No new results were obtained, and the usual manner of solution of pages 31 to 49 leads to these results in a simpler manner. The only reason for the new method is that it is capable of being extended to non-linear systems as well.

FIG. 8.25. Diagram for a single mass m between two springs k with clearances a on each side of the center position (Fig. 8.15a). The maximum amplitude $x_0 - a \mid r_0$.

We start with the system with clearances (Fig. 8.15a), and we can draw the $\dot{x}/\omega_n$- versus x-diagram immediately (Fig. 8.25). First we draw in the points A_1 and A_2 at half clearance distance from the origin. To the right of A_2 and to the left of A_1 we have a simple, linear k-m system, so that Fig. 8.21 applies. We draw in the semicircles in the appropriate spaces in Fig. 8.25. In the region between A_1 and A_2 there is clearance; no spring force nor any other force acts on the mass, which, therefore, moves with constant velocity, i.e., along horizontal lines. Combining two half circles with two pieces of straight horizontal line, we obtain closed curves in Fig. 8.25 signifying periodic motions.

Next we investigate the case of Figs. 8.15b and 8.26: set-up springs with stops. When the mass is to the right of the center O, it feels a spring which would be linear if it only would center in the point P, where its

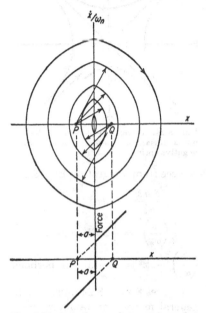

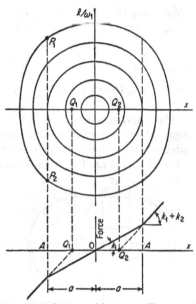

FIG. 8.26. A single mass between two set-up springs with stops. Each spring is originally precompressed through a distance a (Fig. 8.15b).

FIG. 8.27. System with stops. The curves are circles inside and ellipses on Q as centers outside. The ellipses join the circles at P_1, P_2 with a common tangent. An alternate way of representing this case would be with $x/\omega_2 = x/\sqrt{(k_1 + k_2)/m}$ as ordinate; then the inside curves would be ellipses and the outer parts arcs of circles concentric with Q_1 or Q_2. (See Fig. 8.15c.)

extrapolated force would be zero. Hence the velocity-displacement diagram to the right of O consists of circles concentric on point P. Similarly, to the left of O we have circles concentric on point Q.

For the case of Figs. 8.15c and 8.27, where an additional set of springs k_2 comes in at distance a from the center (also called "stops" if k_2 is large), we again consider regions for x smaller than a and larger than a. For $x < a$ the system is strictly linear, and we have concentric circles. They *are* circles, because we plot vertically $\dot{x}/\omega_{n1} = \dot{x}/\sqrt{k_1/m}$, where k_1 is the spring constant actually in force.

For the regions outside $-a < x < a$ we again would have circles, if the ordinate were only $\dot{x}/\sqrt{(k_1 + k_2)/m}$. Since the ordinate differs from this by a constant factor, the curves are elongated circles, or ellipses, with their center at Q, which is the effective center of the spring force for the outside region. The ellipse passing through P_1 and P_2 is such that it joins the inside circle at those two points with a smooth tangent.

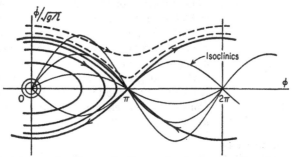

Fig. 8.28. Angular speed versus angle diagram for the large motion of a mathematical gravity pendulum.

As a last example we consider the large 360-deg. motions of a mathematical pendulum, which has two equilibrium positions, one stable, hanging down at $\varphi = 0$, and another unstable, upright at $\varphi = \pi = 180$ deg. The equation is non-linear:

$$\ddot{\varphi} + \frac{g}{l} \sin \varphi$$

$$\dot{\varphi} \frac{d\dot{\varphi}}{d\varphi} + \frac{g}{l} \sin \varphi = 0$$

$$\frac{d\dot{\varphi}}{d\varphi} = -\frac{g}{l} \frac{\sin \varphi}{\dot{\varphi}}$$

Isoclinics for $-\frac{g}{l} \frac{\sin \varphi}{\dot{\varphi}} = $ constant or $\dot{\varphi} = C \sin \varphi$. These are sine waves as shown in Fig. 8.28. If we choose for ordinate $\dot{\varphi}/\omega_n = \dot{\varphi}/\sqrt{g/l}$, we know the details of the large diagram for small regions about $\varphi = 0$, 2π, 4π, . . . and about $\varphi = \pi$, 3π, ... For φ near zero we have the small vibrations of a linear pendulum as in Fig. 8.21. For φ near $\pi = 180$ deg. we have the motions of a linear upside-down pendulum with a negative spring constant (Fig. 8.24). Thus the integral curves near $\varphi = 0$ are circles, and those near $\varphi = \pi$ are 45-deg. hyperbolas. Starting the various sine-wave isoclinics at $\varphi = 0$, we immediately see the slope directions associated with each of them, from the small circles about $\varphi = 0$. Then the sine waves with 45-deg. initial slopes are important at $\varphi = \pi$, because we see that there the slope direction coincides with the slope of the curve itself. Tracing out the integral curves starting at 45 deg. at $\varphi = \pi$, we divide the field into regions. The curves inside this limiting one are closed curves around $\varphi = 0$ or $\varphi = 360$, etc., and begin to resemble circles when close enough to $\varphi = 0$; the curves outside the limiting one are not closed but continuous wavy lines, resembling hyperbolas when close enough to $\varphi = \pi$. The direction of progress in all curves is to the right for positive $\dot{\varphi}$ and to the left for negative $\dot{\varphi}$.

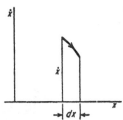

With the general method of isoclinics in the $\dot{x}$- versus x-diagram we can always construct such a diagram for any case of non-linear free vibration with one degree of freedom. It now remains to carry out the second integration on this diagram. Considering an element of curve over the base dx, as in Fig. 8.29, we have

FIG. 8.29. The time consumed in progressing through distance dx is found from $dt = dx/\dot{x}$. This leads to results (8.26) and (8.27).

$$\dot{x} = \frac{dx}{dt} \quad \text{or} \quad dt = \frac{dx}{\dot{x}}$$

In the right hand of this expression dx is the element of distance horizontally, and $\dot{x}$ is the vertical ordinate. Thus we can calculate the time consumed in progressing a distance dx, and in general

$$t = \int \frac{dx}{\dot{x}} \tag{8.26}$$

For a periodic (undamped) motion the $\dot{x}$, x curves are closed, so that for a full period of oscillation, where we run through the closed curve once, we have

$$T = \oint \frac{dx}{\dot{x}} \tag{8.26a}$$

This integral can always be evaluated, although for pronounced non-linear cases it is usually necessary to do it numerically, step by step. For a symmetrical system the four quarter periods all take up the same time, so that the integral simplifies to

$$T = 4 \int_{x=0}^{x=x_{\max}} \frac{dx}{\dot{x}} \quad \text{(symmetrical spring)} \tag{8.27}$$

As an example consider first the linear, undamped case (Fig. 8.21). There

$$(\omega_n \dot{x})^2 + x^2 = \text{constant} = x_{\max}^2$$

or

$$\dot{x} = \frac{1}{\omega_n} \sqrt{x_{\max}^2 - x^2}$$

Substituting this into Eq. (8.27),

$$T = \frac{4}{\omega_n} \int_{x=0}^{x=x_m} \frac{dx}{\sqrt{x_m^2 - x^2}} = \frac{4}{\omega_n} \int_0^{x_m} \frac{d(x/x_m)}{\sqrt{1 - (x/x_m)^2}}$$

$$= \frac{4}{\omega_n} \int_0^1 \frac{dy}{\sqrt{1 - y^2}} = (\text{see handbook}) = \frac{4}{\omega_n} \sin^{-1} y \Big|_0^1$$

$$= \frac{4}{\omega_n} \frac{\pi}{2} = \frac{2\pi}{\omega_n}, \text{ a known result}$$

As a second example take the system with clearances (Fig. 8.25). The integral (8.27) falls into two parts, first the straight line from P to Q, and second the quarter circle from Q to R. Between P and Q the speed is constant $\dot{x} = \omega_n r_0$, where r_0 is the radius of the circle considered. Hence the time required by Eq. (8.26) to go from P to Q is

$$T_{PQ} = \int \frac{dx}{\dot{x}} = \frac{a}{\omega_n r_0} = \frac{a}{x_0 - a} \sqrt{\frac{m}{k}}$$

The time required between Q and R is that or a quarter cycle of a free linear vibration, but to be formal we proceed with Eq. (8.26) anyhow:

$$\left(\frac{\dot{x}}{\omega_n}\right)^2 + (x - a)^2 = \text{constant} = (x_0 - a)^2 = r_0^2$$

$$\dot{x} = \omega_n \sqrt{r_0^2 - (x - a)^2}$$

$$T_{QR} = \frac{1}{\omega_n} \int_a^{x_0} \frac{dx}{\sqrt{r_0^2 - (x - a)^2}} = \frac{1}{\omega_n} \int_a^{x_0} \frac{d(x - a)}{\sqrt{r_0^2 - (x - a)^2}}$$

$$= \frac{1}{\omega_n} \int_0^{r_0} \frac{dy}{\sqrt{r_0^2 - y^2}} = \frac{1}{\omega_n} \int_0^{r_0} \frac{d(y/r_0)}{\sqrt{1 - (y/r_0)^2}}$$

$$= \frac{1}{\omega_n} \int_0^1 \frac{dz}{\sqrt{1 - z^2}} = \frac{1}{\omega_n} \sin^{-1} z \Big|_0^1 = \frac{\pi}{2\omega_n} = \frac{\pi}{2} \sqrt{\frac{m}{k}}$$

The total time from P to R (or the quarter period of the motion) is the sum of these two:

$$\frac{T}{4} = \sqrt{\frac{m}{k}} \frac{\pi}{2} \left(1 + \frac{2}{\pi} \frac{a}{x_0 - a}\right)$$

and the natural frequency ω of the motion with clearances is

$$\omega = \frac{2\pi}{T} = \sqrt{\frac{k}{m}} \left(1 + \frac{2}{\pi} \frac{a}{x_0 - a}\right) \qquad (8.28)$$

This frequency depends on the amplitude x_0; it reduces to the usual $\sqrt{k/m}$ for the case of zero clearance a (see Fig. 8.30).

If the non-linearity is located in the *damping* of the system, the natural frequency is not affected by the amplitude and remains approximately $\sqrt{k/m}$. The only question of interest here is the rate of dying down of the amplitude. An exact solution to this problem can be found by a step-by-step (graphical or numerical) integration of the equation of motion, but this is too laborious. (Only for the simple case of Coulomb dry friction does a simple exact solution exist.)

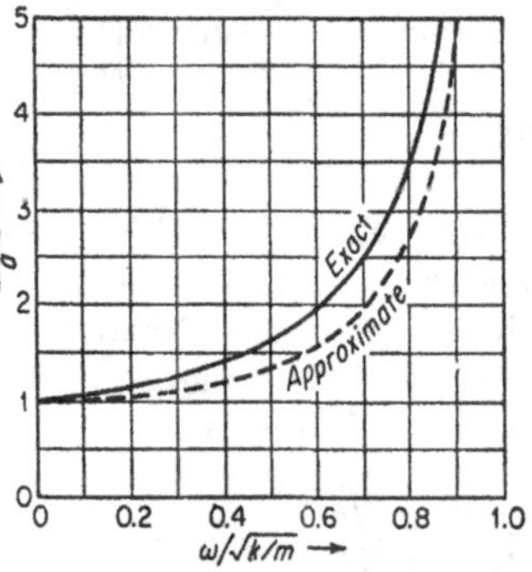

A sufficiently accurate approximation for practical purposes is obtained by calculating the energy spent by the friction force during a cycle and equating this energy to the loss in kinetic energy of the motion. In order to be able to calculate these energy losses, we have to know the shape of the motion, which obviously is not sinusoidal but yet resembles a sinusoid for small values of the damping. The smaller the damping the better is this resemblance, because with a harmonic motion the *large* spring and inertia forces are harmonic and only the *small* damping force causes a deviation from this harmonic motion.

FIG. 8.30. Natural frequency as a function of the maximum amplitude of vibration for a system with clearances (Fig. 8.15a). The exact curve is from (8.28) and the approximate curve is obtained by the method illustrated in Fig. 8.39.

Thus we assume harmonic motion $x = x_0 \sin \omega t$. If the damping force is represented by $f(\dot{x})$, its work per cycle is

$$W = \int f(\dot{x})\, dx = \int_0^T f(\dot{x})\dot{x}\, dt = x_0 \int_0^{2\pi} f(\dot{x}) \cos \omega t\, d(\omega t)$$

The loss in kinetic energy per cycle is

$$\tfrac{1}{2}m\omega^2 x_0^2 - \tfrac{1}{2}m\omega^2(x_0 - \Delta x)^2 = m\omega^2 x_0\, \Delta x - \tfrac{1}{2}m\omega^2(\Delta x)^2 \approx m\omega^2 x_0\, \Delta x$$

Equating the two expressions we find for the decrease in amplitude per cycle

$$\Delta x = \frac{1}{m\omega^2} \int_0^{2\pi} f(\dot{x}) \cos \omega t\, d(\omega t) \tag{8.29}$$

This integral can always be evaluated, even though it may sometimes be necessary to do it graphically.

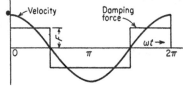

FIG. 8.31. Coulomb damping force.

As an example, consider Coulomb damping, where $f(\dot{x}) = \pm F$. The velocity and the damping force are shown in Fig. 8.31. The integral in (8.29) is seen to consist of four equal parts,

$$4 \int_0^{\frac{\pi}{2}} F \cos \omega t\, d(\omega t) = 4F$$

and the decrement in amplitude per cycle is

$$\Delta x = \frac{4F}{m\omega^2} = 4\frac{F}{k} \cdot \frac{k}{m} \cdot \frac{1}{\omega^2} = 4\frac{F}{k} \tag{8.30}$$

or four times the static deflection of the friction force on the spring. The result is significant in that the amplitude decreases in equal decrements as an arithmetic series, whereas in the case of viscous damping the amplitude decreases in equal percentage ratios as a geometric series (page 40). Incidentally it is of interest to know that (8.30) happens to coincide with the exact solution before mentioned.

8.7. Relaxation Oscillations. A linear vibratory system with negative damping builds up oscillations of infinite amplitude (Fig. 7.2b). Of course, this is physically impossible and in all actual systems the damping becomes positive again for sufficiently large amplitudes, thus limiting the motion. An example of this is the electric transmission line discussed on page 299.

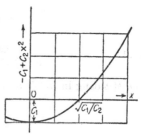

The actual relation between the damping coefficient and the amplitude varies from case to case, but for a general understanding it is useful to write down the simplest possible mathematical expression that will make the damping force negative for small amplitudes x and positive again for larger ones. Such an expression is

Fig. 8.32. Simplest mathematical expression for a non-linear damping coefficient which is negative for small amplitudes and becomes positive for greater amplitudes.

$$\text{Damping force} = -(C_1 - C_2 x^2)\dot{x} \tag{8.31}$$

The damping coefficient as a function of x is shown in Fig. 8.32. It is seen that zero damping occurs at an amplitude $x = \sqrt{C_1/C_2}$.

The differential equation of a single-degree-of-freedom system with this type of non-linear damping is

$$m\ddot{x} - (C_1 - C_2 x^2)\dot{x} + kx = 0 \tag{8.32}$$

Since we shall now give a general discussion of this equation, it is of importance to simplify it as much as possible first, by reducing the number of system characteristics of which there are now four, viz., m, C_1, C_2, and k. To this end we divide by m, and with the notation $k/m = \omega_n^2$ we obtain

$$\ddot{x} - \left(\frac{C_1}{m} - \frac{C_2}{m}x^2\right)\dot{x} + \omega_n^2 x = 0 \tag{8.33}$$

Of the three remaining characteristics, C_1/m, C_2/m, ω_n^2, two can be absorbed by making the variables x and t dimensionless. First consider the time

t, which is measured in seconds. Instead of using this standard unit, we shall now measure time in terms of a unit inherent in the system, for example, $T/2\pi$. This means that for a slowly vibrating system the new time unit is large, while for a rapidly vibrating system it is small. The time is measured in "periods/2π" rather than in "seconds." Let the new time (measured in units of $T/2\pi$) be denoted by t' and the old time (measured in seconds) by t.

Then

$$t' = \frac{t}{T/2\pi} = \omega_n t$$

The new differential coefficients become

$$\frac{d^2x}{dt^2} = \frac{d^2x}{dt'^2} \cdot \frac{t'^2}{t^2} = \omega_n^2 \cdot \frac{d^2x}{dt'^2}$$

and

$$\frac{dx}{dt} = \omega_n \frac{dx}{dt'}$$

Substituting in (8.33) and dividing the equation by ω_n^2,

$$\ddot{x} - \frac{C_1}{m\omega_n}\left(1 - \frac{C_2}{C_1}x^2\right)\dot{x} + x = 0$$

where the dots now signify differentiation with respect to the dimensionless time t'.

There are now only two parameters, $C_1/m\omega_n$ and C_1/C_2. The amplitude x still has the dimension of a length, and in order to make it dimensionless we measure it also in a unit inherent in the equation. A con venient unit is indicated in Fig. 8.32, viz., the amplitude $\sqrt{C_1/C_2}$, for which the positive and negative damping forces balance. Thus we take for our new "dimensionless displacement"

$$y = \frac{x}{\sqrt{C_1/C_2}}$$

which gives the differential equation in the form

$$\ddot{y} - \epsilon(1 - y^2)\dot{y} + y = 0 \tag{8.34}$$

The equation is finally reduced to a single parameter $\epsilon = C_1/m\omega_n$, which has an important physical significance. For harmonic motion this quantity equals the ratio between the maximum negative damping force and the maximum spring force:

$$\frac{C_1}{m\omega_n} = \epsilon = \frac{\text{input force}}{\text{spring force}} \tag{8.35}$$

This can be shown as follows. Let $x = x_0 \sin \omega_n t$, and $\dot{x} = x_0\omega_n \cos \omega_n t$.

From (8.31) the maximum negative damping force in the middle of a stroke ($x = 0$) is $C_1\dot{x}_{max} = C_1x_0\omega_n$. The maximum spring force $kx_{max} = kx_0 = \omega_n^2 mx_0$, so that (8.35) is verified.

In all cases thus far discussed, the input force was much smaller than the spring force, so that ϵ was a small quantity, $\epsilon \ll 1$. This implies a motion practically harmonic and of the natural frequency ω_n. The final amplitude to which the motion will build up can be found from an energy consideration. For amplitudes smaller than this final one, the damping force $F = \epsilon(1 - y^2)\dot{y}$ puts energy into the system; while for amplitudes greater than the final one, the damping dissipates energy. At the final amplitude we have for a full cycle:

$$0 = \int F \, dy = \int_0^{2\pi} F\dot{y} \, dt' = \int_0^{2\pi} \epsilon(1 - y^2)\dot{y}^2 \, dt'$$

The motion is harmonic:

$$y = y_0 \sin \omega_n t = y_0 \sin t'$$

Hence

$$0 = \epsilon \int_0^{2\pi} (1 - y_0^2 \sin^2 t')y_0^2 \cos^2 t' \, dt'$$

or

$$y_0^2 = \frac{\int^{2\pi} \cos^2 t' \, dt'}{\int_0^{2\pi} \sin^2 t' \cos^2 t' \, dt'} = \frac{\pi}{\pi/4} = 4$$

(The evaluation of these integrals is discussed on page 15.) Thus for small values of the parameter ϵ the amplitude of vibration x is

$$x_0 = 2\sqrt{\frac{C_1}{C_2}} \qquad (8.36)$$

or, in words, the amplitude is twice as large as the amplitude at which the damping force just becomes zero. Figure 8.32 shows that this is a reasonable result: the energy put in by the negative damping force in the center part of the motion is neutralized by the energy dissipated by the positive force near the end of the stroke.

So far the introduction of the differential equation (8.32) or (8.34) has not brought us anything new. The importance of these equations is centered rather in the case where the input force or negative damping force is *great* in comparison with the elastic force:

$$\epsilon \gg 1$$

Then the non-linear middle term in (8.34) becomes more important than the other two, so that the assumption of a harmonic motion (which was justified for a small middle term) is untenable. Thus we should expect the motion to be very much distorted, containing a great number of higher harmonics, and we also expect the frequency to differ from ω_n.

Applying the general method of pages 353 to 360 to Eq. (8.34), it becomes

$$\frac{d\dot{y}}{dy} = \epsilon(1 - y^2) - \frac{y}{\dot{y}} \qquad (8.37)$$

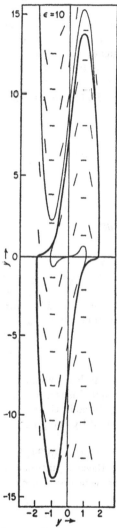

which enables us to fill the $\dot{y}$, y diagram with small directional line stretches at various points. Having this whole field of tangents, a solution can be found by starting from any aribitrary point (*i.e.*, with any arbitrary initial displacement y and velocity $\dot{y}$) and constructing a curve following the tangents. Figure 8.33 ($\epsilon = 10$) shows, for example, that starting at $\dot{y} = 15$ and $y = -2$ the curve goes down, bends up, then goes down again, and thereafter describes a closed figure continuously. Also when starting from rest (*i.e.*, from the origin), it reaches the same closed curve after a short run. As an ordinary steady-state harmonic vibration would be pictured as an ellipse in this diagram, so it is seen that for $\epsilon = 10$ the motion is far from harmonic.

Next transform Fig. 8.33 into the corresponding diagram in terms of $y = f(t')$, as shown in Fig. 8.34. The abscissa of a point in Fig. 8.33 corresponds to the ordinate of Fig. 8.34, whereas the ordinate of that point in Fig. 8.33 is the slope of Fig. 8.34. Thus the construction of Fig. 8.34 from Fig. 8.33 amounts to a second graphical integration.

Our expectations regarding the nature of the motion are fully corroborated by the final result, Fig. 8.34. The motion is seen to be distinctly non-harmonic. The period is not 2π units of time (the unit being $T/2\pi$) but rather 2ϵ units or $2\epsilon \cdot T/2\pi$ sec. This, by virtue of (8.35), is

$$\text{Period} = \frac{2\epsilon}{\omega_n} = \frac{2C_1}{m\omega_n^2} = 2\frac{C_1}{k} \text{ sec.} \qquad (8.38)$$

Fig. 8.33. First integration of Eq. (8.34) for relaxation oscillations in the case that $\epsilon = 10$.

i.e., the period no longer depends on the ratio of mass to spring constant but rather on the ratio of negative damping coefficient to spring constant. The expression (8.38) is twice the *relaxation time* (see page 41) of a system with a positive coefficient C_1. For this reason oscillations of the nature of Fig. 8.34 have been called *relaxation oscillations*.

The result (8.38), as well as the general shape of the vibration, Fig. 8.34, can be made to seem reasonable by a physical analysis as follows.

For $\epsilon = 10$ the damping action is large in comparison to the spring action. Follow the motion in Fig. 8.34 starting from the point A where the amplitude is $x = 2\sqrt{C_1/C_2}$. On account of Fig. 8.32 the damping coefficient at A is positive and remains positive until the amplitude has diminished to one-half its value (point B). Between A and B the velocity will be very small because the weak spring force is opposed by a damping force of which the coefficient is large. Hardly any inertia effect will come in during that time. At the point B the damping reverses, and becomes negative and large, which hurls the mass at a high speed through the point C, where the damping force again reverses. Between B and C the negative damping force has done work on the mass and thus has given it considerable momentum. This momentum is destroyed by the positive damping force from C on, until the mass comes to rest again at D. That the point D should be approximately at $x = 2\sqrt{C_1/C_2}$ seems reasonable from the result (8.36) for the case of harmonic motion.

Since it takes hardly any time to move from B to D we might calculate the period by taking twice the time between A and B. The answer thus found would be slightly too small.

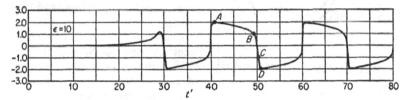

Fig. 8.34. A typical relaxation oscillation, being an integration of Fig. 8.33.

In simplifying the calculation we see in Fig. 8.32 that the damping coefficient between $x = \sqrt{C_1/C_2}$ and $x = 2\sqrt{C_1/C_2}$ can be expressed very well by a straight-line relation.

$$\text{Damping coefficient} = -3C_1 + \frac{3C_1}{\sqrt{C_1/C_2}} \cdot x$$

The damping force is

$$\left(-3C_1 + \frac{3C_1 x}{\sqrt{C_1/C_2}} \right) \dot{x}$$

and this force is opposed only by the spring force $-kx$. Thus the differential equation of the creeping or relaxation motion between A and B is

$$\left(-3C_1 + \frac{3C_1}{\sqrt{C_1/C_2}} x \right) \frac{dx}{dt} = -kx$$

or

$$\frac{3C_1}{k} \left(-\frac{1}{x} + \frac{1}{\sqrt{C_1/C_2}} \right) dx = -dt$$

In integrating this expression we notice that the time progresses from 0 to $T/2$ (half period), while x goes from $2\sqrt{C_1/C_2}$ to $\sqrt{C_1/C_2}$, so that

$$\frac{3C_1}{k} \left(-\log_e x + \frac{x}{\sqrt{C_1/C_2}} \right) \Bigg|_{2\sqrt{C_1/C_2}}^{\sqrt{C_1/C_2}} = -t \Bigg|_{0}^{T/2}$$

After substitution of the limits we find

$$\frac{3C_1}{k}(-\log_e 2 + 1) = \frac{T}{2}$$

or

$$T = 6(1 - \log_e 2)\frac{C_1}{k} = 1.84\frac{C_1}{k}$$

With the slight additional time taken in going from B to D it is seen that (8.38) is verified.

The corresponding results of the graphical integration for the more

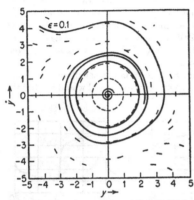

FIG. 8.35. First integration of Eq. (8.34) for a small damping force, $\epsilon = 0.1$.

usual case $\epsilon = 0.1$ are shown in Figs. 8.35 and 8.36.

Relaxation oscillations have been found to occur very often in radio engineering, and the reader is referred to the original papers of Van der Pol for quite a number of applications in that field. In mechanical engineering thus far they have been of little importance.

A case on the border between the electrical and mechanical fields is that of the periodic speed reversals of a separately excited direct-current motor fed by a direct-current series generator driven at a constant speed (Fig. 8.37). The voltage generated in a constant-speed generator is proportional to its magnetic field strength. If there were no magnetic saturation, this field strength would be proportional to the field current i, which in a series machine is the same as the main current. The influence

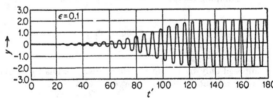

FIG. 8.36. Second integration of Eq. (8.34) for $\epsilon = 0.1$, showing the building up of a non-linear, self-excited vibration to its final amplitude.

of saturation amounts to a less rapid increase of the field strength, and the characteristic of the generator (Fig. 8.38) may be expressed approximately by the relation

$$E_{\text{gen}} = C_1 i - C_2 i^3$$

This generated voltage overcomes first the inductance of its own field coils $\left(L\dfrac{di}{dt}\right)$, second the resistance of the circuit (Ri), and third the countervoltage of the motor. The motor has a *constant* magnetic field and a variable angular speed ω. Its voltage is proportional to the speed, $C_3\omega$. No effect of saturation enters since the field is maintained constant. The voltage equilibrium equation is

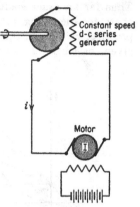

Constant speed
d-c series
generator

$$C_1 i - C_2 i^3 = C_3\omega + L\frac{di}{dt} + Ri \quad (8.39)$$

Another relation between i and ω is obtained from the fact that the energy input per second in the motor is given by its voltage $C_3\omega$ multiplied by its current i. Since the motor drives no load, this energy is used in accelerating its rotating parts of which the moment of inertia is I. The kinetic energy of the motor is $\frac{1}{2}I\omega^2$ and

Motor

$$C_3\omega i = \frac{d}{dt}(\frac{1}{2}I\omega^2) = I\omega\frac{d\omega}{dt}$$

or

$$\frac{d\omega}{dt} = \frac{C_3 i}{I} \quad (8.40)$$

Fig. 8.37. A separately excited motor driven by a series generator has periodic speed reversals of the character shown in Fig. 8.34.

The angular speed ω can be eliminated from (8.39) by differentiating and substituting (8.40), giving

$$C_1\frac{di}{dt} - 3C_2 i^2\frac{di}{dt} - C_3\frac{d\omega}{dt} + L\frac{d^2 i}{dt^2} + R\frac{di}{dt}$$

$$L\frac{d^2 i}{dt^2} - (C_1 - R - 3C_2 i^2)\frac{di}{dt} + \frac{C_3^2}{I}i = 0$$

This equation is equivalent to (8.32). Moreover, the values of $C_1 - R$, C_3, and I in the usual motor are such that

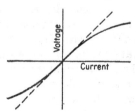

Fig. 8.38. Voltage-current characteristic of a constant-speed series generator.

$\epsilon = \dfrac{C_1 - R}{C_3}\sqrt{\dfrac{I}{L}}$ is much larger than unity. Thus the current i will reverse periodically according to Fig. 8.34, and the velocity of rotation ω will also show periodic reversals on account of Eq. (8.40). By Eq. (8.38) the period of these reversals is

$$T = 2\frac{C_1 - R}{C_3^2}I$$

that is, proportional to the inertia of the motor. If the oscillation were *harmonic* its period would be proportional to the *square root* of the inertia.

8.8. Forced Vibrations with Non-linear Springs. The problem is that of an undamped system with a curved spring characteristic under the influence of a harmonic disturbing force or

$$m\ddot{x} + f(x) = P_0 \cos \omega t \qquad (8.41)$$

Thus far an exact solution to this problem exists only for the simple characteristic of Fig. 8.15b and is so complicated as to be without much practical value. In the following pages an approximate solution will be given, based on the assumption that the motion $x = f(t)$ is sinusoidal

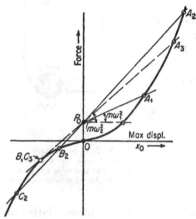

and has the "forced" frequency. This is obviously not true, and the degree of approximation can be estimated only by the seriousness of the deviation from this assumption. Assume then

$$x = x_0 \cos \omega t \qquad (8.42)$$

The inertia force $m\ddot{x}$ is $-mx_0\omega^2 \cos \omega t$, and this force attains its maximum value $-m\omega^2 x_0$ at the same instant that the external force reaches its maximum value P_0 and the spring force its maximum $f(x_0)$. Equation (8.41) is a condition of equilibrium among three forces at

FIG. 8.39. Approximate construction of the amplitudes of forced non-linear vibrations.

any time during the (non-harmonic) motion. Let us satisfy that condition for the harmonic motion (by a proper choice of x_0) at the instant that $x = x_0$. Thus

$$-m\omega^2 x_0 + f(x_0) = P_0$$

or

$$f(x_0) = P_0 + m\omega^2 x_0 \qquad (8.43)$$

At the time when $x = 0$ (in the middle of the stroke), all three forces are zero so that the equilibrium condition is again satisfied. In case $f(x)$ were equal to kx, all three terms of (8.41) would be proportional to $\sin \omega t$, so that (8.41) divided by $\sin \omega t$ would give (8.43) with $f(x_0) = kx_0$ and the equilibrium condition would be satisfied at *all* values of x between 0 and x_0. However, when $f(x) \neq kx$, this is no longer true, and the equilibrium is violated at most points between 0 and x_0. To satisfy the equilibrium at the two points $x = 0$ and $x = x_0$ is the best we can do under the circumstances. Thus the amplitude of the forced vibration will be found approximately from the algebraic equation (8.43).

The most convenient and instructive manner in which this can be done is graphical. Plot the forces vertically and the amplitude x_0 horizontally as in Fig. 8.39. The left side of (8.43) is the (curved) spring characteristic, while the right side of the equation expresses a straight line with the ordinate intercept P_0 and the slope $\tan^{-1}(m\omega^2)$. Where the curve and the straight line intersect, the left-hand force of (8.43) equals the right-hand force, so that equilibrium exists (at the end of a stroke). This determines x_0 as the abscissa of the point of intersection. For slow frequencies (small slopes $m\omega^2$), there is only one such point of intersection A_1, but for greater frequencies and the same force P_0 there are three intersections A_2, B_2, and C_2. In other words, there are then three possible solutions. To see this more clearly, we plot in Fig. 8.40 the ampli-

tude x_0 against the frequency ω for a given constant value of the force P_0, which gives a resonance diagram corresponding to Fig. 2.18, page 44, for the linear case. (Incidentally, Fig. 2.18 can be constructed point by point in an exact manner from Fig. 8.39 with a straight-line characteristic.) It is left for the reader to develop Fig. 8.40 from Fig. 8.39 and in particular to see that for frequencies below BC_3 only one solution exists, and for frequencies

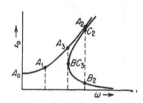

FIG. 8.40. Resonance diagram for a system with a gradually stiffening spring.

above BC_3 three solutions exist; also that the A-branch of the diagram represents motions *in phase* with the force $P_0 \sin \omega t$, while the BC-branch is 180 deg. out of phase with this force. This peculiarity is the same as in Fig. 2.18.

Of the three possible motions A, B, or C, it has been found that C is *unstable*, whereas A and B represent *stable* motions which can be realized by experiment. In order to make this statement seem reasonable, it is

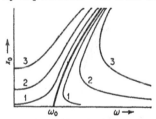

FIG. 8.41. Explains the instability of the C-branch of Fig. 8.40.

necessary to complete the diagram of Fig. 8.40 with curves for other values of P_0, and this is done in Fig. 8.41. The central thick curve is the one for $P_0 = 0$, or, in other words, for the free vibration. It is found by drawing lines with slopes $m\omega^2$ from the origin O (Fig. 8.39) and determining their intersections with the characteristic. For frequencies ω below a certain value ω_0 the slope in Fig. 8.39 is too small

to give any intersection at all. For increasing slope the x_0 becomes greater and greater. For a very small exciting force P_0 we obtain curve 1 of Fig. 8.41, while for greater values of P_0 the curves 2 and 3 result.

Consider a point on the A-branch of one of the curves of Fig. 8.41. If for a given frequency the force P_0 is increased, the amplitude x_0 also increases (we move along a vertical line in Fig. 8.41). The same is true for any point on the B-branch of the curves. But on the C-branch an increase in the force P_0 means a downward motion in Fig. 8.41 (from curve 1 toward curve 2) and this means that an *increase* in the force results in a *decrease* in the amplitude. This cannot happen, however, and what actually takes place is shown in Fig. 8.42, representing the same curve as Fig. 8.40 with the influence of damping taken into account. This damping rounds off the resonance peak in the same manner as with

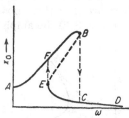

a linear system. If the force amplitude P_0 is kept constant and the frequency ω is gradually increased, the amplitude x_0 suddenly drops from B to C and continues to D. With diminishing ω we pass D, C, and E, where the amplitude suddenly jumps up to F, then continues on to A. The unstable branch BE represents motions that cannot occur.

Fig. 8.42. Discontinuous jumps in the amplitudes of a non-linear system with a gradually stiffening spring.

The characteristic of Fig. 8.39 represents a spring which becomes gradually stiffer with increasing amplitudes. This leads to a natural frequency which increases with the amplitude, as shown by the thick curve bending off to the right in Fig. 8.41. For a spring of *diminishing* stiffness (as, for example, Fig. 8.15b) the natural-frequency curve bends to the left and the unstable C-branch of the curves lies to the left of the central curve. In Fig. 8.43 the upward jump in amplitude happens with *increasing* frequency.

An interesting method of solving this problem accurately by successive approximations for any spring characteristic is due to *Rauscher*. Instead of starting with a given frequency and then solving for the amplitude x_0, as was done in Fig. 8.39, Rauscher starts with an amplitude ratio x_0/P_0 and then solves for the frequency. In Eq. (8.41) the frequency ω is regarded as not fixed, and a first guess at the motion is Eq. (8.42), in which x_0 is given a definite value, while ω is the frequency of the force, as yet floating. Then we may write $P_0 \cos \omega t = P_0 x/x_0$, which transforms the exciting force from

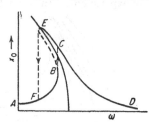

Fig. 8.43. Resonance diagram for a spring in which the stiffness *decreases* with the amplitude.

a time function to a displacement function. The exciting force is now brought to the left-hand side of Eq. (8.41) and combined with the spring force $f(x)$. The problem reduces to one of free vibration, which can be solved by means of the method of the velocity-displacement diagram and a subsequent second integration. The displacement-time curve so obtained will not be the same as the first guess (8.42) for it, but it *will* have the same maximum amplitude x_0. With this new displacement function we enter once more into the differential equation (8.41), transform the exciting force from a t-function to an x-function, and throw it to the left so as to combine it with the spring force. In this manner the third solution for $x = f(t)$ is obtained. These successive solutions for the motion converge very rapidly to the exact one. Being a method of "iteration" it is very closely related to *Stodola's* procedure, discussed on pages 155 to 165. Again, as in that process, if the first guess for the motion happens to be the correct one, the second result will be identical with the first. This can best be shown by applying Rauscher's procedure to the linear case,

$$m\ddot{x} + kx = P_0 \cos \omega t$$

The first guess is $x = x_0 \cos \omega t$, which, if x_0 has a definite value, is the exact motion for some frequency ω. Then

$$P_0 \cos \omega t = \frac{P_0 x}{x_0} \quad \text{and} \quad m\ddot{x} + \left(k - \frac{P_0}{x_0}\right) x = 0$$

This is a free vibration of a linear system like Eq. (2.7) with the solution (8.42) in which the frequency ω is determined by $\omega^2 = \dfrac{k - P_0/x_0}{m}$. This is seen to be the *exact* result (2.23) of page 42. The unusual feature of Rauscher's procedure is that, instead of finding the intersection of the curves of Fig. 8.41 with a vertical line (*i.e.*, solving for x_0 with ω given), the intersection with a horizontal line is determined (*i.e.*, ω is solved for a given x_0) which, of course, is just as good.

The analysis of the electric circuit of Fig. 8.19 follows exactly the same lines except that the inertia force (inductance voltage) has a curved characteristic, whereas the spring force (condenser voltage) follows a straight line.

In the analysis it was assumed that the motion has the same frequency as the force, which would be the case in a linear system. Though this is the only possible motion for *slightly* non-linear systems, it will be seen later (page 377) that for very pronounced non-linearity motions of a frequency 1, 2, 3, 4 . . . times as *slow* as the disturbing frequency ω may be excited.

8.9. Forced Vibrations with Non-linear Damping. The differential equation of this case is

$$m\ddot{x} + f(\dot{x}) + kx = P_0 \sin \omega t \tag{8.44}$$

where $f(\dot{x})$ is not equal to $c\dot{x}$. On account of the presence of the non-linear damping term $f(\dot{x})$, the motion is not harmonic. An exact solution of (8.44) is known only for the case of Coulomb damping, $f(\dot{x}) = \pm F + c\dot{x}$ and even then in a limited region of frequencies only.

In many practical cases the damping is reasonably small, and the curve of motion is sufficiently close to a sinusoid to base an approximate analysis on it. The most general method replaces the term $f(\dot{x})$ by an "equivalent" $c\dot{x}$ and then proceeds to determine the "equivalent damping constant" c in such a manner that with sinusoidal motion the actual damping force $f(\dot{x})$ does the same work per cycle as is done by the equivalent damping force $c\dot{x}$. The value for c thus obtained will not be a true constant but a function of ω and of the amplitude x_0. Therefore, approximately, the system (8.44) can be replaced by a linear one, but the damping constant c has a different numerical value for each value of ω or of x_0.

In carrying out this analysis we first assume for the motion,

$$x = x_0 \sin \omega t$$

The work done per cycle by the equivalent damping force $c\dot{x}$ is $\pi c \omega x_0^2$ as calculated on page 52. For the work per cycle of the general damping force $f(\dot{x})$ we found on page 362:

$$x_0 \int_0^{2\omega} f(\dot{x}) \cos \omega t \, d\omega t$$

Equating the two values we obtain for the equivalent damping constant c:

$$c = \frac{1}{\pi \omega x_0} \int_0^{2\pi} f(\dot{x}) \cos \omega t \, d\omega t \qquad (8.45)$$

The amplitude of the "linearized" Eq. (8.44), as given on page 49, is

$$x_0 = \frac{P_0}{k} \frac{1}{\sqrt{\left(1 - \dfrac{\omega^2}{\omega_n^2}\right)^2 + \left(\dfrac{c\omega}{k}\right)^2}} \qquad (2.28a)$$

In order to calculate the amplitude, the value (8.45) for c has to be substituted in (2.28a).

This general procedure may be applied to any type of damping, even if its law is given merely in curve form, where the integral (8.45) must be evaluated graphically. As an example we shall take the case of dry friction $f(\dot{x}) = \pm F$. On page 362, the value of the integral in (8.45) was found to be $4F$. Hence

$$c \approx \frac{4F}{\pi \omega x_0}$$

indeed an equivalent damping constant depending on both frequency and amplitude. Substituting in (2.28a):

$$x_0 \sqrt{\left(1 - \frac{\omega^2}{\omega_n^2}\right)^2 + \left(\frac{4F}{\pi k x_0}\right)^2} = \frac{P_0}{k}$$

or

$$\left(1 - \frac{\omega^2}{\omega_n^2}\right)^2 x_0^2 = \left(\frac{P_0}{k}\right)^2 - \left(\frac{4F}{\pi k}\right)^2$$

Hence

$$x_0 = \frac{P_0}{k} \cdot \frac{\sqrt{1 - \left(\frac{4}{\pi} \frac{F}{P_0}\right)^2}}{1 - \frac{\omega^2}{\omega_n^2}} \tag{8.46}$$

An *exact* solution of this case also exists. The analysis is too elaborate to be given here in detail, but the results, which do not differ much from

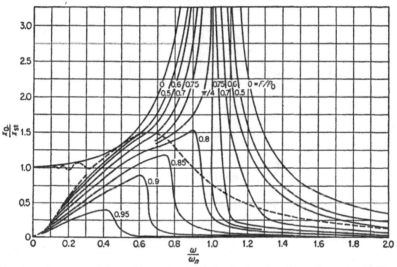

Fig. 8.44. Resonance diagram for a system with dry friction damping. Compare with Fig. 2.22a on page 51.

those of (8.46), are shown in Figs. 8.44 and 8.45. The reader should compare these with Fig. 2.22, page 51.

With Coulomb friction (below the value of $F/P_0 = \pi/4$), the amplitudes at resonance are infinitely large, independent of the damping. At first sight it seems strange that a *damped* vibration can have infinite amplitude. The paradox is explained, however, by considering that the exciting force $P_0 \sin \omega t$ performs work on the system, and, since work is the product of force and displacement, this energy input is proportional to the amplitude. The energy dissipated by damping is also proportional to the amplitude since the friction force F is constant. Thus, if the friction force is small with respect to the exciting force $\left(F < \dfrac{\pi}{4} P_c\right)$, the work input by the latter is greater than the dissipation by the former, no matter how great the amplitude becomes. Thus the amplitude increases

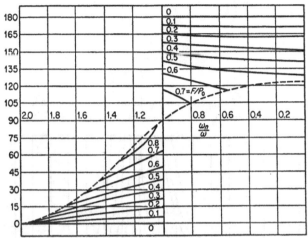

FIG. 8.45. Phase-angle diagram with dry friction damping. Compare with Fig. 2.22b.

without limit, which is another way of saying that it is infinitely large in the steady state. With viscous damping, however, the friction *force* itself is proportional to the amplitude, so that its work dissipation is proportional to the square of the amplitude. Hence, even for a very small friction constant c there will always be a finite amplitude at which the dissipation by damping is equal to the energy input by the exciting force.

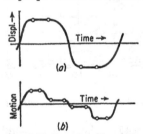

FIG. 8.46. Forced motion with (a) one or (b) two stops per half cycle occurring with great Coulomb damping at slow frequencies.

In connection with the fact that infinite amplitudes occur at resonance with Coulomb damping, the phase angle shows a discontinuous jump at resonance, as can be seen in Fig. 8.45.

For Coulomb friction the (non-linear) friction force is constant, whereas the (linear) inertia and spring forces increase with the amplitude. Thus for large amplitudes the motion will be practically sinusoidal and the approximation (8.46) should be very satisfactory. For smaller amplitudes the curve of motion becomes very much distorted and consequently the approximation for the amplitude is poor. Below the dotted line running through Fig. 8.44 we have motions with one "stop" per half cycle, as shown in Fig. 8.46a. In the blank part in the left-hand lower corner of Fig. 8.44 the motion has more than one stop per half cycle as shown in Fig. 8.46b. No solution could be obtained in that region. For all motions of the types of Fig. 8.46 the approximate formula (8.46) is unreliable. In practice, however, we are interested only in the conditions near resonance, and here the errors of (8.46) are of the order of a few per cent. Thus the general method of (8.45) and (2.28a) is of great practical value. Its consequences for the case of turbulent-air damping, i.e., $f(\dot{x}) = c\dot{x}^2$, have been worked out in the form of diagrams like Figs. 8.44 and 8.45. For further details the reader is referred to the literature.

8.10. Subharmonic Resonance. In this section some cases will be discussed for which the motion differs greatly from a harmonic motion on account of some pronounced non-linearity in the system. It does not matter where this non-linearity appears, whether it be in the spring, in the damping, or in both.

In *linear* systems subjected to an "impure" force disturbance, large amplitudes may be excited at a frequency which is a multiple of the fundamental frequency of the disturbance. The most important technical example of this was discussed in Chap. 5, namely the torsional vibration in internal-combustion engines. The converse of this, i.e., the excitation of large oscillations of a *lower* frequency than ($\frac{1}{2}$, $\frac{1}{3}$, $\frac{1}{4}$ . . . of) the fundamental frequency of the disturbance, never happens in a linear system.

In *non-linear* cases, however, this may occur. Consider, for example, a self-excited relaxation oscillation as in Fig. 8.14. Subject this system to a small harmonic force of a frequency 2, 3, 4 . . . times as fast as the free or natural frequency $\frac{1}{2\epsilon}$. Since the free motion contains all higher harmonics generously, the disturbance (if phased properly) will do work on one of these harmonics and excite it. But this harmonic is an integral part of the whole motion of Fig. 8.34 and will pull all other harmonics with it. The result is that a large motion is excited at a frequency lower than (a submultiple of) the disturbing frequency. This phenomenon is known as "subharmonic resonance" or "frequency demultiplication."

No practical cases of this sort have thus far occurred in mechanical engineering, but in electrical engineering they are of some importance and are beginning to find applications.

Let an electric circuit containing a neon tube, a condenser, a resistance, and a battery be arranged so as to produce a relaxation oscillation of the type of Fig. 8.34, and excite this circuit by a small alternating voltage of constant frequency ω. The natural period T_n of the system (which in this case is not proportional to $\sqrt{LC}$ but to RC) is slowly varied by changing the capacity C. If there were no ω-disturbance, the self-excited period would gradually vary along the dotted line of Fig. 8.47. With the ω-disturbance, however, this does not happen. The system always vibrates at a multiple of the exciting period T_{exc} (*i.e.*, at a submultiple of the exciting *frequency* ω) and picks *that* multiple which is closest to the natural period, as shown in Fig. 8.47. With circuits of this sort, frequency demultiplication up to 200 times has been obtained.

Although the phenomenon was first observed with relaxation oscillations, the explanation given shows that it is not limited to that type of vibration but may occur in any pronounced non-linear system with small "effective" damping. By "effective damping" is meant the total energy dissipation per cycle by the positive and negative damping forces combined. Thus the argument applies to non-linear self-excited vibrations and also to non-linear forced vibrations without any or with very little damping. In the latter case the non-linearity is usually caused by the springs. Two examples will now be considered.

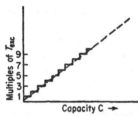

FIG. 8.47. Subharmonic resonance in self-excited relaxation circuit.

Let a cantilever with an iron bob be placed between two permanent magnets (Fig. 8.48a). The "spring" is then made up of two parts, an elastic one (the beam) which is linear, and a magnetic one which is negative and distinctly non-linear. The closer the iron bob approaches to

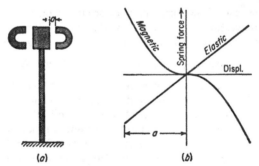

(a)   (b)

FIG. 8.48. (a) Mechanical subharmonic resonant system. The mass can be made to vibrate at its natural frequency by an exciting force of much higher frequency. (b) The magnetic and elastic spring forces acting on the mass of (a).

one of the magnets, the greater the attractive (or negative restoring) force, as shown in Fig. 8.48b. With a combined spring of this sort, the free vibration contains many higher harmonics. Imagine the bob of the cantilever to be subjected to a small alternating force of a frequency which is approximately a multiple of the natural frequency. This force can be realized in many ways, among others by attaching a small unbalanced motor to the bob. The alternating force can then do work on the nth harmonic of the motion and thus keep the system in vibration. In this example no source of energy exists other than the alternating one, and it is seen that the frequency of the alternating source of energy must be a multiple of the natural frequency.

It is not necessary to have an extraneous exciting force acting on the system: subharmonic resonance can be brought about also by a variable spring. The cases discussed in Sec. 8.2 to 8.4 had *linear* springs for which the constant or intensity varied with the time. It was shown there that resonance could occur at higher frequencies than that of the spring variation and also at *half* this frequency but not at any of the *lower* subharmonics ($\frac{1}{3}$, $\frac{1}{4}$, etc.). However, if we have a non-linear spring varying

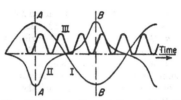

Fig. 8.49. Operation of the system of Fig. 8.48 with alternating current in the magnet windings.

with the time (*i.e.*, a spring for which the stiffness varies with *both* the displacement and the time), these lower subharmonics may be excited. An example of such a system is Fig. 8.48*a*, in which the magnets now consist of soft iron and carry alternating-current windings. The attractive force of such magnets varies not only with the displacement according to Fig. 8.48*b* but also with the time at twice the current frequency. That it is possible for the magnetic forces to do work on the vibration if the phase is proper, is clear from Fig. 8.49. Curve I of that figure represents the motion of the bob, curve II is the spring force of the magnets if there were direct current in them, and curve III shows the intensity variation of the magnets with the time in case the mass were standing still (taken here to be six times as fast as the motion). The actual force exerted by the magnets on the bob is the product of the ordinates of II and III. Just to the left of line AA the magnetic force is against the direction of motion, and just to the right of it the force helps the motion. But III has been placed so that to the left of AA the intensity is small and to the right of AA it is great. The same relations obtain near BB. Thus energy is put into the system. The non-linearity of the system is essential because without it curve II would be sinusoidal and the argument of Fig. 1.16, page 15, would show no energy input. Only the fact that at some distance from either AA or BB the curve II has a negligible ordinate accounts for the energy input.

Under which conditions the "proper phase" between the curves I and III occurs is a question that can be answered only by mathematical analysis. Since this implies a non-linear equation with variable coefficients, it is evident that such an analysis will be extremely difficult.

Problems 217 *to* 233.

PROBLEMS

Chapter 1

1. A force $P_0 \sin \omega t$ acts on a displacement $x = x_0 \sin (\omega t + 30°)$, where $P_0 = 5$ lb., $x_0 = 2$ in., and $\omega = 62.8$ rad./sec.

   a. What is the work done during the first second?

   b. What is the work done during the first $\frac{1}{40}$ sec.?

2. A force $P_0 \sin 3\omega t$ acts on a displacement $x_0 \sin 2\omega t$, so that two force cycles coincide with three motion cycles. Calculate the work done by the force on the motion during (*a*) the first, (*b*) the second, (*c*) the third, and (*d*) the fourth half cycle of the motion.

3. A function has the value P_0 during the time intervals $0 < \omega t < \pi, 2\pi < \omega t < 3\pi$, $4\pi < \omega t < 5\pi$, etc., while the function is zero during the intermediate periods $\pi < \omega t < 2\pi, 3\pi < \omega t < 4\pi$, etc. Find the Fourier components. Note that this function does have an "average value"; determine that value first, subtract it from the curve, and recognize what remains as something you have seen before.

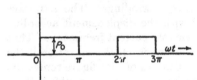

Problem 3.

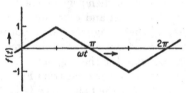

Problem 4.

4. Let a periodic curve $f(t)$ be as shown in the figure. Prove that

$$f(t) = \frac{8}{\pi^2} \left(\sin \omega t - \frac{1}{3^2} \sin 3\omega t + \frac{1}{5^2} \sin 5\omega t - \cdots \right)$$

5. Referring to Fig. 1.18, let the curve to be analyzed consist of a pure sine wave, so that $a_1 = 1$ and all other a's and b's are zero. Sketch the shape of the curve traced on the table E of Fig. 1.18, if the gear B and the scotch crank rotate at equal speeds. The closed curve on E depends on how the two gears are coupled. Show that by displacing them 90 deg. with respect to each other, the table curve varies from a circle to a straight line at 45 deg. Find the area of the circular E-curve and show that $a_1 = 1$ and $b_1 = 0$.

6. Sketch the E-curves of Problem 5 for the case where the scotch crank turns 2, 3, . . . times as fast as B, and show that the area registered by the planimeter is zero in all these cases.

7. Deduce Eq. (1.6) on page 5 by trigonometry.

8. A rectangular curve has the value $+a$ during three-eighths of the time and the value $-a$ during five-eighths of the time, as shown in the figure. Find the Fourier coefficients.

380

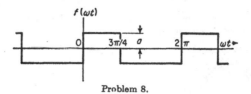

Problem 8.

9. A force is expressed by $P = P_0 \sin (\tfrac{1}{2}\omega t)$ during the time $0 < \omega t < 2\pi$ and is repetitive from there on. (Hence it consists of half sine waves, always above zero.) Find the Fourier coefficients.

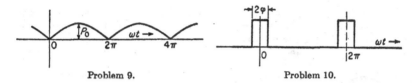

Problem 9. Problem 10.

10. A force has the constant value P_0 during the time intervals $-\varphi < \omega t < +\varphi$, $(2\pi - \varphi) < \omega t < (2\pi + \varphi)$, etc., while it is zero in between $\varphi < \omega t < (2\pi - \varphi)$, etc.

a. Find the Fourier coefficients.

b. Deduce what happens to this solution when P_0 is made greater and φ smaller, indefinitely, but keeping the product $P_0\varphi$ constant.

11. A curve is made up of parabolic arcs as follows. Between $x = -l/2$ and $x = +l/2$ the equation is $y = a(2x/l)^2$. Farther the curve repeats itself by mirroring about the vertical lines $x = -l/2$ and $x = +l/2$. Calculate the Fourier coefficients.

Chapter 2

12. Derive the results (2.28a) and (2.28b) in the manner indicated directly below Eq. (2.27).

13. Derive Eq. (2.23) by an energy method.

14. A uniform bar of total weight W and length l is pivoted at a quarter-length point, while at the other quarter-length point a spring of stiffness k applies at 45 deg. as shown.

a. Find the natural frequency.

b. Turn the figure 90 deg. so that the bar hangs down vertically, and again find the natural frequency.

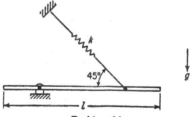

Problem 14.

15. Two solid disks, weighing 50 lb. each, are pinned to a bar at their centers, and also to a side rod after the fashion of the conventional steam-locomotive mechanism. The bar and side rod weigh 50 lb. each. The mechanism is free to roll without slip on a horizontal track. Find the natural frequency of small oscillations about the equilibrium position.

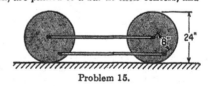

Problem 15.

16. A bar in the form of a 90-deg. bend of total length $2l$ and total mass $2m$ (each arm being m and l) hangs from a pivot in a vertical plane. Gravity is acting. Find the natural frequency of small oscillations.

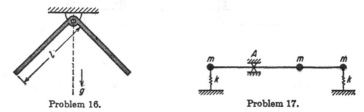

Problem 16.   Problem 17.

17. A weightless bar of length $3l$ carries three distinct equal masses m. It is pivoted at A and carries two springs k. Find the natural frequency of small vibrations. The bar is horizontal so that gravity has no influence.

18. A rotor of weight W and of moment of inertia I about its axis of symmetry is laid with its journals on two guides with radius of curvature R. The radius of the journals is r. When the rotor rolls without sliding, it executes small harmonic vibrations about the deepest point of the track. Find the frequency (energy method, see pages 33 and 37).

Problem 18.

19. The same problem as 18, except that the track is straight ($R = \infty$) and the rotor is unbalanced by a small weight w attached to it at a distance r_1 from the axis.

20. Two cylindrical rolls are located at a distance $2a$ apart; their bearings are anchored and they rotate with a great speed ω in opposite directions. On their tops rests a bar of length l and weight W. Assuming dry friction of coefficient f between the rolls and the bar, the bar will oscillate back and forth longitudinally. (a) Calculate its frequency. (b) If one end of the bar A is pushed into the paper somewhat and B is pulled out, is the equilibrium stable or unstable?

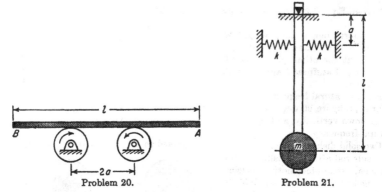

Problem 20.   Problem 21.

21. A pendulum consists of a stiff weightless bar of length l carrying a mass m on its end (see figure). At a distance a from the upper end two springs k are attached to the bar. Calculate the frequency of the vibrations with small amplitude.

22. Turn the figure of Problem 21 upside down. (a) Find the relation between u, m, and l for which the equilibrium is stable. (b) Find the frequency.

23. A weightless bar of length $2l$ is hinged in the middle. At its top end it carries a concentrated weight W and at its bottom end a concentrated weight 2W. Midway between the hinge and the bottom weight there is a spring k. Find the natural frequency of small oscillations in the vertical plane.

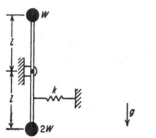

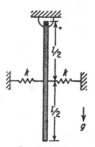

Problem 23.   Problem 24.

24. A bar of total length l and total mass m, uniformly distributed over its length, is pivoted at its top point and is restrained by two springs k at its center point. Find the natural frequency.

25. Calculate the frequency of the stator of Fig. 2.43. The linear stiffness of each of the four springs is k, their average distance from the center of the rotor is a, and the moment of inertia of the stator is I.

26. Calculate the frequency of Problem 25 for the spring system of Fig. 2.44. The beams c are made of steel with a modulus of elasticity E; their dimensions are l_1, l_2, w, and t as indicated in the figure.

27. A stiff weightless horizontal bar of length l is pivoted at one end and carries a mass m at its other end (see figure). It is held by an inextensible string of length h. If the mass is pulled perpendicularly out of the paper and then released, it will oscillate. Calculate the frequency.

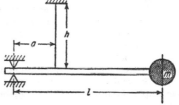

Problem 27.

28. A mass m is attached to the center of a thin wire of cross section A and total length l which is stretched with a large tension of T lb. between two immovable supports. The modulus of elasticity of the wire is E. Calculate the frequency of the vibrations of the mass in a plane perpendicular to the wire.

29. A heavy solid cylinder of diameter D, length l, and mass m can roll over a horizontal surface. Two springs k are attached to the middle of l at a distance a above the center. Calculate the frequency.

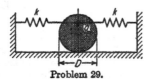

Problem 29.

30. A system consists of a single concentrated mass m attached to the end of a stiff weightless rod of length l. The rod is normally horizontal, and its other end is ball-hinged to a wall. There are two springs k in a horizontal plane attached to the middle of the rod. The mass m is hung from a vertical string of length h. Find the natural frequency of free vibrations.

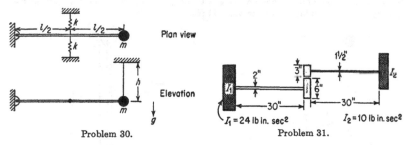

Plan view

Elevation

Problem 30.

$I_1 = 24$ lb in. sec^2 $I_2 = 10$ lb in. sec^2

Problem 31.

31. Find the natural frequency of the geared system shown. The inertia of the two shafts and gears is to be neglected. The shafts are of steel.

32. Find an expression for the linear spring constant k of a steel coil spring of wire diameter d, coil diameter D, and having n turns. Calculate k numerically for $d = 0.1$ in., $D = 1\frac{1}{2}$ in., and $n = 10$.

33. Find the torsional-spring constant of a coil spring, i.e., a coil spring of which the ends are subjected to torques about the longitudinal axis of the spring. Calculate this k numerically for the spring of Problem 32.

34. Find the spring constant k in bending of a coil spring, i.e., the bending moment to be applied to the ends of the spring divided by the angle through which the two ends turn with respect to each other. Calculate this k numerically for the spring of Problem 32.

35. What are the expressions for the linear-spring constants of

a. A cantilever beam of bending stiffness EI with the mass attached to the end l?

b. A beam of total length l on two supports with the mass in the center?

c. A beam of total length l built in at both ends with the mass in the center?

36. Calculate the frequency of the small vertical vibrations of the mass m. The two bars are supposed to be stiff and weightless. The mass is in the center between k_3 and k_4, and k_3 is midway between k_1 and k_2. The mass is guided so that it can move up and down only. It can rotate freely and has no moment of inertia.

Problem 36.   Problem 37.

37. A point on a machine executes simultaneously a horizontal and a vertical vibration of the same frequency. Viewed with the seismic microscope described on page 68 the point will be seen as an ellipse (see figure). By observation, the lengths h and AB are found. (a) Calculate from these the phase angle between the horizontal and vertical motions. What shapes does the ellipse assume for (b) $\varphi =$ zero, and (c) $\varphi = 90$ deg.?

38. a. If in a vibration every maximum amplitude is 2 per cent less than that amplitude in the previous (whole) cycle, what is the damping, expressed in per cent of critical?

  b. If the vibrating system a consists of a weight of 1 lb. and a spring of $k = 10$ lb./in., what is the damping constant numerically and in what units is it expressed?

  39. In the damped spring-bob system shown, the elongation of the spring due to the 20-lb. weight of the bob is $\frac{1}{2}$ in. In free vibrations of the system the amplitude is observed to decay from 0.4 to 0.1 in. in 20 cycles. Calculate the damping constant c in lb.-sec./in.

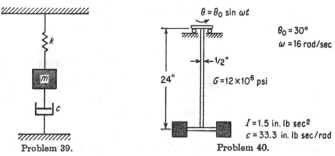

Problem 39. Problem 40.

  40. A mechanical agitator is designed so that the power end rotates in a predetermined simple harmonic motion $\theta = \theta_0 \sin \omega t$. When operating in air, the machine vibrates with negligible damping. Tests show that in normal operation, with the paddle immersed, the fluid damps the motion with a torque of $c = 33.3$ in. lb. sec./rad. and also has the effect of doubling the moment of inertia of the paddle. Find the maximum shear stress in the shaft and the angular displacement amplitude of the paddle for steady-state motion (a) in air and (b) in fluid.

  41. A damped vibrating system consists of a spring of $k = 20$ lb. per inch and a weight of 10 lb. It is damped so that each amplitude is 99 per cent of its previous one (i.e., 1 per cent loss in amplitude per *full* cycle).

  a. Find the frequency by formula and from Fig. 2.9.

  b. Find the damping constant.

  c. Find the amplitude of the force of resonant frequency necessary to keep the system vibrating at 1 in. amplitude.

  d. What is the rate of increase in amplitude if at 1 in. amplitude the exciting force (at resonant frequency) is doubled?

  e. What is the final amplitude to which the system tends under the influence of this doubled force?

  f. Find the amplitude-time relation of this growing vibration.

  42. Find the expression for the steady-state torque, assuming no damping,

  a. In shaft k of Fig. 2.6.

  b. In shaft k_2 of Fig. 2.7.

  43. A torsional pendulum, consisting of a 20-lb. disk, 10 in. diameter, mounted on a steel shaft (30 in. long by $\frac{1}{2}$ in. diameter), is subjected to a simple harmonic alternating torque at a frequency of 7 cycles per second. The shaft material has a yield point of 20,000 lb./in.^{2} in shear. Find the amplitude, M_{ta} of the impressed torque corresponding to a safety factor of 2.

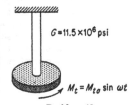

Problem 43.

  44. Find the equivalent spring constant for a mounting for a direct-current bench grinder running at 3,000 r.p.m. and weighing 50 lb.

which will reduce the force transmitted to the bench to one-tenth that which is developed when the grinder is bolted to the bench (assume the bench to be rigid).

45. It is desired to make a seismometer which will measure simple harmonic motion at 5 cycles per second with an error not greater than 10 per cent. What limitation does this impose upon the natural frequency of the instrument?

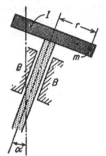

46. A "static balancing machine" (page 233) consists of a bearing B inclined at an angle α with the vertical (see figure). A rotor placed in this bearing has a moment of inertia I and an unbalance m at a distance r from the center. Write the differential equation of the vibrations of the rotor in terms of its angle of rotation φ. Find the natural frequency for small vibrations φ.

47. Find the natural frequency of the small oscillations of a solid half cylinder (the contour consisting of a half circle and a diameter), which rolls without sliding on a horizontal plane.

48. A simple k-m system is at rest. A constant force P is applied to the mass during a stated time interval t_0, after which the force is removed. Find the motion of the mass after removal.

Problem 46.

49. Set up the differential equations of motion of the system of Fig. 2.7; then, by elimination, reduce them to a single differential equation in terms of the variable $\psi = \varphi_1 - \varphi_2/n$, which is an angle that becomes zero if the shafts are not twisted. In this manner verify the statements made on page 30.

50. A weightless, stiff bar is hinged at one end. At a distance l from the hinge there is a mass m, at a distance $2l$ there is a dashpot c, and at a distance $3l$ there is a spring k and an alternating force $P_0 \sin \omega t$. Set up the differential equation. Assuming small damping c (but *not* zero damping), calculate the natural frequency; the amplitude of forced vibration at the spring at the natural frequency and at half natural frequency.

51. A circular solid disk of mass M and radius r is suspended in a horizontal plane from a fixed ceiling by three vertical wires of length l, attached to three equally spaced points on the periphery of the disk.

a. The disk is turned through a small angle about its vertical center line and let go. Calculate the frequency of rotational vibration.

b. The disk is displaced sidewise through a small distance without rotation and let go. Calculate the frequency of the ensuing swinging motion.

52. Prove the statement made on page 59 that there is no phase distortion in a seismographic instrument if the phase-angle diagram Fig. 2.22b is a straight diagonal line passing through the origin.

53. A mass m is suspended from a ceiling by a spring k and a dashpot c. The ceiling has a forced motion $a_0 \sin \omega t$. Calculate the work done by the ceiling on the system per cycle of vibration in the steady state. Write the answer in dimensionless form.

54. In the system of Fig. 2.3 and Fig. 2.22a, the maximum *work* input by the force as a function of frequency is only approximately equal to $\pi P_0 x_0$, where x_0 is the amplitude at $\omega/\omega_n = 1$. The actual maximum work is at a slightly different frequency. Prove that this maximum work can be computed from $\pi P_0 x_0$ by multiplying that quantity by the correction factor

$$\frac{\sqrt{1 - 2(c/c_c)^2}}{1 - (c/c_c)^2}$$

and show that this error is less than 0.1 per cent for a damping as high as $c/c_c = 20$ per cent.

55. In 1940 a large two-bladed windmill, capable of generating 1,250 kw. of electric power was built on Grandpa's Knob near Rutland, Vt. The diameter of the blade circle is 175 ft., the blades rotate at 30 r.p.m. in a plane which is considered vertical for our purpose. The blades are mounted on the "pintle" or cap, which itself can rotate slowly about a vertical axis in order to make the blades face the wind. Since there are only two blades in the rotor, the moment of inertia of the rotor about the vertical pintle axis is very much greater when the blades are pointing horizontally than when they are vertical, 90 deg. further. Let Ω be the constant angular speed of the rotor, ω the very much smaller angular speed of the pintle, and I_{max} and I_{min} the extreme values of inertia about the vertical axis.

a. Assuming the driving mechanism of the pintle motion to be extremely soft torsionally, so that no torque acts on the pintle (except friction, which is to be neglected), find the ratio between the maximum and minimum values of ω.

b. Assuming the pintle drive to be extremely stiff torsionally, so that the pintle motion ω is forcibly uniform, find an expression for the torque in the pintle drive.

56. A helicopter rotor. Find the natural frequency of small up-and-down vibrations of the rotating blade (angular speed Ω). The hub does not move (except to rotate); φ is small; the blade is stiff. Express the answer in terms of a, b, blade mass M, and moment of inertia I_G.

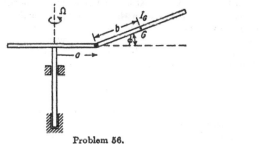

Problem 56.   Problem 57.

57. A uniform bar of weight W and length l is pivoted at its top about a horizontal axis, and the bar hangs down vertically. Now the bar and its hinge are rotated with speed Ω about a vertical axis.

a. Find the position of (dynamic) equilibrium of the bar, expressed as the angle α with respect to the downward vertical as a function of the speed Ω.

b. Find the natural frequency of small oscillations about the vertical position $\alpha = 0$.

c. Find the natural frequency of small oscillations about an arbitrary equilibrium position $\alpha = \alpha_0$ (which occurs, of course, at higher speed Ω).

58. A uniform bar of total length $l = 9$ in. is pivoted without friction at its top. It is excited by a force of $\frac{1}{2}$ lb. maximum value, varying sinusoidally in time (harmonic force) with a frequency of 1 cycle per second. The force acts horizontally at quarter length, *i.e.*, at $2\frac{1}{4}$ in. from the top. Find the "steady-state" forced amplitude of oscillation, expressed in degrees. The bar weighs 4 lb. total.

59. The upper support of the spring-bob system shown moves in simple harmonic motion with an amplitude of 0.05 in. and a frequency equal to the undamped natural frequency of the system.

a. Find the amplitude of the spring force.

b. Find the amplitude of the damping force.

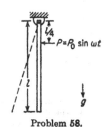

Problem 58.

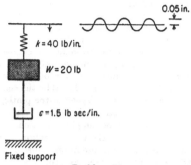

Problem 59.

60. We have discussed two different resonance curves:

a. The case of an exciting force $P_0 \sin \omega t$ of constant amplitude P_0. The resonance curve starts at height 1, ends at height 0 (page 44).

b. The case of an exciting force $A\omega^2 \sin \omega t$, where the amplitude is proportional to ω^2. This resonance curve starts at height 0 and ends at height 1 (Fig. 2.20, page 46). Now discuss the new case of an exciting force $A\omega \sin \omega t$, with an intensity proportional to the first power of the frequency. Find the formula for the resonance curve, bring it to dimensionless form, and find at what height it starts and ends. If you define the static amplitude as the static amplitude at the natural frequency, that is, $x_{stat} = A\omega_n/k$, then find the two frequency ratios ω/ω_n for which the resonance curve has unit height.

61. A single-degree-of-freedom vibrational system of very small damping is operated near resonance. The magnification is so large there and the excitation so small that our measuring instruments are unable to measure the static amplitude OA with sufficient accuracy. All we can measure is the maximum amplitude x_{max}, the amplitudes x_{side} at two spots $2\,\Delta\omega$ apart, the frequency difference $2\,\Delta\omega$, and the natural frequency ω_n. From these the damping ratio must be deduced.

Derive a formula for c/c_c as a function of the above quantities for the simplified case that $\dfrac{c}{c_c}$ is small, that $\Delta\omega/\omega_n$ is small, and the x_{side}/x_{max} is small.

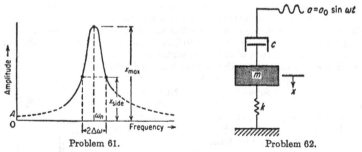

Problem 61.   Problem 62.

62. The upper support of the vibratory system shown is given the motion $a_0 \sin \omega t$. Set up the differential equation for the (absolute) motion of mass, and reduce it to standard form. Write the relation for the dimensionless "resonance diagram." Let $x_{stat} = \dfrac{A\omega_n}{k}$. Sketch the curves for $c/c_c = 0$, $c/c_c = 0.1$, $c/c_c = 1$, and $c/c_c \to \infty$.

63. A reciprocating engine is subjected to a fluctuating exciting force whose frequency is the same as the running speed of the engine. The engine and base weigh 500 lb. and are supported on a vibration-isolating mounting which has an equivalent spring constant of 300 lb./in. and a dashpot which is adjusted so that the damping is 20 per cent of critical.

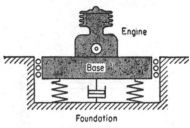

a. Over what speed range in r.p.m. is the amplitude of the fluctuating force transmitted to the foundation actually larger than the excitation?

Problem 63.

b. Over what speed range is the transmitted force amplitude less than 20 per cent of the exciting force amplitude?

Chapter 3

64. Calculate the abscissas and ordinates of the points A, P, and Q in Fig. 3.13.

65. Calculate the natural frequency of the water in the tank system shown.

66. Find the metacentric height of a body made of solid material of specific gravity $\frac{1}{2}$, floating in water, having the shape of a parallelepiped with

a. Square cross section $h \times h$, floating with one of its sides parallel to the water.

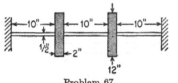

Problem 65. Idealized Frahm tank.

b. Triangular cross section of base b and height h floating with the base down and the point emerging from the water.

c. The same triangular section with the point down.

67. A steel torsion shaft of $\frac{1}{2}$ in. diameter is built into walls at both ends. It carries two flywheels, each at 10 in. from a wall, and, also 10 in. from each other, of steel, 12 in. in diameter and 2 in. thick. Find the two natural frequencies, expressed in v.p.m.

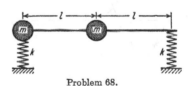

Problem 67.   Problem 68.

68. *a.* Calculate the two natural frequencies of the system shown consisting of a weightless bar of length $2l$, two masses m, and two springs k.

b. Find the location of the "node" or center of rotation of the bar in each of the two natural modes.

69. A weightless string is stretched with a large tension of T lb. between two solid immovable supports. The length of the string is $3l$ and it carries two masses m at distances l and $2l$ from one of the supports. Find the shapes of the natural modes of motion by reasoning along (without mathematics), and then calculate the two natural frequencies (*cf.* Problem 28).

70. A uniform bar of length l and mass m carries at its right end an additional mass, also m. It is mounted on two equal springs k at the extremities. Find the two natural frequencies and the corresponding shapes of vibration.

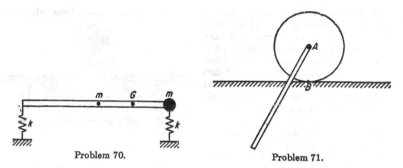

Problem 70.   Problem 71.

71. A system consists of a solid cylinder of radius R and weight W, which rolls without slipping on a horizontal track. To its center is pivoted a uniform bar of total length $3R$ and weight W, equal to that of the cylinder. Set up the equations of the system, and find the natural frequencies of small oscillations.

72. A cylinder of radius r, total mass $2m$, center of gravity G located at distance $r/2$ from the geometric center has a moment of inertia $I_G = mr^2$ about an axis through the center of gravity parallel to the cylinder itself. At a point A at distance $r/2$ from the center and opposite to G, it carries a pin from which can swing a uniform bar of total length $3r$, total mass m, uniformly distributed along the length. The cylinder can roll on the ground with large friction and no slip. There is no friction in the hinge at A. Write the differential equations for the simplified case of *small* angles φ and θ, and find the natural frequencies.

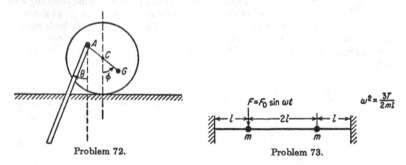

Problem 72.   Problem 73.

73. A weightless string of total length $4l$ carries two masses, m each, at the quarter-length points and nothing at the mid-point. The tension in the string is T. The first mass is excited with a force of F_0 lb., varying at a forced frequency $\omega^2 = 3T/2ml$. Find the steady-state forced motions at both masses.

74. In the undamped vibration absorber of Fig. 3.6 let the mass ratio M/m be 5, and let the damper be tuned to the main system so that also $K/k = 5$. Further let the external force P_0 be absent. Find the two natural modes of motion, *i.e.*, the ratio between the amplitudes of M and m at the natural frequencies. Also calculate those frequencies.

75. Let the system of Problem 74 be provided with a dashpot across the damper spring, having a damping constant of 5 per cent of "critical" ($c = \sqrt{4km}/20$). Assuming that the natural modes of motion calculated in Problem 74 are not appreciably altered by this small amount of damping, calculate the rate of decay in each of the two natural motions.

76. The period of roll of the "Conte di Savoia" (see page 112) is 25 sec., the metacentric height is 2.2 ft., and the weight of the ship is 45,000 tons. Calculate

a. Its moment of inertia about the roll axis.

b. Its maximum angular momentum when rolling 10 deg. to either side.

The characteristics of each one of the three gyroscopes installed on board the ship are:

> Gyro moment of inertia, $4.7 \times 10^6/32.2$ ft. lb. sec.^{2}
> Gyro speed, 800 r.p.m.

Let these three gyroscopes precess from $\psi = -30$ deg. to $\psi = +30$ deg., and let this happen during a time (say 2 sec.) which is short in comparison with a half period of the ship's roll. Let this precession take place at the middle of a roll always in a sense to cause positive damping.

c. Find the rate of decay of a rolling motion of the ship, assuming that no damping action exists other than that of the gyroscopes.

77. An automobile has main springs which are compressed 4 in. under the weight of the body. Assume the tires to be infinitely stiff. The car stands on a platform which is first at rest and then is suddenly moved downward with an acceleration $2g$.

a. How far does the platform move before the tires leave it?

b. Assuming the car to have a speed of 30 m.p.h., draw the profile of the road which would correspond to the $2g$-accelerated platform. This question has meaning for *front* wheels only.

78. The car of Problem 77 runs over a road surface consisting of sine waves of 1 in. amplitude (*i.e.*, having 2 in. height difference between crests and valleys) and with distances of 42 ft. between consecutive crests. There are no shock absorbers.

a. Find the critical speed of the car.

b. Find the amplitude of vertical vibration of the chassis at a forward speed of 40 m.p.h.

79. A double pendulum consists of two equal masses m, hanging on weightless strings of length l each. In addition to gravity, there are two mechanical springs of stiffness k. The equilibrium position is a vertical line. Set up the differential equations of motion carefully and calculate the two natural frequencies. (Small angles.)

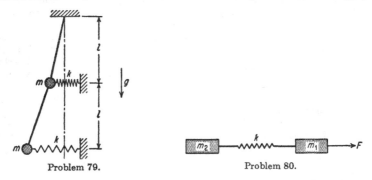

Problem 79.   Problem 80.

80. Starting surges in the cables of an elevator in a high building. Long (up to 1,000-ft.) elevator cables are quite flexible longitudinally. The hoisting machinery usually is located high in the building, and when the cab is below, the starting torque of the machine can reach it only through this flexibility. Hence the machine starts first, stretching the cable, and the cab starts later. Idealize this situation as two masses and a spring, m_1 being the equivalent mass of the hoisting engine (i.e., the mass which at the periphery of the hoisting drum matches the moment of inertia of the engine), m_2 being the mass of the cab, and k the spring constant of the cables. Let F be the applied force (i.e., the hoisting torque divided by the drum radius).

First consider the case that F comes on suddenly, that is, $F = 0$ for $t < 0$ and $F = F_0$ for $t > 0$. Set up the differential equations, and from them deduce a single differential equation for the cable stretch only, not containing the motions of the masses m_1 or m_2 individually. Solve, and find the cable force as a function of time. To this must be added the constant static force $m_2 g$. Hence the cab will accelerate upward in spurts, the maximum acceleration being twice the average. Plot $F = f(t)$ to see this clearly.

81. In order to improve the performance of the elevator of the previous problem, it is proposed to let the torque come up gradually, rising linearly with the time from zero to F_0 during the interval from $t = 0$ to $\omega_n t = 2\pi$ (a full cycle of the natural motion). Thereafter the force remains constant, equal to F_0. Again set up the differential equation for the cable stretch only (not containing x_1 and x_2 individually), and prove that now the operation is very smooth and the cable tension never rises above its steady-state value.

82. Generalize the previous problems. The hoisting engine force still grows linearly with the time from zero to its maximum value F_0, but, instead of doing this during time 0 (as in Problem 80) or during time $\omega_n t = 2\pi$ (as in Problem 81), it grows during the time $\omega_n t = \alpha$.

Prove that the ratio of the maximum cable force to the force F_0 is expressed by

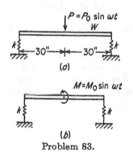

$$\frac{F_{max}}{F_0} = 1 + \frac{1}{\alpha} \sqrt{2 - 2 \cos \alpha}$$

Plot this relation up to $\alpha = 6\pi$, and show that it fits the result of the two previous problems as special cases.

83. The symmetrical system shown is an idealization of a balancing machine frame and consists of a uniform slender and rigid rod of weight $\overline{W} = 100$ lb. supported by a spring at each end, for each of which the spring constant is $k = 400$ lb./in.

Problem 83.

a. Find the amplitude of the force in each spring due to a simple harmonic alternating force P_0 of 20 lb amplitude impressed upon the bar at a frequency of 12 cycles/sec.

b. Solve the same problem when the impressed force is replaced by a couple M_0 having the same frequency and an amplitude of 1,000 in.-lb.

84. Find the mechanical impedance Z of the top point of a spring k, from which is suspended a mass m.

85. Find the mechanical impedance Z of a ceiling from which is suspended a mass m by means of a spring k and a dashpot c parallel to the spring.

86. Find the mechanical impedance of a ceiling from which is suspended a mass m through a spring k. In addition there is a dashpot c between the mass and the ground.

87. The system shown in the figure is a special case of Fig. 3.27 in the text. Find the amplitude of motion of the dashpot and, from it, the force transmitted to the ground.

Problem 87.

88. Derive Eq. (3.49) (page 121) of the text.

89. Find the mechanical impedance Z of a mass M which is attached to the ground through a spring K.

90. An engine m_1, excited by $P_0 \sin \omega t$, is to be mounted through a protecting spring k on a foundation with the impedance of the previous problem. Find the transmissibility.

91. In the centrifugal pendulum of Fig. 3.10b let Ω be the speed of rotation of the disk, a the distance from the disk center to the center of swing of the pendulum, b the distance from the swing center to the center of gravity of the pendulum, and finally k the radius of gyration of the pendulum mass about its swing center. Find the natural frequency and try to design a pendulum that will swing back and forth three times per revolution.

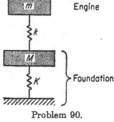

Problem 90.

92. Prove that the most favorable damping in the viscous Lanchester damper (curve 3 of Fig. 3.14c, page 101) is given by

$$\frac{c}{2m\Omega} = [2(1 + \mu)(2 + \mu)]^{-\frac{1}{2}}$$

93. A three-bladed airplane propeller is idealized as three flat massless cantilever springs, spaced 120 deg. apart and carrying concentrated masses m at their ends, at a

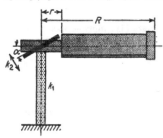

Problem 93.

distance R away from the shaft center. They are built in at a distance r from the shaft center into a hub having a moment of inertia I, with a definite angle α between the blade plane and the plane of the entire propeller (see figure). Let the spring constant of each blade in its limber direction be k_2 and let the blade be infinitely stiff against bending in its stiff direction (90 deg. from the limber direction). Let the hub be mounted on a shaft of torsional stiffness k_1. Find the two natural frequencies of the non-rotating system (the "blade frequency" and the "hub frequency"), as a function of the blade angle α, and find in particular whether the blade frequency is raised or lowered with increasing blade angle α.

94. The same as Problem 93, this time the shaft k_1 is stiff against torsion, but flexible against extension. The hub therefore can vibrate linearly in the shaft direction. Let k_1 mean the extensional spring constant of the shaft and let the inertia of the hub be expressed by its mass M rather than its moment of inertia.

95. The same as Problem 93, but this time the blade stiffness in its own plane is no longer considered infinite. Let the stiffness of one blade in its stiff plane be k_2 and in its limber plane k_3; let k_1 as before be the torsional stiffness of the shaft. For simplicity let $I = 0$.

96. A combination of Problems 94 and 95, the blade having stiffnesses k_3 and k_2, the shaft being stiff in torsion and having k_1 in extension, and the hub mass M being zero for simplicity.

97. A mass m is suspended at distance l below the ceiling by two equal springs k arranged symmetrically at an angle α. This angle α is the angle under the static influence of gravity with the springs carrying the weight.   Find

a. The natural frequency of up-and-down motion.

b. The natural frequency of sidewise motion.

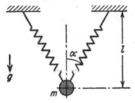

Problem 97.   Problem 98.

98. A uniform disk of weight W and radius r rolls without sliding on a plane table. At its center it carries a hinge with a weightless pendulum of length l and a concentrated weight w at its end.   Find the natural frequencies for motion in the plane of the paper.

Chapter 4

99. Derive Eq. (4.11) by working out the determinant (4.5).

100. A simple massless beam of bending stiffness EI and length $4l$, simply supported at its ends, carries a mass m at a distance l from each of the supports.   Find:

a. The three influence numbers.

b. The two natural frequencies.

c. The two natural modes of motion.

101. A flexible weightless beam of section EI and length l is simply supported at its two ends and carries two equal masses m, each at $\frac{1}{4}l$ and at $\frac{1}{2}l$ from one of the ends. Calculate the two frequencies by the method of influence numbers.

102. In Fig. 3.1, let $m_1 = m$, $m_2 = 5m$, $k_1 = k$, $k_2 = 3k$, and $k_3 = 7k$.   Let a force $P_0 \sin \omega t$ be acting on m_1.   Find:

a. The frequency ω of P_0 at which m_1 does not move.

b. The amplitude of m_2 at this frequency.

Solve this problem without the use of large formulas by a physical consideration, as suggested in Fig. 4.4.

103. Derive Eq. (4.22).

104. Check the various frequencies shown in Fig. 4.19.

105. By Rayleigh's method find the natural frequency of a string with tension **T** and length $3l$, carrying masses m at distances l and $2l$ from one end.   The mass of the string itself is $3m$.

106. A beam EI on two supports, of length l and of mass μ_1 per unit length (total mass $m = \mu_1 l$) carries a concentrated mass M in the middle.   Find the natural frequency by Rayleigh's method, and in particular find what fraction of m should be added to M in order to make the simple formula (2.10) applicable.

107. The same as Problem 106, but for a beam of total mass m, clamped solidly at both ends, and carrying a mass M at its center.

108. A weightless cantilever spring of length $2l$ and bending stiffness EI carries two concentrated weights, each of mass m, one at the free end $2l$ and the other at the center l.   Calculate the two natural frequencies.

109. A pulley of moment of inertia $2mr^2$ and radius r carries two equal masses m, supported from the string by two different springs k and $2k$.   The masses are supposed to move up and down only, and the pulley can rotate freely.   Set up the equations, and find the natural frequencies of this system.

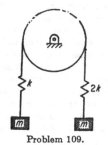

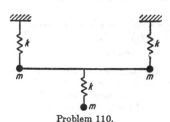

Problem 109.   Problem 110.

110. The system shown has three degrees of freedom if we consider only small vertical motions of the masses.

a. Use either Newton's laws or influence coefficients to derive a set of equations from which the natural frequencies and natural modes can be determined.

b. The system has three natural modes. Find any one of these modes and its corresponding natural frequency. Any method may be used, including the educated guess.

111. A weightless cantilever beam of bending stiffness EI and total length l carries two masses m, one at its end and one at its middle. Calculate the lowest natural frequency by Rayleigh's method, assuming that the mode shape is a parabola $y = ax^2$.

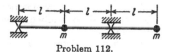

Problem 111.   Problem 112.

112. A beam, weightless, has constant stiffness EI, total length $3l$, and concentrated masses m and m as shown. Find the natural configurations and frequencies.

113. The massless cantilever beam of uniform stiffness EI is fastened at the left end to a rod which has a vertical motion of $a_0 \sin \omega_f t$. Two unequal masses are placed as indicated in the sketch.

a. Find the influence coefficients $\alpha_{11}, \alpha_{12}, \alpha_{22}$.

b. Find the natural frequencies of the system.

c. Find the normal modes of vibration of the system.

d. Find the forcing frequency ω_f for which mass (2) acts as an "absorber" for mass (1).

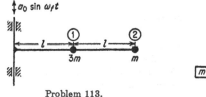

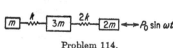

Problem 113.   Problem 114.

114. A three-mass system m, $3m$, and $2m$ is connected by two springs k and $2k$. It is excited at the mass $2m$ by a force $P_0 \sin \omega t$. Set up the differential equations, and solve for the motion of the excited mass $2m$. Sketch the resonance diagram, and calculate the locations of the three or four principal points in it.

115. Three unequal weights are symmetrically mounted on a weightless beam, as shown. The influence coefficients are $\alpha_{11} = \alpha_{33} = 800$; $\alpha_{22} = 1,400$; $\alpha_{12} = \alpha_{21} = $

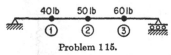

Problem 115.

$\alpha_{23} = \alpha_{32} = 1,000$; $\alpha_{13} = \alpha_{31} = 600$. All units are 10^{-6} in./lb. Find the first natural frequency, using the iteration (Stodola's) method.

116. A 1-in.-diameter steel shaft supporting three 40-lb. weights, which may be regarded as mass particles, is supported in self-aligning bearings, so that the critical-speed problem is the same as the free-vibration problem for the system regarded as a simply supported beam. The distributed mass of the beam has already been accounted for in the 40-lb. weights of the mass particles. The influence coefficients are found to be $\alpha_{11} = \alpha_{33} = 1.711$; $\alpha_{22} = 3.06$; $\alpha_{12} = \alpha_{21} = \alpha_{23} = \alpha_{32} = 2.11$; $\alpha_{13} = \alpha_{31} = 1.339$. All units are 10^{-3} in./lb.

a. Find the first natural frequency (or critical speed) by the iteration method.

b. Find the second natural frequency (this is easy, because the second-mode shape is known from symmetry, and the iteration method degenerates to a single step).

c. Find the third natural frequency, using the orthogonality condition to determine the third-mode shape.

W_1 W_2 W_3 $W_1 = W_2 = W_3 = 40$ lb
|←—15"—→|←—15"—→|←—15"—→|←—15"—→| $E = 30 \times 10^6$

Problem 116.

117. Modify the previous problem by changing the weight W_2 from 40 to 60 lb., thus making the system unsymmetrical. Find the three frequencies. This time the exact method, by direct solution of the differential equation, involves less work than the iteration method.

118. The frequency formula (29) on page 432 is expressed in general terms. Show that it can be transformed to

$$\text{v.p.m.} = \frac{30}{\pi} a_n \sqrt{\frac{Eg}{\gamma} \cdot \frac{k}{l^2}}$$

where k is the radius of gyration of the beam section and γ the weight per unit volume of beam material. Apply this to three frequent cases: a beam of rectangular section bh; a beam of solid circular section of diameter d; a thin-walled pipe of diameter d and wall thickness t, all made of steel. Show that the formula reduces to

$$\text{v.p.m.} = 560,000 \frac{h}{l^2} a_n \quad \text{for the rectangle of height } h$$

$$\text{v.p.m.} = 480,000 \frac{d}{l^2} a_n \quad \text{for the solid circular bar}$$

$$\text{v.p.m.} = 685,000 \frac{d}{l^2} a_n \quad \text{for the thin-walled pipe}$$

(All dimensions are in inches.) Note that neither the wall thickness of the tube nor the width of the rectangular beam have any influence on the frequency.

119. Refer to Fig. 4.28 on page 153. Calculate the first cantilever frequency by Rayleigh's method based on the following assumed shapes:

a. A parabola.

b. A mixture of a parabola and a cubic parabola $(y = Cx^3)$, so adjusted that the bending moment or curvature at the free end is zero.

120. *a.* In a system consisting of either a weightless string or a weightless shaft, supported at both ends and loaded with n discrete weights, assume the Rayleigh curve for the first mode to be equal to the static-deflection curve due to gravity. Prove that Rayleigh's method then leads to the formula

$$\omega_n = \sqrt{g \frac{W_1 y_1 + W_2 y_2 + W_3 y_3}{W_1 y_1^2 + W_2 y_2^2 + W_3 y_3^2}}$$

where y_1, y_2, y_3 are the static deflections. In calculating the strain energy note that it is the same as that due to the gravity forces; and, instead of evaluating an integral over the length of the beam, find the work done by these forces as they deform the beam.

b. Use the above formula to check the first frequency of Problem 116.

121. For a uniform beam on two end supports, the natural shape is a half sine wave. Suppose this fact is not known and the natural frequency is to be determined by the Stodola method. Assume for a first deflection curve a parabola $y = y_0(1 - 4x^2/l^2)$ with the center as origin. Enter with this expression into the Stodola process, and carry out the integrations by calculation (not graphically). Determine the first approximation to the natural frequency in this manner.

122. A beam of uniform cross section EI, mass per unit length μ or ρA, and total length $3l$ is supported at the points $x = 0$ and $x = 2l$.

a. Write an approximate shape, fit for starting the Rayleigh method, by assuming 180 deg. worth of sine wave tipped down at 0: $y = \sin \pi \frac{x}{3l} - ax$. Calculate the slope a in this expression so as to satisfy the condition at $x = 2l$. What is the bending moment at the end $x = 3l$?

b. Assume an algebraic expression

$$y = x + ax^2 + bx^3 + cx^4$$

and calculate a, b, and c so as to satisfy the conditions of zero bending moment at $x = 0$ and at $x = 3l$, and also the condition at $x = 2l$.

c. Find the frequency by Rayleigh's method, using either expression a or b, whichever you dislike least.

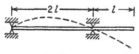 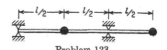

Problem 122.   Problem 123.

123. A uniform beam of total length $3l/2$ has supports at one end and at two-thirds of its length. The mass of the beam itself is m, and in addition it carries concentrated masses $m/3$ at the overhung end and in the middle of the span. Bending stiffness is EI.

a. Set up a "Rayleigh" shape of vibrational deformation in algebraic form, satisfying the condition of zero curvature at the two extreme ends.

b. With this formula find the frequency by Rayleigh's method.

124. A uniform bar is freely supported at its two quarter-length points. Find the natural frequency by Rayleigh's method. It is suggested that you use an algebraic polynomial trial expression which fits the end conditions.

125. A beam of length l on two supports has zero height at the supports and a linearly increasing height toward a maximum at the center. For half the beam let $\mu_1 = C_1 x$ and $EI = C_2 x^3$.

Take for the Rayleigh deflection curve an ordinary parabola, and find the natural frequency.

126. A cantilever beam of length l has a constant bending stiffness EI, but a mass variable along its length. It is μ_0 per running inch at the built-in end, it is $2\mu_0$ per running inch at the free end; and it varies linearly from μ_0 to $2\mu_0$ along the length. Taking a parabola as an approximation of the first mode of motion, calculate

  a. The frequency by Rayleigh's method.

  b. By Stodola's method find the next better approximation for the deflection curve, and estimate the frequency from it.

127. Apply Stodola's method to the problem of the natural frequency of a simple, uniform cantilever beam. Assume for the first approximation of the shape $y = Cx^2$. Find the next iteration by integration, and identify the end deflections of the two shapes, thus determining the frequency.

128. A ship's propeller shaft has a length of 200 ft. between the engine and the propeller. The shaft diameter is 12 in. The propeller has the same moment of inertia as a solid steel disk of 4 ft. diameter and 6 in. thickness. The modulus of shear of the shaft is $G = 12 \times 10^6$ lb. per square inch. If the shaft is supposed to be clamped at the engine, find the natural frequency of torsional vibration, taking account of the inertia of the shaft by means of Rayleigh's method (steel weighs 0.28 lb. per cubic inch).

129. The coil springs of automobile-engine valves often vibrate so that the individual coils move up and down in the direction of the longitudinal axis of the spring. This is due to the fact that a coil spring considered as a "bar" with distributed mass as well as flexibility can execute longitudinal vibrations as determined by Eq. (4.21a). Find the equivalents for μ_1 and AE in (4.21a) in terms of the coil diameter D, wire diameter d, number of turns per inch n_1, modulus of shear G, mass per turn of spring m_1.

Calculate the first natural period of such a spring of total length $l(n = n_1l)$ clamped on both sides.

130. A cantilever beam of total length $2l$ has a stiffness EI and a mass per unit length μ_1 along a part l adjacent to the clamped end, whereas the other half of it has a stiffness $5EI$ and a unit mass $\mu_1/2$. Find the fundamental frequency by Rayleigh's method.

131. A small ⅛-hp. motor frame has the following characteristics (Fig. 4.34): $\alpha = 220$ deg., $R = 2.75$ in.; $I = 0.0037$ in.⁴; $E = 27.10^6$ lb./in.²; $\mu = 0.00052$ lb. sec.²/in.². Find the fundamental frequency.

132. A mass hangs on a coil spring (Fig. 2.3 without damping or excitation). If the mass of the spring itself is not negligible with respect to the end mass, calculate what percentage of the spring mass has to be added to the end mass if the natural frequency is to be found from $\omega^2 = k/m$

  a. By Rayleigh's method.

  b. By the exact theory.

133. A uniform bar of length l, bending stiffness EI, and mass per unit length μ_1 is freely supported on two points at distance $l/6$ from each end. Find the first natural frequency by Rayleigh's method.

134. A ship drive, such as that discussed with reference to Fig. 5.18, consists of a propeller weighing 50,000 lb. and a line shaft of 19 in. diameter and 188 ft. length, on the other end of which there is a large gear weighing again 50,000 lb. The gear is driven by pinions and steam turbines which have no influence on the longitudinal vibrations of the system. On the inboard side of the main gear the thrust is taken by a Kingsbury thrust bearing, the supporting structure of which has a stiffness in the longitudinal direction of the shaft of 2.5×10^6 lb./in.   The propeller has four blades and consequently gives four longitudinal impulses to the shaft per revolution.

Calculate the two critical speeds of the installation, considering it as a two-degree-of-freedom system, distributing the shaft mass equally to the propeller and to the gear mass.

135. Solve Problem 134 by the exact method, assuming the shaft mass to be uniformly distributed, and find the numerical answer for the lowest critical speed.   The data of Problem 134 are taken from an actual case.   The vibration was eliminated by stiffening the thrust bearing supports.

136. To calculate by Rayleigh's method the antisymmetrical, three-noded frequency of a free-free bar of length $2l$, assume for the curve a sine wave extending from -180 deg. to $+180$ deg., with a base line rotated through a proper angle about the mid-point, so that it intersects the sine curve in two points besides the center point.

a. Determine the slope of the base line so as to satisfy the condition that the angular momentum about the center remains zero during the vibration.

b. Calculate the frequency with the curve so found.

137. The potential energy of a membrane, such as is shown in Fig. 4.37, is calculated by multiplying the tension T by the increment in area of each element caused by the elastic deformation.

a. If the deformation has rotational symmetry about the central axis (as shown in Fig. 4.37), derive that this energy is

$$ Pot = \int \frac{T}{2} \left(\frac{dy}{dr} \right)^2 dA $$

b. Assume for the deformation a sinusoid of revolution and calculate the frequency by Rayleigh's method.

138. In connection with the numerical Stodola or "iteration" method, discussed on page 163, carry out the following calculations:

a. Starting with the first assumption for the second mode, $a_1 = 1.000$; $a_2 = 0.500$; $a_3 = -0.750$; carry out the various steps without eliminating the first mode, and observe that gradually the solution converges to the first mode and not to the second.

b. In figuring the third mode start with an assumption such as $a_1 = a_3 = 1.000$, $a_2 = -1.000$, and eliminate from this solution the first and second harmonic contents by means of Eq. 3.9b.   Note that the shape so obtained is the exact solution.

Chapter 5

139. A single-cylinder engine weighs complete 300 lb.; its reciprocating weight is 10 lb., and the rotating weight is 5 lb.   The stroke $2r = 5$ in., and the speed is 500 r.p.m.

a. If the engine is mounted floating on *very* weak springs, what is the amplitude of vertical vibration of the engine?

b. If the engine is mounted solidly on a solid foundation, what is the alternating force amplitude transmitted?

Assume the connecting rod to be infinitely long.

140. Construct the piston-acceleration curve (Fig. 5.3) for an engine with a very short connecting rod, $l/r = 3$.

141. Sketch one full cycle of the inertia-torque variation [Eq. (5.15)] for an engine with $l/r = 3$.

142. Prove the four propositions on inertia balance stated on page 182.   Find also the balance properties of a three-cylinder two-cycle (0-120-240) engine.

143. A 4-cylinder engine has all 4 cylinders in one plane, on a crank shaft of 2 cranks in line, 90 deg apart.　Find:

a. The amount of necessary counterweight at A or A' in order to reduce the primary inertia force of one crank and pair of pistons to a force of constant magnitude rotating in a direction opposite to that of the crank shaft.

b. The secondary inertia force of one crank.

c. The necessary counterwight and its angular location at B and C (gears rotating at 1:1 speed opposite to the crank shaft) in order to balance for primary forces and moments.

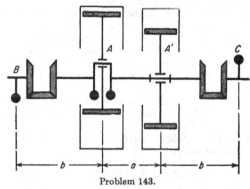

Problem 143.

144. The figure shows a "wobble-plate" engine.　A number of stationary cylinders are equally spaced angularly around the central shaft.　By properly proportioning the inertia of the piston and piston rods in relation to the inertia of the wobble plate, the engine can be balanced perfectly.　For purposes of this analysis the pistons and rods may be assumed to have a uniformly distributed mass around the axis of rotation.　The wobble plate is assumed to be a disk of total weight W_{disk}, uniformly distributed over its circular area of radius R_{disk}.　The total weight of all pistons and connecting rods is W_{pis}, supposedly concentrated on a circle of radius R_{pis} from the x-axis. Find the relation between these variables for which perfect balance is accomplished.

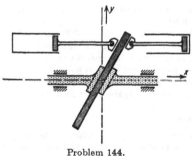

Problem 144.

145. The torsional amplitudes of any engine at slow speeds are very large but the crank shaft stresses associated with it are small.　In order to visualize this condition, consider a two-disk system I_1, I_2, connected by a shaft k, with a torque $T_0 \sin \omega t$ acting on disk I_1 only.　Calculate and plot:

a. The amplitude of the engine I_1 as a function of frequency.

b. The shaft torque as a function of frequency.

146. Find the first natural frequency of a four-cylinder engine driving an electric generator of the following characteristics:

$$I_{1,2,3,4} \text{ of the cranks, pistons, etc.} = 50 \text{ lb. in. sec.}^2 \text{ each}$$
$$I_5 \text{ of flywheel-generator assembly} = 1,000 \text{ lb. in. sec.}^2$$
$$k_1 = k_2 = k_3 = k_4 = 10^7 \text{ in. lb./rad.}$$

147. The torsional system shown consists of three flywheels with moments of inertia 200, 100, and 1,000 lb. in. sec.^{2}, respectively. The flywheels are connected by two shafts, stiffness of each shaft being 10^6 in.-lb./rad. Make a reasonable guess for the first natural frequency, and with it perform one Holzer calculation.

From the result of this calculation, deduce whether your guess for the natural frequency was too high or too low.

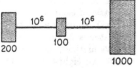

Problem 147.

148. An engine consists of four cylinders in line, each having an equivalent moment of inertia of 100 lb. in. sec.^{2} It drives a generator of $I = 2,000$ lb. in. sec.^{2} through a connecting shaft of stiffness $k = 10^4$ in.-lb./rad. The three crank throws between the four cylinders have a stiffness of $k = 10^5$ in.-lb./rad. each. Find the first and second natural frequencies by Holzer's method.

149. The sketch shows a six-cylinder (four-cycle) truck engine. Estimate the natural frequency, and make one Lewis calculation with it. Call that answer good, but state whether it is too large or too small.

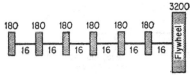

Problem 149.

Find the location of the third- and sixth-order critical speeds. Explain whether or not these are "major" speeds and why.

All inertias are WR^2 values, expressed in lb./in.^{2} All flexibilities are in millions of in.-lb./rad.

150. A six-cylinder engine coupled to a heavy generator has cylinder inertias of 50 lb. in. sec.^{2} and crank throw stiffnesses of $k = 60 \times 10^6$ in.-lb./rad. The generator inertia is 2,000 lb. in. sec.^{2}

Make a good guess at the first natural frequency, and express the answer in v.p.m.

151. An eight-cylinder in-line engine has cylinder inertias of 5,000 lb. in. sec.^{2} each and throw stiffnesses of 1 million in.-lb. rad. each. It is coupled through a shaft of $k = 2 \times 10^6$ to a flywheel of 50,000.

a. Find the first natural frequency by Lewis's method.

b. Estimate the second frequency.

152. a. An eight-cylinder engine drives a heavy propeller through a flexible shaft. Find an approximate value for the first mode ω^2, suitable for a first try in the Holzer table.

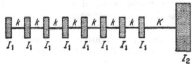

$k = 10^8$ in. lb /rad $I_1 = 1000$ lb in. sec^2
$K = 5 \times 10^6$ in. lb /rad $I_2 = 20,000$ lb in. sec^2

Problem 152.

b. Find an approximation for the second mode ω^2, suitable for a first try either by Lewis's method or by a Holzer table.

c. Let the crank shaft be 0, 180, 90, 270, 270, 90, 180, 0, and let the firing order be 1 6 2 5 8 3 7 4. For the first mode, which of the following three orders of vibration is serious?

$$\text{Order } \tfrac{1}{2}, 2, 4$$

d. Same question as (c), but now for the second mode. Assume that all critical speeds are within the running range of the engine.

153. A Diesel engine by itself has a rotatory inertia of $WR^2 = 200,000$ lb. in.^{2}, and a total stiffness of the entire crank shaft of 5 million in.-lb./rad. On one end the engine is directly connected to a flywheel of $WR^2 = 1$ million lb. in.^{2} Make a good guess at the natural frequency of the system, and check it with one Lewis calculation. As a result of this make a second guess, but do not follow the process any further. Express the second guess in terms of v.p.m.

154. A six-cylinder Diesel engine is directly coupled to a generator. The moment of inertia of each crank is I, and that of the generator is $5I$. The torsional stiffness of the shafting between two cylinders is k, and that between the last cylinder and the generator is $2k$ (equivalent shaft length one-half of that between cylinders). Using Lewis's method, find the first two natural frequencies of this system.

155. An eight-cylinder in-line Diesel engine drives two generators through gears. The eight cylinders have each an inertia of 1,000 lb. in. sec.²; the stiffness of one crank throw is 10×10^6 in.-lb./rad.; the generator inertias are 15,000 and 5,000 lb. in. sec.², and their drive shafts have stiffnesses of 10×10^6 and 20×10^6 in.-lb./rad. as indicated.

a. Describe briefly how you go about combining Lewis's method with Holzer's method for a branched system such as this one.

b. Estimate the first frequency, and make the first Lewis-Holzer calculation.

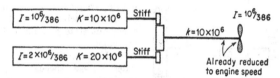

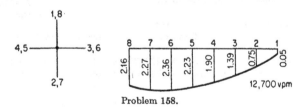

Problem 156.

156. Two engines, one twice as inert as the other, running at the same speed, drive a common propeller through a reduction gear. The stiffnesses and inertias are as shown in the figure. Find the first three natural frequencies.

157. *a.* Explain briefly why in a four-cycle engine torsional vibrations occur of integer and half-integer orders, and why the orders appearing in a two-cycle engine are integer only.

b. A 9-cylinder 2-cycle engine has the firing order 1 8 6 4 2 9 7 5 3. Sketch the vector diagrams (equal magnitude of vectors) for the orders 1 to 9. How many of these diagrams are different?

c. Assuming that in the first mode of motion all cylinders have the same amplitude, what orders of vibration from 1 to 9 are most serious?

d. Assuming that in the second mode of motion the cylinder amplitudes are 4, 3, 2, 1, 0, −1, −2, −3, and −4 deg. for the consecutive cylinders 1 to 9, which of the orders 1 to 9 are serious?

Problem 158.

158. A four-cycle Diesel engine consists of 16 cylinders on an eight-throw crank shaft in two banks of 8 cylinders each with a 60-deg. V-angle between. The crank diagram is shown, as is also the configuration of one of the natural modes of the system, and the firing order is 1 6 2 5 8 3 7 4. The engine has to run in a speed range of 2,400 r.p.m. on down. Consider four possible critical speeds, for which the p-factors (ratio of harmonic alternating torque to mean torque) are shown below:

Order	5½	6	6½	7
p-factor	0.38	0.28	0.20	0.14

Find which one of the four above critical speeds is the most dangerous one, and list all four speeds in order of their severity.

159. A General Motors two-cycle Diesel (type 12-567, six cylinders in line, two banks at 45-deg. V) drives an Eliott generator at 720 r.p.m. All inertias are lb. in. sec.^{2}; all stiffnesses, in.-lb./rad. Make rough guesses for the frequencies of the first and the second modes of this

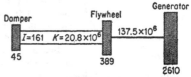

Problem 159.

system. Based on these guesses, find what order critical speed is near the running speed of 720 r.p.m. in the first (one noded) mode and in the second (two-noded) mode.

160. The same question as Problem 159 for a General Motors 16-278A type two-cycle Diesel, eight cylinders in line, two banks at 40 deg. V, driving a generator at 720 r.p.m. It looks like this:

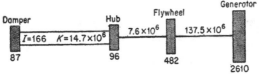

Problem 160.

161. The contribution toward the moment of inertia of a single crank throw by the crank itself is usually of the same order of magnitude as the contribution by the moving parts, i.e., the piston and connecting rod. In the problem below assume this to be the case, i.e., the I of a crank is twice that caused by the piston and connecting rod.

A truck engine (The Autocar Co., model 3770HV) has the following characteristics:

Number of cylinders in line:	6
Piston weight:	2 lb. 10 oz.
Rot. end connecting rod weight:	3 lb. 0
Rec. end connecting rod weight:	1 lb. 4 oz.
Crank radius:	2½ in.
₵ distance between cranks:	5⅛ in.
Shaft diameter:	3¼ in.
Crank-pin diameter:	2⅜ in.

The length of the crankpin is equal to the length of the shaft in the main bearings. The engine, as usual with automobile installations, is "free" at the front end and carries a "large" flywheel at the rear end. The system behind this flywheel is "very soft," i.e., the drive shaft between engine and wheels has a value of k very much smaller than the crank shaft.

Estimate the lowest natural frequency for the purpose of starting either a Lewis or a Holzer calculation.

162. A centrifugal pendulum damper of order $n = 3$ is to be designed for an engine at 800 r.p.m. The available radial distance from the center of the shaft to the point of suspension of the pendulum $R = 4\frac{1}{2}$ in. The alternating torque that the damper has to produce is 15 ft.-lb., and the allowable angle of its relative swing is 0.1 radian. What is the required weight of the pendulum?

163. A centrifugal pendulum consists of a point mass m on a weightless string of length a, attached at a point distant a from the center of rotation. Angular speed of the disk is Ω.

a. Find the natural frequency of *small* vibrations.

b. If the pendulum string is held at position B 90 deg. from the radial direction and released from that position, at what speed will it fly through the radial position A? The angular speed Ω is held constant.

Problem 163.

164. a. Sketch the steam-torque curve of a double-acting steam cylinder of which the inlet valve is open for one-fourth revolution after the dead-center position. During the next quarter revolution the steam expands according to pv = constant. The engine works without compression.

b. Sketch the combined torque curve of an engine made up of three such cylinders on a 120, 240, 360 deg. crank shaft and also the combined torque curve of a six-cylinder Diesel engine based on Fig. 5.19a. Compare the two.

165. Draw the four fundamental star diagrams for the engine (Fig. 5.27) for each of the four firing orders that are possible with the given crank shaft (see figure).

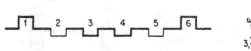

Problem 165. Given crankshaft.

166. Discuss the star diagrams for the eight-cylinder engine (0, 180, 90, 270, 270, 90, 180, 0) without considering the elastic curve. How many fundamental diagrams are there, and to which orders of vibration do they belong?

167. The turbine ship drive of Fig. 5.18, page 195, is excited by the four-bladed propeller only, the intensity of the exciting torque being 0.075 times the mean torque. Assume a propeller damping corresponding to twice the slope of the diagram, Fig. 5.28, and assume that diagram to be a parabola. Neglect damping in other parts of the installation.

a. Calculate the amplitude at resonance at the propeller.

b. From the Holzer calculations of pages 196 and 197 find the resonant torque amplitudes in the shafts 2-3 and 3-4.

c. At what propeller r.p.m. does this critical condition occur?

168. Problem 167 determines the resonant amplitude of the ship drive, Fig. 5.18. The resonance curve about that critical condition is found by calculating the undamped resonance curve and sketching in the damped one. Points on the undamped curve are determined by calculating a Holzer table for neighboring frequencies and by interpreting the "remainder torque" of these tables as a forced propeller-exciting torque. The mean propeller torque is 6,300,000 in. lb. at the rated speed of 90 r.p.m. and is proportional to the square of the speed. Find the amplitudes of forced vibration at the propeller for $\omega^2 = 145$ and $\omega^2 = 215$, and from the results sketch the resonance curve.

169. An aircraft engine consists of two six-cylinder-in-line blocks arranged parallel to each other and coupled to each other at each end by three identical spur gears, so that the two blocks run at equal speeds in the same direction. One set of natural modes has nodes at both ends of each block.

a. What modification has to be made in the first line of the ordinary Holzer table to accommodate the node at one end?

b. What is the Holzer criterion at the other end?

c. What is the Θ in Lewis's method (page 193)?

d. Calculate the lowest natural frequency of a system of six equal inertias I, coupled to each other and to two solid walls at either end by seven identical shafts of stiffness k.

170. A variation (due to Chilton) of the damper of Fig. 5.37 consists of a steel block of weight W with a hardened cylindrical bottom that can roll on a hard cylindrical guide (see figure). The two radii of curvature R_1 and R_2 are large and their difference $\Delta R = R_2 - R_1$ is small. The distance between the center of gravity G and the contact point is a, and the radius of gyration about G is ρ.

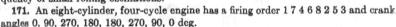

a. Calculate the natural frequency of small rolling oscillations in a gravity field g.

b. The assembly rotates with speed Ω about a center O; the distance $OG = r_0$, and gravity is neglected. Calculate the frequency of small rolling oscillations.

Problem 170.

171. An eight-cylinder, four-cycle engine has a firing order 1 7 4 6 8 2 5 3 and crank angles 0, 90, 270, 180, 180, 270, 90, 0 deg.

a. Sketch the vector diagrams for the various orders of vibration without considering the magnitude of the vectors.

b. If at a certain mode the Holzer amplitudes are as follows:

No. 1, 1.000;   No. 2, 0.900;   No. 3, 0.800; etc.

down to No. 8, 0.300, and, if the $3\frac{1}{2}$ order harmonic torque is 100,000 in. lb., find the work input per cycle at the resonance of this order if cylinder No. 1 vibrates ± 1 deg.

c. If the above mode occurs with a value $\omega^2 = 2,000$ in the Holzer table, what is the critical r.p.m. of order $3\frac{1}{2}$?

d. What is the most dangerous r.p.m. of this engine?

e. What is the state of balance of this engine?

172. An idealized single-acting steam engine with constant pressure during the entire stroke (no cutoff) and an infinitely long connecting rod has a torque-angle diagram consisting of 180 deg. of sine wave, then 180 deg. of zero torque, etc.; the torque never becomes negative. Find the harmonic torque components by a Fourier analysis, in terms of the mean torque $T_{max/\pi}$.

Chapter 6

173. Consider a simple beam on two plain end supports of total length l and of uniform bending stiffness EI. Consider this beam itself weightless, but loaded with a single central mass m. Then we know that the natural frequency for bending vibrations is $\omega_n^2 = 48EI/ml^3$. Substitute this exact frequency in Holzer's extended method (page 230), and verify that the end bending moment is zero, as it should be. Then perform two more such calculations with the values $\omega^2 ml^3/EI = 40$ and again 60, one below and one above the true frequency. What are the "remainder" bending moments for these two cases?

174. Find the critical speed in revolutions per minute of the system shown in Fig. 6.1 in which the disk is made of solid steel with a diameter of 5 in. and a thickness of 1 in. The total length of the steel shaft between bearings is 20 in., and its diameter is $\frac{1}{2}$ in. The bearings have equal flexibility in all directions, the constant for either one of them being $k = 100$ lb./in.

175. The same as Problem 174 except that the bearings have different vertical and horizontal flexibilities: $k_{hor} = 100$ lb./in. and $k_{vert} = 200$ lb./in. for *each* of the bearings.

176. On a horizontal platform are two small motors A and B, their shafts parallel and horizontal, at distance $2a$ apart. The motors are unbalanced, each producing a rotating centrifugal force. The motors rotate at equal speeds in the same direction and their two centrifugal forces are equal in magnitude, but one of them runs by the constant angle α ahead of the other.

a. If C denotes the instantaneous intersection of the two centrifugal forces through A and B, prove that the locus of C is a circle passing through A and B with its center somewhere on the perpendicular bisecting AB.

b. Using this result, prove that the resultant of the two centrifugal forces is a single rotating force rotating about a fixed point located on the perpendicular bisector of AB at a distance $a \tan \alpha/2$ from AB.

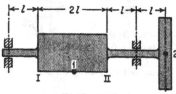

Problem 177.

177. The figure shows a machine with a rigid overhung rotor. The initial unbalance consists of 1 in. oz. in the center of the main rotor and of 2 in. oz. on the overhung disk, 90 deg. away from the first unbalance. Find the corrections in the planes I and II.

178. A rotor is being balanced in the machine of Fig. 6.10, pivoted about the fulcrum F_1. The following amplitudes of vibration are observed at the critical speed:

1. 14 mils for the rotor without additional weights.
2. 10 mils with 3 oz. placed in location 0 deg.
3. 22 mils with 3 oz. placed in location 90 deg.
4. 22 mils with 3 oz. placed in location 180 deg.

Find the weight and location of the correction (Fig. 6.11).

179. A short rotor or flywheel has to be balanced. Observations of the vibration at one of the bearings are made in four runs as follows:

> Run 1; rotor "as is":   amplitude 6 mils
> Run 2; with 1 lb. at 0 deg.:   amplitude 5 mils
> Run 3; with 1 lb. at 180 deg.:   amplitude 10 mils
> Run 4; with 1 lb. at 90 deg.:   amplitude $10\frac{1}{2}$ mils

Find the amount and angular location of the necessary correction weight.

180. In the balancing process we make the following observations:

$a_0 =$ amplitude of vibration of the unbalanced rotor "as is."

$a_1 =$ amplitude with an additional one-unit correction at the location 0 deg.

$a_2 =$ same as a_1 but now at 180 deg.

The ideal rotor, unbalanced only with a unit unbalance (and thus *not* containing the original unbalance), will have a certain amplitude which we cannot measure. Call that amplitude x. Let the unknown location of the original unbalance be φ.

Solve x and φ in terms of a_0, a_1, and a_2, and show that in this answer there is an ambiguity in sign. Thus *four* runs are necessary to determine completely the diagram of Fig. 6.11.

181. In a Thearle balancing machine (page 238), the total mass of the rotating parts is M, the eccentricity e, the mass of each of the balls at the ends of the arms is

m, and the arm radius r. Find the angle α which the arms will include in their equilibrium position when released about the resonant speed.

182. A steel disk of 5 in. diameter and 1 in. thickness is mounted in the middle of a shaft of a total length of 24 in. simply supported on two rigid bearings (Fig. 6.1). The shaft diameter is $\frac{1}{2}$ in.; it is made of steel also. The shaft has filed over its entire length two flat spots (Fig. 6.17a), so that the material taken away on either side amounts to $\frac{1}{500}$ part of the cross section (total loss in cross section $\frac{1}{250}$). Find the primary and secondary critical speeds.

Find the amplitude of the secondary alternating force, and calculate the unbalance which would cause an equal force at the primary critical speed.

183. A weightless shaft of total length l between bearings (simple supports) carries at its center a disk of diametral mass moment of inertia I_d. The disk is keyed on at a (small) angle φ_0. When rotating at constant angular speed ω, the centrifugal forces tend to diminish the angle φ_0 to a new value $\varphi_0 - \varphi$. Find the ratio φ/φ_0 as a function of the speed ω.

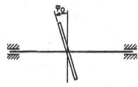

Problem 183.

184. Consider a simply supported shaft with a central disk. One possible mode of vibration involves no displacement of the center of the disk, but only an angular displacement of it. The shaft stiffness constant at the disk is $k = 12EI/l$, being the moment required at the center of the shaft to produce unit angle φ at that location.

a. Find the natural frequency of this motion for a non-rotating disk ($\Omega = 0$), and call this frequency ω_0.

b. Find the natural frequency (or frequencies) ω for the case of a disk rotating with angular speed Ω.

c. Plot this result in a diagram ω/ω_0 versus Ω/ω_0.

185. Calculate the critical speed of an overhung (cantilever) disk of 10 in. diameter weighing 20 lb., the shaft being of steel, 10 in. long and $\frac{1}{2}$ in. diameter.

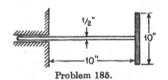

Problem 185.   Problem 186.

186. A shaft of length $2l$ and bending stiffness EI is supported on two bearings as shown. The bearings allow the shaft to change its angle freely but prevent any deflection at those two points. The disk at the end has a moment of inertia I_p about its axis of rotation. (Thus I is measured in in.^{4}, and I_p in lb. in. sec.^{2}.) The mass of the disk is m. Find the critical speed.

187. A shaft of stiffness EI, length l, and negligible mass is freely supported at both ends. At one end it carries a disk of mass m and moment of inertia I_d about the diameter. This disk is very close to the bearing: assume it to be at the bearing, so that it can tilt, but not displace its center of gravity.

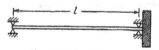

Problem 187.

a. For the case of no rotation, find the natural frequency.

b. Try to find the whirling critical speed, but do not spend more than 3 minutes trying.

c. Set up the general frequency equation in terms of a rotational speed Ω and a whirl ω; make this equation dimensionless in terms of K- and R-functions.

$$K = \frac{\omega}{\sqrt{\frac{EI}{I_d l}}}; \quad R = \frac{\Omega}{\sqrt{\frac{EI}{I_d l}}}$$

d. Solve it; plot it; verify answer *a*, and give the answer to question *b*.

188. A shaft EI of total length l is supported "freely" at its two ends At a quarter length between the end bearings the shaft carries a disk of mass m and of diametral moment of inertia I_d. Find

a. The non-rotating natural frequency.

b. The whirl frequency.

189. A shaft of total length l on end bearings carries two disks at the quarter-length points. The disks have mass m and inertia I_d; the shaft stiffness is EI.

a. Set up the equations for the whirling shaft, where the whirl frequency is equal to the r.p.m. (forward synchronous precession).

b. Make the frequency equation dimensionless in terms of

$$\text{The critical-speed function} \quad K^2 = \frac{ml^3\omega^2}{EI}$$

$$\text{The disk effect} \quad D = \frac{I_d}{ml^2}$$

c. Find the whirling speed for the following three cases:

$$D = 0, \quad D = \infty, \quad \text{and} \quad D = \tfrac{1}{12}$$

d. Suppose the shaft to be *non-rotating* now. For this different problem the analysis is almost the same. Reason out what change this makes in the original set-up, and find what small change occurs in the final K, D equation. Find the frequencies again for $D = 0, D = \infty, D = \tfrac{1}{12}$.

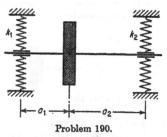

Problem 190.

190. A solid disk of mass M and radius R is keyed to a stiff and weightless shaft, supported by springs k_1 and k_2 at distances a_1 and a_2. The nearer spring is the stiffer one, so that $k_1 a_1 = k_2 a_2$. The shaft rotates at speed Ω. Calculate the natural frequencies of the system and plot them in the form ω/ω_a against Ω/ω_a,

where $\omega_a^2 = (k_1 + k_2)/M$ and $\omega_b^2 = (k_1 a_1^2 + k_2 a_2^2)/\tfrac{1}{4}MR^2 = 4\omega_a^2$.

191. Calculate the abscissas and ordinates of several points on the curves of Fig. 6.43 by means of Eq. (6.21).

192. A cantilever shaft has a stiffness EI over a length l and is completely stiff over an additional distance l_1. The stiff part has a total mass m while the flexible part is supposedly massless. Calculate the natural frequencies as a function of l_1/l between the values $0 < l_1/l < 1$, and plot the result in a curve.

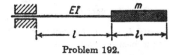

Problem 192.

193. In a laboratory experiment one small electric motor drives another through a long coil spring (n turns, wire diameter d, coil diameter D). The two motor rotors have inertias I_1 and I_2 and are distance l apart.

a. Calculate the lowest torsional natural frequency of the set-up.

b. Assuming the ends of the spring to be "built in" to the shafts, calculate the r.p.m. of the assembly at which the coil spring bows out at its center, due to whirling.

194. Consider the transverse free vibration of a pipe containing a flowing liquid. If the liquid does not move, the beam analysis of page 148 applies, with μ_1 replaced by $\mu_{\text{pipe}} + \mu_{liq}$.

a. When the liquid flows through the pipe with a velocity V_0, show that the differential equation of motion is

$$EI \frac{\partial^4 y}{\partial x^4} + \mu_{liq} V_0^2 \frac{\partial^2 y}{\partial x^2} + 2\mu_{liq} V_0 \frac{\partial^2 y}{\partial x \, \partial t} + (\mu_{liq} + \mu_{\text{pipe}}) \frac{\partial^2 y}{\partial t^2} = 0$$

b. Assume, for simplicity, that the third term in the result for (*a*) is of minor importance as compared to the others. Solve the differential equation for this simplified case and show that the lowest natural frequency for simply supported ends is

$$\omega^2 = \omega_0^2 - \frac{\pi^2 \mu_{liq}}{\mu_{\text{pipe}} + \mu_{liq}}$$

where ω_0 is the natural frequency for $V_0 = 0$.

c. The result for (*b*) indicates that ω decreases with increasing V_0; that is, the stiffness decreases. Find for what speed $V_{0,\text{crit}}$ the lateral stiffness disappears and the problem is reduced to a *static* one. For the case of $\mu_{\text{pipe}} \ll \mu_{liq}$ express $V_{0,\text{crit}}$ in half wave lengths of pipe (*l*) per cycle of vibration when the liquid does not move.

195. The drive of an aerodynamic wind tunnel consists of a driving motor I_1 coupled to a large fan I_2 which drives the air through the wind tunnel. The torsional elasticity between the motor and the fan is k, and the tunnel is idealized as an organ pipe of length l and cross section A. The coupling between the fan and the air column is expressed by two constants C_1 and C_2 with the following meaning: C_1 is a pitch constant relating the air displacement in the tunnel to the angular displacement of the fan: $\xi = -C_1 \varphi_2$. The second constant C_2 relates the forward torque on the fan to the pressure variation in the air column at the fan, so that the fan torque variation is $C_2 d\xi/dx$. In both these expressions x is the distance along the tunnel measured from left to right and ξ is the alternating component of the air displacement, positive to the right.

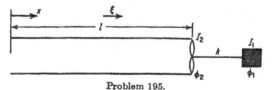

Problem 195.

Set up the differential equations of the system and from it deduce the frequency equation.

Chapter 7

196. Test the stability of the following frequency equations:
(*a*) $s^3 + 5s^2 + 3s + 2 = 0$.
(*b*) $s^4 + 8s^3 + 10s^2 + 5s + 7 = 0$.
(*c*) $s^4 - 2s^3 + 5s^2 - 3s + 2 = 0$.

197. The landing gear of an airplane consists of two wheels whose axes are rigidly attached to the fuselage and a third trailing wheel which is castored, *i.e.*, can swivel about a vertical axis. Necessarily the center of gravity of the airplane is located so that its projection falls within the triangle formed by the three wheels.

a. Prove that, if this gear is rolling over the ground with the two wheels forward as usual, the operation is unstable, *i.e.*, a small angular deviation of the rear wheel will increase and the plane will execute a "ground loop."

b. Prove that if the castored wheel is located in front of the two steady wheels, as with a so-called "tricycle," landing gear, the operation is stable.

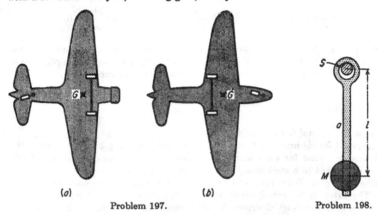

(*a*) (*b*)

Problem 197. Problem 198.

198. A pendulum with a light rod a and a heavy weight of mass M at a distance l from the point of support is hanging on a round shaft S. If the shaft S is rotating at a large angular velocity ω and the friction torque on the shaft is T_0, find:

a. The equilibrium position of the pendulum in terms of the angle α_0 with the vertical.

Discuss the small vibrations which the pendulum may execute about this equilibrium position for the following three cases:

b. The friction torque T_0 is absolutely constant.

c. T_0 increases slightly with increasing velocity of slip.

d. T_0 decreases slightly with increasing velocity of slip.

199. A weight W rests on a table with the coefficient of friction f. A spring k is attached to it with one end while the motion of the other end is prescribed by

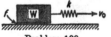

Problem 199.

$$t < 0 \quad v = 0$$
$$t > 0 \quad v = v_0$$

or, in words, at the time $t = 0$ the spring end suddenly starts moving with a constant velocity v_0. Discuss the motion and construct displacement-time diagrams of the mass for the three cases b, c, and d of Problem 198.

200. A certain cross section has a diagram (Fig. 7.21) with the following curves:

$$\text{Lift} = L_0 \sin 2\alpha$$
$$\text{Drag} = D_0 - \frac{D_0}{2} \cos 2\alpha$$

If a piece of such a section is mounted in the apparatus of Fig. 7.17 in the position $\alpha = 90$ deg., for what ratio L_0/D_0 does instability start?

201. The figure shows a Watt's governor with the dimensions l, a, m, M, and k. At standstill the spring k is such that the angle α of the flyball arms is 30 deg. At the full rotational speed Ω the angle α is 45 deg.

    a. Express k in terms of the other variables.

    b. Calculate the natural frequency at standstill.

    c. Calculate the natural frequency while the governor is rotating with speed Ω.

202. Transform Eq. (7.25) into a relation between four dimensionless variables: one frequency ratio $f = \omega_g/\omega_e$, two damping ratios $C_e = (c/c_e)_{eng}$ and $C_2 = (c/c_e)_{gov}$, and a dimensionless feedback or coupling quantity F. Plot the results so found on a diagram for one certain value of f; C_e = ordinate, C_g = abscissa, and Γ = parameter for the various curves. Interpret these graphs.

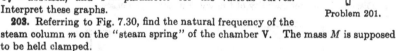

Problem 201.

203. Referring to Fig. 7.30, find the natural frequency of the steam column m on the "steam spring" of the chamber V. The mass M is supposed to be held clamped.

204. A system consists of an engine I_1, driving a shaft k_1. At the other end of k_1 is attached a fluid-flywheel coupling (page 217), the "driver" of which has an inertia I_2. The "follower" is attached to a piece of driven machinery of inertia I_3. Set up the differential equations of motion, using Eq. (5.37a), write the frequency equation, and find whether the system is or is not capable of self-excited oscillation.

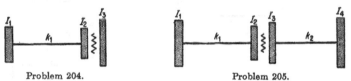

Problem 204.   Problem 205.

205. The same as Problem 204, only the "follower" of inertia I_3 drives a shaft k_2, at the other end of which is a flywheel I_4.

206. A windmill rotor of moment of inertia I_1 drives a driven machine of inertia I_2 through a shaft of torsional stiffness k. The wind torque on the rotor near the operating speed is $T_0 + C_1\omega$, where T_0 is the steady or average torque and ω is the instantaneous angular speed of the windmill. C_1 is a constant which may be either positive (torque increasing with speed) or negative (torque decreasing with speed). The driven machine presents a counter torque which is constant $= T_0$, independent of its speed. There is a governor in the system which keeps it from running away, but the governor is so slow that it does not budge when torsional vibrations take place in the system. Find whether the system is stable or unstable against torsional vibration, and give the answer for the two cases C_1 positive and C_1 negative.

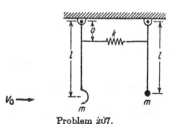

Problem 207.

207. A system consists of two pendulums, each of mass m concentrated in the bob, the bars of length l being weightless. They are coupled by a spring k at distance a from the top. One of the bobs is hemispherical. The wind blows with a speed V_0 and exerts a force $F = CV_{rel}^2$ on the hemispherical bob only. Find whether the small oscillations about the equilibrium position are stable or unstable.

208. A chamber has two openings as shown in the figure. At A a steady stream of air comes into it. At B air leaks out of it past a valve. When the valve is in the equilibrium position, the in and outflow balance, so that the pressure in the vessel remains constant. The equations of the system can be written as

$$m\ddot{x} + kx = Ap$$
$$\dot{p} = Cx$$

Here A is the area of the piston, x is the valve position, p is the pressure in the tank, and C is a constant, implying that the amount of air leaking out at B is proportional to the valve setting x.

  a. Discuss the sign of the constant C for the two cases shown in (a) and (b) of the figure.

  b. Discuss the flutter stability of the valve in these two cases.

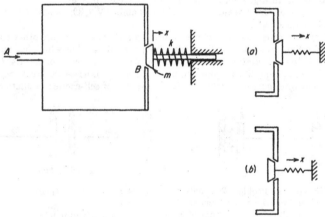

Problem 208.

209. In Problem 208 add a dashpot c to the piston in parallel with the spring, and then answer (b) again.

210. A piston of mass m in a cylinder is held by a spring k. The piston acts as a valve for the air flow that enters the cylinder through a series of many very small holes. The air leaks away through a small leak. At the equilibrium state, there is no motion of the piston, and the amount of air coming into the piston equals that escaping through the leak. Let

  x = upward displacement of piston, measured from the equilibrium position
  A = area of piston
  p = pressure in cylinder above the static-equilibrium pressure
  $C_1 x$ = volume of air entering cylinder per second (above amount entering at equilibrium)
  $C_2 p$ = volume of air leaking out per second (above normal leak at equilibrium)

Set up the equations, and derive the conditions of stability. Assume the air can be considered incompressible.

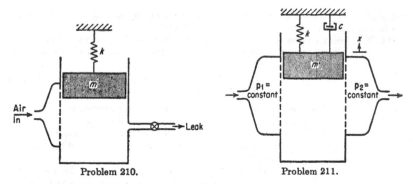

Problem 210.   Problem 211.

211. A more general case of Problem 210 with variable exit and entrance areas and taking account of compressibility of the gas.   Let

x = upward displacement of piston from static-equilibrium position
p_0, ρ_0 = pressure and density of gas in cylinder at equilibrium
V_0 = volume of cylinder at equilibrium
A = area of piston
p, ρ = pressure and density in cylinder at any time
$p' = p - p_0$, pressure in cylinder above p_0
$w_1 = C_1 x - C_2 p'$, mass rate of flow of gas into cylinder (above amount entering at static equilibrium)
$w_2 = C_3 x + C_4 p'$, mass rate of flow of gas out of cylinder (above amount leaving at static equilibrium)

Assume that $\dfrac{p}{\rho^\gamma}$ = constant describes the relation obeyed by the gas in the cylinder. Derive the conditions for stability.   Does the compressibility decrease the stability? Is the dashpot c necessary for stability?

212. Before the disaster of the Tacoma bridge was definitely diagnosed as caused by Kármán vortices, theories were proposed basing the failure on a flutter phenomenon.   In idealizing this situation, assume the following differential equations:

$$m\ddot{x} + k_x x = A V^2 \left(\theta - \frac{\dot{x}}{V}\right)$$

$$I\ddot{\theta} + k_\theta \theta = a \cdot A V^2 \left(\theta - \frac{\dot{x}}{V}\right)$$

Problem 212.

where $\theta - \dot{x}/V$ is the apparent angle of attack, A is a coefficient, so that the right hand member of the top equation is the lift, and a is a moment arm (usually about a quarter span) giving the wind torque on the bridge.

Discuss this with Routh's criteria, and find a wind speed above which the bridge will flutter.

213. A composite body consists of two masses m_1 and m_2, connected together by a weightless connection of length l between the two individual centers of gravity.   The masses have moments of inertia I_1 and I_2 about their own centers of gravity.   Find the moment of inertia of the combination about the composite G, and prove that $ab < \rho_G^2$.   Only when I_1 and I_2 both are zero (point masses connected by a weightless rod) is $ab = \rho_G^2$. Thus if an aircraft landing gear can be idealized as above, the wheel will be unstable in shimmy [see Eq. (7.42), page 334].

214. Two point masses m_1 and m_2 are connected by a uniform rod of mass m_3. Find the location of the center of gravity G of the combination and the moment of inertia of the entire assembly about G. Prove that

$$ab > \rho^2$$

except when $m_3 = 0$, when $ab = \rho^2$. Thus, if this assembly represents an aircraft

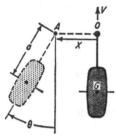

landing gear, the wheel will be stable against shimmy. However, the idealization of Problem 213 is much closer to the truth than this on account of the shape of the landing wheel itself.

215. In an aircraft landing gear let O be the point where the unstressed vertical swivel axis of the wheel meets the ground. The vertical strut is flexible with constant k, and during shimmy it bends out sidewise by an amount x. Let A be the point where the bent swivel axis intersects a plane at distance r above the ground. Let θ be the shimmy angle during landing speed V. Assume that point

Problem 215.

O moves forward in a straight line, i.e., that the front-wheel shimmy does not push the airplane aside. Rubber tires have been found to experience a sidewise force from the ground which is proportional to the angle between the vertical plane of the wheel and the direction of *relative* forward motion.

a. Show that this (small) angle is expressed by $\theta + \dfrac{\dot{x} + a\dot{\theta}}{V}$.

b. Assuming that only the wheel has mass, show that the equations of (small) motion of the wheel are

$$C\left(\theta + \frac{\dot{x} + a\dot{\theta}}{V}\right) + kx = -m(\ddot{x} + a\ddot{\theta})$$

$$kxa = I_G\ddot{\theta}$$

c. Find under what circumstances these equations represent a stable, nonshimmying wheel.

216. In the simple set-up for a shimmying aircraft landing gear (Fig. 7.41), add a torsional dashpot in the swivel axis, giving a torque $c\phi$ at point B in that figure. Complete the analysis of page 333 with this new term, and prove that the condition for stable operation is

$$c > \frac{mV}{l}(\rho_G^2 - ab)$$

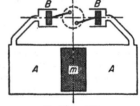

Problem 217.

which contains Eq. (7.42) as a special case for zero damping.

Chapter 8

217. In the center of the cylinder AA of cross section A (see figure) a piston of mass m can slide without friction. The pistons BB are moving back and forth in opposite phase and change the pressure of the air in the cylinder A between 95 and 105 per cent of atmospheric pressure. Assume that this change in pressure takes place isothermally or that $pv =$ const. The volume of one half of A together with its pipe and the cylinder B is V. Find the frequency or frequencies of motion of BB at which the mass m is in unstable equilibrium. Give a general discussion with the aid of Fig. 8.11.

218. A pendulum consists of a uniform bar of 5 in. length and ½ lb. weight. The base is given an alternating harmonic motion in a vertical direction with an amplitude $e = 0.5$ in. At what speed of the driving motor will the pendulum become stable in an upright position? Assume the curve of Fig. 8.12 to be a parabola passing through the origin and through the point $y = 0.5$ and $x = -0.1$.

219. Consider the free vibration of a simple system consisting of a mass on a non-linear spring of characteristic $f(x)$, given as a graphical curve. Show that the iso-clinics in the sense of page 344 are curves similar to the spring characteristic, and construct the $\dot{x}$- versus x-diagram. Show that it reduces to Fig. 8.21 for the linear case $f(x) = kx$.

220. Find the natural frequency for the system shown as a function of maximum amplitude x_0.

There are no forces in the springs at the center static-equilibrium position.

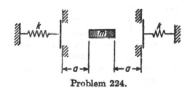

Problem 220.

221. Using the method of Fig. 8.39, find the approximate natural frequency for free vibrations of the system shown in Fig. 8.15a. This result and the exact answer are plotted in Fig. 8.30.

222. Calculate and plot the natural frequency of the system Fig. 8.15b as a function of the amplitude. Do this by the exact method of Eq. (8.26) as well as by the approximate method of Fig. 8.39.

223. A single-degree-of-freedom system consists of a mass m and a non-linear spring which has a stiffness k for deflections up to a, both ways, and has a constant force, independent of x, for values of x greater than a, both ways.

Calculate the natural frequency of free vibrations as a function of amplitude x_0, and in particular find the frequency for $x_0 = 2a$.

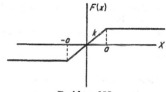

Problem 223.   Problem 224.

224. Find the natural frequency for the system shown. The springs are initially in compression by an amount F_0.

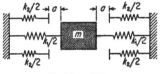

Problem 225.

225. Do Problem 222 for the system of Fig. 8.15c repeated here.

226. Do Problem 220 for initial clearance a on each side of the mass.

227. Find a few of the slopes drawn in Fig. 8.33, and from that figure construct one cycle of Fig. 8.34.

228. Consider a modified Van der Pol equation for relaxation oscillations:

$$\ddot{y} - \epsilon(1 - y^4)\dot{y} + y = 0$$

Here the fourth power of y occurs instead of the square as in the usual Van der Pol equation. For almost-sinusoidal vibrations, i.e., for small ϵ, find the value of the dimensionless amplitude y to which the system ultimately will go in a steady state.

229. For a simple free vibrating system with linear spring and mass subjected to dry friction (Coulomb damping), show that the equation of motion is

$$m\ddot{x} + kx = \begin{cases} -F \text{ for } \dot{x} > 0 \\ +F \text{ for } \dot{x} < 0 \end{cases}$$

Construct and describe the $\dot{x}/\omega_n$- versus x-diagram, and from it deduce the amplitude decrement per cycle [Eq. (8.30)].

230. A belt with a horizontal section moves at constant velocity V_0 driven by a pair of pulleys. On the horizontal section is a weight W, attached to a spring k. The coefficient of dry friction between the weight and the belt is f_m when moving and f_s when standing still relative to each other, and $f_s = 2f_m$. Set up a velocity-displacement diagram for this case, and note that the trajectories in this diagram consist exclusively of straight lines and circular arcs. Find the "limit cycles" or steadystate motions, and show how these steady states are reached when starting from different points in the diagram.

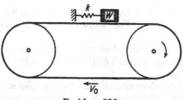

Problem 230.

231. Give a discussion and derive a result corresponding to Eq. (8.46) for the forced vibrations of a system with a damping proportional to the square of the velocity ($F = \pm c\dot{x}^2$).

232. Consider the velocity-displacement diagram ($\dot{x}/\omega_n$ versus x) for a system with a linear mass and spring but with non-linear damping. Prove that the total energy of the system (kinetic plus potential) is proportional to the square of the radius to any point. Thus for an undamped system (Fig. 8.21) to energy is constant, but for a damped system (Fig. 8.22, 8.23, or 8.33) the energy changes continuously. Where does this energy come from?

233. Prove that in the most general velocity-displacement diagram ($\dot{x}$ versus x) the acceleration at any point P is given by the "subnormal," i.e., by the segment cut out from the horizontal x-axis by two lines: the normal to the integral curve at P and the normal dropped from P on the horizontal axis.

ANSWERS TO PROBLEMS

1. (a) -157.0 in.-lb.   (b) $+0.40$ in.-lb.

2. (a) $\dfrac{6}{5} P_0 x_0$.   (b) $\dfrac{6}{5} P_0 x_0$.   (c) $\dfrac{-6}{5} P_0 x_0$.   (d) $\dfrac{-6}{5} P_0 x_0$.

3. $f(t) = \dfrac{P_0}{2} + \dfrac{2P_0}{\pi} \left(\sin \omega t + \dfrac{1}{3} \sin 3\omega t + \dfrac{1}{5} \sin 5\omega t + \cdots \right)$

8. When $n = 1, 3, 9, \ldots,$ $b_n = + \dfrac{a\sqrt{2}}{\pi n}$.

$n = 5, 7, 13, \ldots,$ $b_n = - \dfrac{a\sqrt{2}}{\pi n}$.

$n = 2, 10, 18, \ldots,$ $b_n = - \dfrac{2a}{\pi n}$.

$n = 4, 8, 12, \ldots,$ $b_n = 0$.

$n = 1, 7, 9, \ldots,$ $a_n = + \dfrac{2a}{\pi n} \left(\dfrac{2+\sqrt{2}}{2} \right)$.

$n = 3, 5, 11, \ldots,$ $a_n = + \dfrac{2a}{\pi n} \left(\dfrac{2-\sqrt{2}}{2} \right)$.

$n = 2, 6, 10, \ldots,$ $a_n = + \dfrac{2a}{\pi n}$.

$n = 4, 12, 20, \ldots,$ $a_n = + \dfrac{4a}{\pi n}$.

$n = 8, 16, 24, \ldots,$ $a_n = 0$.

9. $b_0 = \dfrac{2P_0}{\pi}, b_n = - \dfrac{4}{4n^2-1} \dfrac{P_0}{\pi}; a_n = 0$.

10. (a) $a_n = 0, b_0 = \dfrac{P_0\varphi}{\pi}; b_n = \dfrac{2P_0}{n\pi} \sin n\varphi$.   (b) $a_n = 0, b_0 = \dfrac{P_0\varphi}{\pi}; b_n = \dfrac{2P_0\varphi}{\pi}$.

11. $y = \dfrac{a}{3} + \dfrac{4a}{\pi^2} \displaystyle\sum_{n=1}^{n=\infty} \dfrac{(-1)^n}{n^2} \cos \dfrac{2n\pi x}{l}$.

14. (a) $\omega_n^2 = \dfrac{6}{7} \dfrac{kg}{W}$.   (b) $\omega^2 = \dfrac{6}{7} \dfrac{kg}{W} + \dfrac{12}{7} \dfrac{g}{l}$.

15. $\omega_n^2 = 5.22$ rad./sec.^{2}

16. $\omega_n^2 = \dfrac{3g}{2\sqrt{2}\,l}$.

17. $\omega_n^2 = \dfrac{5k}{6m}$.

18. $\omega^2 = \dfrac{r^2}{R - r} \cdot \dfrac{W}{\dfrac{Wr^2}{g} + I}$.

19. $\omega^2 = r_1 \dfrac{W}{\dfrac{Wr^2}{g} + I}$.

20. (a) $\omega^2 = \dfrac{fg}{a}$ (b) Unstable.

21. $\omega^2 = \dfrac{g}{l} + \dfrac{2ka^2}{ml^2}$ (see theorem on page 270).

22. (a) $a^2 > \dfrac{mgl}{2k}$.   (b) $\omega^2 = -\dfrac{g}{l} + \dfrac{2ka^2}{ml^2}$.

23. $\omega_n^2 = \dfrac{1}{12}\dfrac{kg}{W} + \dfrac{2g}{3}$.

24. $\omega_n^2 = \dfrac{3}{2}\left(\dfrac{g}{l} + \dfrac{K}{m}\right)$.

25. $\omega^2 = \dfrac{4ka^2}{I}$.

26. $\omega^2 = \dfrac{2Ea^2wt^3}{Il_1^2(2l_1 + 3l_2)}$.

27. $\omega^2 = \dfrac{a}{l} \cdot \dfrac{g}{h}$.

28. $\omega^2 = \dfrac{4T}{ml}$.

29. $\omega^2 = \dfrac{4}{3}\dfrac{k}{m}\left(1 + \dfrac{2a}{D}\right)^2$.

30. $\omega_n^2 = \dfrac{g}{h} + \dfrac{k}{2m}$.

31. $\omega_n = 153$ rad./sec.

32. (a) $k = \dfrac{Gd^4}{8nD^3}$.   (b) $k = 4.45$ lb. per inch.

33. (a) $k = EI/l$, where EI is the bending stiffness and l is the total length πDn of the spring.

     (b) $k = 3.13$ in.-lb. per radian.

34. (a) $k = \dfrac{\pi d^4}{32l} \cdot \dfrac{E}{\left(1 + \dfrac{E}{2G}\right)}$.   (b) $k = 2.78$ in.-lb. per radian

35. (a) $\dfrac{3EI}{l^3}$.   (b) $\dfrac{48EI}{l^3}$.   (c) $\dfrac{192EI}{l^3}$.

36. $\omega^2 = \dfrac{4}{m\left[\dfrac{1}{4k_1} + \dfrac{1}{4k_2} + \dfrac{1}{k_3} + \dfrac{1}{k_4}\right]}$.

37. (a) $\sin \varphi = \dfrac{AB}{h}$.   (b) Straight line through origin.

     (c) Ellipse with vertical and horizontal major axes.

38. (a) 0.318%.   (b) 0.00102 lb.-sec./in.

39. 0.0309 lb.-sec./in.

40. (a) $9,340$ lb./in.^{2}; $34.3°$.   (b) $25,800$ lb./in.^{2}; $38.9°$.

41. (a) $\omega = 27.8$ radians per second or $f = 4.42$ cycles per second

     (b) $c = 0.0023$ lb. in.$^{-1}$ sec.   (c) $P_0 = 0.064$ lb.

     (d) One per cent per cycle at the beginning; slower later on.

     (e) 2 in.   (f) $x = 2 - e^{-(ct/2m)}$

42. (a) Torque $= T_0 \cdot \dfrac{I_2}{I_1 + I_2} \cdot \dfrac{1}{1 - \dfrac{\omega^2}{\omega_n^2}}$, where $\omega_n^2 = \dfrac{k(I_1 + I_2)}{I_1 I_2}$.

(b) Same as (a), except that I_2 becomes $n^2 I_2$ and $\dfrac{1}{k}$ becomes $\dfrac{1}{k_1} + \dfrac{1}{n^2 k_2}$.

43. 115.5 in.-lb.   **44.** 1,161 lb./in.   **45.** $f < 1.51$ cycles/sec.

46. (a) $I \ddot{\varphi} + mgr \sin \alpha \sin \varphi = 0$.   (b) $\omega^2 = \dfrac{mgr \sin \alpha}{I}$.   **47.** $\omega^2 = \dfrac{8g}{r(9\pi - 16)}$.

48. $x = \dfrac{P}{k} [\cos \omega_n(t - t_0) - \cos \omega_n t]$ where t starts upon application of the load.

49. $\dfrac{I_1 I_2 n^2}{I_1 + I_2 n^2} \ddot{\psi} + \dfrac{k_1 k_2 n^2}{k_1 + k_2 n^2} \psi = T_0 \cdot \dfrac{I_2 n^2}{I_1 + I_2 n^2} \sin \omega t$.

50. $\omega^2 = \dfrac{9k}{m} - \dfrac{4c^2}{m^2}$.   (b) $x = \dfrac{3P_0}{4c} \sqrt{\dfrac{m}{k}}$.   (c) $x_0 = \dfrac{4}{3} \dfrac{P_0}{k} \left[1 - \dfrac{32}{81} \dfrac{c^2}{km} \right]$

51. (a) $\omega^2 = \dfrac{2g}{l}$.   (b) $\omega^2 = \dfrac{g}{l}$.

53. $\dfrac{\text{Work/cycle}}{ka_0^2} = 2\pi \dfrac{c}{c_c} \cdot \dfrac{\omega}{\omega_n} \cdot \left(\dfrac{y}{a_0} \right)^2$ where y is the relative motion (across the dashpot), described by Eq. (2.28a) in which the force $P_0 = m\omega^2 a_0$.

55. (a) $\omega_{\max}/\omega_{\min} = I_{\min}/I_{\max}$.
(b) Torque $= \omega \Omega (I_{\max} - I_{\min}) \sin 2\Omega t$, very large, so that (a) is the practical alternative.

56. $\omega_n^2 = \Omega^2 \dfrac{abM + I_G + Mb^2}{b^2 M + I_G}$.

57. (a) $\cos \alpha = \dfrac{3g}{2l\Omega^2}$.   (b) $\omega_n^2 = \dfrac{3g}{2l} - \Omega^2$.   (c) $\omega_n^2 = \dfrac{3g}{2l} \cos \alpha - \Omega^2 \cos 2\alpha$.

58. $9\frac{1}{4}$ deg. from vertical.   **59.** (a) 2.77 lb.   (b) 2.00 lb.

60. $\dfrac{\omega}{\omega_n} = 0.62, 1.62$.   **61.** $\dfrac{c}{c_c} = \dfrac{\Delta\omega}{\omega_n} \dfrac{x_{\text{side}}}{x_{\max}}$

63. (a) 205 r.p.m.   (b) 427 r.p.m.   **65.** $\omega^2 = \dfrac{2g}{\pi R}$.

66. (a) $-\dfrac{h}{12}$ (unstable).   (b) $\dfrac{b^2}{h \cdot 6\sqrt{2}} - h \dfrac{2 - \sqrt{2}}{3}$.   (c) Same as (b).

67. 477 v.p.m.; 828 v.p.m.

68. $\omega_1^2 = 0.76k/m$ with the node at $2.62l$ to right of left mass.
$\omega_2^2 = 5.24k/m$ with the node at $0.38l$ to right of left mass.

69. $\omega_1^2 = \dfrac{T}{ml}$. $\omega_2^2 = \dfrac{3T}{ml}$.

70. $\omega_1^2 = 0.73 \dfrac{k}{m}$, node at $0.17l$ to the left of the left end.

$\omega_2^2 = 3.27 \dfrac{k}{m}$, node at $0.14l$ to the left of the right end.

71. $\omega_1^2 = 0.$ $\omega_2^2 = \dfrac{2g}{3R}.$ **72.** $\dfrac{\omega^2}{g/r} = \dfrac{19}{33} \pm \dfrac{\sqrt{229}}{33}.$ **73.** $x_1 = 0; x_2 = -2l\dfrac{F_0}{T}$

74. $\omega_1^2 = 0.64k/m$ with $x_1/x_2 = +0.36.$
$\omega_2^2 = 1.56k/m$ with $x_1/x_2 = -0.56.$

75. First mode: 10 per cent per cycle decay in amplitude.
Second mode: 24 per cent per cycle decay in amplitude.

76. (a) 3.1×10^9 ft.-lb. sec.^{2} (b) 1.38×10^8 ft.-lb. sec.
(c) Arithmetic decay; roll angle diminishes by 2.7 deg. each half cycle of roll.

77. 4.37 in.   **78.** (a) 45 m.p.h. (b) 4.75 in.   **79.** $\omega^2 = \dfrac{k}{m} + \dfrac{g}{l}(2 \pm \sqrt{2}).$

80. $F_{\text{cable}} = \dfrac{m_2}{m_1 + m_2}F_0(1 - \cos \omega_n t) + m_2 g,$ where $\omega_n^2 = \dfrac{k(m_1 + m_2)}{m_1 m_2}.$

81. $F_{\text{cable}} = \dfrac{m_2}{m_1 + m_2}F_0\dfrac{1}{2\pi}(\omega_n t - \sin \omega_n t) + m_2 g$ during growing period;

$F_{\text{cable}} = F_0\dfrac{m_2}{m_1 + m_2}$ thereafter.

83. (a) 11.8 lb. compression (or tension) in both.
(b) 43.3 lb. compression in one, tension in other.

84. $Z = \dfrac{k}{-\dfrac{k}{m\omega^2} + 1}.$ **85.** $Z = \dfrac{k + j\omega c}{1 - \dfrac{k + j\omega c}{m\omega^2}}.$ **86.** $Z = \dfrac{k}{1 - \dfrac{k}{m\omega^2 - j\omega c}}$

87. $x_0 = \dfrac{P}{\sqrt{(m\omega^2)^2 + \left[\omega c\left(1 - \dfrac{m\omega^2}{k}\right)\right]^2}};$ force is $c\omega x_0$

89. $Z = k\left(1 - \dfrac{\omega^2}{\omega_n^2}\right).$ This is the classical resonance curve.

90. Transmissibility $= \dfrac{1}{1 - m_1\omega^2\left(\dfrac{1}{k} + \dfrac{1}{K - M\omega^2}\right)}.$ **91.** $\omega = \Omega\sqrt{\dfrac{ab}{k^2}}.$

93. $\omega^4[1 + 3mR^2 \cos^2 \alpha/I] - \omega^2[\omega_1^2 + \omega_2^2 + 3k_2R^2/I] + \omega_1^2\omega_2^2 = 0.$
where $\omega_1^2 = k_1/I$ and $\omega_2^2 = 3k/3m.$

94. $\omega^4[1 + 3m \sin^2 \alpha/M] - \omega^2[\omega_1^2 + \omega_2^2 + 3k_2/M] + \omega_1^2\omega_2^2 = 0.$
where $\omega_1^2 = k_1/M$ and $\omega_2^2 = 3k_2/3m.$

95. $\omega^4[1 + 3k_2R^2 \cos^2 \alpha/k_1 + 3k_3R^2 \sin^2 \alpha/k_1]$
$\qquad\qquad - \omega^2[\omega_1^2 + \omega_2^2 + 3k_2k_3R^2/mk_1] + \omega_1^2\omega_2^2 = 0.$
where $\omega_1^2 = 3k_2/3m$ and $\omega_2^2 = 3k_3/3m.$

96. $\omega^4[1 + 3k_2 \sin^2 \alpha/k_1 + 3k_3 \cos^2 \alpha/k_1] - \omega^2\left(\omega_1^2 + \omega_2^2 + \dfrac{3k_2k_3}{mk_1}\right) + \omega_1^2\omega_2^2 = 0.$

In the last four problems solutions are simple and physically lucid for $\alpha = 0$ and $\alpha = 90$ deg.   Check up what these frequencies are.   Increasing blade angle α means increased "coupling" between the two modes, and this always causes the two uncoupled frequencies to spread apart (see Figs. 3.2 and 3.7). Therefore the blade frequency is either raised or lowered by an increase in α depending on whether it is higher or lower than the engine frequency to start with.

97. (a) $\omega_n^2 = \dfrac{2k}{m} \cos^2 \alpha + \dfrac{g}{l} \sin^2 \alpha = \dfrac{2k}{m} \cos^2 \alpha \left(1 + \dfrac{\delta_{stat}}{l} \sin^2 \alpha \right).$

$\quad$ (b) $\omega_n^2 = \dfrac{2k}{m} \sin^2 \alpha + \dfrac{g}{l} \cos^2 \alpha.$

Note that the answer to Problem 14a is in error by omitting a gravity term. Deduce the correct answer by using the results of this problem.

98. $\omega_1^2 = 0$

$\quad \omega_2^2 = \dfrac{g}{l} \left[1 + \dfrac{2}{3} \dfrac{w}{W} \right].$

100. (a) $\alpha_{11} = \alpha_{22} = \dfrac{3l^3}{4EI}; \ \alpha_{12} = \dfrac{7l^3}{12EI}.$ $\quad$ (b) $\omega^2 = \dfrac{1}{m(\alpha_{11} \pm \alpha_{12})}$ with $\dfrac{x_1}{x_2} = \pm 1.$

101. $\alpha_{11} = \dfrac{16}{768} \dfrac{l^3}{EI} \quad \alpha_{12} = \dfrac{11}{708} \dfrac{l^3}{EI} \quad \alpha_{22} = \dfrac{9}{768} \dfrac{l^3}{EI}$

$\quad \omega_1 = 5.63 \sqrt{\dfrac{EI}{ml^3}} \quad \omega_2 = 28.5 \sqrt{\dfrac{EI}{ml^3}}.$

102. (a) $\omega = \sqrt{\dfrac{2k}{m}}.$ $\quad$ (b) $x_2 = \dfrac{P_0}{7k}.$

105. $\omega^2 = \dfrac{6}{11} \dfrac{T}{ml}$ assuming a shape consisting of three straight stretches of string.

106. Half the mass of the beam has to be added to the central mass. Curve assumed is half a sine wave.

107. Three-eighths of the beam mass is effective. Assumed curve is a full (360-deg.) sine wave, vertically displaced.

108. $\omega_1 = 0.59 \sqrt{\dfrac{EI}{ml^3}}.$ $\quad \omega^2 = 3.89 \sqrt{\dfrac{EI}{ml^3}}.$ $\qquad$ **109.** $\omega_1^2 = 0; \ \omega_{2,3}^2 = \left(\dfrac{9}{4} \pm \sqrt{\dfrac{17}{16}} \right) \dfrac{k}{m}.$

110. $\omega_n^2 = \dfrac{k}{2m}, \dfrac{k}{m}, \dfrac{2k}{m}.$ $\quad$ **111.** $\omega_n^2 = \dfrac{64}{17} \dfrac{EI}{ml^3}.$ $\quad$ **112.** $\omega_n^2 = 0.933 \dfrac{EI}{ml^3}; \ 10.29 \dfrac{EI}{ml^3}.$

113. (a) $\alpha_{11} = \dfrac{l^3}{3EI}; \ \alpha_{22} = \dfrac{8}{3} \dfrac{l^3}{EI}; \ \alpha_{12} = \dfrac{5}{6} \dfrac{l^3}{EI}.$ $\quad$ (b) $\omega_1^2 = \dfrac{2}{7} \dfrac{EI}{ml^3}; \ \omega_2^2 = 6 \dfrac{EI}{ml^3}.$

$\quad$ (c) $\left(\dfrac{y_1}{y_2} \right)_{\omega_1} = \dfrac{1}{3}; \ \left(\dfrac{y_1}{y_2} \right)_{\omega_2} = -1.$ $\quad$ (d) $\omega^2 = \dfrac{6}{11} \dfrac{EI}{ml^3}.$

114. $A = -\dfrac{\dfrac{P}{2m} \left(\omega^4 - \dfrac{2k}{m} \omega^2 + \dfrac{2}{3} \dfrac{k^2}{m^2} \right)}{\dfrac{k}{m} \left(\dfrac{k}{m} - \omega^2 \right) \left(\dfrac{2k}{m} - \omega^2 \right)}.$

115. $\omega_1 = 52.3$ rad./sec.

116. $\omega_{1,2,3} = 40; 160; 367$ rad./sec. $\qquad$ **117.** $\omega_{1,2,3} = 37.6, 143, 355$ rad./sec.

119. (a) $\omega_1 = 4.48 \sqrt{\dfrac{EI}{\mu_1 l^4}}$; (b) $\omega_1 = 3.57 \sqrt{\dfrac{EI}{\mu_1 l^4}}$ (exact $= 3.52$).

121. $y = y_0 \left(\dfrac{x^4}{24} - \dfrac{x^6}{90 l^2} - \dfrac{5 l^2 x^2}{96} + 0.0106 l^4 \right).$

$\omega_1 = 9.71 \sqrt{\dfrac{EI}{\mu_1 l^4}}$ (exact $= \pi^2 = 9.87$).

122. (a) $a = \dfrac{\sqrt{3}}{4l}.$ (b) $a = 0,\ b = \dfrac{-3}{8 l^2},\ c = \dfrac{1}{16 l^2}.$ (c) $\omega_1 = 1.56 \sqrt{\dfrac{EI}{\mu_1 l^4}}.$

123. (a) $y = \dfrac{x}{l} - \dfrac{3}{2} \dfrac{x^3}{l^3} + \dfrac{1}{2} \dfrac{x^4}{l^4}.$ (b) $\omega_1 = 4.07 \sqrt{\dfrac{EI}{m l^3}}.$ **124.** $\omega_1 = 22.22 \sqrt{\dfrac{EI}{\mu_1 l^4}}.$

125. $\omega_1^2 = 10.9 \dfrac{C_2}{C_1 l^2}.$ **126.** (a) $\omega_1^2 = 10.9 \dfrac{EI}{\mu_0 l^4}.$ (b) $\omega_1^2 = 7.6 \dfrac{EI}{\mu_0 l^4}.$

127. $\omega_1^2 = 13.85 \dfrac{EI}{\mu l^4}$ (exact $= 12.39$).

128. $\omega^2 = 2,960$ rad.^{2}/sec.^{2} $f = 8.66$ cycles per second.

129. Equivalent of μ_1 is $m_1 n_1$; equivalent of AE is $\dfrac{G d^4}{8 n_1 D^3}$; $\omega^2 = \dfrac{\pi^2 G d^4}{8 n_1^2 D^3 m_1 l^2}.$

130. $\omega^2 = 2.80 \dfrac{EI}{\mu_1 l^4}$ for a quarter cosine wave; the coefficient 2.80 becomes 1.35 if the stiff half of the beam does not bend so that the deflection curve is one-eighth cosine wave and a piece of straight line.

131. 745 cycles per second.

132. (a) Assuming straight line deformation one-third of spring mass is to be added to end mass.

(b) Frequency determined by the transcendental equation

$$\frac{\omega^2}{k/m} = \frac{\omega/\sqrt{k/m_s}}{\tan \omega/\sqrt{k/m_s}}.$$

For $m_s \ll m$, and retaining the first two terms of the Taylor series development we find again that one-third of the spring mass is to be added to the end mass.

133. $\omega^2 = \dfrac{\pi^4}{\dfrac{3}{2} - \dfrac{4}{\pi}} \cdot \dfrac{EI}{\mu_1 l^4}$ assuming curve a sine wave passing through $\dfrac{l}{6}$ and $\dfrac{5l}{6}.$

134. 132 r.p.m. and 376 r.p.m.

135. $\tan pl = \dfrac{p(q + r)}{q r - p^2}$ where $p^2 = \dfrac{\mu_1 \omega^2}{AE}$; $q = \dfrac{M_1 \omega^2}{AE}$; $r = \dfrac{M_2 \omega^2 - k}{AE}.$

Solve by trial and error assuming values for ω^2. Plot left and right side of equation against ω^2 and get intersection of the two curves. The first critical speed is at 127.5 r.p.m.

136. (a) $y = \sin \dfrac{\pi x}{l} - \dfrac{3}{\pi} \dfrac{x}{l} \cdots$ (b) $\omega^2 = \dfrac{\pi^4}{1 - 6/\pi^2} \cdot \dfrac{EI}{\mu l^4} = 15\,75 \dfrac{EI}{\mu l^4}$.

The exact solution, listed on p. 432 as a_2 of the hinged-free beam, has a factor 15.4.

137. $\omega^2 = \dfrac{\pi^2}{4} \cdot \dfrac{\pi^2 + 4}{\pi^2 - 4} \cdot \dfrac{T}{\mu R^2} = 5.80 \dfrac{T}{\mu R^2}$

The exact answer, involving Bessel's functions, has a factor 5.74.

139. (a) $\frac{1}{8}$ in. vertically. (b) 267 lb. vertically.

142. Primary and secondary forces balanced; both moments unbalanced.

143. (a) $\frac{1}{2} W_{rec}$, or half the rec. wt. of a crank, i.e., one piston and fraction of one rod.

(b) Zero.

(c) $W_{rec} \sqrt{\dfrac{(a+b)^2 + b^2}{(a+2b)^2}}$; $\alpha_A = 0$; $\alpha_{A1} = 90°$;

$\alpha_B = 180° + \tan^{-1} \dfrac{b}{a+b}$

$\alpha_C = 180° + \tan^{-1} \dfrac{a+b}{b}$.

144. $\dfrac{W_{pis} R_{pis}^2}{W_{disk} R_{disk}^2} = \dfrac{\cos^2 \alpha}{2}$.

145. (a) $\dfrac{\varphi_1}{T_0/k} = \dfrac{(\omega/\omega_n)^2 - I_1/I_1 + I_2}{1 - (\omega/\omega_n)^2} \cdot \left(\dfrac{\omega_n}{\omega}\right)^2 \cdot \dfrac{I_2}{I_1 + I_2}$,

showing $\varphi_1 \to \infty$ for $\omega = 0$.

(b) $\dfrac{k(\varphi_1 - \varphi_2)}{T_0} = \dfrac{I_2/I_1 + I_2}{1 - (\omega/\omega_n)^2}$

146. $\omega_1 = 168$ radians per second.

147. $\omega_1^2 = 2,800$ rad./sec.^{2} **148.** $\omega_1^2 = 27.4$; $\omega_2^2 = 629.3$ rad./sec.^{2}

149. 16,800 v.p.m.; r.p.m.$_6 = 2,800$; r.p.m.$_2 = 5,600$. **150.** 2,800 v.p.m.

151. (a) $\Theta = 110°$; $\omega = 3.45$. (b) $\Theta = 270°$; $\omega = 8.48$.

152. (a) $\omega^2 = 730$. (b) $\omega^2 = 16,000$. (c) 4. (d) $\frac{1}{2}$.

153. 1,570 v.p.m. **154.** $\omega_1 = 0.350 \sqrt{\dfrac{k}{I}}$; $\omega_2 = 0.825 \sqrt{\dfrac{k}{I}}$.

155. $\omega_1 = 21.3$ rad./sec. **156.** $\omega_1 = 67.5$; $\omega_2 = 97.5$; $\omega_3 = 188$ rad./sec.

157. (b) 1, 8; 2, 7; 3, 6; 4, 5; 9. (c) 9. (d) 2 and 7. **158.** $5\frac{1}{2}$, 6, $6\frac{1}{2}$, 7.

159. 3,810 v.p.m.; 6,630 v.p.m. **160.** 1,215 v.p.m.; 5,530 v.p.m.

161. $\omega_1 = 2,080$ rad./sec. **162.** 4.9 lb.

163. (a) $\omega_n = \Omega$; (b) $V = a\Omega \sqrt{2}$.

166. Five fundamental diagrams:

(1) For orders $\frac{1}{2}$, $3\frac{1}{2}$, $4\frac{1}{2}$, $7\frac{1}{2}$, etc.

(2) For orders 1, 3, 5, 7, etc.

(3) For orders $1\frac{1}{2}$, $2\frac{1}{2}$, $5\frac{1}{2}$, $6\frac{1}{2}$, etc.

(4) For orders 2, 6, 10, etc.

(5) Majors 4, 8, 12, etc.

167. (a) 0.0047 radian. (b) 293,000 in.-lb.; 41,300 in.-lb. (c) 31.6 r.p.m.

168. -0.00359 radian. $+0.00423$ radian.

169. (a) $\beta = 0$; first shaft torque arbitrarily assumed.　　(b) End $\beta = 0$.

(c) $\Theta = \tau = 180$ deg.　　(d) $\omega^2 = \dfrac{\pi^2}{42} \cdot \dfrac{k}{I}$.

170. (a) $\dfrac{g}{\rho^2 + a^2}\left(\dfrac{R_1R_2 - a\,\Delta R}{\Delta R}\right)$.　　(b) $\dfrac{\Omega^2 r_G}{\rho^2 + a^2}\left[\dfrac{R_1R_2 - a\,\Delta R - a^2\,\Delta R/r_G}{\Delta R}\right]$.

171. (b) 4,000 in.-lb.　　(c) 122.0 r.p.m.　　(d) 107 r.p.m.　　(e) Balanced.

172. $a_1 = \dfrac{\pi}{2}$; all other a's are zero.

$b_n = \dfrac{2}{n^2 - 1}$ for even n; $b_n = 0$ for odd n.

Order:　　　　　　　　1　2　3　4　5　6
Per cent of mean torque: 157　66.7　0　13.3　0　5.7

173. $+0.35\dfrac{EI}{l}$; $-1.3\dfrac{EI}{l}$.　　　　**174.** 990 r.p.m.

175. 990 r.p.m. horizontally and 1,260 r.p.m. vertically.

177. Counting angles from the $+1$ unbalance toward the $+2$ unbalance (at which $\varphi = 90$ deg.), the corrections are:
In Plane I: 2.06 in. oz. at 104 deg.
In Plane II: 4.03 in. oz. at 263 deg.

178. 4.2 oz. at 315 deg.　　　　**179.** 1.15 lb. at 306 dry.

180. $x^2 = \dfrac{a_1^2 + a_2^2}{2_4} - a_0^2$; $\cos\varphi = \dfrac{a_1^2 - a_2^2}{4a_0 x}$.

Ambiguity between $+\varphi$ and $-\varphi$.

181. $\alpha = 2\cos^{-1}\dfrac{Me}{2mr}$.

182. Primary speed, 1440 r.p.m.; secondary speed, 720 r.p.m. Secondary force amplitude is 0.044 lb., corresponding to an unbalance of 7.1×10^{-4} in.-lb.

183. $\dfrac{\varphi}{\varphi_0} = \dfrac{1}{1 + \dfrac{12EI}{I_d\omega^2 l}}$.

184. (a) $\omega_0^2 = \dfrac{12EI}{I_d l}$; (b) $\dfrac{\omega}{\omega_0} = \dfrac{\Omega}{\omega_0} \pm \sqrt{1 + \left(\dfrac{\Omega}{\omega}\right)^2}$.　　**185.** 804 r.p.m.

186. $K^2 = \dfrac{12}{7}\left[\left(2 - \dfrac{1}{D}\right) + \sqrt{\dfrac{1}{D^2} - \dfrac{9}{4D} + 4}\right]$

where K and D are the abbreviations used in Eq. (6.14).

187. (a) $\omega^2 = \dfrac{3EI}{I_d l}$, (b) Does not exist, (d) $K = R \pm \sqrt{R^2 + 3}$.

188. (a) $\dfrac{K^2}{64} - \dfrac{K}{64}\left(\dfrac{28}{9} + \dfrac{1}{4D}\right) + \dfrac{1}{3D} = 0$.　　(b) $\dfrac{K^2}{64} - \dfrac{K}{64}\left(\dfrac{28}{9} - \dfrac{1}{4D}\right) - \dfrac{1}{3D} = 0$.

189. (c) $\dfrac{ml^3\omega^2}{EI} = 48; 192; 96.$ (d) $\dfrac{ml^3\omega^2}{EI} = 48; 192; 25.6$ and $358.$

190. With x of the disk center and φ of the shaft there are two differential equations, the latter having the gyroscopic term $\frac{1}{2}MR^2\Omega\omega\varphi$, where ω is the angular speed of the forward whirl of shaft center line. The frequency equation falls apart into two quadratics:

$$\left[\left(\frac{\omega}{\omega_a}\right)^2 - 1\right]\left[\left(\frac{\omega}{\omega_a}\right)^2 - \left(\frac{\omega}{\omega_a}\right)\cdot 2\frac{\Omega}{\omega_a} - \left(\frac{\omega_b}{\omega_a}\right)^2\right] = 0$$

so that two roots are $\omega/\omega_a = \pm 1$ independent of Ω (a forward and backward whirl with the shaft parallel to itself) and two other roots, one a forward whirl the frequency of which increases with Ω and a backward whirl of a frequency decreasing with Ω.

192. If plotted as: $\omega^2/EI/m \left(l + \dfrac{l_1}{2}\right)^3 = f(l_1/l)$ the curve is nearly straight at an ordinate falling from 3 to 2.90 between the points $l_1/l = 0$ and 1. A second critical with a node somewhere in the stiff part has a very high frequency running from ∞ at $l_1/l = 0$ to 567 at $l_1/l = 1$ in the same diagram.

193. (a) $\omega^2 = \dfrac{Ed^4(I_1 + I_2)}{64DnI_1I_2}.$ (b) $22.4\sqrt{\dfrac{Ed^2}{8\pi^2D^2n^2l^2\rho_{steel}(1 + E/2G)}}.$

194. (c) $2l$, one full wave length per cycle.

195. $I_1\ddot{\varphi}_1 + k(\varphi_1 - \varphi_2) = 0.$

$I_2\ddot{\varphi}_2 + k(\varphi_2 - \varphi_1) - C_1C_2\dfrac{\omega}{c}\tan\dfrac{\omega l}{c}\cdot\varphi_2 = 0.$

$-k^2 + (-I_1\omega^2 + k)\cdot\left(-I_2\omega^2 + k - C_1C_2\dfrac{\omega}{c}\tan\dfrac{\omega l}{c}\right) = 0.$

in which $c = \sqrt{AE/\mu}$, the velocity of sound; see p. 137.

196. (a) Stable. (b) Unstable. (c) Unstable.

198. (a) $\sin\alpha_0 = \dfrac{T_0}{Mgl}.$ (b) Undamped vibrations of frequency $\omega^2 = g\cos\alpha_0/l.$

(c) Damped vibrations; same frequency. (d) Increasing vibration.

199. (b) $x = v_0t^* - v_0\sqrt{\dfrac{m}{k}}\cdot\sin\sqrt{\dfrac{k}{m}}\,t^*$, where $t^* = t - \dfrac{fW}{kv}.$

(c) Damped oscillations about $x = v_0t.$*

(d) Oscillations with increasing amplitude which lead to a motion with periodic stops of the mass.

200. $\dfrac{L_0}{D_0} > \dfrac{3}{4}.$

201. (a) $k = \dfrac{m\Omega^2}{\sqrt{6}-2}\left(1 + \dfrac{a\sqrt{2}}{l}\right).$ (b) $\omega^2 = \dfrac{k}{M + 2m}.$

(c) $\omega^2 = \dfrac{k(\sqrt{6}-1) - \frac{1}{2}m\Omega^2}{m + M}.$

202. For complete solution with curves see *Trans. A.I.E.E.*, 1933, p. 340.

203. $\omega^2 = \dfrac{a^2E}{mV}.$

204. $I_1\ddot{\varphi}_1 + k_1(\varphi_1 - \varphi_2) = 0.$

$I_2\ddot{\varphi}_2 + k_1(\varphi_2 - \varphi_1) + \dfrac{\Delta m}{T}(r_B^2\ddot{\varphi}_2 + r_A^2\ddot{\varphi}_3) = 0.$

$I_3\ddot{\varphi}_3 + \dfrac{\Delta m}{T}(r_A^2\ddot{\varphi}_3 - r_B^2\ddot{\varphi}_2) = 0.$

$s^3 + s^2\left(\dfrac{r_A^2}{I_3} + \dfrac{r_B^2}{I_2}\right)\dfrac{\Delta m}{T} + s\left(\dfrac{k_1}{I_2} + \dfrac{k_1}{I_1}\right) + \dfrac{\Delta m}{T}\left[\dfrac{k_1 r_A^2}{I_2 I_3} + \dfrac{k_1 r_A^2}{I_1 I_3} + \dfrac{k_1 r_B^2}{I_1 I_2}\right] = 0.$

The system is stable.

205. $I_1\ddot{\varphi}_1 + k_1(\varphi_1 - \varphi_2) = 0$

$I_2\ddot{\varphi}_2 + k_1(\varphi_2 - \varphi_1) + \dfrac{\Delta m}{T}r_B^2\ddot{\varphi}_2 - \dfrac{\Delta m}{T}r_A^2\ddot{\varphi}_3 = 0$

$I_3\ddot{\varphi}_3 + k_2(\varphi_3 - \varphi_4) + \dfrac{\Delta m}{T}r_A^2\ddot{\varphi}_3 - \dfrac{\Delta m}{T}r_B^2\ddot{\varphi}_2 = 0$

$I_4\ddot{\varphi}_4 + k_2(\varphi_4 - \varphi_3) = 0$

$s^5 + A_4 s^4 + A_3 s^3 + A_2 s^2 + A_1 s + A_0 = 0$

where $A_4 = \dfrac{\Delta m}{T}\left[\dfrac{r_A^2}{I_3} + \dfrac{r_B^2}{I_2}\right];\ A_3 = \dfrac{k_1}{I_1} + \dfrac{k_1}{I_2} + \dfrac{k_2}{I_3} + \dfrac{k_2}{I_4}$

$A_2 = \dfrac{\Delta m}{T}\left(\dfrac{r_A^2 k_1}{I_2 I_3} + \dfrac{r_A^2 k_1}{I_1 I_3} + \dfrac{r_B^2 k_1}{I_1 I_2} + \dfrac{r_A^2 k_2}{I_3 I_4} + \dfrac{r_B^2 k_2}{I_2 I_4} + \dfrac{r_B^2 k_2}{I_2 I_3}\right)$

$A_1 = \dfrac{k_1 k_2}{I_2 I_3} + \dfrac{k_1 k_2}{I_1 I_4} + \dfrac{k_1 k_2}{I_2 I_4} + \dfrac{k_1 k_2}{I_1 I_3}$

$A_0 = \dfrac{\Delta m}{T}\left(\dfrac{k_1 k_2 r_A^2}{I_2 I_3 I_4} + \dfrac{k_1 k_2 r_A^2}{I_1 I_3 I_4} + \dfrac{k_1 k_2 r_B^2}{I_1 I_2 I_4} + \dfrac{k_1 k_2 r_B^2}{I_1 I_2 I_3}\right).$

The system is stable.

206. $C_1 < 0$ stable; $C_1 > 0$ unstable.

207. $0 > -\left(\dfrac{ka}{ml}\right)^2$ always stable.   **208.** Unstable for both cases.

209. Stable if $\dfrac{ck}{m} > -AC.$ **210.** $kC_2 > AC_1.$

211. Only condition for stability is $(C_2 + C_4)k > (C_1 - C_3)A.$

212. $V = \sqrt{\dfrac{k\theta}{aA}}.$ **215.** Stable if $k > \dfrac{C}{a}.$

217. Unstable frequencies are $\omega = \alpha\sqrt{\dfrac{2A^2 p}{MV}}$ where $\alpha = 2,\ 1,\ \frac{2}{3},\ \frac{2}{4},\ \frac{2}{5}$, etc., and p is atmospheric pressure $= 14.6$ lb. per square inch. The slope of the line in the diagram of Fig. 8.11 is 0.10.

218. 1,085 r.p.m.   **220.** $\omega_n = 2\sqrt{\dfrac{k_1}{m}}\dfrac{1}{\sqrt{\dfrac{k_1}{k_2}} + 1}.$ **221.** $\omega_n = \sqrt{\dfrac{k}{m}}\left(1 - \dfrac{a}{x_0}\right)^{1/2}.$

222. Exact: $\omega = \sqrt{\dfrac{k}{m}}\cdot\dfrac{\pi/2}{\cos^{-1}\left(\dfrac{1}{1 + \dfrac{kx_0}{F}}\right)}.$ Approximate: $\omega = \sqrt{\dfrac{k}{m} + \dfrac{F}{mx_0}}.$

223. $\omega_n = \sqrt{\dfrac{k}{m}}\dfrac{\pi/2}{\tan^{-1}\sqrt{\dfrac{1}{2\left(\dfrac{x_0}{a}-1\right)}}+\sqrt{2\left(\dfrac{x_0}{a}-1\right)}}.$

$\omega_n\Big]_{\frac{x_0}{a}=2} = 0.773\sqrt{\dfrac{k}{m}}.$

224. Exact: $\omega_n = \dfrac{\pi}{2}\sqrt{\dfrac{k}{m}}\left[\dfrac{1}{\cos^{-1}\dfrac{1}{1+\dfrac{k(x_0-a)}{F}}+\dfrac{1}{\left(\dfrac{x_0}{a}-1\right)^2+\dfrac{2F}{k}\left(\dfrac{x_0}{a}-1\right)}}\right].$

Approximate: $\omega_n = \sqrt{\dfrac{k}{m}}\sqrt{1-\dfrac{a}{x_0}+\dfrac{F}{k}}.$

225. Exact:

$\omega_n = \dfrac{\pi}{2}\sqrt{\dfrac{k_1}{m}}\dfrac{1}{\sqrt{\dfrac{k_1}{k_1+k_2}}\cos^{-1}\dfrac{k_1/k_2}{\dfrac{x_0}{a}\left(\dfrac{k_1}{k_2}+1\right)-1}+\sin^{-1}\dfrac{1}{\sqrt{\dfrac{k_2}{k_1}\left(\dfrac{x_0}{a}-1\right)^2+\left(\dfrac{x_0}{a}\right)^2}}}.$

Approximate: $\omega_n = \sqrt{\dfrac{k_1}{m}}\sqrt{1+\dfrac{k_2}{k_1}\left(1-\dfrac{a}{x_0}\right)}.$

226. $\omega_n = \dfrac{2\sqrt{\dfrac{k_1}{m}}}{1+\sqrt{\dfrac{k_1}{k_2}}+\dfrac{\left(1+\sqrt{\dfrac{k_2}{k_1}}\right)\left(1+\dfrac{k_2}{k_1}\right)}{\dfrac{x_0}{a}-1}},$

where x_0 is defined as $\dfrac{1}{2}$ of the total motion.

228. $y_{\text{steady state}} = 1.68.$

230. Limit cycle is circle about a center $x = f_m\dfrac{W}{k}$, $\dfrac{\dot{x}}{\omega_n} = 0$, passing through the point $x = f_s\dfrac{W}{k}$, $\dot{x}/\omega_n = V_0/\omega_n$, and cut off at the top by the straight line $\dfrac{\dot{x}}{\omega_n} = \dfrac{V_0}{\omega_n^2}$.

The frequency is $\omega_n^2 = \dfrac{kg}{W}$.

231. $x_0^2 = \left(\dfrac{P_0}{k}\right)^2\cdot\dfrac{3\pi}{8D}\left[-\dfrac{3\pi}{16D}\left(1-\dfrac{\omega^2}{\omega_n^2}\right)^2+\sqrt{\dfrac{9\pi^2}{256D^2}\left(1-\dfrac{\omega^2}{\omega_n^2}\right)^4+1}\right]$

where $D = c\omega^2 P_0/k^2$ is a dimensionless variable involving the damping constant c, the damping force being $c\dot{x}^2$.

APPENDIX

A COLLECTION OF FORMULAS

I. Linear Spring Constants

("Load" per inch deflection)

Coil dia. D; wire dia. d; n turns $\qquad k = \dfrac{Gd^4}{8nD^3}$ (1)

Cantilever $\qquad k = \dfrac{3EI}{l^3}$ (2)

Cantilever $\qquad k = \dfrac{2EI}{l^2}$ (3)

Beam on two supports; centrally loaded $\qquad k = \dfrac{48EI}{l^3}$ (4)

Beam on two supports; load off center $\qquad k = \dfrac{3EIl}{l_1^2 l_2^2}$ (5)

Clamped-clamped beam; centrally loaded $\qquad k = \dfrac{192EI}{l^3}$ (6)

Circular plate, thickness t; centrally loaded; circumferential edge simply supported $\qquad k = \dfrac{16\pi D}{R^2}\dfrac{1 + \mu}{3 + \mu}$ (7)

in which the plate constant is

$$D = \dfrac{Et^3}{12(1 - \mu^2)}$$ (7a)

μ = Poisson's ratio ≈ 0.3

Circular plate; circumferential edge clamped $\qquad k = \dfrac{16\pi D}{R^2}$ (8)

Two springs in series $\qquad k = \dfrac{1}{1/k_1 + 1/k_2}$ (9)

II. Rotational Spring Constants

("Load" per radian rotation)

Twist of coil spring; wire dia. d; coil dia. D; n turns $\qquad k = \dfrac{Ed^4}{64nD}$ (10)

429

Bending of coil spring

$$k = \frac{Ed^4}{32nD} \cdot \frac{1}{1 + E/2G} \tag{11}$$

Spiral spring; total length l; moment of inertia of cross section I

$$k = \frac{EI}{l} \tag{12}$$

Twist of hollow circular shaft, outer dia. D, inner dia. d, length l

$$k = \frac{GI_p}{l} = \frac{\pi}{32} \frac{G(D^4 - d^4)}{l} \tag{13}$$

For steel $\quad k = 1.18 \times 10^6 \times \dfrac{D^4 - d^4}{l}$

Cantilever

$$k = \frac{EI}{l} \tag{14}$$

Cantilever

$$k = \frac{2EI}{l^2} \tag{15}$$

Beam on two simple supports; couple at center

$$k = \frac{12EI}{l} \tag{16}$$

Clamped-clamped beam; couple at center

$$k = \frac{16EI}{l} \tag{17}$$

III. Natural Frequencies of Simple Systems

End mass M; spring mass m, spring stiffness k

$$\omega_n = \sqrt{k/(M + m/3)} \tag{18}$$

End inertia I; shaft inertia I_s, shaft stiffness k

$$\omega_n = \sqrt{k/(I + I_s/3)} \tag{19}$$

Two disks on a shaft

$$\omega_n = \sqrt{\frac{k(I_1 + I_2)}{I_1 I_2}} \tag{20}$$

Cantilever; end mass M; beam mass m, stiffness by formula (2)

$$\omega_n = \sqrt{\frac{k}{M + 0.23m}} \tag{21}$$

Simply supported beam; central mass M; beam mass m; stiffness by formula (4)

$$\omega_n = \sqrt{\frac{k}{M + 0.5m}} \tag{22}$$

Massless gears, speed of $I_2 n$ times as large as speed of I_1

$$\omega_n = \sqrt{\frac{1}{\dfrac{1}{k_1} + \dfrac{1}{n^2 k_2}} \times \dfrac{I_1 + n^2 I_2}{I_1 \cdot n^2 I_2}} \tag{23}$$

$$\omega_n^2 = \frac{1}{2}\left(\frac{k_1}{I_1} + \frac{k_2}{I_2} + \frac{k_1 + k_2}{I_2}\right) \pm$$
$$\pm \frac{1}{2}\sqrt{\left(\frac{k_1}{I_1} + \frac{k_2}{I_2} + \frac{k_1 + k_2}{I_2}\right)^2 - 4\frac{k_1 k_2}{I_1 I_2 I_3}(I_1 + I_2 + I_3)} \tag{24}$$

IV. Uniform Beams

(Longitudinal and torsional vibration)

Longitudinal vibration of cantilever: A = cross section, E = modulus of elasticity. μ_1 = mass per unit length, $n = 0,1,2,3$ = number of nodes

$$\omega_n = \left(n + \frac{1}{2}\right) \pi \sqrt{\frac{AE}{\mu_1 l^2}} \qquad (25)$$

For steel and l in inches this becomes

$$f = \frac{\omega_n}{2\pi} = (1 + 2n)\frac{51,000}{l}$$

cycles per second $\qquad (25a)$

Longitudinal vibration of beam clamped (or free) at both ends

$$\omega_n = n\pi \sqrt{\frac{AE}{\mu_1 l^2}} \quad n = 1, 2, 3, \ldots (26)$$

For steel, l in inches

$$f = \frac{\omega_n}{2\pi} = \frac{102,000}{l} \text{ cycles/sec.} \quad (26a)$$

Torsional vibration of beams

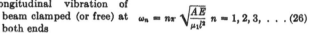

Same as (25) and (26); replace tensional stiffness AE by torsional stiffness C ($= GI_p$ for circular cross section); replace μ_1 by the moment of inertia per unit length $i = \dfrac{I_{bar}}{l}$

Organ pipe open at one end, closed at the other

For air at 60°F., l in inches:

$$f = \frac{\omega_n}{2\pi} = (1 + 2n)\frac{3,300}{l}$$

cycles/sec. $\quad (27a)$

$$n = 0, 1, 2, 3, \ldots$$

Water column in rigid pipe closed at one end (l in inches)

$$f = \frac{\omega_n}{2\pi} = (1 + 2n)\frac{14,200}{l}$$

cycles/sec. $\quad (27b)$

$$n = 0, 1, 2, 3, \ldots$$

Organ pipe closed (or open) at both ends (air at 60°F.)

$$f = \frac{n6,600}{l} \text{ cycles/sec.} \quad (28a)$$

$$n = 1, 2, 3, \ldots$$

Water column in rigid pipe closed (or open) at both ends

$$f = \frac{n28,400}{l} \text{ cycles/sec.} \quad (28b)$$

$$n = 1, 2, 3, \ldots$$

For water columns in non-rigid pipes

$$\frac{f_{non-rigid}}{f_{rigid}} = \frac{1}{\sqrt{1 + \dfrac{300,000D}{E_{pipe}t}}}$$

E_{pipe} = elastic modulus of pipe, lb./in.²

D, t = pipe diameter and wall thickness, same units

V. Uniform Beams

(Transverse or bending vibrations)

The same general formula holds for all the following cases,

$$\omega_n = a_n \sqrt{\frac{EI}{\mu_1 l^4}} \tag{29}$$

where EI is the bending stiffness of the section, l is the length of the beam, μ_1 is the mass per unit length $= W/gl$, and a_n is a numerical constant, different for each case and listed below

Cantilever or "clamped-free" beam	$a_1 =$	3.52
	$a_2 =$	22.0
	$a_3 =$	61.7
	$a_4 =$	121.0
	$a_5 =$	200.0

Simply supported or "hinged-hinged" beam	$a_1 =$	$\pi^2 =$	9.87
	$a_2 =$	$4\pi^2 =$	39.5
	$a_3 =$	$9\pi^2 =$	88.9
	$a_4 =$	$16\pi^2 =$	158.
	$a_5 =$	$25\pi^2 =$	247.

"Free-free" beam or floating ship	$a_1 =$	22.0
	$a_2 =$	61.7
	$a_3 =$	121.0
	$a_4 =$	200.0
	$a_5 =$	298.2

"Clamped-clamped" beam has same frequencies as "free-free"	$a_1 =$	22.0
	$a_2 =$	61.7
	$a_3 =$	121.0
	$a_4 =$	200.0
	$a_5 =$	298.2

"Clamped-hinged" beam may be considered as half a "clamped-clamped" beam for even a-numbers	$a_1 =$	15.4
	$a_2 =$	50.0
	$a_3 =$	104.
	$a_4 =$	178.
	$a_5 =$	272.

"Hinged-free" beam or wing of autogyro may be considered as half a "free-free" beam for even a-numbers	$a_1 =$	0
	$a_2 =$	15.4
	$a_3 =$	50.0
	$a_4 =$	104.
	$a_5 =$	178.

(See Problem 118.)

VI. Rings, Membranes, and Plates

Extensional vibrations of a ring, radius r, weight density γ:

$$\omega_n = \frac{1}{r} \sqrt{\frac{Eg}{\gamma}} \tag{30}$$

Bending vibrations of ring, radius r, mass per unit length μ_1, in its own plane with

n full "sine waves" of disturbance along circumference

$$\omega_n = \frac{n(n^2 - 1)}{\sqrt{1 + n^2}} \sqrt{\frac{EI}{\mu_1 r^4}} \tag{31}$$

Circular membrane of tension T, mass per unit area μ_1, radius r

$$\omega_n = a_{cd} \sqrt{\frac{T}{\mu_1 r^2}} \tag{32}$$

The constant a_{cd}, is shown below, the subscript c denotes the number of nodal circles, and the subscript d the number of nodal diameters:

d \ c	1	2	3
0	2.40	5.52	8.65
1	3.83	7.02	10.17
2	5.11	8.42	11.62
3	6.38	9.76	13.02

Membrane of any shape of area A roughly of equal dimensions in all directions, fundamental mode:

$$\omega_n = \text{const.} \sqrt{\frac{T}{\mu_1 A}} \tag{33}$$

circle.......................... const. $= 2.40\pi = 4.26$
square.......................... const. $= 4.44$
quarter circle.................. const. $= 4.55$
2×1 rectangle.............. const. $= 4.97$

Circular plate of radius r, mass per unit area μ_1; the "plate constant D" defined by Eq. (7a), p. 429:

$$\omega_n = a \sqrt{\frac{D}{\mu_1 r^4}} \tag{34}$$

For free edges, 2 perp. nodal diameters............ $a = 5.25$
For free edges, one nodal circle, no diameters....... $a = 9.07$
Clamped edges, fundamental mode................ $a = 10.21$
Free edges, clamped at center, umbrella mode....... $a = 3.75$

Rectangular plate, all edges simply supported, dimensions l_1 and l_2:

$$\omega_n = \pi^2 \left(\frac{m^2}{l_1^2} + \frac{n^2}{l_2^2} \right) \sqrt{\frac{D}{\mu_1}} \qquad m = 1, 2, 3, \ldots ; n = 1, 2, 3, \ldots \tag{35}$$

Square plate, all edges clamped, length of side l, fundamental mode:

$$\omega_n = \frac{36}{l^2} \sqrt{\frac{D}{\mu_1}} \tag{36}$$

INDEX

434

A CATALOG OF SELECTED
DOVER BOOKS
IN SCIENCE AND MATHEMATICS

Astronomy

BURNHAM'S CELESTIAL HANDBOOK, Robert Burnham, Jr. Thorough guide to the stars beyond our solar system. Exhaustive treatment. Alphabetical by constellation: Andromeda to Cetus in Vol. 1; Chamaeleon to Orion in Vol. 2; and Pavo to Vulpecula in Vol. 3. Hundreds of illustrations. Index in Vol. 3. 2,000pp. 6¹/₈ x 9¹/₄.

Vol. I: 0-486-23567-X
Vol. II: 0-486-23568-8
Vol. III: 0-486-23673-0

EXPLORING THE MOON THROUGH BINOCULARS AND SMALL TELE-SCOPES, Ernest H. Cherrington, Jr. Informative, profusely illustrated guide to locating and identifying craters, rills, seas, mountains, other lunar features. Newly revised and updated with special section of new photos. Over 100 photos and diagrams. 240pp. 8¹/₄ x 11.　　0-486-24491-1

THE EXTRATERRESTRIAL LIFE DEBATE, 1750–1900, Michael J. Crowe. First detailed, scholarly study in English of the many ideas that developed from 1750 to 1900 regarding the existence of intelligent extraterrestrial life. Examines ideas of Kant, Herschel, Voltaire, Percival Lowell, many other scientists and thinkers. 16 illustrations. 704pp. 5³/₈ x 8¹/₂.　　0-486-40675-X

THEORIES OF THE WORLD FROM ANTIQUITY TO THE COPERNICAN REVOLUTION, Michael J. Crowe. Newly revised edition of an accessible, enlightening book re-creates the change from an earth-centered to a sun-centered conception of the solar system. 242pp. 5³/₈ x 8¹/₂.　　0-486-41444-2

ARISTARCHUS OF SAMOS: The Ancient Copernicus, Sir Thomas Heath. Heath's history of astronomy ranges from Homer and Hesiod to Aristarchus and includes quotes from numerous thinkers, compilers, and scholasticists from Thales and Anaximander through Pythagoras, Plato, Aristotle, and Heraclides. 34 figures. 448pp. 5³/₈ x 8¹/₂.　　0-486-43886-4

A COMPLETE MANUAL OF AMATEUR ASTRONOMY: TOOLS AND TECHNIQUES FOR ASTRONOMICAL OBSERVATIONS, P. Clay Sherrod with Thomas L. Koed. Concise, highly readable book discusses: selecting, setting up and maintaining a telescope; amateur studies of the sun; lunar topography and occultations; observations of Mars, Jupiter, Saturn, the minor planets and the stars; an introduction to photoelectric photometry; more. 1981 ed. 124 figures. 25 halftones. 37 tables. 335pp. 6¹/₂ x 9¹/₄.　　0-486-42820-8

AMATEUR ASTRONOMER'S HANDBOOK, J. B. Sidgwick. Timeless, comprehensive coverage of telescopes, mirrors, lenses, mountings, telescope drives, micrometers, spectroscopes, more. 189 illustrations. 576pp. 5³/₈ x 8¹/₄. (Available in U.S. only.)　　0-486-24034-7

STAR LORE: Myths, Legends, and Facts, William Tyler Olcott. Captivating retellings of the origins and histories of ancient star groups include Pegasus, Ursa Major, Pleiades, signs of the zodiac, and other constellations. "Classic."—Sky & Telescope. 58 illustrations. 544pp. 5³/₈ x 8¹/₂.　　0-486-43581-4

Chemistry

THE SCEPTICAL CHYMIST: THE CLASSIC 1661 TEXT, Robert Boyle. Boyle defines the term "element," asserting that all natural phenomena can be explained by the motion and organization of primary particles. 1911 ed. viii+232pp. $5^3/_8$ x $8^1/_2$.
0-486-42825-7

RADIOACTIVE SUBSTANCES, Marie Curie. Here is the celebrated scientist's doctoral thesis, the prelude to her receipt of the 1903 Nobel Prize. Curie discusses establishing atomic character of radioactivity found in compounds of uranium and thorium; extraction from pitchblende of polonium and radium; isolation of pure radium chloride; determination of atomic weight of radium; plus electric, photographic, luminous, heat, color effects of radioactivity. ii+94pp. $5^3/_8$ x $8^1/_2$.
0-486-42550-9

CHEMICAL MAGIC, Leonard A. Ford. Second Edition, Revised by E. Winston Grundmeier. Over 100 unusual stunts demonstrating cold fire, dust explosions, much more. Text explains scientific principles and stresses safety precautions. 128pp. $5^3/_8$ x $8^1/_2$.
0-486-67628-5

MOLECULAR THEORY OF CAPILLARITY, J. S. Rowlinson and B. Widom. History of surface phenomena offers critical and detailed examination and assessment of modern theories, focusing on statistical mechanics and application of results in mean-field approximation to model systems. 1989 edition. 352pp. $5^3/_8$ x $8^1/_2$.
0-486-42544-4

CHEMICAL AND CATALYTIC REACTION ENGINEERING, James J. Carberry. Designed to offer background for managing chemical reactions, this text examines behavior of chemical reactions and reactors; fluid-fluid and fluid-solid reaction systems; heterogeneous catalysis and catalytic kinetics; more. 1976 edition. 672pp. $6^1/_8$ x $9^1/_4$.
0-486-41736-0 $31.95

ELEMENTS OF CHEMISTRY, Antoine Lavoisier. Monumental classic by founder of modern chemistry in remarkable reprint of rare 1790 Kerr translation. A must for every student of chemistry or the history of science. 539pp. $5^3/_8$ x $8^1/_2$.
0-486-64624-6

MOLECULES AND RADIATION: An Introduction to Modern Molecular Spectroscopy. Second Edition, Jeffrey I. Steinfeld. This unified treatment introduces upper-level undergraduates and graduate students to the concepts and the methods of molecular spectroscopy and applications to quantum electronics, lasers, and related optical phenomena. 1985 edition. 512pp. $5^3/_8$ x $8^1/_2$.
0-486-44152-0

A SHORT HISTORY OF CHEMISTRY, J. R. Partington. Classic exposition explores origins of chemistry, alchemy, early medical chemistry, nature of atmosphere, theory of valency, laws and structure of atomic theory, much more. 428pp. $5^3/_8$ x $8^1/_2$. (Available in U.S. only.)
0-486-65977-1

GENERAL CHEMISTRY, Linus Pauling. Revised 3rd edition of classic first-year text by Nobel laureate. Atomic and molecular structure, quantum mechanics, statistical mechanics, thermodynamics correlated with descriptive chemistry. Problems. 992pp. $5^3/_8$ x $8^1/_2$.
0-486-65622-5

ELECTRON CORRELATION IN MOLECULES, S. Wilson. This text addresses one of theoretical chemistry's central problems. Topics include molecular electronic structure, independent electron models, electron correlation, the linked diagram theorem, and related topics. 1984 edition. 304pp. $5^3/_8$ x $8^1/_2$.
0-486-45879-2

Engineering

DE RE METALLICA, Georgius Agricola. The famous Hoover translation of greatest treatise on technological chemistry, engineering, geology, mining of early modern times (1556). All 289 original woodcuts. 638pp. 6¾ x 11.   0-486-60006-8

FUNDAMENTALS OF ASTRODYNAMICS, Roger Bate et al. Modern approach developed by U.S. Air Force Academy. Designed as a first course. Problems, exercises. Numerous illustrations. 455pp. 5⅜ x 8½.   0-486-60061-0

DYNAMICS OF FLUIDS IN POROUS MEDIA, Jacob Bear. For advanced students of ground water hydrology, soil mechanics and physics, drainage and irrigation engineering and more. 335 illustrations. Exercises, with answers. 784pp. 6⅛ x 9¼.   0-486-65675-6

THEORY OF VISCOELASTICITY (SECOND EDITION), Richard M. Christensen. Complete consistent description of the linear theory of the viscoelastic behavior of materials. Problem-solving techniques discussed. 1982 edition. 29 figures. xiv+364pp. 6⅛ x 9¼.   0-486-42880-X

MECHANICS, J. P. Den Hartog. A classic introductory text or refresher. Hundreds of applications and design problems illuminate fundamentals of trusses, loaded beams and cables, etc. 334 answered problems. 462pp. 5⅜ x 8½.   0-486-60754-2

MECHANICAL VIBRATIONS, J. P. Den Hartog. Classic textbook offers lucid explanations and illustrative models, applying theories of vibrations to a variety of practical industrial engineering problems. Numerous figures. 233 problems, solutions. Appendix. Index. Preface. 436pp. 5⅜ x 8½.   0-486-64785-4

STRENGTH OF MATERIALS, J. P. Den Hartog. Full, clear treatment of basic material (tension, torsion, bending, etc.) plus advanced material on engineering methods, applications. 350 answered problems. 323pp. 5⅜ x 8½.   0-486-60755-0

A HISTORY OF MECHANICS, René Dugas. Monumental study of mechanical principles from antiquity to quantum mechanics. Contributions of ancient Greeks, Galileo, Leonardo, Kepler, Lagrange, many others. 671pp. 5⅜ x 8½.   0-486-65632-2

STABILITY THEORY AND ITS APPLICATIONS TO STRUCTURAL MECHANICS, Clive L. Dym. Self-contained text focuses on Koiter postbuckling analyses, with mathematical notions of stability of motion. Basing minimum energy principles for static stability upon dynamic concepts of stability of motion, it develops asymptotic buckling and postbuckling analyses from potential energy considerations, with applications to columns, plates, and arches. 1974 ed. 208pp. 5⅜ x 8½.   0-486-42541-X

BASIC ELECTRICITY, U.S. Bureau of Naval Personnel. Originally a training course; best nontechnical coverage. Topics include batteries, circuits, conductors, AC and DC, inductance and capacitance, generators, motors, transformers, amplifiers, etc. Many questions with answers. 349 illustrations. 1969 edition. 448pp. 6½ x 9¼.   0-486-20973-3

ROCKETS, Robert Goddard. Two of the most significant publications in the history of rocketry and jet propulsion: "A Method of Reaching Extreme Altitudes" (1919) and "Liquid Propellant Rocket Development" (1936). 128pp. 5³/₈ x 8¹/₂.   0-486-42537-1

STATISTICAL MECHANICS: PRINCIPLES AND APPLICATIONS, Terrell L. Hill. Standard text covers fundamentals of statistical mechanics, applications to fluctuation theory, imperfect gases, distribution functions, more. 448pp. 5³/₈ x 8¹/₂.   0-486-65390-0

ENGINEERING AND TECHNOLOGY 1650–1750: ILLUSTRATIONS AND TEXTS FROM ORIGINAL SOURCES, Martin Jensen. Highly readable text with more than 200 contemporary drawings and detailed engravings of engineering projects dealing with surveying, leveling, materials, hand tools, lifting equipment, transport and erection, piling, bailing, water supply, hydraulic engineering, and more. Among the specific projects outlined-transporting a 50-ton stone to the Louvre, erecting an obelisk, building timber locks, and dredging canals. 207pp. 8³/₈ x 11¹/₄.   0-486-42232-1

THE VARIATIONAL PRINCIPLES OF MECHANICS, Cornelius Lanczos. Graduate level coverage of calculus of variations, equations of motion, relativistic mechanics, more. First inexpensive paperbound edition of classic treatise. Index. Bibliography. 418pp. 5³/₈ x 8¹/₂.   0-486-65067-7

PROTECTION OF ELECTRONIC CIRCUITS FROM OVERVOLTAGES, Ronald B. Standler. Five-part treatment presents practical rules and strategies for circuits designed to protect electronic systems from damage by transient overvoltages. 1989 ed. xxiv+434pp. 6¹/₈ x 9¹/₄.   0-486-42552-5

ROTARY WING AERODYNAMICS, W. Z. Stepniewski. Clear, concise text covers aerodynamic phenomena of the rotor and offers guidelines for helicopter performance evaluation. Originally prepared for NASA. 537 figures. 640pp. 6¹/₈ x 9¹/₄. 0-486-64647-5

INTRODUCTION TO SPACE DYNAMICS, William Tyrrell Thomson. Comprehensive, classic introduction to space-flight engineering for advanced undergraduate and graduate students. Includes vector algebra, kinematics, transformation of coordinates. Bibliography. Index. 352pp. 5³/₈ x 8¹/₂.   0-486-65113-4

HISTORY OF STRENGTH OF MATERIALS, Stephen P. Timoshenko. Excellent historical survey of the strength of materials with many references to the theories of elasticity and structure. 245 figures. 452pp. 5³/₈ x 8¹/₂.   0-486-61187-6

ANALYTICAL FRACTURE MECHANICS, David J. Unger. Self-contained text supplements standard fracture mechanics texts by focusing on analytical methods for determining crack-tip stress and strain fields. 336pp. 6¹/₈ x 9¹/₄.   0-486-41737-9

STATISTICAL MECHANICS OF ELASTICITY, J. H. Weiner. Advanced, self-contained treatment illustrates general principles and elastic behavior of solids. Part 1, based on classical mechanics, studies thermoelastic behavior of crystalline and polymeric solids. Part 2, based on quantum mechanics, focuses on interatomic force laws, behavior of solids, and thermally activated processes. For students of physics and chemistry and for polymer physicists. 1983 ed. 96 figures. 496pp. 5³/₈ x 8¹/₂.   0-486-42260-7

Mathematics

FUNCTIONAL ANALYSIS (Second Corrected Edition), George Bachman and Lawrence Narici. Excellent treatment of subject geared toward students with background in linear algebra, advanced calculus, physics and engineering. Text covers introduction to inner-product spaces, normed, metric spaces, and topological spaces; complete orthonormal sets, the Hahn-Banach Theorem and its consequences, and many other related subjects. 1966 ed. 544pp. $6^1/_8$ x $9^1/_4$. 0-486-40251-7

DIFFERENTIAL MANIFOLDS, Antoni A. Kosinski. Introductory text for advanced undergraduates and graduate students presents systematic study of the topological structure of smooth manifolds, starting with elements of theory and concluding with method of surgery. 1993 edition. 288pp. $5^3/_8$ x $8^1/_2$. 0-486-46244-7

VECTOR AND TENSOR ANALYSIS WITH APPLICATIONS, A. I. Borisenko and I. E. Tarapov. Concise introduction. Worked-out problems, solutions, exercises. 257pp. $5^5/_8$ x $8^1/_4$. 0-486-63833-2

AN INTRODUCTION TO ORDINARY DIFFERENTIAL EQUATIONS, Earl A. Coddington. A thorough and systematic first course in elementary differential equations for undergraduates in mathematics and science, with many exercises and problems (with answers). Index. 304pp. $5^3/_8$ x $8^1/_2$. 0-486-65942-9

FOURIER SERIES AND ORTHOGONAL FUNCTIONS, Harry F. Davis. An incisive text combining theory and practical example to introduce Fourier series, orthogonal functions and applications of the Fourier method to boundary-value problems. 570 exercises. Answers and notes. 416pp. $5^3/_8$ x $8^1/_2$. 0-486-65973-9

COMPUTABILITY AND UNSOLVABILITY, Martin Davis. Classic graduate-level introduction to theory of computability, usually referred to as theory of recurrent functions. New preface and appendix. 288pp. $5^3/_8$ x $8^1/_2$. 0-486-61471-9

AN INTRODUCTION TO MATHEMATICAL ANALYSIS, Robert A. Rankin. Dealing chiefly with functions of a single real variable, this text by a distinguished educator introduces limits, continuity, differentiability, integration, convergence of infinite series, double series, and infinite products. 1963 edition. 624pp. $5^3/_8$ x $8^1/_2$. 0-486-46251-X

METHODS OF NUMERICAL INTEGRATION (SECOND EDITION), Philip J. Davis and Philip Rabinowitz. Requiring only a background in calculus, this text covers approximate integration over finite and infinite intervals, error analysis, approximate integration in two or more dimensions, and automatic integration. 1984 edition. 624pp. $5^3/_8$ x $8^1/_2$. 0-486-45339-1

INTRODUCTION TO LINEAR ALGEBRA AND DIFFERENTIAL EQUATIONS, John W. Dettman. Excellent text covers complex numbers, determinants, orthonormal bases, Laplace transforms, much more. Exercises with solutions. Undergraduate level. 416pp. $5^3/_8$ x $8^1/_2$. 0-486-65191-6

RIEMANN'S ZETA FUNCTION, H. M. Edwards. Superb, high-level study of landmark 1859 publication entitled "On the Number of Primes Less Than a Given Magnitude" traces developments in mathematical theory that it inspired. xiv+315pp. $5^3/_8$ x $8^1/_2$. 0-486-41740-9

CALCULUS OF VARIATIONS WITH APPLICATIONS, George M. Ewing. Applications-oriented introduction to variational theory develops insight and promotes understanding of specialized books, research papers. Suitable for advanced undergraduate/graduate students as primary, supplementary text. 352pp. 5³/₈ x 8¹/₂.
0-486-64856-7

MATHEMATICIAN'S DELIGHT, W. W. Sawyer. "Recommended with confidence" by *The Times Literary Supplement*, this lively survey was written by a renowned teacher. It starts with arithmetic and algebra, gradually proceeding to trigonometry and calculus. 1943 edition. 240pp. 5³/₈ x 8¹/₂.
0-486-46240-4

ADVANCED EUCLIDEAN GEOMETRY, Roger A. Johnson. This classic text explores the geometry of the triangle and the circle, concentrating on extensions of Euclidean theory, and examining in detail many relatively recent theorems. 1929 edition. 336pp. 5³/₈ x 8¹/₂.
0-486-46237-4

COUNTEREXAMPLES IN ANALYSIS, Bernard R. Gelbaum and John M. H. Olmsted. These counterexamples deal mostly with the part of analysis known as "real variables." The first half covers the real number system, and the second half encompasses higher dimensions. 1962 edition. xxiv+198pp. 5³/₈ x 8¹/₂.
0-486-42875-3

CATASTROPHE THEORY FOR SCIENTISTS AND ENGINEERS, Robert Gilmore. Advanced-level treatment describes mathematics of theory grounded in the work of Poincaré, R. Thom, other mathematicians. Also important applications to problems in mathematics, physics, chemistry and engineering. 1981 edition. References. 28 tables. 397 black-and-white illustrations. xvii + 666pp. 6¹/₂ x 9¹/₄.
0-486-67539-4

COMPLEX VARIABLES: Second Edition, Robert B. Ash and W. P. Novinger. Suitable for advanced undergraduates and graduate students, this newly revised treatment covers Cauchy theorem and its applications, analytic functions, and the prime number theorem. Numerous problems and solutions. 2004 edition. 224pp. 6¹/₂ x 9¹/₄.
0-486-46250-1

NUMERICAL METHODS FOR SCIENTISTS AND ENGINEERS, Richard Hamming. Classic text stresses frequency approach in coverage of algorithms, polynomial approximation, Fourier approximation, exponential approximation, other topics. Revised and enlarged 2nd edition. 721pp. 5³/₈ x 8¹/₂.
0-486-65241-6

INTRODUCTION TO NUMERICAL ANALYSIS (2nd Edition), F. B. Hildebrand. Classic, fundamental treatment covers computation, approximation, interpolation, numerical differentiation and integration, other topics. 150 new problems. 669pp. 5³/₈ x 8¹/₂.
0-486-65363-3

MARKOV PROCESSES AND POTENTIAL THEORY, Robert M. Blumental and Ronald K. Getoor. This graduate-level text explores the relationship between Markov processes and potential theory in terms of excessive functions, multiplicative functionals and subprocesses, additive functionals and their potentials, and dual processes. 1968 edition. 320pp. 5³/₈ x 8¹/₂.
0-486-46263-3

ABSTRACT SETS AND FINITE ORDINALS: An Introduction to the Study of Set Theory, G. B. Keene. This text unites logical and philosophical aspects of set theory in a manner intelligible to mathematicians without training in formal logic and to logicians without a mathematical background. 1961 edition. 112pp. 5³/₈ x 8¹/₂. 0-486-46249-8

INTRODUCTORY REAL ANALYSIS, A.N. Kolmogorov, S. V. Fomin. Translated by Richard A. Silverman. Self-contained, evenly paced introduction to real and functional analysis. Some 350 problems. 403pp. 5⅜ x 8½. 0-486-61226-0

APPLIED ANALYSIS, Cornelius Lanczos. Classic work on analysis and design of finite processes for approximating solution of analytical problems. Algebraic equations, matrices, harmonic analysis, quadrature methods, much more. 559pp. 5⅜ x 8½. 0-486-65656-X

AN INTRODUCTION TO ALGEBRAIC STRUCTURES, Joseph Landin. Superb self-contained text covers "abstract algebra": sets and numbers, theory of groups, theory of rings, much more. Numerous well-chosen examples, exercises. 247pp. 5⅜ x 8½. 0-486-65940-2

QUALITATIVE THEORY OF DIFFERENTIAL EQUATIONS, V. V. Nemytskii and V.V. Stepanov. Classic graduate-level text by two prominent Soviet mathematicians covers classical differential equations as well as topological dynamics and ergodic theory. Bibliographies. 523pp. 5⅜ x 8½. 0-486-65954-2

THEORY OF MATRICES, Sam Perlis. Outstanding text covering rank, nonsingularity and inverses in connection with the development of canonical matrices under the relation of equivalence, and without the intervention of determinants. Includes exercises. 237pp. 5⅜ x 8½. 0-486-66810-X

INTRODUCTION TO ANALYSIS, Maxwell Rosenlicht. Unusually clear, accessible coverage of set theory, real number system, metric spaces, continuous functions, Riemann integration, multiple integrals, more. Wide range of problems. Undergraduate level. Bibliography. 254pp. 5⅜ x 8½. 0-486-65038-3

MODERN NONLINEAR EQUATIONS, Thomas L. Saaty. Emphasizes practical solution of problems; covers seven types of equations. ". . . a welcome contribution to the existing literature. . . ."—*Math Reviews.* 490pp. 5⅜ x 8½. 0-486-64232-1

MATRICES AND LINEAR ALGEBRA, Hans Schneider and George Phillip Barker. Basic textbook covers theory of matrices and its applications to systems of linear equations and related topics such as determinants, eigenvalues and differential equations. Numerous exercises. 432pp. 5⅜ x 8½. 0-486-66014-1

LINEAR ALGEBRA, Georgi E. Shilov. Determinants, linear spaces, matrix algebras, similar topics. For advanced undergraduates, graduates. Silverman translation. 387pp. 5⅜ x 8½. 0-486-63518-X

MATHEMATICAL METHODS OF GAME AND ECONOMIC THEORY: Revised Edition, Jean-Pierre Aubin. This text begins with optimization theory and convex analysis, followed by topics in game theory and mathematical economics, and concluding with an introduction to nonlinear analysis and control theory. 1982 edition. 656pp. 6⅛ x 9¼. 0-486-46265-X

SET THEORY AND LOGIC, Robert R. Stoll. Lucid introduction to unified theory of mathematical concepts. Set theory and logic seen as tools for conceptual understanding of real number system. 496pp. 5⅜ x 8¼. 0-486-63829-4

TENSOR CALCULUS, J.L. Synge and A. Schild. Widely used introductory text covers spaces and tensors, basic operations in Riemannian space, non-Riemannian spaces, etc. 324pp. 5⅜ x 8¼.　　　　　　　　　　　　　　　　0-486-63612-7

ORDINARY DIFFERENTIAL EQUATIONS, Morris Tenenbaum and Harry Pollard. Exhaustive survey of ordinary differential equations for undergraduates in mathematics, engineering, science. Thorough analysis of theorems. Diagrams. Bibliography. Index. 818pp. 5⅜ x 8½.　　　　　　　　　　　　　　　0-486-64940-7

INTEGRAL EQUATIONS, F. G. Tricomi. Authoritative, well-written treatment of extremely useful mathematical tool with wide applications. Volterra Equations, Fredholm Equations, much more. Advanced undergraduate to graduate level. Exercises. Bibliography. 238pp. 5⅜ x 8½.　　　　　　　　　　　　　　　0-486-64828-1

FOURIER SERIES, Georgi P. Tolstov. Translated by Richard A. Silverman. A valuable addition to the literature on the subject, moving clearly from subject to subject and theorem to theorem. 107 problems, answers. 336pp. 5⅜ x 8½.　　0-486-63317-9

INTRODUCTION TO MATHEMATICAL THINKING, Friedrich Waismann. Examinations of arithmetic, geometry, and theory of integers; rational and natural numbers; complete induction; limit and point of accumulation; remarkable curves; complex and hypercomplex numbers, more. 1959 ed. 27 figures. xii+260pp. 5⅜ x 8½.
　　　　　　　　　　　　　　　　　　　　　　0-486-42804-8

THE RADON TRANSFORM AND SOME OF ITS APPLICATIONS, Stanley R. Deans. Of value to mathematicians, physicists, and engineers, this excellent introduction covers both theory and applications, including a rich array of examples and literature. Revised and updated by the author. 1993 edition. 304pp. 6⅛ x 9¼.　0-486-46241-2

CALCULUS OF VARIATIONS, Robert Weinstock. Basic introduction covering isoperimetric problems, theory of elasticity, quantum mechanics, electrostatics, etc. Exercises throughout. 326pp. 5⅜ x 8½.　　　　　　　　　　　　0-486-63069-2

THE CONTINUUM: A CRITICAL EXAMINATION OF THE FOUNDATION OF ANALYSIS, Hermann Weyl. Classic of 20th-century foundational research deals with the conceptual problem posed by the continuum. 156pp. 5⅜ x 8½.　　0-486-67982-9

CHALLENGING MATHEMATICAL PROBLEMS WITH ELEMENTARY SOLUTIONS, A. M. Yaglom and I. M. Yaglom. Over 170 challenging problems on probability theory, combinatorial analysis, points and lines, topology, convex polygons, many other topics. Solutions. Total of 445pp. 5⅜ x 8½. Two-vol. set.
　　　　　　　　　　　Vol. I: 0-486-65536-9　Vol. II: 0-486-65537-7

INTRODUCTION TO PARTIAL DIFFERENTIAL EQUATIONS WITH APPLICATIONS, E. C. Zachmanoglou and Dale W. Thoe. Essentials of partial differential equations applied to common problems in engineering and the physical sciences. Problems and answers. 416pp. 5⅜ x 8½.　　　　　　　　0-486-65251-3

STOCHASTIC PROCESSES AND FILTERING THEORY, Andrew H. Jazwinski. This unified treatment presents material previously available only in journals, and in terms accessible to engineering students. Although theory is emphasized, it discusses numerous practical applications as well. 1970 edition. 400pp. 5⅜ x 8½.　0-486-46274-9

Math—Decision Theory, Statistics, Probability

INTRODUCTION TO PROBABILITY, John E. Freund. Featured topics include permutations and factorials, probabilities and odds, frequency interpretation, mathematical expectation, decision-making, postulates of probability, rule of elimination, much more. Exercises with some solutions. Summary. 1973 edition. 247pp. 5⅜ x 8½.
0-486-67549-1

STATISTICAL AND INDUCTIVE PROBABILITIES, Hugues Leblanc. This treatment addresses a decades-old dispute among probability theorists, asserting that both statistical and inductive probabilities may be treated as sentence-theoretic measurements, and that the latter qualify as estimates of the former. 1962 edition. 160pp. 5⅜ x 8½.
0-486-44980-7

APPLIED MULTIVARIATE ANALYSIS: Using Bayesian and Frequentist Methods of Inference, Second Edition, S. James Press. This two-part treatment deals with foundations as well as models and applications. Topics include continuous multivariate distributions; regression and analysis of variance; factor analysis and latent structure analysis; and structuring multivariate populations. 1982 edition. 692pp. 5⅜ x 8½.
0-486-44236-5

LINEAR PROGRAMMING AND ECONOMIC ANALYSIS, Robert Dorfman, Paul A. Samuelson and Robert M. Solow. First comprehensive treatment of linear programming in standard economic analysis. Game theory, modern welfare economics, Leontief input-output, more. 525pp. 5⅜ x 8½.
0-486-65491-5

PROBABILITY: AN INTRODUCTION, Samuel Goldberg. Excellent basic text covers set theory, probability theory for finite sample spaces, binomial theorem, much more. 360 problems. Bibliographies. 322pp. 5⅜ x 8½.
0-486-65252-1

GAMES AND DECISIONS: INTRODUCTION AND CRITICAL SURVEY, R. Duncan Luce and Howard Raiffa. Superb nontechnical introduction to game theory, primarily applied to social sciences. Utility theory, zero-sum games, n-person games, decision-making, much more. Bibliography. 509pp. 5⅜ x 8½.
0-486-65943-7

INTRODUCTION TO THE THEORY OF GAMES, J. C. C. McKinsey. This comprehensive overview of the mathematical theory of games illustrates applications to situations involving conflicts of interest, including economic, social, political, and military contexts. Appropriate for advanced undergraduate and graduate courses; advanced calculus a prerequisite. 1952 ed. x+372pp. 5⅜ x 8½.
0-486-42811-7

FIFTY CHALLENGING PROBLEMS IN PROBABILITY WITH SOLUTIONS, Frederick Mosteller. Remarkable puzzlers, graded in difficulty, illustrate elementary and advanced aspects of probability. Detailed solutions. 88pp. 5⅜ x 8½.
0-486-65355-2

PROBABILITY THEORY: A CONCISE COURSE, Y. A. Rozanov. Highly readable, self-contained introduction covers combination of events, dependent events, Bernoulli trials, etc. 148pp. 5⅜ x 8¼.
0-486-63544-9

THE STATISTICAL ANALYSIS OF EXPERIMENTAL DATA, John Mandel. First half of book presents fundamental mathematical definitions, concepts and facts while remaining half deals with statistics primarily as an interpretive tool. Well-written text, numerous worked examples with step-by-step presentation. Includes 116 tables. 448pp. 5⅜ x 8½.
0-486-64666-1

Math—Geometry and Topology

ELEMENTARY CONCEPTS OF TOPOLOGY, Paul Alexandroff. Elegant, intuitive approach to topology from set-theoretic topology to Betti groups; how concepts of topology are useful in math and physics. 25 figures. 57pp. 5⅜ x 8½.　　0-486-60747-X

A LONG WAY FROM EUCLID, Constance Reid. Lively guide by a prominent historian focuses on the role of Euclid's Elements in subsequent mathematical developments. Elementary algebra and plane geometry are sole prerequisites. 80 drawings. 1963 edition. 304pp. 5⅜ x 8½.　　0-486-43613-6

EXPERIMENTS IN TOPOLOGY, Stephen Barr. Classic, lively explanation of one of the byways of mathematics. Klein bottles, Moebius strips, projective planes, map coloring, problem of the Koenigsberg bridges, much more, described with clarity and wit. 43 figures. 210pp. 5⅜ x 8½.　　0-486-25933-1

THE GEOMETRY OF RENÉ DESCARTES, René Descartes. The great work founded analytical geometry. Original French text, Descartes's own diagrams, together with definitive Smith-Latham translation. 244pp. 5⅜ x 8½.　　0-486-60068-8

EUCLIDEAN GEOMETRY AND TRANSFORMATIONS, Clayton W. Dodge. This introduction to Euclidean geometry emphasizes transformations, particularly isometries and similarities. Suitable for undergraduate courses, it includes numerous examples, many with detailed answers. 1972 ed. viii+296pp. 6⅛ x 9¼.　　0-486-43476-1

EXCURSIONS IN GEOMETRY, C. Stanley Ogilvy. A straightedge, compass, and a little thought are all that's needed to discover the intellectual excitement of geometry. Harmonic division and Apollonian circles, inversive geometry, hexlet, Golden Section, more. 132 illustrations. 192pp. 5⅜ x 8½.　　0-486-26530-7

THE THIRTEEN BOOKS OF EUCLID'S ELEMENTS, translated with introduction and commentary by Sir Thomas L. Heath. Definitive edition. Textual and linguistic notes, mathematical analysis. 2,500 years of critical commentary. Unabridged. 1,414pp. 5⅜ x 8½. Three-vol. set.
　　　　Vol. I: 0-486-60088-2　Vol. II: 0-486-60089-0　Vol. III: 0-486-60090-4

SPACE AND GEOMETRY: IN THE LIGHT OF PHYSIOLOGICAL, PSYCHOLOGICAL AND PHYSICAL INQUIRY, Ernst Mach. Three essays by an eminent philosopher and scientist explore the nature, origin, and development of our concepts of space, with a distinctness and precision suitable for undergraduate students and other readers. 1906 ed. vi+148pp. 5⅜ x 8½.　　0-486-43909-7

GEOMETRY OF COMPLEX NUMBERS, Hans Schwerdtfeger. Illuminating, widely praised book on analytic geometry of circles, the Moebius transformation, and two-dimensional non-Euclidean geometries. 200pp. 5⅜ x 8¼.　　0-486-63830-8

DIFFERENTIAL GEOMETRY, Heinrich W. Guggenheimer. Local differential geometry as an application of advanced calculus and linear algebra. Curvature, transformation groups, surfaces, more. Exercises. 62 figures. 378pp. 5⅜ x 8½.　　0-486-63433-7

History of Math

THE WORKS OF ARCHIMEDES, Archimedes (T. L. Heath, ed.). Topics include the famous problems of the ratio of the areas of a cylinder and an inscribed sphere; the measurement of a circle; the properties of conoids, spheroids, and spirals; and the quadrature of the parabola. Informative introduction. clxxxvi+326pp. 5⅜ x 8½.   0-486-42084-1

A SHORT ACCOUNT OF THE HISTORY OF MATHEMATICS, W. W. Rouse Ball. One of clearest, most authoritative surveys from the Egyptians and Phoenicians through 19th-century figures such as Grassman, Galois, Riemann. Fourth edition. 522pp. 5⅜ x 8½.   0-486-20630-0

THE HISTORY OF THE CALCULUS AND ITS CONCEPTUAL DEVELOP-MENT, Carl B. Boyer. Origins in antiquity, medieval contributions, work of Newton, Leibniz, rigorous formulation. Treatment is verbal. 346pp. 5⅜ x 8½.   0-486-60509-4

THE HISTORICAL ROOTS OF ELEMENTARY MATHEMATICS, Lucas N. H. Bunt, Phillip S. Jones, and Jack D. Bedient. Fundamental underpinnings of modern arithmetic, algebra, geometry and number systems derived from ancient civilizations. 320pp. 5⅜ x 8½.   0-486-25563-8

THE HISTORY OF THE CALCULUS AND ITS CONCEPTUAL DEVELOP-MENT, Carl B. Boyer. Fluent description of the development of both the integral and differential calculus—its early beginnings in antiquity, medieval contributions, and a consideration of Newton and Leibniz. 368pp. 5⅜ x 8½.   0-486-60509-4

GAMES, GODS & GAMBLING: A HISTORY OF PROBABILITY AND STATISTICAL IDEAS, F. N. David. Episodes from the lives of Galileo, Fermat, Pascal, and others illustrate this fascinating account of the roots of mathematics. Features thought-provoking references to classics, archaeology, biography, poetry. 1962 edition. 304pp. 5⅜ x 8½. (Available in U.S. only.)   0-486-40023-9

OF MEN AND NUMBERS: THE STORY OF THE GREAT MATHEMATICIANS, Jane Muir. Fascinating accounts of the lives and accomplishments of history's greatest mathematical minds—Pythagoras, Descartes, Euler, Pascal, Cantor, many more. Anecdotal, illuminating. 30 diagrams. Bibliography. 256pp. 5⅜ x 8½.   0-486-28973-7

HISTORY OF MATHEMATICS, David E. Smith. Nontechnical survey from ancient Greece and Orient to late 19th century; evolution of arithmetic, geometry, trigonometry, calculating devices, algebra, the calculus. 362 illustrations. 1,355pp. 5⅜ x 8½. Two-vol. set.   Vol. I: 0-486-20429-4   Vol. II: 0-486-20430-8

A CONCISE HISTORY OF MATHEMATICS, Dirk J. Struik. The best brief history of mathematics. Stresses origins and covers every major figure from ancient Near East to 19th century. 41 illustrations. 195pp. 5⅜ x 8½.   0-486-60255-9

Physics

OPTICAL RESONANCE AND TWO-LEVEL ATOMS, L. Allen and J. H. Eberly. Clear, comprehensive introduction to basic principles behind all quantum optical resonance phenomena. 53 illustrations. Preface. Index. 256pp. $5^3/_8$ x $8^1/_2$.　　0-486-65533-4

QUANTUM THEORY, David Bohm. This advanced undergraduate-level text presents the quantum theory in terms of qualitative and imaginative concepts, followed by specific applications worked out in mathematical detail. Preface. Index. 655pp. $5^3/_8$ x $8^1/_2$.
0-486-65969-0

ATOMIC PHYSICS (8th EDITION), Max Born. Nobel laureate's lucid treatment of kinetic theory of gases, elementary particles, nuclear atom, wave-corpuscles, atomic structure and spectral lines, much more. Over 40 appendices, bibliography. 495pp. $5^3/_8$ x $8^1/_2$.
0-486-65984-4

A SOPHISTICATE'S PRIMER OF RELATIVITY, P. W. Bridgman. Geared toward readers already acquainted with special relativity, this book transcends the view of theory as a working tool to answer natural questions: What is a frame of reference? What is a "law of nature"? What is the role of the "observer"? Extensive treatment, written in terms accessible to those without a scientific background. 1983 ed. xlviii+172pp. $5^3/_8$ x $8^1/_2$.
0-486-42549-5

AN INTRODUCTION TO HAMILTONIAN OPTICS, H. A. Buchdahl. Detailed account of the Hamiltonian treatment of aberration theory in geometrical optics. Many classes of optical systems defined in terms of the symmetries they possess. Problems with detailed solutions. 1970 edition. xv + 360pp. $5^3/_8$ x $8^1/_2$.　　0-486-67597-1

PRIMER OF QUANTUM MECHANICS, Marvin Chester. Introductory text examines the classical quantum bead on a track: its state and representations; operator eigenvalues; harmonic oscillator and bound bead in a symmetric force field; and bead in a spherical shell. Other topics include spin, matrices, and the structure of quantum mechanics; the simplest atom; indistinguishable particles; and stationary-state perturbation theory. 1992 ed. xiv+314pp. $6^1/_8$ x $9^1/_4$.　　0-486-42878-8

LECTURES ON QUANTUM MECHANICS, Paul A. M. Dirac. Four concise, brilliant lectures on mathematical methods in quantum mechanics from Nobel Prize-winning quantum pioneer build on idea of visualizing quantum theory through the use of classical mechanics. 96pp. $5^3/_8$ x $8^1/_2$.　　0-486-41713-1

THIRTY YEARS THAT SHOOK PHYSICS: THE STORY OF QUANTUM THEORY, George Gamow. Lucid, accessible introduction to influential theory of energy and matter. Careful explanations of Dirac's anti-particles, Bohr's model of the atom, much more. 12 plates. Numerous drawings. 240pp. $5^3/_8$ x $8^1/_2$.　　0-486-24895-X

ELECTRONIC STRUCTURE AND THE PROPERTIES OF SOLIDS: THE PHYSICS OF THE CHEMICAL BOND, Walter A. Harrison. Innovative text offers basic understanding of the electronic structure of covalent and ionic solids, simple metals, transition metals and their compounds. Problems. 1980 edition. 582pp. $6^1/_8$ x $9^1/_4$.
0-486-66021-4

HYDRODYNAMIC AND HYDROMAGNETIC STABILITY, S. Chandrasekhar. Lucid examination of the Rayleigh-Benard problem; clear coverage of the theory of instabilities causing convection. 704pp. $5^{3}/_{8}$ x $8^{1}/_{4}$. 0-486-64071-X

INVESTIGATIONS ON THE THEORY OF THE BROWNIAN MOVEMENT, Albert Einstein. Five papers (1905–8) investigating dynamics of Brownian motion and evolving elementary theory. Notes by R. Fürth. 122pp. $5^{3}/_{8}$ x $8^{1}/_{2}$. 0-486-60304-0

THE PHYSICS OF WAVES, William C. Elmore and Mark A. Heald. Unique overview of classical wave theory. Acoustics, optics, electromagnetic radiation, more. Ideal as classroom text or for self-study. Problems. 477pp. $5^{3}/_{8}$ x $8^{1}/_{2}$. 0-486-64926-1

GRAVITY, George Gamow. Distinguished physicist and teacher takes reader-friendly look at three scientists whose work unlocked many of the mysteries behind the laws of physics: Galileo, Newton, and Einstein. Most of the book focuses on Newton's ideas, with a concluding chapter on post-Einsteinian speculations concerning the relationship between gravity and other physical phenomena. 160pp. $5^{3}/_{8}$ x $8^{1}/_{2}$. 0-486-42563-0

PHYSICAL PRINCIPLES OF THE QUANTUM THEORY, Werner Heisenberg. Nobel Laureate discusses quantum theory, uncertainty, wave mechanics, work of Dirac, Schroedinger, Compton, Wilson, Einstein, etc. 184pp. $5^{3}/_{8}$ x $8^{1}/_{2}$. 0-486-60113-7

ATOMIC SPECTRA AND ATOMIC STRUCTURE, Gerhard Herzberg. One of best introductions; especially for specialist in other fields. Treatment is physical rather than mathematical. 80 illustrations. 257pp. $5^{3}/_{8}$ x $8^{1}/_{2}$. 0-486-60115-3

AN INTRODUCTION TO STATISTICAL THERMODYNAMICS, Terrell L. Hill. Excellent basic text offers wide-ranging coverage of quantum statistical mechanics, systems of interacting molecules, quantum statistics, more. 523pp. $5^{3}/_{8}$ x $8^{1}/_{2}$. 0-486-65242-4

THEORETICAL PHYSICS, Georg Joos, with Ira M. Freeman. Classic overview covers essential math, mechanics, electromagnetic theory, thermodynamics, quantum mechanics, nuclear physics, other topics. First paperback edition. xxiii + 885pp. $5^{3}/_{8}$ x $8^{1}/_{2}$.
0-486-65227-0

PROBLEMS AND SOLUTIONS IN QUANTUM CHEMISTRY AND PHYSICS, Charles S. Johnson, Jr. and Lee G. Pedersen. Unusually varied problems, detailed solutions in coverage of quantum mechanics, wave mechanics, angular momentum, molecular spectroscopy, more. 280 problems plus 139 supplementary exercises. 430pp. $6^{1}/_{2}$ x $9^{1}/_{4}$.
0-486-65236-X

THEORETICAL SOLID STATE PHYSICS, Vol. 1: Perfect Lattices in Equilibrium; Vol. II: Non-Equilibrium and Disorder, William Jones and Norman H. March. Monumental reference work covers fundamental theory of equilibrium properties of perfect crystalline solids, non-equilibrium properties, defects and disordered systems. Appendices. Problems. Preface. Diagrams. Index. Bibliography. Total of 1,301pp. $5^{3}/_{8}$ x $8^{1}/_{2}$. Two volumes. Vol. I: 0-486-65015-4 Vol. II: 0-486-65016-2

WHAT IS RELATIVITY? L. D. Landau and G. B. Rumer. Written by a Nobel Prize physicist and his distinguished colleague, this compelling book explains the special theory of relativity to readers with no scientific background, using such familiar objects as trains, rulers, and clocks. 1960 ed. vi+72pp. $5^{3}/_{8}$ x $8^{1}/_{2}$. 0-486-42806-0

A TREATISE ON ELECTRICITY AND MAGNETISM, James Clerk Maxwell. Important foundation work of modern physics. Brings to final form Maxwell's theory of electromagnetism and rigorously derives his general equations of field theory. 1,084pp. 5⅜ x 8½. Two-vol. set.   Vol. I: 0-486-60636-8   Vol. II: 0-486-60637-6

MATHEMATICS FOR PHYSICISTS, Philippe Dennery and Andre Krzywicki. Superb text provides math needed to understand today's more advanced topics in physics and engineering. Theory of functions of a complex variable, linear vector spaces, much more. Problems. 1967 edition. 400pp. 6½ x 9¼.   0-486-69193-4

INTRODUCTION TO QUANTUM MECHANICS WITH APPLICATIONS TO CHEMISTRY, Linus Pauling & E. Bright Wilson, Jr. Classic undergraduate text by Nobel Prize winner applies quantum mechanics to chemical and physical problems. Numerous tables and figures enhance the text. Chapter bibliographies. Appendices. Index. 468pp. 5⅜ x 8½.   0-486-64871-0

METHODS OF THERMODYNAMICS, Howard Reiss. Outstanding text focuses on physical technique of thermodynamics, typical problem areas of understanding, and significance and use of thermodynamic potential. 1965 edition. 238pp. 5⅜ x 8½.
0-486-69445-3

THE ELECTROMAGNETIC FIELD, Albert Shadowitz. Comprehensive under- graduate text covers basics of electric and magnetic fields, builds up to electromagnetic theory. Also related topics, including relativity. Over 900 problems. 768pp. 5⅜ x 8¼.
0-486-65660-8

GREAT EXPERIMENTS IN PHYSICS: FIRSTHAND ACCOUNTS FROM GALILEO TO EINSTEIN, Morris H. Shamos (ed.). 25 crucial discoveries: Newton's laws of motion, Chadwick's study of the neutron, Hertz on electromagnetic waves, more. Original accounts clearly annotated. 370pp. 5⅜ x 8½.   0-486-25346-5

EINSTEIN'S LEGACY, Julian Schwinger. A Nobel Laureate relates fascinating story of Einstein and development of relativity theory in well-illustrated, nontechnical volume. Subjects include meaning of time, paradoxes of space travel, gravity and its effect on light, non-Euclidean geometry and curving of space-time, impact of radio astronomy and space-age discoveries, and more. 189 b/w illustrations. xiv+250pp. 8⅜ x 9¼.   0-486-41974-6

THE VARIATIONAL PRINCIPLES OF MECHANICS, Cornelius Lanczos. Philosophic, less formalistic approach to analytical mechanics offers model of clear, scholarly exposition at graduate level with coverage of basics, calculus of variations, principle of virtual work, equations of motion, more. 418pp. 5⅜ x 8½.   0-486-65067-7

Paperbound unless otherwise indicated. Available at your book dealer, online at www.doverpublications.com, or by writing to Dept. GI, Dover Publications, Inc., 31 East 2nd Street, Mineola, NY 11501. For current price information or for free catalogues (please indicate field of interest), write to Dover Publications or log on to www.doverpublications.com and see every Dover book in print. Dover publishes more than 400 books each year on science, elementary and advanced mathematics, biology, music, art, literary history, social sciences, and other areas.